10th MATBUD'2023
Scientific-Technical Conference

10th MATBUD'2023 Scientific-Technical Conference

Editors

Katarzyna Mróz
Tomasz Tracz
Tomasz Zdeb
Izabela Hager

MDPI • Basel • Beijing • Wuhan • Barcelona • Belgrade • Manchester • Tokyo • Cluj • Tianjin

Editors
Katarzyna Mróz
Cracow University of Technology
Poland

Tomasz Tracz
Cracow University of Technology
Poland

Tomasz Zdeb
Cracow University of Technology
Poland

Izabela Hager
Cracow University of Technology
Poland

Editorial Office
MDPI
St. Alban-Anlage 66
4052 Basel, Switzerland

This is a reprint of articles from the Proceedings published online in the open access journal *Materials Proceedings* (ISSN 2673-4605) (available at: https://www.mdpi.com/2673-4605/13/1).

For citation purposes, cite each article independently as indicated on the article page online and as indicated below:

LastName, A.A.; LastName, B.B.; LastName, C.C. Article Title. *Journal Name* **Year**, *Volume Number*, Page Range.

ISBN 978-3-0365-7094-5 (Hbk)
ISBN 978-3-0365-7095-2 (PDF)

Cover image courtesy of Katarzyna Mróz

Contents

Preface to "10th MATBUD'2023 Scientific-Technical Conference"

The 10th MATBUD'2023 Conference on 'Building Materials Engineering and Innovative Sustainable Materials' was jointly organized by the interdisciplinary research group of the Faculty of Civil Engineering and Faculty of Civil Engineering and Physics at Cracow University of Technology in Poland and the Nantes Université IUT Saint Nazaire GeM Institute in France.

The auspices of the Section of Building Materials Engineering, which is a part of the Committee of Civil Engineering of the Polish Academy of Sciences and the Construction Commission, Cracow Branch of the Polish Academy of Sciences, were extended to the organization of the conference.

The event took place within the framework of the Welcome to Poland project financed by the Polish National Agency for Academic Exchange (NAWA): "MATBUD'2023—Developing international scientific cooperation in the field of building materials engineering" BPI_WTP_2021_1_00002.

The scientific committee worked under the chairmanship of Prof. Lech Czarnecki (Honorary Chairman) and Prof. Izabela Hager (Chairwoman; the project coordinator).

We would like to express our gratitude to Prof. Lech Czarnecki, the Honorary Chairman of the Scientific Committee, who has played a key role in MATBUD events from the outset, generously sharing his time and expertise with us. Our heartfelt thanks go to him and all members of the Scientific Committee for their voluntary paper review and valuable suggestions during the preparation of this conference.

We would like to express our heartfelt appreciation to Professor Jacek Śliwiński, the originator and architect of these cyclic meetings, who has devoted his time and effort to organizing MATBUD events with unwavering commitment since its first edition in 1996, and has also provided strong support for the organization of these 10 events.

This year, we received about 78 paper submissions, which were reviewed by national and international experts. In total, 44 of the received papers were accepted by the Scientific Committee for oral presentations at the conference and for publication in the Proceedings.

The primary focus of this conference was several areas of building materials engineering: development, processing, and the evaluation of properties and applications. MATBUD'2023 mainly emphasized cement and alkali-activated binders, aspects of durability, and innovative sustainable materials. Research in materials engineering is presently oriented towards finding waste management solutions for structural material production and developing cement composites regarding their manufacturing (e.g., 3D printing) and design (e.g., fiber-reinforced concretes).

The conference hosted participants from over 45 research units worldwide.

The organizing committee would like to express great gratitude to all the authors and participants for their valuable contribution to the conference.

Katarzyna Mróz, Tomasz Tracz, Tomasz Zdeb, and Izabela Hager
Editors

Fundusze Europejskie Wiedza Edukacja Rozwój · Rzeczpospolita Polska · NAWA · Unia Europejska Europejski Fundusz Społeczny

materials proceedings

MDPI

Editorial

Statement of Peer Review †

Katarzyna Mróz *, Tomasz Tracz, Tomasz Zdeb and Izabela Hager

Chair of Building Materials, Faculty of Civil Engineering, Cracow University of Technology, 24 Warszawska Str., 31-155 Cracow, Poland; tomasz.tracz@pk.edu.pl (T.T.); tomasz.zdeb@pk.edu.pl (T.Z.); izabela.hager@pk.edu.pl (I.H.)

* Correspondence: katarzyna.mroz@pk.edu.pl; Tel.: +48-12-628-23-71

† Presented at the 10th MATBUD'2023 Scientific-Technical Conference "Building Materials Engineering and Innovative Sustainable Materials", Cracow, Poland, 19–21 April 2023.

In submitting conference proceedings to *Materials Proceedings*, the volume editors of the proceedings certify to the publisher that all papers published in this volume have been subjected to peer review administered by the volume editors. Reviews were conducted by expert referees to the professional and scientific standards expected of a proceedings journal.

- Type of peer review: single-blind;
- Conference submission management system: web-based portal, email
- Number of submissions sent for review: 51
- Number of submissions accepted: 44
- Acceptance rate (number of submissions accepted/number of submissions received): 0.86
- Average number of reviews per paper: 2
- Total number of reviewers involved: 38
- Any additional information on the review process: All submissions were reviewed for linguistic correctness and formatting by one of the volume editors prior the peer review.

Conflicts of Interest: The authors declare no conflict of interest.

Citation: Mróz, K.; Tracz, T.; Zdeb, T.; Hager, I. Statement of Peer Review. *Mater. Proc.* **2023**, *13*, 45. https://doi.org/10.3390/materproc2023013045

Published: 17 March 2023

materials proceedings

MDPI

Proceeding Paper

Fiber-Reinforced Alkali-Activated Materials Based on Waste Materials †

Martin Mildner *, Jan Fořt and Robert Černý

Department of Materials Engineering and Chemistry, Faculty of Civil Engineering, Czech Technical University in Prague, Thákurova 7/2077, 166 29 Prague, Czech Republic

* Correspondence: martin.mildner@fsv.cvut.cz

† Presented at the 10th MATBUD'2023 Scientific-Technical Conference "Building Materials Engineering and Innovative Sustainable Materials", Cracow, Poland, 19–21 April 2023.

Abstract: The adverse effects associated with a rise in global temperature require substantial advances in various industries, the building industry in particular, with an emphasis on sustainability and circular economy measures. Research effort on the design of alkali-activated materials with sufficient engineering properties is thus on the rise, as these materials form a possible way to replace cementitious binders in the future. This paper deals with the description of an alternative material without the use of cementitious binders. The alkaline activation of a blended precursor composed of a finely ground granulated blast furnace slag and metashale, activated using waste alkalis from industrial production is studied. In addition, this material was reinforced using 25 mm long fibers of a waste fiberglass reinforcement fabric to improve the mechanical properties. This research confirmed the suitability of using a range of waste or secondary raw materials to produce new materials which then have lower environmental impacts.

Keywords: alkali activation; metashale; blast furnace slag; waste alkali activator; waste fibers

1. Introduction

Cement composites have become irreplaceable elements in the construction industry over the last few decades due to their properties and low production costs. However, their production involves the consumption of significant quantities of raw materials and primary energy for manufacturing. Moreover, the production and transport of cement place a high CO_2 burden on the environment [1,2]. Despite all drawbacks, cement is the most widely used building material in the world due to its superior engineering properties. In terms of their environmental impact and material properties, alkali-activated materials (AAM) seem to be a suitable option and they also match the development direction of building materials due to the valorization of various by-products or industrial waste products [3].

AAMs are formed by the alkaline activation of aluminosilicates (precursors) using highly alkaline activators. The precursors are usually obtained as waste materials or secondary raw materials from industrial production (blast furnace slag, silica fume, metashale and metakaolin), but natural precursors (rice husk, palm oil ash) can be used as well. Through the utilization of various waste or natural precursors, the production of binders and the associated release of CO_2 are noticeably reduced. As activators, water glass or hydroxide are the most commonly used [4–6]. Apart from being aggressive substances, their production, which is very energy-demanding, also represents a substantial burden on the natural environment and human health. For this reason, waste alkalis, which can also be obtained as waste raw materials from industrial production, should be used instead [7]. The selection and use of the precursors and activators significantly influence the final properties of the AAM, such as the mechanical, thermal or electrical properties, as well as their fire and frost resistance [8,9]. In regards to the properties of AAM, some of them can

check for updates

Citation: Mildner, M.; Fořt, J.; Černý, R. Fiber-Reinforced Alkali-Activated Materials Based on Waste Materials. *Mater. Proc.* **2023**, *13*, 1. https://doi.org/10.3390/materproc2023013001

Academic Editors: Katarzyna Mróz, Tomasz Tracz, Tomasz Zdeb and Izabela Hager

Published: 13 February 2023

also be modified by adding fiber reinforcement to the material's matrix. The use of fibers can also lead to a reduction in the cracks arising from debonding, sliding and pull-out [10].

Glass fibers are synthetic amorphous silicates that are commonly available at low cost. Their main advantage is their high flexural strength [11]. Pernica et al. [12] studied a Na geopolymer E-glass composite and observed an increase in the flexural strength from 175.9 to 255.2 MPa. Sankar and Kriven [11] observed the behavior of geopolymer-reinforced E-glass lazy weaves and observed an increase in the flexural tensile strength as well.

This study aimed to assess the potential of waste or secondary raw materials as the main sources for the production of new alternative building materials. In addition to the waste precursor, the alkaline activator represents a waste material from industrial cleaning operations and can also be viewed as waste material. In other words, this research contemplated the production of alternative AAMs based on waste materials, with a significantly reduced environmental footprint. Thus, the utilization of selected waste products can be seen as beneficial in terms of the environment and cost savings achieved by avoiding costly disposal. To improve the mechanical properties and reduce the shrinkage of the material, waste glass fibers were utilized.

2. Experiments

2.1. Materials

The precursor used to produce the alkali-activated material was a combination of two aluminosilicates materials: finely ground granulated blast furnace slag (produced by Kotouč Štramberk, Czech Republic) and metashale (produced by České Lupkové Závody, A.S., Czech Republic). Finely ground granulated blast furnace slag (GGBFS) is a secondary raw material from iron production. Metashale (RON D460 HR) is produced by heat treatment of the waste material from metashale Mefisto L05 production. Both precursors are shown in Figure 1 and their chemical composition is presented in Table 1.

(a)

(b)

Figure 1. (a) Finely ground granulated blast furnace slag (GGBFS); (b) metashale (RON D460 HR).

Table 1. Chemical composition of the precursors used.

Precursor	SiO_2	Al_2O_3	Fe_2O_3	TiO_2	MgO	CaO	K_2O	MnO	SO_3	LOI
GGBFS	39.1	9.8	0.5	0.3	8.7	38.8	0.7	0.9	0.6	0.6
RON D460 HR	49.1	47.3	0.9	1.6	0.1	0.2	0.5	—	—	0.3

Waste alkalis (WAA) based on sodium hydroxide from the industrial processes of glass production (AGS Glass, Teplice, Czech Republic) were used as an alkaline activator. These alkalis are in solid form in the form of pellets (see Figure 2), from which an activating solution of the appropriate molarity is produced. For the purpose of this research, an 8 M solution of WAA was used.

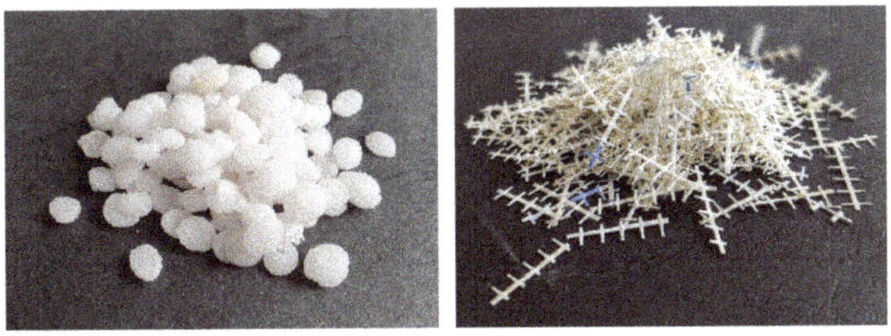

(a) (b)

Figure 2. (**a**) Solidified waste alkalis based on sodium hydroxide (WAA); (**b**) fibers from the waste fiberglass reinforcement fabric (WFF).

Fibers from waste fiberglass reinforcement fabric (WFF) were used to reinforce the alkali-activated composite. This fabric is very important for ensuring high quality and a strong final layer in the interior plaster. The fabric was cut into individual fibers 25 mm in length (see Figure 2). The producer reported that the minimal tensile strength of the fiber is from 2000 to 2200 N/50 mm, while the elongation is 3.5%.

A mixture of fine standard aggregate in three fractions (0.08–0.5, 0.5–1.0 and 1.0–2.0 mm) was used for the mortar samples.

The experimental program was divided into two sets of sample mixtures. One set of samples was without WFF; the other had WFF fibers. The procedure for mixing the WFF was as follows. The mixture was mixed one step at a time. First, the dry components (a mixture of the respective precursors and fine aggregates) were mixed and then an 8 M solution of WAA was added. At the very end, WFF fibers were inserted in batches to avoid the fibers clumping in the matrix.

The individual mixtures differed in the percentage of the respective precursors, i.e., GGBFS and RON, which changed after 20%. The composition of the studied mixtures is shown in Table 2.

Table 2. Composition of 1 m^3 of the studied mixtures.

Mixture Designation	GGBFS (kg)	RON (kg)	Fibers WFF (kg)	Fine Aggregate (kg) of Fractions			Solution of 8 M WAA (kg)
				0.08–0.5 mm	0.5–1.0 mm	1.0–2.0 mm	
(a) Composition with fibers							
80S+20R	460	115	5.1	380	380	380	296
60S+40R	345	230					
40S+60R	230	345					
20S+80R	115	460					
(b) Composition without fibers							
80S+20R	460	115	—	382	382	382	297
60S+40R	345	230					
40S+60R	230	345					
20S+80R	115	460					

2.2. Experimental Methods

2.2.1. Chemical and Mineralogical Properties

The phase composition of the measured mixtures was studied by X-ray diffraction (XRD). Diffractograms were recorded using a Malvern PANalytical Aeris diffractometer equipped with a CoKα source operating at 7.5 mA and 40 kV. The incident beam'd path consisted of iron beta-filter, Soller slits at 0.04 rad and a divergence slit at 1/2°.

For measurement by XRF analysis, a sequential WD-XRF spectrometer (ARL 9400 XP, Thermo ARL) was used, which was equipped with a Rh anode tipped X-ray tube (type 4 GN) fitted with a 50 μm Be window.

2.2.2. Basic Material Properties

The bulk density and matrix density of each mixture in the hardened state were determined. Bulk density was determined on dried samples by measuring them with a digital caliper and then weighing them. The matrix density was determined using a helium pycnometer (Pycnomatic ATC, Thermo Scientific, Waltham, MA, USA). The open porosity of the samples was calculated from these values.

2.2.3. Mechanical Properties

Flexural and compressive tensile strengths were tested after 28 days of curing under laboratory conditions (20 °C and 50% RH). Three prismatic specimens of 40 × 40 × 160 mm from each mix were tested each time for flexural tensile strength and then for compressive strength on the resulting fractures. An ED60 hydraulic press was used for the measurements.

2.2.4. Analyzing the Microstructure of the Surface

Scanning electron microscopy (SEM) was used to document where the matrix material interfaced with the blended fibers. A test sample measuring 40 × 40 × 160 mm was cut using a saw into 1 mm slices, which were then used for SEM. A Phenom XL (ThermoScientific) was used with a magnification range of 80 to 100,000×.

3. Results and Discussion

3.1. Basic Material Properties

The basic characteristics of the material such as the bulk density, matrix density and the open porosity of the material calculated from them are shown in Table 3. There was no clear trend in the open porosity values of the different mixtures. The results were more or less similar for all measured mixtures and oscillated around the average value. The mixture labeled 80S+20R is slightly out of line, having the highest measured value of 19.82%. This is probably due to measurement uncertainty.

Table 3. Basic material properties of the mixtures.

Mixture Designation	Bulk Density (kg/m^3)	Matrix Density (kg/m^3)	Open Porosity (%)
80S+20R	2025	2525	19.82
60S+40R	2022	2495	18.97
40S+60R	2039	2500	18.44
20S+80R	2026	2506	19.18

3.2. Chemical and Mineralogical Properties

Alkaline activation increased the amorphous matter content and also caused the formation of a zeolitic structure, diffracting at low angles. As the slag content increased, the amount of amorphous matter increased as well (see Figure 3). It may be observed that the phase composition of alkali-activated S+R mixtures (Table 4) reflects the gradually changing composition of the mixed precursors.

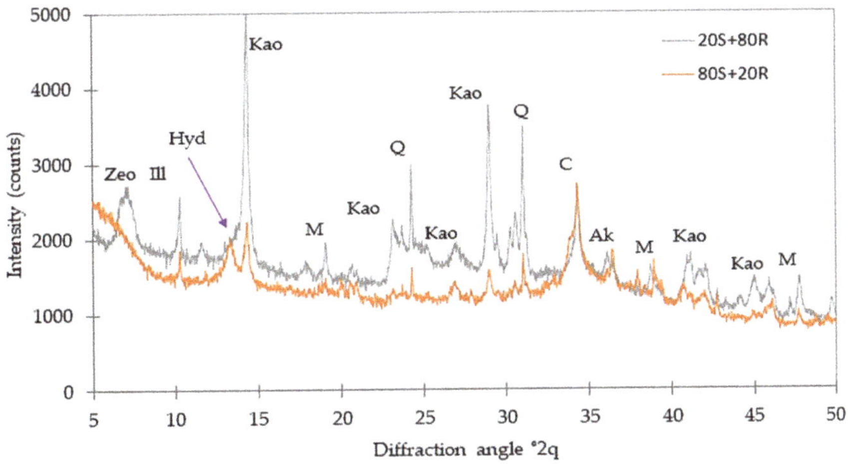

Figure 3. Comparison of two extreme activated products: 20S+80R and 80S+20R. Explanatory notes: Kao, kaolinite; Ill, illite; Zeo, zeolitic structure; Hyd, hydrotalcite; Ak, akermanite; C, calcite; Q, quartz; M, mullite.

Table 4. Phase composition of the mixtures (XRD analysis).

Minerals	Formula	Individual Phase Content (%)			
		80S+20R	60S+40R	40S+60R	20S+80R
Amorphous phase	—	80	75	63	61
Calcite	$CaCO_3$	3	4	5	4
Quartz	SiO_2	0	0	2	3
Akermanite	$Ca_2Mg(Si_2O_7)$	5	4	3	2
Merwinite	$Ca_3Mg(SiO_4)_2$	3	2	2	0
Kaolinite	$Al_2(OH)_4Si_2O_5$	2	6	11	13
Illite	$K_{0.65}Al_{2.0}[Al_{0.65}Si_{3.35}O_{10}](OH)_2$	1	2	5	6
Mullite	$Al_6Si_2O_{13}$	1	3	5	8
Zeolite	—	0	0	0	1
Hydrotalcite	$Mg_6Al_2CO_3(OH)_{16}\cdot 4H_2O$	5	4	4	2

The amount of R-related phases (kaolinite, illite, mullite) decreased, while the amount of slag-related minerals (akermanite, merwinite) increased with the slag content. Interestingly, the calcite content was about the same in all samples. This might indicate that akermanite also undergoes some kind of alteration in an alkaline environment. The zeolite-like structure was clearly apparent only in the 20S+80R; the lower R content caused its merge in low-angle intensity to increase. Hydrotalcite was observed as clear crystalline product of activation; its content increased with the slag dosage (magnesium and carbonate anions both come from slag).

3.3. Mechanical Properties

The mortar specimens (the samples were 40 × 40 × 160 mm in dimension) were tested for basic mechanical properties, i.e., flexural tensile strength and compressive strength. The mechanical properties were investigated both on specimens produced without the addition of WFF and on separate specimens with WFF added. Both strengths (flexural and compressive tensile) were tested over a standard period of 28 days.

Figure 4 below shows the results of the individual flexural tensile strength tests. As expected, the specimens without waste fibers showed lower values for these strengths. The highest values were achieved by the 60S+40R blend in both cases (without and with WFF). The flexural tensile strength of the samples with WFF fibers was 8.96 MPa, an improvement of nearly 27% in these strengths.

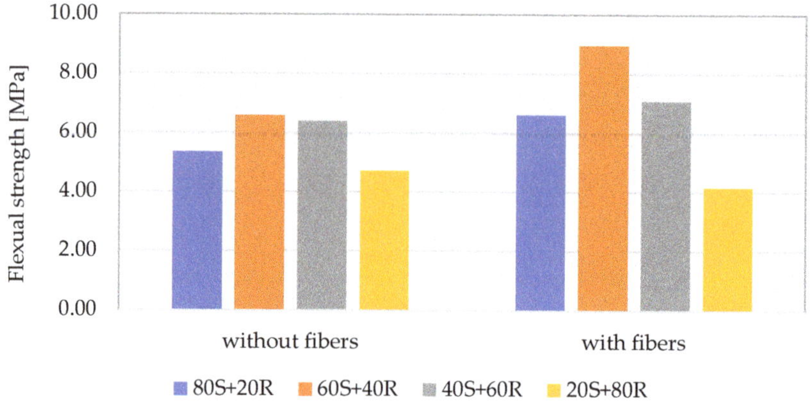

Figure 4. Comparison of the flexural strength of samples without and with WFF fibers.

For the mixture labeled 20S+80R, the final strengths deteriorated and decreased after the addition of fibers. This phenomenon was caused by the poor precursor ratio. A large amount of RON D460 HR metashale did not produce good strengths under alkaline activation.

The compressive strength (Figure 5) of the specimens was tested on fractions of the specimens after the flexural tensile strength test. Again, the compressive strength of the fiber specimens without WFF was highest for the mixture marked 60S+40R, which achieved 46.9 MPa. There was a slight increase in the compressive strength after the addition of WFF fibers. For the 60S+40R blend, the increase was below 10 MPa.

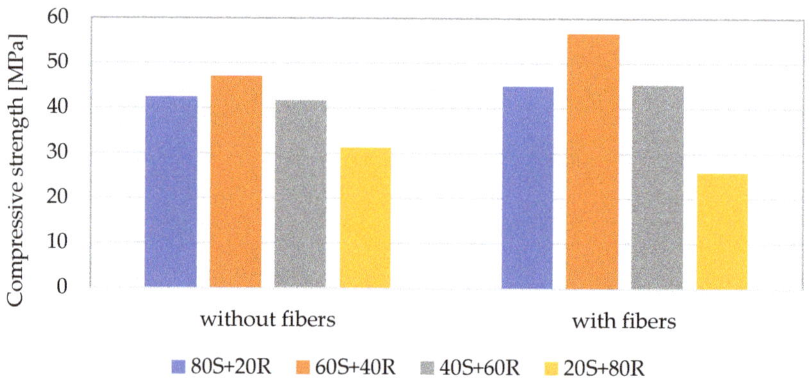

Figure 5. Comparison of the compressive strength of samples withoutand with WFF fibers.

The best blend in terms of mechanical properties appeared to be a blend of the alkali-activated composite with 60% GGFBS and 40% RON D460 HR within the precursor blend. The fibers helped to improve the structure and thus increased the flexural and compressive tensile strengths of the alkali-activated material.

3.4. Scanning Electron Microscopy (SEM)

Scanning electron microscopy (SEM) was used to investigate the surface of the alkali-activated composite. The aim was to capture the interface between the WFF fibers and the sample matrix, and to map any surface changes or micro-cracks in the material (see Figure 6).

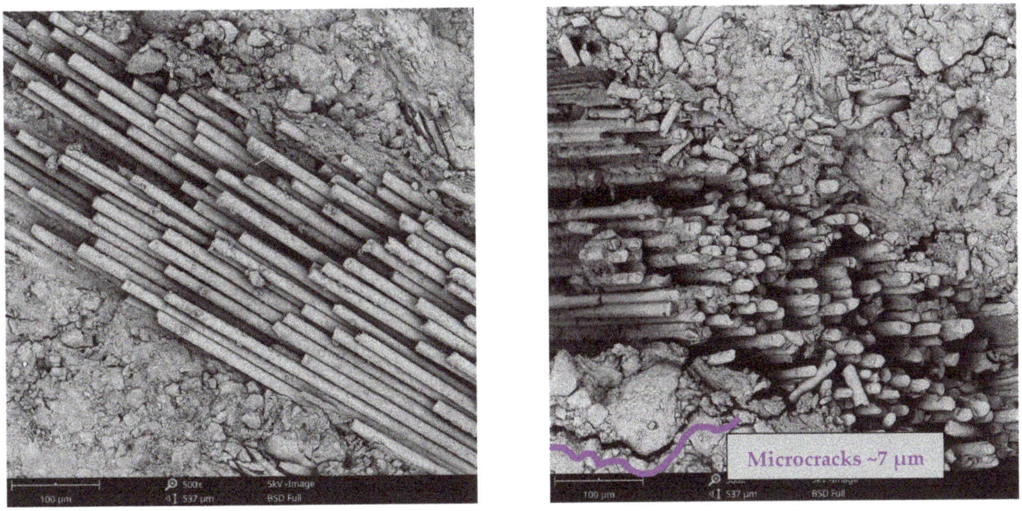

(**a**) (**b**)

Figure 6. (**a**) The interface between the fiber and the matrix. (**b**) Structure of the material with microcrack-trapping fibers.

In Figure 6a, a single strand of the fibers can be seen, in which the individual glass fibers can be seen due to the cut. The quality of the interconnection between the fibers and the sample matrix can be seen. This resulted in a stronger and more durable material structure. In Figure 6b, the ability of the fibers to prevent the propagation of microcracks (shown in purple in the figure) is more evident. The size of the microcracks in the samples was around 7 µm. The largest microcrack captured was 20 µm.

4. Conclusions

This study considered the utilization of various waste materials in the form of upcycling to design environmentally friendly reinforced materials that may be viewed as an alternative to cementitious materials. For this purpose, a blended precursor composed of waste metashale and ground granulated blast furnace slag was activated by a waste cleaning solution and reinforced by waste fiberglass. The results obtained indicated the successful integration of the WFF into the material's matrix without any distinct segregation or the formation of microcracks on the interface. Increasing the metashale from 20% to 40% resulted in a positive effect on the mechanical properties, specifically from 42 MPa to 47 MPa in compressive strength without fibers and from 45 MPa to 57 MPa with fibers. However, a further increase in the metashale content reduced both the flexural and compressive strength. The most beneficial effect of applying fibers was seen for the combination of 60% slag and 40% metashale. In general, the mechanical strength of the designed materials reached a sufficient level and can be deemed to be a viable way for future follow-up work aiming to preciselydetermine the optimal ratio of the components.

Author Contributions: Conceptualization, J.F. and R.Č.; methodology, J.F. and R.Č.; software, M.M.; validation, J.F. and R.Č.; formal analysis, J.F. and R.Č.; investigation, J.F. and M.M.; resources, J.F.; data curation, M.M.; writing—original draft preparation, M.M. and J.F.; writing—review and editing, J.F.; visualization, M.M.; supervision, J.F.; project administration, R.Č.; funding acquisition, R.Č. All authors have read and agreed to the published version of the manuscript.

Funding: This research was funded by the support of M.era-Net Call 2021, Project No. 9262 and financial support from Technology Agency of Czech Republic under project TH80020002 and by the Czech Technical University in Prague under project No. SGS22/137/OHK1/3T/11.

Institutional Review Board Statement: Not applicable.

Informed Consent Statement: Not applicable.

Data Availability Statement: Not applicable.

Acknowledgments: The publication cost of this paper was covered by funds from the Polish National Agency for Academic Exchange (NAWA): "MATBUD 2023—Developing international scientific cooperation in the field of building materials engineering" BPI/WTP/2021/1/00002, MATBUD 2023.

Conflicts of Interest: The authors declare no conflict of interest.

References

1. Zhang, H.; Kodur, V.; Cao, L.; Qi, S. Fiber Reinforced Geopolymers for Fire Resistance Applications. *Procedia Eng.* **2014**, *71*, 153–158. [CrossRef]
2. Refaat, M.; Mohsen, A.; Nasr, E.A.R.; Kohail, M. Minimizing energy consumption to produce safe one-part alkali-activated materials. *J. Clean. Prod.* **2021**, *323*, 129137. [CrossRef]
3. Korniejenko, K.; Łach, M. Geopolymers reinforced by short and long fibers—Innovative materials for additive manufacturing. *Curr. Opin. Chem. Eng.* **2020**, *28*, 167–172. [CrossRef]
4. Parathi, S.; Nagarajan, P.; Pallikkara, S.A. Ecofriendly geopolymer concrete: A comprehensive review. *Clean Technol. Environ. Policy* **2021**, *23*, 1701–1713. [CrossRef]
5. Davidovits, J. Geopolymers and geopolymer new materials. *J. Therm. Anal.* **1989**, *35*, 429–444. [CrossRef]
6. Davidovits, J. *Geopolymer Chemistry and Applications*, 5th ed.; Institut Géopolymère: Saint-Quentin, France, 2020.
7. Fořt, J.; Mildner, M.; Keppert, M.; Černý, R. Waste solidified alkalis as activators of aluminosilicate precursors: Functional and environmental evaluation. *J. Build. Eng.* **2022**, *54*, 104598. [CrossRef]
8. Silva, G.; Kim, S.; Bertolotti, B.; Nakamatsu, J.; Aguilar, R. Optimization of a reinforced geopolymer composite using natural fibers and construction wastes. *Constr. Build. Mater.* **2020**, *258*, 119697. [CrossRef]
9. Liew, Y.M.; Kamarudin, H.; Al Bakri, A.M.; Luqman, M.; Nizar, I.K.; Ruzaidi, C.M.; Heah, C.Y. Processing and characterization of calcined kaolin cement powder. *Constr. Build. Mater.* **2012**, *30*, 794–802. [CrossRef]
10. Ranjbar, N.; Mehrali, M.; Mehrali, M.; Alengaram, U.J.; Zamin, M. Cement and Concrete Research Graphene nanoplatelet-fly ash based geopolymer composites. *Cem. Concr. Res.* **2015**, *76*, 222–231. [CrossRef]
11. Sankar, K.; Kriven, W.M. Geopolymer reinforced with E-glass leno weaves. *J. Am. Chem. Soc.* **2017**, *100*, 2492–2501. [CrossRef]
12. Pernica, D.; Reis, P.N.B.; Ferreira, J.A.M.; Louda, P. Effect of test conditions on the bending strength of a geopolymer-reinforced composite. *J. Mater. Sci.* **2010**, *45*, 744–749. [CrossRef]

materials proceedings

Proceeding Paper

Potential Use of Construction Waste for the Production of Geopolymers: A Review †

Liga Radina *, Andina Sprince, Leonids Pakrastins, Rihards Gailitis and Gita Sakale

Institute of Structural Engineering, Riga Technical University, Kipsalas 6A, LV-1048 Riga, Latvia
* Correspondence: liga.radina@rtu.lv; Tel.: +371-26356332

† Presented at the 10th MATBUD'2023 Scientific-Technical Conference "Building Materials Engineering and Innovative Sustainable Materials", Cracow, Poland, 19–21 April 2023.

Abstract: In recent decades, geopolymer concrete has often been viewed as an alternative to traditional concrete. Although its comparatively lower production of greenhouse gas emissions during a lifecycle is usually mentioned at the top of the list of benefits, the possibility of using various waste materials in its production is a clear advantage as well. This literature review summarizes and analyses the existing information on the different available construction wastes for the production of geopolymer and foamed geopolymer concrete and analyzes the curing conditions, constituents in the aluminosilicate precursor, mechanical properties, and the activator type. As part of the literature review, the use of autoclaved aerated concrete and brick wastes has been evaluated. Autoclaved concrete has been chosen because it is a typical low-strength, cement-based construction material and demolition waste that is currently disposed of in landfills, making it quite a challenge for direct use as a supplementary cementitious material. On the other hand, brick waste, one of the most common construction wastes, can be feasibly used in the form of brick dust. This literature review uses data from randomly selected studies.

Keywords: construction waste; geopolymer; porous geopolymers; waste clay bricks; autoclaved aerated concrete waste

check for updates

Citation: Radina, L.; Sprince, A.; Pakrastins, L.; Gailitis, R.; Sakale, G. Potential Use of Construction Waste for the Production of Geopolymers: A Review. *Mater. Proc.* **2023**, *13*, 2. https://doi.org/10.3390/materproc2023013002

Academic Editors: Katarzyna Mróz, Tomasz Tracz, Tomasz Zdeb and Izabela Hager

Published: 13 February 2023

1. Introduction

The construction industry is the single largest energy consumer in the European Union (EU), as well as one of the largest carbon dioxide (CO_2) emitters (mainly from construction, demolition, and renovation) [1]. Over the past two decades, countless studies have been conducted on ways to improve the current situation. One of the researched areas relates to the development of alternatives for traditional concrete, which is a widely used material for construction. Demand for ordinary Portland cement-based concrete (hereafter referred to as OPC-based concrete) is continuously increasing [2]. Unfortunately, OPC-based concrete requires a considerable amount of ingredients and resources which are acquired through mining and exhaustive processing, causing immense quantities of greenhouse gas emissions to be sent into the atmosphere [3,4]. This review examines geopolymer as a promising alternative because it not only allows for a reduction in CO_2 emissions and water consumption in comparison with OPC-based concrete production (the former by 80% [5]), but it can also be produced effectively using various waste materials [6]. A majority of studies about geopolymer concrete focus on using high amorphous precursors, such as fly ash, silica fume, and ground granulated blast-furnace slag [2,7], which are industrial byproducts. These are abundantly available in some countries. However, in many, they are either unavailable or available only in small, limited quantities [8]; therefore, it is always useful to search for new, suitable, and preferably local materials for geopolymer production. By analyzing the literature, regulatory acts, and studies, it was found that there is a large amount of construction waste in the European Union (and also in Latvia) which is currently being disposed of in landfills or reused only in small quantities. Therefore, within the framework of this review, much attention is paid to widely available and local construction waste (clay brick waste and aerated concrete waste) in order to evaluate the possibilities of their use for the production of geopolymers.

2. Review Methodology

Data from peer-reviewed journal articles were mainly used to analyze the existing information on the production of geopolymers using various construction wastes, including their mechanical and thermal properties and the porosity of samples. The related literature was searched for in Scopus and Web of Science databases using the search keywords "autoclaved aerated concrete waste", "clay brick waste", and "construction waste". Once the search results were obtained, only the articles from the last 6 years were selected for use. The articles, which were used in the introduction section, had no restrictions related to their publication date.

3. Autoclaved Aerated Concrete Waste

Autoclaved aerated concrete (AAC) is a combination of silica sand, lime, cement, water, and an expansion agent. It is relatively lightweight and has lower thermal conductivity and lower shrinkage in comparison to traditional concrete [9]. Disposal of AAC waste to landfills may cause contaminating leaching and pH changes in the water and surrounding soil [10]; therefore, instead of hiding AAC waste from our sight and pretending that the problem is fixed, an approach where AAC waste is reused and recycled should be chosen.

It is important to understand that AAC waste is produced not only during the demolition and construction works, but also during AAC production and transportation, as it is easily damaged [9,11]. In their article, He X. et al. reported that 3–5% of the entire AAC production is waste [12], while Zou D., using the situation in China as an example, wrote that AAC waste will account for about 40% of the various types of building material waste by 2025 [13].

In terms of production volumes, according to the European Autoclaved Aerated Concrete Association, there are more than 100 AAC manufacturing plants in 18 countries, which are producing 15 million cubic meters of AAC every year, while the global AAC market was valued at more than 4445 million EUR in 2021 and is expected to reach more than 8255 million EUR by 2030 [13]. This means that the wider production will be followed by an increasingly large amount of waste, and that the conventional disposal of AAC waste, including backfilling and stacking, will not meet the increasing demand for AAC waste disposal [12]. Additionally, landfill capacities are limited, therefore it is expected that landfill fees will increase, which could lead to the development of more complex and cost-intensive recycling processes [14].

Recycling of AAC Waste

In order to eliminate the above-mentioned environmental problems, several scientists have studied AAC waste-recycling methods. Extensive research has been conducted on AAC waste as an adsorbent material that can solidify harmful chemical components due to its porous structure [13,15] and its incorporation into the mortar, which is one of the most widely used building materials [16]. Because AAC waste is rich in silicate, it may partially replace sand, which is necessary for making mortar; thus, not only does it reduce AAC waste, but it also saves the natural river sand resources. According to Lam's study [17], AAC waste can replace up to 25% of natural sand in new AAC fabrication, yielding materials whose main properties (compressive strength, bulk density volume, and drying shrinkage) meet the technical requirements of ASTM C 1693—standard specification for autoclaved aerated concrete.

Several researchers have been focusing on the chemical properties of AAC waste. They have reported that this type of waste could produce sodium aluminosilicate, aluminosilicate zeolites, or replace cementitious materials. However, the preparation procedures require not only a specific environment, but also a special chemical treatment, which significantly limits this type of AAC waste utilization [11,12].

Scientists' interest in using AAC waste as a supplementary cementitious material is significant. This can be explained by the low strength and reactivity of hydrated AAC particles, though there may be some residual unhydrated phases inside the AAC waste [18]. Within the framework of the study, He X. et al. attempted to find a possible approach to using AAC waste as a cement substitute in building materials. He used a mechanical grinding of the AAC waste in the water environment to promote particle refinement and ion dissolution promotion. This experiment showed that AAC waste could be efficiently used as an alternative cementitious material in cement and concrete after the wet-milling treatment, which notably improved the particle fineness, distribution homogeneity, pH value, and other properties of the AAC waste slurry [12].

Although there are studies on the use of AAC waste for the replacement of sand, which is also a component of geopolymers, or on the development of alternative cementitious materials, during the preparation of this review, no articles were found on the possibilities of using AAC waste in the production of geopolymers.

High amorphous precursors, such as fly ash, silica fume, and ground granulated blast-furnace slag, are unavailable locally in Latvia; therefore, it was essential to evaluate the possibilities of using locally available construction and demolition waste for the production of geopolymers. The literature analysis revealed that clay brick waste (CBW) is much more promising for this purpose.

4. Clay Brick Waste

CBW constitutes a major part of the solid waste generated by construction and demolition activities worldwide. Its disposal results in both the pollution of the environment and the occupation of large areas of land. However, the actuality of the problem can be entirely understood when considering the following data: firstly, construction and demolition (hereafter C&D) activities in the Europe Union are responsible for generating more than 850 million tons of C&D waste per year [19]. In China, this amount exceeds 1.5 billion tons of C&D waste per year, which has led to severe environmental and social problems [20]. In the United States of America, 600 million tons of C&D debris was generated in 2018, which is more than twice the amount, and is generated as municipal solid waste [21].

Secondly, research data indicate that CBW accounts for an average of 30% of total C&D waste in the EU [22]. According to Zhu L. and Zhu Z. [23], CBW from demolished brick walls accounted for approximately 54% of C&D waste in Spain; however, it must be understood that CBW is obtained not only as a result of the demolition, as a large amount of broken clay bricks is obtained after the firing activities, transportation, and also construction with this material.

In their article [24], L. M. Beleuk a Moungam et al. mentioned a brick factory in Cameroon where the annual volume of bricks produced is almost 4000 tons; however, 17% of the production is broken. Thirdly, a significant part of CBW is deposited in landfills or reclamation sites, which are expensive and inefficient. The distances between demolition sites and disposal areas are increasing, negatively affecting transportation costs. In addition, landfills and reclamation sites are limited; however, CBW occupies significant areas, damaging the soil structure [21,23]. Therefore, many scientists are looking for alternatives, both in terms of providing effective waste management practices, thus ensuring a cleaner and greener environment, and also searching for different ways to successfully reuse and recycle the already existing CBW [25–27].

4.1. Geopolymer Production Using CBW

Considering the various studies conducted on the methods of using CBW to produce new building materials, as well as the limited volume of reviews, the use of CBW is analyzed only for geopolymers, without mentioning other building materials.

The most extensively investigated precursor materials are slag, fly ash, silica fume, and metakaolin [2,28,29]. However, recent studies [22,30] have confirmed the feasibility of using low-amorphous aluminosilicates for the synthesis of geopolymers. Waste clay bricks, on the other hand, are excellent low-amorphous aluminosilicates for geopolymer production due to their chemical and mineralogical properties, allowing one to obtain samples with sufficient strength [8]. Clay, obtained from CBW, has the natural advantage of being already calcined at up to a high temperature of 950 °C [31] during the manufacturing process. The combined water in clay minerals evaporates, creating disordered amorphous phases of alumina and silica, which, in turn, allows us to look at CBW as an environmentally friendly and low-cost raw material for the production of geopolymers [8]. In addition, CBW can be used to produce geopolymers with or without widely used precursor materials such as fly ash and slag. More information on CBW as a precursor material and relevant studies is summarized in Table 1. It also provides information about components, the curing conditions, activator type, and compressive strength.

Table 1. CBW-based geopolymer material properties and curing conditions.

Constituents in Aluminosilicate Precursor	Activator Type	Curing Conditions, °C	Compressive Strength, MPa	Ref.
CBW (with the addition of a reduced amount (30%) of PC)	NaOH	25 ± 3, 80% humidity	2 ± 0.5 (7 days) 7 ± 0.5 (28 days)	[32]
CBW only	Na_2SiO_3	25 ± 3	3.3 (7 days)	[2]
			5.4 (28 days) 10.5 (90 days)	
CBW + metakaolin	NaOH + Na_2SiO_3	40 for 20 h, later—ambient temperature	7 ± 0.75	[33]
CBW + fly ash	Na_2SiO_3	25 ± 3	7.5 (7 days)	[2]
			22.4 (28 days)	
CBW + dolomite-concrete powder	NaOH	80 for 24 h, later—40, >95% humidity	15.0	[34]
CBW + Natural Pozzolana	NaOH + Na_2SiO_3	80	37.0 ± 3 (7 days)	[22]
CBW + metakaolin	NaOH + Na_2SiO_3	Ambient cond. for 24 h, then 4–5 h 60	39.0 (7 days) 43.0 (28 days)	[35]
Waste brick powder	NaOH + Na_2SiO_3	20	41.9 (28 days)	[36]
C&D-based masonry waste (CBW, roof tile, hollow brick), glass	NaOH	95–115	42.0 (2 days) 45.0 (3 days)	[37]
CBW + slag	Na_2SiO_3	25 ± 3	68.3 (7 days)	[2]
			81.8 (28 days) 88.5 (90 days)	
Waste brick powder	NaOH + Na_2SiO_3	60 and 80, 50% humidity	-	[38]

Upon analyzing the information available in Table 1, it can be seen that more strength is obtained for geopolymers that use CBW in combination with slag. High-strength samples cannot be obtained using only CBW at ambient conditions, as the compressive strength of the samples, in this case, does not exceed 7 MPa, even after 28 days. The activator type for geopolymers, which are produced using CBW, does not differ from activators which are used in other geopolymers. In most cases, it is NaOH together with Na_2SiO_3; however, in some studies, they have been used alone. These conclusions are also confirmed by J. Migunthanna, who states in her study [2] that CBW-only one-part geopolymers and CBW-only two-part geopolymers are not capable of achieving high compressive strength in ambient curing conditions. She evidences this with the low degree of reaction of CBW, suggesting that elevated curing temperatures are more suitable for 100% CBW-based geopolymers [2,22]. It should be noted that Table 1 includes information on only a few studies wherein CBW has been used to produce geopolymers. Due to the limited volume of the review, its purpose was to show the vast possibilities of how this kind of C&D waste, alone or in combination with other raw materials, can be used in the production of geopolymers. Secondly, Table 1 deliberately includes studies with different curing temperatures to show that the creation of geopolymers is possible not only at temperatures of 60 °C and higher, but also in ambient conditions.

Many studies have determined the best production conditions of geopolymers, thus providing good geopolymerization [35]. Nonetheless, this analysis is complicated by the fact that, not only must the aforementioned curing conditions be observed, but also different precursor materials and activators, alkaline solution concentration [39], and particle size. Therefore, more information on the curing temperature and activator type is provided in Sections 4.2 and 4.3.

4.2. Curing Temperature

Curing conditions have a significant impact on the process of geopolymerization, which is a process of forming an amorphous or semi-crystalline polymeric structure consisting of sialate (Si-O-Al) and siloxo (Si-O-Si) bonds as a result of heterogeneous reactions of powder aluminosilicate oxides dissolved in a high-alkaline solution medium. It is reported that the curing temperature from 25 °C to 145 °C and curing time from 2 h to 24 h highly affects the dissolution of the precursor material [35]. According to Udawattha et al. [40], the recommended curing temperature is 50–80 °C. It is also confirmed by Chen K. et al. in their study [32], showing that the dissolution rate and geopolymerization increase with an increase in temperature. This is explained by the fact that an increase in temperature also increases the average kinetic energy of reactant molecules. Chen K. with his colleagues further emphasized that the control of the temperature may not only significantly affect the geopolymerization process, but might also affect the hardened characterizations of geopolymers [39,40]. Moreover, the results of Yener's and Karaaslan's research on the curing time and temperature effect on the properties of pumice-based geopolymers has shown that a curing treatment at 60 °C and 75 °C up to 168 h increased the strength of the geopolymer samples by almost two-fold compared to the 24-h heat-curing time and approximately 13 times compared to ambient curing [35].

However, one of the biggest disadvantages of geopolymers is the fact that, in the case of fly ash-based and other geopolymers, curing must be conducted at a relatively high temperature due to the poor hydration reactivity. The heat-curing process leads to high costs and energy consumption. It creates a barrier for the broad application of fly ash-based geopolymers in building processes due to the formulation of in situ cast concrete [41]; therefore, during the development of the review, special attention was paid to the possibility of using ambient conditions during the curing time. Despite the apparent advantages of using a relatively high temperature, several studies have confirmed that the curing of a fly ash-based geopolymer at an ambient temperature could be significantly accelerated after adding a small proportion of slag [42] or OPC with high CaO content [41]. Moreover, J. Migunthanna et al. have proven in their research [2] that geopolymers from binary blends of CBW with other aluminosilicate precursors such as slag and fly ash show good compressive strength also at ambient curing conditions. Of course, a more thorough analysis of the existing research is needed. However, it is already clear that geopolymers containing CBW can be successfully manufactured even if curing temperatures are not elevated.

4.3. Activator Type

Such alkaline activators as sodium hydroxide, sodium silicate, KOH, potassium silicate, and sodium metasilicate are widely used to produce geopolymers [43,44]. There are two types of activators, namely, solid and liquid. Liquid activators are usually used in two-part geopolymers, whereas solid activators are usually used in one-part geopolymers. The use of solid activators lowers the cost of materials and environmental footprint, and transportation becomes easier [45].

Unfortunately, the aqueous activators, which are used in two-part geopolymer production, are highly corrosive and hazardous, therefore it is difficult to use them on an industrial scale [2].

The information gathered in Table 1 confirms a general trend that NaOH, sodium silicate Na_2SiO_3, and their combination are mainly used as activators for geopolymers in the production in which CBW is used. It should be noted that Na_2SiO_3 is produced via the direct fusion of pure silica with soda ash in a furnace at a temperature of approximately 1400 °C [2]. This process is highly energy consuming, and CO_2 is emitted not only when furnaces are fired using oil and gas, but also during the chemical reaction. Of course, this activator is not used in large quantities, therefore the environmental impact is small; however, this is a factor to consider when choosing a suitable activator.

5. Conclusions Remarks and Research Perspective

The conducted review provides clear evidence of how urgent the problem of C&D waste is and confirms the interest of researchers in the development of various building materials using such widespread wastes as CBW and AAC waste. Although, during the planning of the review, the possibility of using both AAC waste and CBW for the production of geopolymers was foreseen, a thorough analysis of the literature allowed CBW to be prioritized.

Even though scientists have studied geopolymers for more than two decades, many factors must be evaluated and analyzed when designing geopolymer compositions. It is necessary to choose suitable curing conditions, precursor materials and activators, and an alkaline solution concentration when trying to reduce the necessary resources and environmental impact.

Particular attention should be paid to the effect of CBW particle size, as the results widely vary. Some articles provide information that the compressive strength of geopolymers, produced using only CBW or CBW in combination with fly ash, is only slightly affected by the particle size change. At the same time, other studies indicate more than a 70% increase in compressive strength after decreasing CBW particle size. In contrast, while using CBW with slag, CBW particle size significantly affected the properties of the geopolymers.

Future research will focus on developing geopolymers using CBW and slag and on providing ambient curing conditions. This direction seems promising and not so resource intensive. This is not only an opportunity to develop geopolymers and reduce the C&D waste simultaneously, it is also a way to reduce the necessary amount of high amorphous precursors such as fly ash, silica fume, and ground granulated blast-furnace slag that are unavailable locally in Latvia.

Author Contributions: L.R.: methodology, conceptualization, formal analysis, and writing—original draft preparation. A.S.: validation, conceptualization, methodology, validation, formal analysis, writing—review and editing, and supervision. L.P.: writing—review and editing, supervision, project administration, and funding acquisition. R.G.: validation, writing—review and editing G.S.: validation, writing—review and editing, and supervision. All authors have read and agreed to the published version of the manuscript.

Funding: This publication was supported by Riga Technical University's Doctoral Grant program.

Institutional Review Board Statement: Not applicable.

Informed Consent Statement: Not applicable.

Data Availability Statement: Not applicable.

Acknowledgments: This research was supported by the Latvian Council of Science within the scope of the project "Foamed Geopolymer Made by Additive Manufacturing for the Construction Technology Applications (3D-FOAM)" No. project9608. This research was supported by the European Social Fund within Project No 8.2.2.0/20/I/008 "Strengthening of PhD students and academic personnel of Riga Technical University and BA School of Business and Finance in the strategic fields of specialization" of the Specific Objective 8.2.2. "To Strengthen Academic Staff of Higher Education Institutions in Strategic Specialization Areas" of the Operational Programme "Growth and Employment. Publication cost of this paper was covered with founds of the Polish National Agency for Academic Exchange (NAWA): "MATBUD'2023 - Developing international scientific cooperation in the field of building materials engineering" BPI/WTP/2021/1/00002, MATBUD'2023.

Conflicts of Interest: The authors declare no conflict of interest.

References

1. An Official European Commission Website. In Focus: Energy Efficiency in Buildings. Available online: https://commission.europa.eu/news/focus-energy-efficiency-buildings-2020-02-17_en (accessed on 11 December 2022).
2. Migunthanna, J.; Rajeev, P.; Sanjayan, J. Investigation of waste of clay brick as partial replacement of geopolymer binders for rigid pacement application. *Constr. Build. Mater.* **2021**, *305*, 124787. [CrossRef]
3. Shah, K.W.; Huseien, G.F.; Xiong, T. Functional nanomaterials and their applications toward smart and green buildings. In *New Materials in Civil Engineering*; Pijush, S., Dookie, K., Nagesh, R.I., Sandeep, C., Eds.; Butterworth-Heinemann: Oxford, UK, 2020; Chapter 11; pp. 154–196.
4. Saez-Perez, M.; Duran-Suarez, J.A.; Castro-Gomes, J.P. Improving the behavior of green concrete geopolymers using different hemp preservation conditions (fresh and wet). *Minerals* **2022**, *12*, 1530. [CrossRef]
5. Davidovits, J. Geopolymer cement to minimize carbon-dioxide greenhouse-warming. *Ceram. Trans.* **1993**, *37*, 165–182.
6. Akbari, H.; Heller, T.; Shin, S.; Yang, X.; Kolay, P.K.; Kumar, S.; Mohanty, M. Geopolymer based concrete to reduce carbon footprint of the construction industry. *Min. Eng.* **2013**, *65*, 57–62.
7. Figiela, B.; Brudny, K.; Lin, W.T.; Korniejenko, K. Investigation of mechanical properties and microstructure of construction and demolition waste-based geoopolymers. *J. Compos. Sci.* **2022**, *6*, 191. [CrossRef]
8. Migunthanna, J.; Rajeev, P.; Sanjayan, J. Waste clay bricks as a geopolymer binder for pavement construction. *Sustainability* **2022**, *14*, 6456. [CrossRef]
9. Rahman, R.A.; Fazlizan, A.; Thongtha, A.; Asim, N.; Noorashikin, M.S. The physical and mechanical properties of autoclaved aerated concrete (AAC) with recycled AAC as a partial replacement for sand. *Buildings* **2022**, *12*, 60.
10. Straits Research. Autoclaved Aerated Concrete Market Growth, Trends and Forecast to 2030. Available online: https://straitsresearch.com/report/autoclaved-aerated-concrete-market (accessed on 10 December 2022).
11. Iucolano, F.; Campanile, A.; Caputo, D.; Liduori, B. Sustainable management of autoclaved aerated concrete wastes in gypsum composites. *Sustainability* **2021**, *13*, 3961. [CrossRef]
12. He, X.; Zheng, Z.; Yang, J.; Su, Y.; Wang, T.; Strnadel, B. Feasibility of incorporating autoclaved aerated concrete waste for cement replacement in sustainable building materials. *J. Clean. Prod.* **2020**, *250*, 119455. [CrossRef]
13. Zou, D.; Que, Z.; Cui, W.; Wang, X.; Guo, Y.; Zhang, S. Feasibility of recycling autoclaved aerated concrete waste for partial sand replacement in mortar. *J. Build. Eng.* **2022**, *52*, 104481. [CrossRef]
14. Zou, D.; Que, Z.; Su, D.; Liu, T.; Zhou, A.; Li, Y. Sustainable use of recycled autoclaved aerated concrete waste as internal curing materials in ultra—High performance concrete. *J. Clean. Prod.* **2022**, *373*, 133910. [CrossRef]
15. Volk, R.; Steins, J.J.; Stemmermann, P.; Schultmann, F. Comparison of different post-demolition autoclaved aerated concrete (AAC) recycling options. *IOP Conf. Ser. Earth Environ. Sci.* **2022**, *1078*, 012074. [CrossRef]
16. Castellar, J.A.; Formosa, J.; Chimenos, J.M.; Canals, J.; Bosch, M.; Rosell, J.R.; Silva, H.P.D.; Morató, J.; Brix, H.; Arias, C.A. Crushed autoclaved aerated concrete (CAAC), a potential reactive filler medium for enhancing phosphorus removal in nature-based solutions-preliminary batch studies. *Water* **2019**, *11*, 1442. [CrossRef]
17. Lam, N.N. Recycling of AAC waste in the manufacture of autoclaved aerated concrete in Vietnam. *Int. J. GEOMATE* **2021**, *20*, 128–134. [CrossRef]
18. Qin, L.; Gao, X. Recycling of waste autoclaved aerated concrete powder in Portland cement by accelerated carbonation. *Waste Manag.* **2019**, *89*, 254–264. [CrossRef] [PubMed]
19. Saez, P.V.; Osmani, M. A diagnosis of construction and demolition Waste generation and recovery practice in the European Union. *J. Clean. Prod.* **2019**, *241*, 118400. [CrossRef]
20. Huang, B.; Wang, X.; Kua, H.; Geng, Y.; Bleischwitz, R.; Ren, J. Construction and demolition waste management in China through the 3R principle. *Resour. Conserv. Recycl.* **2018**, *129*, 36–44. [CrossRef]

21. An Official Website of the United States Government—EPA, United States Environmental Protection Agency. Sustainable Management of Construction and Demolition Materials. Available online: https://www.epa.gov/smm/sustainable-management-construction-and-demolition-materials (accessed on 9 December 2022).
22. Silva, G.; Castaneda, D.; Kim, S.; Castaneda, A.; Bertolotti, B.; Ortega-San-Martin, L.; Nakamatsu, J.; Aguilar, R. Analysis of the production conditions of geopolymer matrices from natural pozzolana and fired clay brick wastes. *Constr. Build. Mater.* **2019**, *215*, 633–643. [CrossRef]
23. Zhu, L.; Zhu, Z. Reuse of clay brick waste in mortar and concrete. *Adv. Mater. Sci. Eng.* **2020**, *2020*, 6326178. [CrossRef]
24. Beleuk a Moungam, L.M.; Mohamed, H.; Kamseu, E.; Billong, N.; Melo, U.C. Properties of geopolymers made from fired clay bricks wases and rice husk ash (RHA)—Sodium Hydroxide (NaOH) activator. *Mater. Sci. Appl.* **2017**, *8*, 537–552.
25. Olofinnade, O.M.; Ede, A.N.; Booth, C.A. *Sustainability of Waste Glass Powder and Clay Brick Powder as Cement Substitute in Green Concrete*; Hussain, C.M., Ed.; Handbook of Environmental Materials Management; Springer: Cham, Switzerland, 2018; pp. 1–22.
26. Scheinherrova, L.; Dolezelova, M.; Vimmrova, A.; Vejmelkova, E.; Jerman, M.; Pommer, V.; Cerny, R. Fired clay brick waste as low and eco-friendly pozzolana active filler in gypsum-based binders. *J. Clean. Prod.* **2022**, *368*, 133142. [CrossRef]
27. Abdellatief, M.; Elermam, W.E.; Alanazi, H.; Tahwia, A.M. Production and optimization of sustainable cement brick incorporating clay brick wastes using response surface method. *Ceram. Int.* 2022. *in press*. Available online: https://www.sciencedirect.com/science/article/pii/S0272884222041645 (accessed on 9 December 2022).
28. Luhar, I.; Luhar, S.A. Comprehensive review of fly ash—Based geopolymer. *J. Compos. Sci.* **2022**, *6*, 219. [CrossRef]
29. Wu, X.; Shen, Y.; Hu, L. Performance of geopolymer concrete activated by sodium silicate and silica fume activator. *Case Stud. Constr. Mater.* **2022**, *17*, e01513. [CrossRef]
30. Abadel, A.A.; Albidah, A.S.; Altheeb, A.H.; Alrshoudi, F.A.; Abbas, H.; Al-Salloum, Y.A. Effect of molar ratios on strength, microstructure & embodied energy of metakaolin geopolymer. *Adv. Concr. Constr.* **2021**, *11*, 127–140.
31. Hanein, T.; Thienel, K.C.; Zunino, F.; Marsh, A.T.; Maier, M.; Wang, B.; Canut, M.; Juenger, M.C.; Ben Haha, M.; Avet, F.; et al. Clay calcination technology: State-of-the-art review by the RILEM TC 282-CCL. *Mater. Struct.* **2022**, *55*, 3. [CrossRef]
32. Robayo-Salazar, R.A.; Meia-Arcila, J.M.; de Gutierrez, R.M. Eco-efficient alkali-activated cement based on red clay brick wastes suitable for the manufacturing of building materials. *J. Clean. Prod.* **2017**, *166*, 242–252. [CrossRef]
33. Rovnanik, P.; Rovnaniková, P.; Vysvaril, M.; Grzeszczyk, S.; Janowska-Renkas, E. Rheological properties and microstructure of binary waste red brick powder/metakaolin geopolymer. *Constr. Build. Mater.* **2018**, *188*, 924–933.
34. Ouda, A.S.; Gharieb, M. Development the properties of brick geopolymer pastes using concrete waste incorporating dolomite aggregate. *J. Build. Eng.* **2020**, *27*, 100919.
35. Yener, E.; Karaaslan, C. Curing time and temperature effect on the resistance to wet-dry cycles of fly ash added pumice based geopolymer. *J. Cem. Based Compos.* **2020**, *2*, 19–25. [CrossRef]
36. Fort, J.; Vejmelkova, E.; Konakova, D.; Alblova, N.; Cachova, M.; Keppert, M.; Rovnanikova, P.; Cerny, R. Application of waste brick powder in alkali activated aluminosilicates: Functional and environmental aspects. *J. Clean. Prod.* **2018**, *194*, 714–725.
37. Ulugöl, H.; Kul, A.; Yıldırım, G.; Şahmaran, M.; Aldemir, A.; Figueira, D.; Ashour, A. Mechanical and microstructural characterization of geopolymers from assorted construction and demolition waste-based masonry and glass. *J. Clean. Prod.* **2021**, *280*, 124358.
38. Fort, J.; Novotny, R.; Vejmelkova, E.; Trnik, A.; Rovnanikova, P.; Keppert, M.; Pommer, V.; Cerny, R. Characterization of geopolymers prepared using powdered brick. *J. Mater. Res. Technol.* **2019**, *8*, 6253–6261.
39. Chen, K.; Cheng, Y.; Yu, M.; Liu, L.; Wang, Y.; Zhang, J. Predicting the Geopolymerization Process of Fly-Ash-Based Geopolymer Using Machine Learning. *Buildings* **2022**, *12*, 1792.
40. Udawattha, C.D.; Lakmini, A.V.R.D.; Halwatura, R.U. Fly ash-based geopolymer mud concrete block. In Proceedings of the Moratuwa Engineering Research Conference (MERCon)/4th International Multidisciplinary Engineering Research Conference, Katubedda, Sri Lanka, 30 May–1 June 2018; pp. 583–588.
41. Zhao, J.; Wang, K.; Wang, S.; Wang, Z.; Yang, Z.; Shumuye, E.D.; Gong, X. Effect of elevated temperature on mechanical properties of high-volume fly ash-based geopolymer concrete, mortar and paste cured at room temperature. *Polymers* **2021**, *13*, 1473. [CrossRef] [PubMed]
42. Fang, G.; Ho, W.K.; Tu, W.; Zhang, M. Workability and mechanical properties of alkali-activated fly ash-slag concrete cured at ambient temperature. *Constr. Build. Mater.* **2018**, *172*, 476–487.
43. Hajimohammadi, A.; Ngo, T.; Mendis, P. Enhancing the strength of pre-made foams for foam concrete applications. *Cem. Concr. Compos.* **2018**, *87*, 164–171. [CrossRef]
44. Kyung, W.K.; Hyung, M.L.; Seog-Young, Y.; Hyunseok, K. Fast-curing geopolymer foams with an enhanced pore homogeneity derived by hydrogen peroxide and sodium dodecyl sulfate surfactant. *Minerals* **2022**, *12*, 821.
45. Ghazy, M.; Abd Elaty, M.; Mostafa, S. Properties of one-part versus two-part geopolymers composites-a review. *Am. J. Eng. Res.* **2022**, *11*, 1–14.

materials proceedings

Proceeding Paper

A Study of Fly Ash-Based Geopolymers with Basalt Flour Addition †

Barbara Kozub *, Krzysztof Miernik and Szymon Gądek

Department of Materials Engineering, Faculty of Materials Engineering and Physics, Cracow University of Technology, Al. Jana Pawła II 37, 31-864 Cracow, Poland
* Correspondence: barbara.kozub@pk.edu.pl
† Presented at the 10th MATBUD'2023 Scientific-Technical Conference "Building Materials Engineering and Innovative Sustainable Materials", Cracow, Poland, 19–21 April 2023.

Abstract: The purpose of this study is to evaluate the effect of basalt flour addition, replacing quartz sand, and its proportion on fly ash-based geopolymers' properties. As a base material, F-grade fly ash was used. The activation process was carried out using a 10 mol solution of sodium hydroxide and an aqueous solution of sodium silicate. The tests included measurements of density, compressive and flexural strength, abrasion resistance, and observation of the microstructure of geopolymers. The results of the study showed that basalt flour significantly increases compressive strength and causes a slight increase in flexural strength—by about 106% and 11%, respectively—and it allows for the reduction of the size of voids and the share of porosity in the structure of the tested geopolymers. Basalt flour has an application potential in geopolymer materials to make them more useful in construction.

Keywords: fly ash; quartz sand; basalt flour; geopolymer; abrasion resistance; strength properties

1. Introduction

Geopolymers, by definition, are included in the third-generation lime cement and ordinary Portland cement. They are amorphous synthetic alkali aluminosilicates belonging to the group of inorganic polymers, the properties of which depend mainly on the type of base material used, the type and amount of activator used during their production, as well as external factors of the polycondensation process, such as temperature and heating time. In the face of global trends and challenges, developing binders for the manufacture of geopolymer concrete has become a hot topic in construction science [1–5].

Geopolymers are widely regarded as substitute materials for Portland cement. This is due to their good strength and thermal properties, as well as their very good corrosion resistance. These materials are widely used in various industries [6].

One of the most commonly used raw materials in geopolymer production is fly ash. Standard EN-450-1:2012 [7] defines fly ash as fine-grained dust resulting from the combustion of coal dust, which consists mainly of glassy, spheroidal particles.

Basalts, as basic minerals, include intermediate plagioclase feldspars, as well as augite pyroxene with or without olivine, magnetite, and variable amounts of glass. Natural large deposits of basaltic rocks, which are used as aggregate for concrete, are found in Egypt. During the process of crushing and grinding basalt rock, large amounts of basalt powder are produced as a byproduct. Numerous works have investigated the activity of basalt and the possibilities of its use. The reactivity of basalt mainly depends on the surface area of its particles, the content of silica and alteration minerals, as well as the content of volcanic glass [8–13].

The purpose of this study is to investigate the possibility of using basalt flour as a substitute raw material for quartz sand during the production of fly ash-based geopolymers. The effect of the addition of basalt flour and its proportion on the strength properties of

check for updates

Citation: Kozub, B.; Miernik, K.; Gądek, S. A Study of Fly Ash-Based Geopolymers with Basalt Flour Addition. *Mater. Proc.* **2023**, *13*, 3. https://doi.org/10.3390/materproc2023013003

Academic Editors: Katarzyna Mróz, Tomasz Tracz, Tomasz Zdeb and Izabela Hager

Published: 13 February 2023

the produced geopolymers was evaluated. The scope of the research conducted includes density measurements, compressive and flexural strength tests, abrasion resistance, and microstructure studies using a scanning electron microscope (SEM).

2. Material

The base material from which the test samples were made was fly ash from coal combustion (Figure 1a) from the Skawina Combined Heat and Power Plant (Skawina, Poland). The ash used is classified as class F. Obtained by XRD analysis on an X-ray diffractometer from Panalytical Aeris (Malvern PANalytical, Lelyweg 1, Almelo, the Netherlands), the proportion of phases comprising the fly ash used in the study is shown in Figure 2. The main components are quartz and mullite, with a small proportion of hematite and magnetite, not exceeding 4%. Quantitative analysis was carried out using the Rietveld method in HighScore Plus software (version: 4.8, Malvern PANalytical B.V, Almelo, the Netherlands) with a PDF-4+ database.

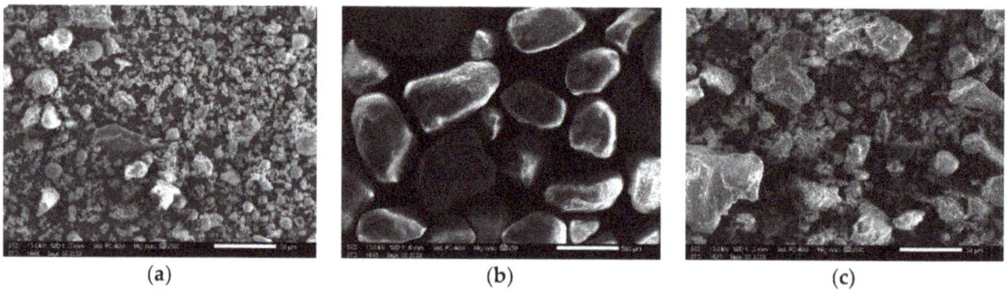

Figure 1. SEM micrographs of (**a**) fly ash, (**b**) quartz sand, (**c**) basalt flour.

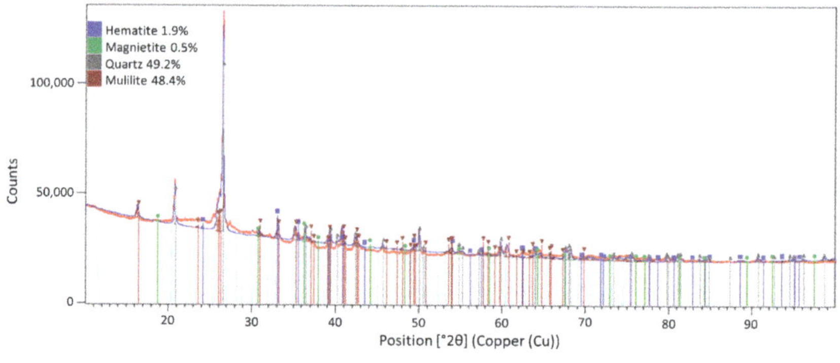

Figure 2. Diffractogram for fly ash.

Materials added in different proportions to the fly ash were quartz sand (Figure 1b) and basalt flour (Figure 1c). Basalt is a volcanic rock that is formed from magma melted in the Earth's mantle. Basalt flour is produced as a result of processing during the extraction of the raw material. The oxide compositions of the fly ash, quartz sand, and basalt powder used for the tests, which were determined with the use of the JEOL JSM-820 scanning electron microscope (IXR Inc., Austin, TX, USA) with the EDS attachment, are shown in Table 1.

Table 1. Oxide composition of raw materials.

Oxides [wt.%]	Fly Ash	Basalt Flour	Quartz Sand
Silica (SiO_2)	53.83 ± 0.38	51.89 ± 0.35	94.24 ± 0.49
Aluminum oxide (Al_2O_3)	28.12 ± 0.24	17.40 ± 0.19	4.81 ± 0.11
Calcium oxide (CaO)	3.16 ± 0.10	10.33 ± 0.16	0.95 ± 0.12
Magnesium oxide (MgO)	2.15 ± 0.07	5.10 ± 0.10	-
Sodium oxide (Na_2O)	2.06 ± 0.07	4.17 ± 0.09	-
Iron oxide (FeO)	6.73 ± 0.22	9.69 ± 0.25	-
Potassium oxide (K_2O)	3.94 ± 0.10	1.42 ± 0.06	-

The summary plots of particle size distribution and cumulative curves for ash, sand, and basalt meal are shown in Figure 3a,b, respectively. Particle size distribution tests were carried out using an instrument from Anton Paar GmbH (Graz, Austria).

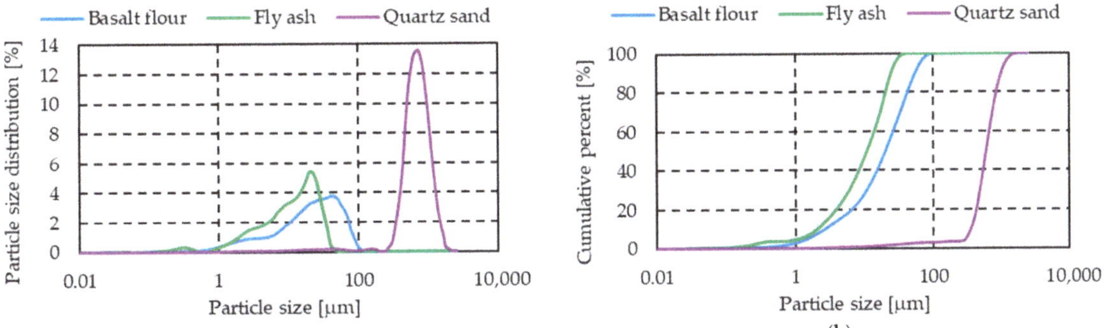

Figure 3. (**a**) The particle size distribution curves for fly ash, quartz sand, and basalt flour; (**b**) the cumulative curves for fly ash, quartz sand, and basalt flour.

3. Sample Preparation

The first step in the preparation of the geopolymer mortar was to mix dry ingredients in a GEOLAB cement mortar mixer (GEOLAB, Warsaw, Poland), which were previously weighed in the appropriate proportions for each mixture according to Table 2, which shows the composition of the mixtures used for the geopolymers, along with sample determinations.

Table 2. Percentage of individual dry components in the mixtures used to produce geopolymers, along with designations.

Mixture No.	Fly Ash [wt.%]	Basalt Flour [wt.%]	Quartz Sand [wt.%]	Sample ID
Mixture I	50	50	-	BF_0
Mixture II	50	37.5	12.5	BF_12.5
Mixture III	50	25	25	BF_25
Mixture IV	50	12.5	37.5	BF_37.5
Mixture V	50	-	50	BF_50

Then, an alkaline 10 mol solution of sodium hydroxide and an aqueous solution of sodium silicate R-145, whose mass ratio was 1:2.5, was prepared and used to activate the prepared mixtures. The dry ingredients with the addition of an alkaline solution were mixed until a uniform paste was obtained. The ratio of dry ingredients to the solution was 0.4. The prepared geopolymer paste was poured into molds, which were then cured in an SLW 750 STD laboratory oven (POL-EKO-APARATURA, Wodzislaw Slaski, Poland) at 75 °C for 24 h. The samples were seasoned under laboratory conditions for 28 days.

4. Methods

The density of the produced geopolymers was measured using a geometric method—measurements were made of the volume and weight of the samples made for compressive strength tests.

Compressive strength and flexural strength measurements were carried out in accordance with PN-EN 196-1:2016-07 [14]. The tests were performed on a Matest 3000 kN testing machine (Matest, Treviolo, Italy) equipped with heads for compressive and flexural strength measurements. For compressive strength tests, six perpendicular specimens of $50 \times 50 \times 50$ mm^3 each were prepared. In contrast, four $50 \times 50 \times 200$ mm^3 specimens were made for flexural strength testing. The specimens were made in accordance with PN-EN 12390-1:2021-12.

The abrasion resistance of the produced geopolymers was determined using the Boehme method. The test was carried out in accordance with PN-EN 14157:2017-11 [15]. The samples prepared for the measurements had dimensions of $71 \times 71 \times 71$ mm^3. The abrasive used during the test was alumina (20 g was used for each sample pass). The clamping force of the sample to the disc was equal to 294 N. For each specimen (three specimens were abraded for each compound), 16 cycles were performed for 20 rotations of the disc per cycle. After the cycle was completed, the sample was rotated 90° and another cycle was started. One of the methods described in the standard for determining abrasion resistance is the method for measuring sample height loss, which was used in this study. This method involves measuring the height of the sample before and after the test and determining their difference. Based on the literature [16], the classification of the tested geopolymers in terms of abrasion resistance was made.

The standard deviation was determined for all obtained results, which was plotted on graphs in the form of error bars.

The microstructure of the fabricated geopolymers was also observed using a JEOL JSM-820 scanning electron microscope (IXR Inc., Austin, TX, USA). A JOEL DII-29030SCTR vacuum coating machine (IXR Inc., Austin, TX, USA) was used to sputter gold onto the samples.

5. Results and Discussion

Analyzing the results of density measurements of the tested geopolymers (Figure 4), it can be observed that the replacement of quartz sand with basalt flour has a slight effect on the obtained values—a slight increase in density can be observed with an increase in the share of basalt meal in the mixture. This is related to the approximate densities of sand and basalt flour. The obtained results do not differ from the values of densities of fly ash-based geopolymers presented in the literature [2,4,17].

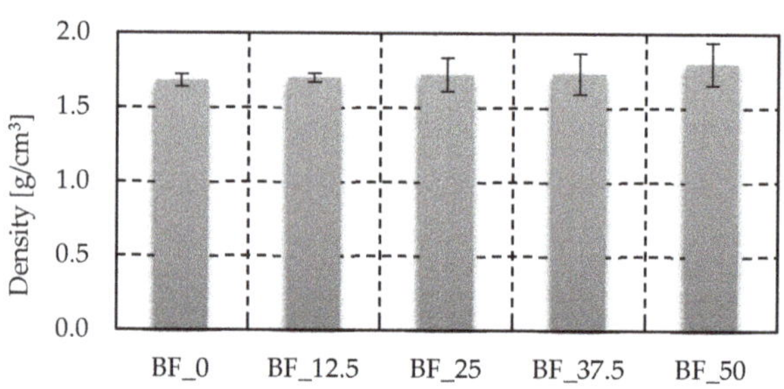

Figure 4. The density of the tested geopolymers.

As can be observed in Figure 5a,b, replacing quartz sand with basalt flour has a favorable effect on the strength properties of the produced geopolymers. As the proportion of basalt meal in the mix increases, there is a significant increase in compressive strength and a gentle increase in flexural strength. The highest values of compressive strength and flexural strength were obtained for the sample which was made from a mixture consisting of fly ash and basalt flour in a ratio of 1:1. The recorded increase in these values relative to the reference sample, made from a mixture of fly ash and sand in a ratio of 1:1, was about as high as 106% and 11% for compressive strength and flexural strength, respectively. It is also worth noting that for samples with 37.5% and 50% basalt meal, the resulting compressive strengths are higher than the compressive strengths of average concretes used in residential and commercial construction, typically ranging between 17 MPa and 28 MPa [18]. The particles of basalt powder are small, so they can fill the voids in the geopolymer structure, which is confirmed by SEM images, and thus effectively improve the strength properties of the tested composites. In addition, the particles of basalt powder dispersed in the geopolymer structure can contribute to passivation and stress dissipation, which allows the delay of the appearance of plastic deformation and the appearance of cracks, which increases the bending strength [19,20].

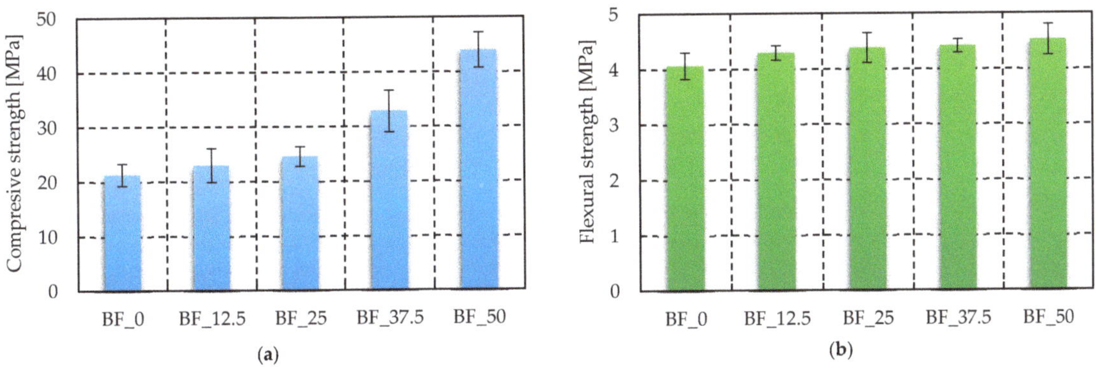

Figure 5. Results for tested geopolymers of (**a**) compressive strength test, (**b**) flexural strength.

The obtained results of compressive strength are also higher in comparison with the results from measurements of compressive strength of geopolymers produced on the basis of basalt meal activated with aqueous sodium hydroxide solution, as presented by Saray et al. [21].

In their work, Venyite et al. [22] studied geopolymers produced based on a mixture of metakaolin, calcined laterite, and basalt flour activated with a 6 mol sodium hydroxide solution and cured at room temperature. The authors showed that the incorporation of basalt into geopolymers based on calcined laterite and metakaolin resulted in compressive strengths of 41.14, 34.46, 40.46, and 24.93 MPa for 20, 30, 40, and 50 wt.% basalt addition.

Beskopylny et al. [23] studied, among other things, the statistical compressive and flexural strengths of fine-grained geopolymer concrete with different types of stone flours. They obtained compressive strength values between 34.1 MPa and 52.2 MPa, and values between 4 MPa and 6.7 MPa for flexural strength for the produced geopolymers.

Figure 6 shows the values of height loss for all tested geopolymers after abrasion tests. Analyzing the results obtained, it can be concluded that all the geopolymers produced are very hard-to-wear materials, as evidenced by the height loss not exceeding the value of 2.5 mm. The highest abrasion resistance was demonstrated by samples made from a mixture consisting of fly ash and basalt flour in a ratio of 1:1, for which the measured average value of sample height loss was 0.81 mm.

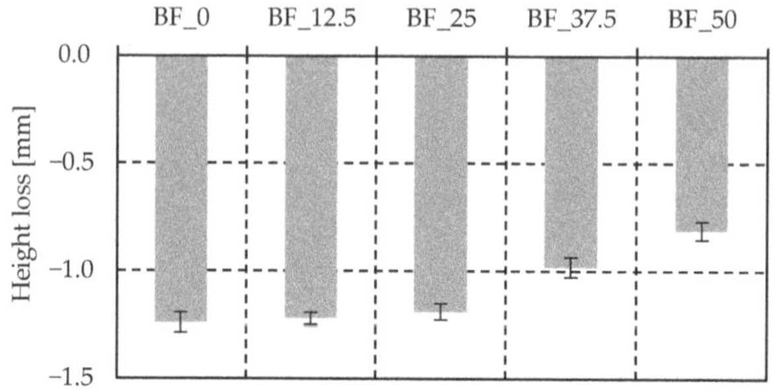

Figure 6. The height loss of tested geopolymers due to abrasion test.

Similar results for a geopolymer made from a mixture based on fly ash and quartz sand, also activated with a 10 mol sodium hydroxide solution, were obtained by Bazan et al. [24], where the average value of sample height loss after abrasion tests was 0.9 ± 0.02 mm, which also classifies the tested material as very difficult to abrade.

Figure 7 shows microphotographs of the structures of the produced geopolymers. For all samples, typical features of the structure of geopolymers produced based on fly ash [25] can be observed, including unreacted spheroidal fly ash particles and dissolved fly ash particles. In addition, it is possible to observe the presence of sand and basalt particles, present in the geopolymer gel in the samples for the production of which they were used. During the study of the structure of the tested geopolymer composites, the influence of the proportion of basalt flour on porosity could be observed—with the increase in the proportion of basalt flour in the mixture, the size of the voids, as well as their share in the structure, decreases.

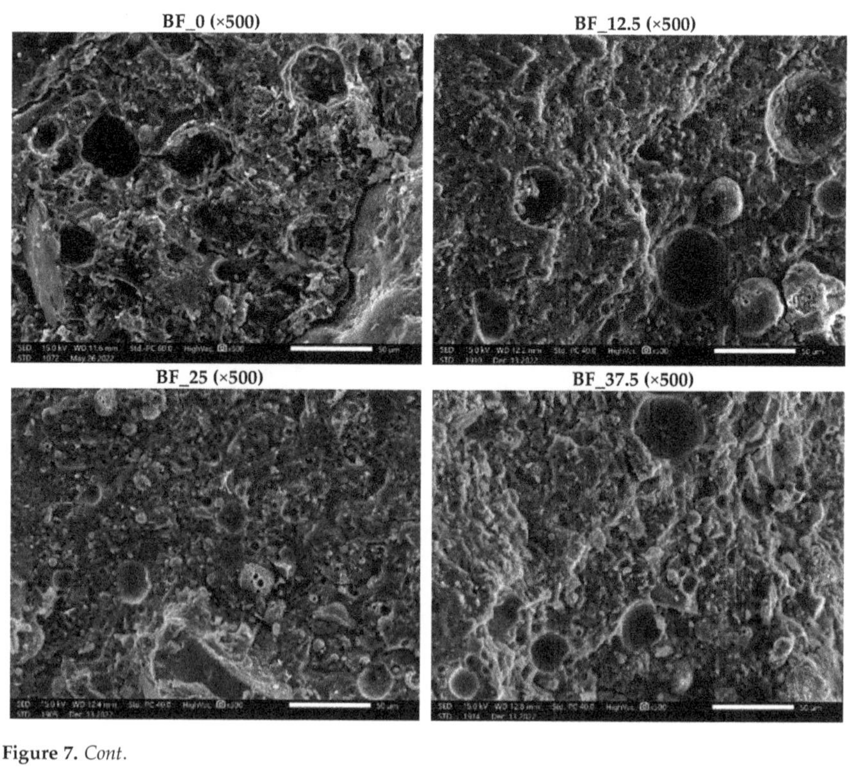

Figure 7. *Cont.*

BF_50 (×500)

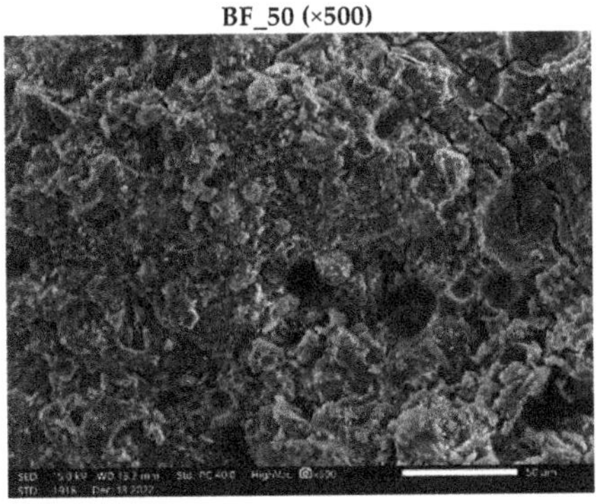

Figure 7. The microstructure SEM photography of produced geopolymers.

6. Conclusions

The study investigated the possibility of using basalt flour as a substitute raw material for quartz sand during the production of fly ash-based geopolymers. The effect of the addition of basalt flour and its proportion on the durability properties of the produced geopolymers was evaluated. Based on the analysis of the obtained test results, the following conclusions can be formulated:

- Basalt flour can be successfully used as a substitute for quartz sand in the production of geopolymer mortars;
- Replacement of quartz sand with basalt flour has a slight effect on the density of geopolymers—a small increase in density can be observed as the proportion of basalt flour in the mixture increases;
- The use of an additive in the form of basalt flour allows a significant improvement in compressive strength and a slight increase in flexural strength, by about 106% and 11%, respectively, compared to geopolymers made based on fly ash with the addition of quartz sand in a ratio of 1:1;
- The addition of basalt powder allows for the reduction of the size of voids and the porosity in the structure of the tested geopolymers.

Basalt flour can be successfully used as a replacement for quartz river sand during the production of fly ash-based geopolymers which have application potential in construction.

Author Contributions: Conceptualization, B.K.; methodology, B.K.; validation, B.K.; formal analysis, B.K. and K.M.; investigation, B.K.; resources, B.K.; data curation, B.K.; writing—original draft preparation, B.K. and S.G.; writing—review and editing, B.K., S.G. and K.M.; visualization, B.K.; supervision, B.K. All authors have read and agreed to the published version of the manuscript.

Funding: The publication cost of this paper was covered with funds from the Polish National Agency for Academic Exchange (NAWA): "MATBUD'2023—Developing international scientific cooperation in the field of building materials engineering" BPI/WTP/2021/1/00002, MATBUD'2023.

Institutional Review Board Statement: Not applicable.

Informed Consent Statement: Not applicable.

Data Availability Statement: Not applicable.

Acknowledgments: This work has been supported by the National Centre for Research and Development in Poland in the framework of the project SMART-G Smart Geopolymers (ERA-MIN2-3/SMART-G/1/2022).

Conflicts of Interest: The authors declare no conflict of interest.

References

1. Mugahed Amran, Y.H.; Alyousef, R.; Alabduljabbar, H.; El-Zeadani, M. Clean production and properties of geopolymer concrete: A review. *J. Clean. Prod.* **2020**, *251*, 119679. [CrossRef]
2. Kozub, B.; Bazan, P.; Gailitis, R.; Korniejenko, K.; Mierzwiński, D. Foamed Geopolymer Composites with the Addition of Glass Wool Waste. *Materials* **2021**, *14*, 4978. [CrossRef] [PubMed]
3. Provis, J.L.; Van Deventer, J.S.J. *Geopolymers: Structure, Processing, Properties and Industrial Applications*; Woodhead Publishing Limited: Cambridge, UK, 2009.
4. Kozub, B.; Castro-Gomes, J. An Investigation of the Ground Walnut Shells' Addition Effect on the Properties of the Fly Ash-Based Geopolymer. *Materials* **2022**, *15*, 3936. [CrossRef]
5. Davidovits, J. *Geopolymer Chemistry and Applications*, 4th ed.; Institut Geopolymere: Saint-Quentin, France, 2008.
6. Hu, W.; Nie, Q.; Huang, B.; Shu, X.; He, Q. Mechanical and microstructural characterization of geopolymers derived from red mud and fly ashes. *J. Clean. Prod.* **2018**, *186*, 799–806. [CrossRef]
7. *PN-EN 450-1:2012*; Popiół Lotny do Betonu—Część 1: Definicje, Specyfikacje i Kryteria Zgodności. Polish Committee for Standardization: Warsaw, Poland, 2014. (In Polish)
8. Korkanc, M.; Tugrul, A. Evaluation of Selected Basalts from the Point of Alkali-Silica Reactivity. *Cem. Concr. Res.* **2005**, *35*, 505–512. [CrossRef]
9. Çopuroğlu, O.; Andiç-Çakir, Ö.; Broekmans, M.A.; Kühnel, R. Mineralogy, Geochemistry and Expansion Testing of an Alkali-Reactive Basalt from Western Anatolia, Turkey. *Mater. Charact.* **2009**, *60*, 756–766. [CrossRef]
10. Korkanç, M.; Tuğrul, A. Evaluation of Selected Basalts from Niğde, Turkey, as Source of Concrete Aggregate. *Eng. Geol.* **2009**, *75*, 291–307. [CrossRef]
11. Unčík, S.; Kmecová, V. The Effect of Basalt Powder on the Properties of Cement Composites. *Procedia Eng.* **2013**, *65*, 51–56. [CrossRef]
12. Laibao, L.; Yunsheng, Z.; Wenhua, Z.; Zhiyong, L.; Lihua, Z. Investigating the Influence of Basalt as Mineral Admixture on Hydration and Microstructure Formation Mechanism of Cement. *Constr. Build. Mater.* **2013**, *48*, 434–440. [CrossRef]
13. Marfil, S.A.; Maiza, P.J.; Bengochea, A.L.; Sota, J.D.; Batic, O.R. Relationships between SiO_2, Al_2O_3, Fe_2O_3, CaO, K_2O, and Expansion in the Determination of the Alkali Reactivity of Basaltic Rocks. *Cem. Concr. Res.* **1998**, *28*, 189–196. [CrossRef]
14. *PN-EN 196-1:2016-07*; Metody badania cementu—Część 1: Oznaczanie wytrzymałości. Polish Committee for Standardization: Warsaw, Poland, 2018.
15. *PN-EN 14157:2017-11*; Test methods for natural stone—Determination of abrasion resistance. Polish Committee for Standardization: Warsaw, Poland, 2017.
16. Plinninger, R.J. *Klassifizierung und Prognose von Werkzeugverschleiß bei konventionellen Gebirgslösungsverfahren im Festgestein*; Münchner Geologische Hefte, Reihe B: Angewandte Geologie, B17; Ludwig-Maximilians-Universität München/Institut für Allgemeine und Angewandte Geologie: München, German, 2002.
17. Sefiu, A.R.; Zhang, B.; Rohiverth, G.; Tiju, T.; Yang, M. Geopolymer for use in heavy metals adsorption, and advanced oxidative processes: A critical review. *J. Clean. Prod.* **2019**, *213*, 42.
18. Gerald, B.; Neville, P.E. *Concrete Manual: Chapter 3: The strength of concrete; Based on the 2015 IBC® and ACI 318-14*; International Code Council: Washington, DC, USA, 2015; ISBN 978-1-60983-618-4.
19. Punurai, W.; Kroehong, W.; SaptamongkoL, A.; Chindaprasirt, P. Mechanical properties, microstructure and drying shrinkage of hybrid fly ash-basalt fiber geopolymer paste. *Constr. Build. Mater.* **2018**, *186*, 62–70. [CrossRef]
20. Li, M.; Gong, F.; Wu, Z. Study on mechanical properties of alkali-resistant basalt fiber reinforced concrete. *Constr. Build. Mater.* **2020**, *245*, 118424. [CrossRef]
21. El-Shahte, M.; Saraya1, I.; El-Fadaly, E. Preliminary Study of Alkali Activation of Basalt: Effect of NaOH Concentration on Geopolymerization of Basalt. *J. Mater. Sci. Chem. Eng.* **2017**, *5*, 58–76.
22. Venyite, P.; Makone, E.C.; Kaze, R.C.; Nana, A.; Nemaleu, J.G.N.; Kamseu, E.; Melo, U.C.; Leonelli, C. Effect of Combined Metakaolin and Basalt Powder Additions to Laterite-Based Geopolymers Activated by Rice Husk Ash (RHA)/NaOH Solution. *Silicon* **2022**, *14*, 1643–1662. [CrossRef]
23. Beskopylny, A.; Shcherban, E.; Stel'makh, S.; Mailyan, L.; Meskhi, B.; El'shaeva, D. The Influence of Composition and Recipe Dosage on the Strength Characteristics of New Geopolymer Concrete with the Use of Stone Flour. *Appl. Sci.* **2022**, *12*, 613. [CrossRef]
24. Bazan, P.; Kozub, B.; Korniejenko, K.; Gailitis, R.; Sprince, A. Tribo-Mechanical Behavior of Geopolymer Composites with Wasted Flax Fibers. *IOP Conf. Ser. Mater. Sci. Eng.* **2021**, *1190*, 012030. [CrossRef]
25. Temuujin, J.; Minjigmaa, A.; Lee, M.; Chen-Tan, N.; Van Riessen, A. Characterisation of class F fly ash geopolymer pastes immersed in acid and alkaline solutions. *Cem. Concr. Compos.* **2011**, *33*, 1086–1091. [CrossRef]

materials proceedings

MDPI

Proceeding Paper

Influence of the Size of Milled Coal Gangue Particles on the Mechanical Properties of Geopolymers †

Beata Figiela [1,*], Kinga Korniejenko [1], Alperen Bulut [2], Baran Şahin [2], Göksu Azizağaoğlu [3], Kinga Pławecka [1] and Barbara Kozub [1]

1 Faculty of Material Engineering and Physics, Cracow University of Technology, Jana Pawła II 37, 31-8646 Cracow, Poland

2 Faculty of Engineering, İzmir Institute of Technology, Gülbahçe Kampüsü, Izmir 35430, Turkey

3 Faculty of Engineering and Natural Sciences, Bursa Technical University, Bursa 16310, Turkey

* Correspondence: beata.figiela@pk.edu.pl

† Presented at the 10th MATBUD'2023 Scientific-Technical Conference "Building Materials Engineering and Innovative Sustainable Materials", Cracow, Poland, 19–21 April 2023.

Abstract: Geopolymers are inorganic materials resulting from the synthesis of silicon and aluminum in a polycondensation reaction. In this study, coal mine waste material from the Wieczorek mine in the Śląskie Voivodeship was used to produce geopolymers. The material was prepared, crushed and milled beforehand due to its large dimensions. The material was subjected to sieve analysis, which allowed to distinguish three fractions. The next step was thermal activation of the obtained powder grain sizes. After thermal activation, the material was combined with an alkaline solution to prepare geopolymers. Photographs of the microstructure were taken in order to determine the chemical composition of the geopolymer and to study the phase composition. The best compressive and bending strengths were exhibited by geopolymer samples with particle sizes ranging below 200 µm—19 MPa and 5.7 MPa, respectively.

Keywords: geopolymer; coal gangue; particle size; circular economy

Citation: Figiela, B.; Korniejenko, K.; Bulut, A.; Şahin, B.; Azizağaoğlu, G.; Pławecka, K.; Kozub, B. Influence of the Size of Milled Coal Gangue Particles on the Mechanical Properties of Geopolymers. *Mater. Proc.* **2023**, *13*, 4. https://doi.org/10.3390/materproc2023013004

Academic Editors: Katarzyna Mróz, Tomasz Tracz, Tomasz Zdeb and Izabela Hager

Published: 13 February 2023

1. Introduction

Geopolymers are called alkaline-activated materials (due to the first phase of geopolymer formation), and are created by combining aluminosilicate materials with alkaline compounds—strongly basic sodium, potassium [1,2] or acid phosphates [3,4]. Aluminum and silicon are obtained from minerals (most often metakaolin [5], volcanic tuffs and waste materials, such as fly ash or slags) [6–8]. Geopolymerization takes place at temperatures not exceeding 100 °C [9,10]. Geopolymers are characterized by good mechanical properties [11], including compressive strength, fire resistance, corrosion resistance, binding of heavy metal elements, [12,13] these properties predispose geopolymeric materials to construction applications, immobilization of hazardous materials or securing waste landfills [14]. Geopolymer materials are generally expected to replace Portland cement, the long-term use of which has led to a high carbon footprint and environmental problems [15,16]. Geopolymers are environmentally friendly compared to conventional cement, and their production results in approximately six times lower CO_2 emissions to the atmosphere than the manufacturing of cement [17,18].

Coal shales are materials produced during coal mining [19]. Their amount increases year by year alongside energy consumption. One of the largest coal mining countries is China, where the annual production is over 70 million tons, resulting in the creation of huge amounts of deposited coal shales [20]. So far, they are not used in large quantities, hence the need to find an appropriate application for this waste [21]. The main components of gangue include quartz, kaolinite and illite, all three of which contain silicon and aluminum compounds [22]. One of the possible applications for coal shales is to use them in geopolymerization [23].

This study aimed to observe the effects of different particle sizes of milled coal gangue on the mechanical properties of geopolymers. In order to better understand the potential of coal gangue in the geopolymerization process, various components of the process were examined, e.g., the influence of thermal activation on the reactivity of the raw material, the influence of particle size [24] especially on fly ash-based geopolymers [25,26] or the influence of the type of alkaline activator on the strength properties [27].

2. Materials and Methods

The base raw material for the production of geopolymers was obtained from the Wieczorek mine (Poland). The first stage of work with this material required crushing it in a jaw crusher and grinding, as the coal shales were rather large. The obtained sample of the material was subjected to a quantitative analysis of the phase composition using the company's X-ray diffractometer PANanalitycal Almelo. The study showed the presence the following minerals in the material: quartz—42.5%, muscovite—12.5%, kaolinite—36.5%, and illite—9.0% (see: Table 1). The quantitative analysis of the phase composition given in Table 1 shows approximate values. The amorphous phase was not determined in the material.

Table 1. Percentage of the minerals in the material.

Identified Phase	Quartz	Kaolinite	Illite	Muscovite
Chemical Formula	SiO_2	$KAl_2(Si_3Al)O_{10}(OH,F)_2$	$Al_2Si_2O_5(OH)_4$	$(K,H_3O)Al_2Si_3AlO_{10}(OH)_2$
Percentage (%)	42.5%	35.6%	9.0%	12.6%

The crushing stage was followed by sieve analysis of the obtained material and divided the material into specific fractions; below 200 µm, in the range of 200 µm to 900 µm and from 900 µm to 1.25 mm. The raw material was heated in a chamotte furnace at 700 °C for 24 h. Calcination is a thermal method of activating a material to remove the carbon content and create the appropriate microstructure. The loss of ignition was 5.2% (LOI).

After grinding and calcination of the material, an analysis of the particle size of the obtained material was carried out. Figure 1 shows the results obtained with this method. The main particle size under 200 µm for the material obtained by sieve analysis was 21 µm, the overall particle size did not exceed 70 µm. For the material in the size range of 200–900 µm, the average particle size was 409 µm, whereas less than 2% of the particles were of the size within the range of 2.2 µm to 125 µm. In the highest size range of 900 µm–1.25 mm, the average grain size of the raw material was 924 µm. The wide distribution of particles for W 200–900 µm was caused by the error of sieve analysis. Some of the undersized particles were not able to pass through the sieve due to the agglomeration of larger particles, while some of them became stuck in the mesh of the sieve.

The finished material with a different range of grain fractions was mixed with a prepared 10 M alkaline solution.

An alkaline solution of 10 M sodium hydroxide with aqueous sodium silicate was prepared. For this purpose, a weighed amount of technical sodium hydroxide flakes was mixed with tap water and stirred until the component dissolved. Then, aqueous sodium silicate was added to the mixture at a ratio of 1:2.5. The resulting solution was allowed to stand for 24 h to stabilize the concentration under laboratory conditions. After this time, the solution and material from the mine were placed in the bowl of a low-speed mixer for approximately 10 min. In this way, three geopolymer mixes were produced (Table 2). The resulting mass was poured into a set of rectangular molds, then to remove air bubbles, the molds were placed on a vibrating table. The last stage of preparing the samples entailed placing them in a laboratory dryer at the temperature of 75 °C for 24 h. On the second day, after taking the molds out of the dryer, the samples were demolded and then seasoned under laboratory conditions for 28 days.

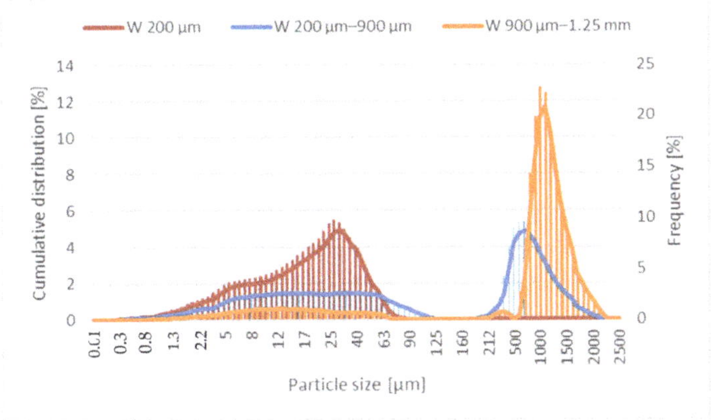

Figure 1. Diagram of particle distribution of W 200 µm, W 200–900 µm and W 900 µm–1.25 mm.

Table 2. Composition of the prepared geopolymers.

Sample	Alkaline Solution 10 M	Mining Raw Material Particle Size
G 900 µm–1.25 mm	Technical sodium flakes, tap water, sodium silicate	900 µm–1.25 mm
G 200 µm–900 µm	Technical sodium flakes, tap water, sodium silicate	200 µm–900 µm
G 200 µm	Technical sodium flakes, tap water, sodium silicate	200 µm

The mineral phase was identified using X-ray diffraction on a PANanalitycal Almelo. The test was carried out on powdered material using a Cu lamp. Analysis of the identified phases was performed with the use of High Score Plus software.

The particle size analyses were performed with using a Particle Size Analyzer AntonPaar GmbH (Graz, Austria). Compressive strength tests were carried out on a testing machine MATEST 3000 kN with a speed of 0.05 MPa/s, according to PN-EN 196-1:2016-07 standard "Cement test methods—Part 1: Strength determination ". The tests were carried out on cubic samples with dimensions of 50 mm × 50 mm × 50 mm. The flexural strength tests were carried out similarly to the compressive strength tests, according to the PN-EN 196-1:2016-07 on the MATEST 3000 kN testing machine with a speed of 0.05 MPa/s. These tests were carried out on 50 mm × 50 mm × 200 mm. The distance between the support points was 150 mm. The observation of the microstructure of the geopolymers was made using scanning electron microscope (SEM) of the JEOL JSM 820 type with EDS on the breakthroughs of the samples after the compressive strength tests. The preparation of samples for observation entailed sputtering a thin layer of gold by a vacuum sprayer JEOL—JEE-4X.

3. Results

3.1. Compressive Strength

The results for compressive strength are shown in Figure 2. The test was carried out on four samples for each variant. The graph shows the average value obtained from these measurements.

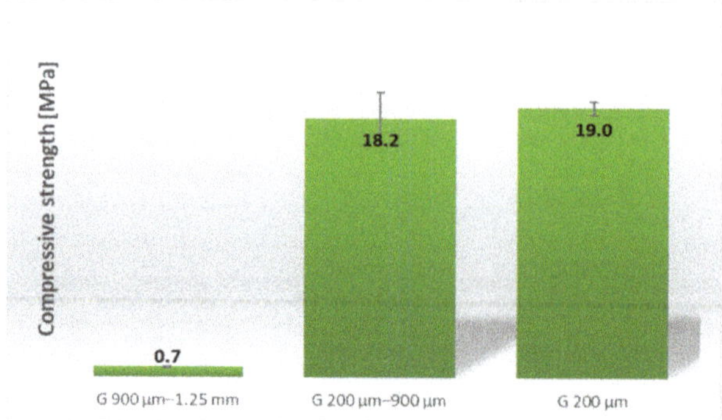

Figure 2. Results from the compressive strength test.

The best compressive strength was achieved by geopolymers with particle sizes of less than 200 µm—19 MPa. A slightly lower compressive strength was obtained by geopolymers from the base material fraction of 200 µm to 900 µm—18.2 MPa. A significant decrease in strength was noted for geopolymers whose particle sizes of the raw material were in the range of 900 µm to 1.25 mm—0.7 MPa.

3.2. *Flexural Strength*

The results for flexural strength are shown in Figure 3. The test was carried out on three samples for each variant. The graph shows the mean value obtained from these measurements.

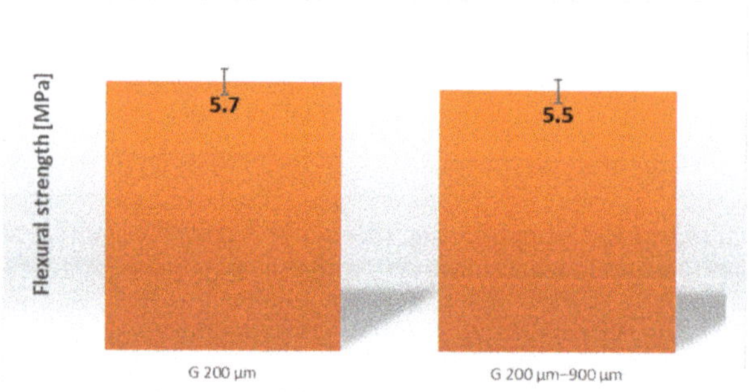

Figure 3. Results from the flexural strength test.

In the case of bending strength, samples with particle sizes of the raw material above 900 µm obtained a result below the reading sensitivity of the testing machine, hence no value was recorded for this type of sample. Again, the best strength was achieved with geopolymers with particle sizes below 200 µm, while this value was comparable to the result for geopolymers with grain sizes of 200–900 µm.

3.3. *SEM Observation*

Figure 4 shows a SEM image for sample G 200–900 µm.

The observed microstructure is typical for geopolymers. A significant number of pores ranging from 2 to 3 µm were observed. The EDS analysis identified compounds of silicon, aluminum, sodium, calcium, potassium, chlorine and iron in the selected areas 1–3 (Figure 5).

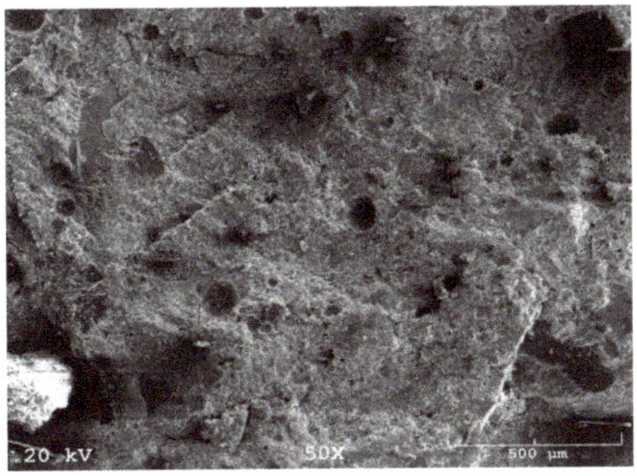

Figure 4. SEM image of geopolymer G 200–900 µm at 50× magnification.

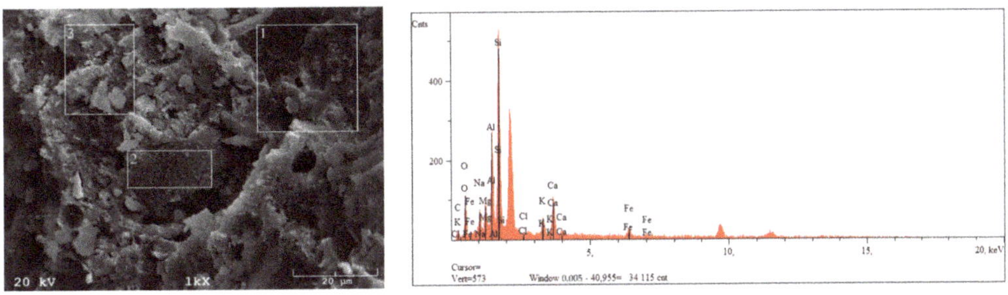

(**a**) (**b**)

Figure 5. Sample G 200–900 (**a**) with marked areas of measurements, (**b**) results of the EDS analysis.

4. Conclusions

The obtained results show the influence of precursor particle size for geopolymerization using a sodium activator on mechanical properties. The best result for mechanical strength, both for compression and flexural strength, was obtained for geopolymers where particle sizes were below 200 µm—19 MPa and 5.7 MPa, respectively. A slightly worse result, about 4% lower, was obtained for samples with particle sizes in the range of 200–900 µm. The worst results were obtained from samples with the largest particle sizes from 900 µm to 1.25 mm. A large amount of fine-fraction particles in the geopolymer G 200–900 µm, below 100 µm, can affect the high strength of the geopolymer, giving the effect of a filler. The high mechanical strength of G 200 µm resulted from the presence of very small particles below 90 µm, which facilitate the solubility of SiO_2 and Al_2O_3 in the NaOH solution [24,28]. Grinding precursor particles to sizes below 900 µm allows improved strength properties to be achieved, and at the same time saving energy, without the need to further grind particles to smaller than 200 µm.

Coal gangue from the mining industry, for example Wieczorek, is a material that is still being researched and has the potential to be used in geopolymerization.

It is an environmentally friendly solution, especially if the method for its preparation as a raw material to produce geopolymers is energy-saving.

Author Contributions: Conceptualization, B.F. and K.K.; methodology, B.F.; validation, K.K., formal analysis, B.F.; investigation, A.B.; B.Ş.; G.A.; K.P.; resources; data curation, B.F.; B.K.; writing—original draft preparation, B.F.; writing—review and editing, B.K.; supervision, K.K.; All authors have read and agreed to the published version of the manuscript.

Funding: This research received no external funding.

Institutional Review Board Statement: Not applicable.

Informed Consent Statement: Not applicable.

Data Availability Statement: Not applicable.

Acknowledgments: This work was supported by the Polish National Centre for Research and Development under the LIDER Grant no: LIDER/31/0168/L-10/18/NCBR/2019. The publication cost of this paper was covered with funds from the Polish National Agency for Academic Exchange (NAWA): "MATBUD'2023—Developing international scientific cooperation in the field of building material engineering" BPI/WTP/2021/1/00002, MATBUD'2023.

Conflicts of Interest: The authors declare no conflict of interest.

References

1. Le, V.S.; Louda, P.; Tran, H.N.; Nguyen, P.D.; Bakalova, T.; Buczkowska, K.E.; Dufkova, I. Study on Temperature-Dependent Properties and Fire Resistance of Metakaolin-Based Geopolymer Foams. *Polymers* **2020**, *12*, 2994. [CrossRef]
2. Bazan, P.; Kozub, B.; Łach, M.; Korniejenko, K. Evaluation of Hybrid Melamine and Steel Fiber Reinforced Geopolymers Composites. *Materials* **2020**, *13*, 5548. [CrossRef]
3. Vanitha, N.; Revathi, T.; Sivasakthi, M.M.; Jeyalakshmi, R. Microstructure properties of poly(phospho-siloxo) geopolymeric network with metakaolin as sole binder reinforced with n-SiO_2 and n-Al_2O_3. *J. Solid State Chem.* **2022**, *312*, 123188. [CrossRef]
4. Silva, G.; Kim, S.; Aguilar, R.; Nakamatsu, J. Natural fibers as reinforcement additives for geopolymers—A review of potential eco-friendly applications to the construction industry. *Sustain. Mat. Technol.* **2020**, *23*, e00132. [CrossRef]
5. Geng, J.; Zhou, M.; Li, Y.; Chen, Y.; Han, Y.; Wan, S.; Zhou, X.; Hou, H. Comparison of red mud and coal gangue blended geopolymers synthesized through thermal activation and mechanical grinding preactivation. *Constr. Build. Mater.* **2017**, *153*, 185–192. [CrossRef]
6. Bai, T.; Song, Z.; Wang, H.; Wu, Y.; Huang, W. Performance evaluation of metakaolin geopolymer modified by different solid wastes. *J. Clea. Prod.* **2019**, *226*, 114–121. [CrossRef]
7. Le, C.H.; Louda, P.; Buczkowska, K.E.; Dufkova, I. Investigation on Flexural Behavior of Geopolymer-Based Carbon Textile/Basalt Fiber Hybrid Composite. *Polymers* **2021**, *13*, 751. [CrossRef] [PubMed]
8. Kumar, R.; Kumar, S.; Mehrotra, S.P. Towards sustainable solutions for fly ash through mechanical activation. *Resour. Conserv. Recycl.* **2007**, *52*, 157–179. [CrossRef]
9. Davidovits, R.; Pelegris, C.; Davidovits, J. *Standardized Method in Testing Commercial Metakaolins for Geopolymer Formulations*; Technical Paper #26-MK-Testing; Geopolymer Institute Library: Saint-Quentin, France, 2019.
10. Kozub, B.; Bazan, P.; Mierzwiński, D.; Korniejenko, K. Fly-Ash-Based Geopolymers Reinforced by Melamine Fibers. *Materials* **2021**, *14*, 400. [CrossRef] [PubMed]
11. Fan, F.; Liu, Z.; Xu, G.; Peng, H.; Cai, C.S. Mechanical and thermal properties of fly ash based geopolymers. *Constr. Build. Mater.* **2018**, *160*, 66–81. [CrossRef]
12. Sandanayake, M.; Gunasekara, C.; Law, D.; Zhang, G.; Setunge, S.; Wanijuru, D. Sustainable criterion selection framework for green building materials–An optimisation based study of fly-ash geopolymer concrete. *Sustain. Mat. Techn.* **2020**, *25*, e00178. [CrossRef]
13. Huang, G.; Ji, Y.; Li, J.; Hou, Z.; Dong, Z. Improving strength of calcinated coal gangue geopolymer mortars via increasing calcium content. *Constr. Build. Mater.* **2018**, *166*, 760–768. [CrossRef]
14. Lin, K.L.; Lo, K.W.; Cheng, T.W.; Lin, W.T.; Lin, Y.W. Influence 411 of SiC Sludge on the Microstructure of Geopolymers. *Materials* **2020**, *13*, 2203. [CrossRef] [PubMed]
15. Wang, Y.-S.; Alrefaei, Y.; Dai, J.-G. Silico-Aluminophosphate and Alkali-Aluminosilicate Geopolymers: A Comparative Review. *Front. Mater.* **2019**, *6*, 106. [CrossRef]
16. Korniejenko, K.; Lin, W.-T.; Šimonová, H. Mechanical Properties of Short Polymer Fiber-Reinforced Geopolymer Composites. *J. Compos. Sci.* **2020**, *4*, 128. [CrossRef]
17. Mikuła, J.; Korniejenko, K. (Eds.) *Innovative, Cost Effective and Eco-Friendly Fibre-Based Materials for Construction Industry*; Cracow, Wydawnictwo Politechniki Krak: Kraków, Poland, 2015.
18. Davidovits, J. Environmentally Driven Geopolymer Cement Applications. In Proceedings of the Geopolymer 2002 Conference, Melbourne, Australia, 18 June 2002.
19. Klojzy-Karczmarczyk, B.; Mazurek, J. Propozycje rozszerzenia działań celem zagospodarowania materiałów odpadowych z górnictwa węgla kamiennego. *Zesz. Nauk. Inst. Gospod. Surowcami Miner. I Energią Pol. Akad. Nauk.* **2017**, *98*, 151–166.
20. Li, J.; Wang, J. Comprehensive utilization and environmental risks of coal gangue: A review. *J. Cleaner Prod.* **2019**, *239*, 117946. [CrossRef]
21. Li, C.; Wanb, J.; Suna, H.; Li, L. Investigation on the activation of coal gangue by a new compound method. *J. Hazard. Mater.* **2010**, *179*, 515–520. [CrossRef]
22. Yhang, Y.I.; Ling, T.C. Reactivity activation of waste coal gangue and its impact on the properties of cementbased materials—A review. *Constr. Build. Mater.* **2020**, *234*, 117424.

23. Zhang, W.; Dong, C.; Huang, P.; Sun, Q.; Li, M.; Chai, J. Experimental Study on the Characteristics of Activated Coal Gangue and Coal Gangue-Based Geopolymer. *Energies* **2020**, *13*, 2504.
24. Li, Z.; Gao, Y.; Zhang, J.; Zhang, C.; Chen, J.; Liu, C. Effect of particle size and thermal activation on the coal gangue based geopolymer. *Mater. Chem. Phys.* **2021**, *267*, 124657. [CrossRef]
25. Nath, S.; Kumar, S. Role of particle fineness on engineering properties and microstructure of fly ash derived geopolymer. *Construct. Build. Mater.* **2020**, *233*. [CrossRef]
26. Aateef, N.; Assi, E.; Deaver, E.; Ziehl, P. Effect of source and particle size distribution on the mechanical and microstructural properties of fly Ash-Based geopolymer concrete. *Constr. Build. Mater.* **2018**, *167*, 372–380.
27. Heng, Y.; Ma, H.; Chen, H.; Wang, J.; Shi, J.; Li, Z.; Yu, M. Preparation and characterization of coal gangue geopolymers. *Constr. Build. Mater.* **2018**, *187*, 318–326.
28. Mucsi, G.; Kumar, S.; Csőke, B.; Kumar, R.; Molnár, Z.; Rácz, A.; Mádai, F.; Debreczeni, A. Control of geopolymer properties by grinding of land filled fly ash. *Int. J. Miner. Process.* **2015**, *143*, 50–58. [CrossRef]

materials proceedings

Proceeding Paper

The Influence of Casting Static Compaction Pressure on Carbonated Reactive Magnesia Cement (CRMC)-Based Mortars †

Erick Grünhäuser Soares [1,*], João Castro-Gomes [1] and Manuel Magrinho [2]

1 Centre of Materials and Building Technologies (C–MADE), Department of Civil Engineering and Architecture, University of Beira Interior (UBI), 6201-001 Covilhã, Portugal
2 Centre of Materials and Building Technologies (C–MADE), Department of Chemistry, University of Beira Interior (UBI), 6201-001 Covilhã, Portugal
* Correspondence: e.grunhauser@ubi.pt
† Presented at the 10th MATBUD'2023 Scientific-Technical Conference "Building Materials Engineering and Innovative Sustainable Materials", Cracow, Poland, 19–21 April 2023.

Abstract: The current study evaluates the influence of the static compaction pressure applied during the casting process on Carbonated Reactive Magnesia Cement-based mortars. For this purpose, mortars, embodying biomass fly ash as filler, were designed and moulded through static compaction pressures of 10, 30, 50, and 70 MPa. The moulded specimens were submitted to an accelerated carbonation curing period of 24 h under controlled conditions. The devised mortars were evaluated through compressive strength tests, and their microstructure was assessed through Mercury Intrusion Porosimetry (MIP), Thermogravimetry and Derivative Thermogravimetry (TG-DTG), and Fourier-transform Infrared Spectroscopy (FTIR) analyses. The results showed that the increment in the static compaction pressure during the specimens' casting process not only led the mortars to reduce their porosity by up to ~30% and increase their compressive strength by up to ~58% (from 19.8 MPa to 31.2 MPa) but also that such a change seems to hinder the CO_2 diffusion into the specimens' core, thus resulting in a lower content of carbonated products. In addition, the MIP analyses demonstrated that the static compaction pressure applied in the mortar casting process changes the pores' characteristics, while TG-DTG and FTIR analyses provided evidence that the devised mortars were carbonated to a certain degree. Therefore, this work demonstrated that Carbonated Reactive Magnesia Cement-based mortars are highly influenced by the static compaction pressure applied during the casting process, at least up to a certain value.

Keywords: Carbonated reactive magnesia cement; CO_2 mineralisation; influencing factor; static compaction pressure; biomass fly ash

Citation: Grünhäuser Soares, E.; Castro-Gomes, J.; Magrinho, M. The Influence of Casting Static Compaction Pressure on Carbonated Reactive Magnesia Cement (CRMC)-Based Mortars. *Mater. Proc.* **2023**, *13*, 5. https://doi.org/10.3390/materproc2023013005

Academic Editors: Katarzyna Mróz, Tomasz Tracz, Tomasz Zdeb and Izabela Hager

Published: 13 February 2023

1. Introduction

Carbonatable binders have emerged as one of the alternative binders to Portland Cement-based materials that could assist in the global trend of reducing CO_2 emissions [1]. Carbonated Reactive Magnesia Cement (CRMC), which is based on the transition of Magnesia to Hydrated Magnesium Carbonates (HMCs), such as Dypingite ($Mg_5(CO_3)_4(OH)_2{\cdot}5H_2O$), Hydromagnesite ($Mg_5(CO_3)_4(OH)_2{\cdot}4H_2O$), and Nesquehonite ($MgCO_3{\cdot}3H_2O$), through Accelerated Carbonation Curing (ACC) [2–5], is among this group of binders.

However, the success of the ACC of CRMC-based materials is highly dependent on a wide range of factors, such as the conditions to which the fresh CRMC-based material is exposed (e.g., CO_2 environment, curing temperature, period of exposition, and relative humidity), the raw materials' properties (e.g., Magnesia source, Magnesia calcination history, and type of aggregates used), the blend design (e.g., water content, pH, additives used, and the use of Magnesia replacement), and both geometry and porosity of the fresh material [6]. In addition to the previously mentioned carbonation-curing influencing factors, the compaction pressure on the specimens' casting process of carbonatable binders is another factor that may significantly influence the compressive strength obtained in this group of binders, mainly due to its influence on the porosity of the material [7]. However, it seems that this approach has not yet been investigated regarding CaO-based materials.

For this purpose, an exploratory approach was developed aiming to investigate how a CRMC-based mortar exposed to ACC behaves when modifications in the static compaction pressure of the casting process are made. A total of 16 specimens were produced and tested for four different static compaction pressures (i.e., 10, 30, 50, and 70 MPa). Moreover, the mixture, the production method, and the ACC conditions were defined based on previous studies [8–10], which used a reactive compound (MgO), a waste-based material as filler, and sand as fine aggregate to produce CRMC-based mortars that were moulded through static compaction and cured under pressurized ACC for 24 h under controlled conditions.

With the aim of understanding the CRMC-based mortar development, the devised mortars were then evaluated through compressive strength tests, Mercury Intrusion Porosimetry (MIP) analysis, Thermogravimetry and Derivative Thermogravimetry (TG-DTG), and Fourier-transform Infrared Spectroscopy (FTIR) analyses.

2. Materials and Methods

2.1. Materials

Reactive Magnesia (r-MgO), with the commercial name "Calcined Magnesite 92/200", was supplied by Richard Baker Harrison Ltd., Liverpool, England. Its reactivity was recorded as 48 s [11], thus classifying it as medium-reactive magnesia [12]. Biomass fly ash (BFA) was used as it was received and consisted of residual energy production material from a burning forestry source, which was supplied by Central de Biomassa do Fundão, Unipessoal LDA, Fundão, Portugal. The fine aggregate used was river sand (RS), supplied by Tabal-Sepor Areias e Argamassas LDA, Salvaterra de Magos, Portugal, which presented a medium particle diameter (D_{50}) of 1.08 mm and a coefficient of uniformity of 3.02. It is important to note that RS was employed in the devised materials to represent their true condition aside from facilitating the CO_2 diffusion into the specimens' core although the use of fine aggregates is known to cause quartz contamination that, in turn, may lead to awkwardness in phase identification.

The raw materials' oxide composition is displayed in Table 1, which was estimated through Energy-dispersive X-ray spectroscopy (EDX) analysis (S-3400N Spectrometer, Hitachi, Tokyo, Japan). The raw materials' physical properties, namely true density, Blaine fineness, and Loss on Ignition (LOI), are shown in Table 2. The true density was determined through a helium gas displacement pycnometer equipment (AccuPyc 1330, Micromeritics, Norcross, GA, USA). The Blaine fineness of the powders (r-MgO and BFA) was determined using a Blaine air permeability apparatus (BSA1, Acmel Labo, Champlan, France). The LOI was obtained through TG-DTG analysis (SDT Q-50, TA Instruments, New Castle, DE, USA).

Table 1. Oxide composition.

Raw Material	Oxide Composition											
	Na_2O	MgO	Al_2O_3	SiO_2	P_2O_5	SO_3	Cl^-	K_2O	CaO	TiO_2	MnO	Fe_2O_3
r-MgO	-	93.3	-	1.6	-	-	-	-	3.5	-	-	1.6
BFA	1.6	5.7	20.3	35.1	2.4	2.7	0.5	8.9	14.9	0.7	0.7	6.3
RS	3.3	0.7	13.2	76.5	-	-	-	3.4	1.4	-	-	1.6

Table 2. Physical properties.

Raw Material	Physical Properties		
	True Density (g/cm^3)	Blaine Fineness (cm^2/g)	LOI (%) at 1000 °C
r-MgO	3.0	6263	10.4
BFA	2.3	4029	7.5
RS	2.6	-	3.7

2.2. Mixture Design and Production of Specimens

The mixture labels are presented in Table 3, and they consist of a binder-to-sand volume ratio of 1:3, a water-to-solids-in-the-binder mass ratio of 0.41, where the solids in the binder are composed of r-MgO and BFA in a volume ratio of 2:3. Four cubic-shaped specimens with 40 mm edge were produced for each mixture label based on the methodology described in recent studies [8–10]. A pressurised ACC chamber of approximately 75 litres was used, and an initial partial vacuum pressure of 0.2 bar was applied. Afterwards, the following ACC conditions were provided: CO_2 concentration > 99%; CO_2 partial pressure = 0.7 bar; T = 60 ± 2 °C; relative humidity (RH) > 99%; and AAC period = 24 h. Following the AAC period, the CRMC-based specimens were immediately stored in lab room conditions (20 ± 2 °C and an RH of 60 ± 5%) for 24 h to cool them before compressive strength testing.

Table 3. Mixture labels.

Mixture Label	Raw Material (vol. %)			w/b	Static Compaction Pressure (MPa)
	r-MgO	BFA	RS		
M.P-10	10.0	15.0	75.0	0.41	10
M.P-30					30
M.P-50					50
M.P-70					70

2.3. Assessment of the CRMC-Based Mortars Developed

Compressive strength tests were carried out through uniaxial loading in triplicate, and their average was reported. A compression machine with digital readout and self-centering platens, operated at a constant loading rate of 1.35 kN/s, was used (ADR Touch 3000 BS EN, ELE International, Leighton Buzzard, UK). The tested material was collected for further microstructural investigations (TG-DTG and FTIR analyses). MIP analyses were performed using a mercury porosimeter (AutoPore IV 9500, Micromeritics, Norcross, GA, USA), with maximum and minimum applied pressures of ~34,000 psia and ~0.5 psia, respectively, thus corresponding to a minimum pore size of 5 nm and a maximum pore size of 345 μm. A mercury surface tension of 480 mM/m was applied. Mercury-solid contact angles for intrusion and extrusion were defined as 130° and 104°, respectively. Specimen fragments with a mass range of 1.50–1.90 g were obtained by sawing the specimens that were not subjected to the compressive strength test to acquire more accurate data for MIP analysis [13]. Before being tested, these fragments were stored in a glass desiccator containing silica gel for seven days to ensure the removal of moisture and ensure the effectiveness of the test. TG-DTG analyses were performed from ambient temperature (20 ± 2 °C) to 1000 °C at a heating rate of 20 °C/min under nitrogen flow (SDT Q-50, TA Instruments, New Castle, DE, USA). The material tested consisted of ~7 mg of particles passing through a 63 μm sieve, which were first submitted to a drying stage of 24 h at 60 °C to avoid the overlapping effect of free water release with the dehydration step in the TG-DTG curves. FTIR data were obtained by recording the infrared spectrum from 600 to 1600 cm^1 using a FTIR Spectrometer apparatus (Nicolet iS10 with a Smart ATR accessory, Thermo Fisher Scientific, Waltham, MA, USA).

3. Results and Discussion

The average compressive strength results of each mixture label are shown in Figure 1 and Table 4. The specimens reached a compressive strength of 19.8 MPa to 31.2 MPa. It was noticed that a change in only one parameter of the casting process can enhance the compressive strength results by ~58% when comparing M.P-10 with M.P-70. However, it was also observed that the degree of enhancement in compressive strength reduces as the static compaction pressure of the casting process increases. Thus, when comparing the M.P-50 with M.P-70, the compressive strength gain was only 1.5%. Therefore, it must be highlighted that the static compaction pressure applied in the casting process plays an important role in the strength development of CRMC-based materials but this strength gain may be almost negligible for static compaction pressures above 50 MPa. Similar behaviour was observed in steel slag-based CO_2-cured pastes [7]. Furthermore, the calculated standard deviation of the compressive strength results indicates that the materials devised for this study exhibit a high degree of homogeneity.

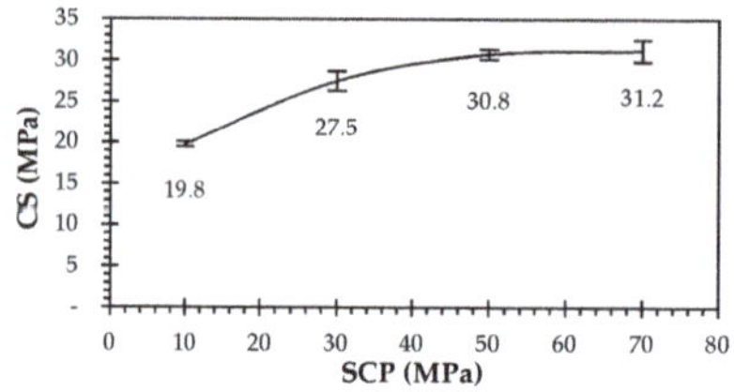

Figure 1. Static compaction pressure (SCP) vs. compressive strength (CS).

Table 4. Compressive strength results.

Parameter		Mixture Label M.P-10	M.P-30	M.P-50	M.P-70
Compressive strength (MPa)		19.79	27.50	30.75	31.21
Standard deviation (MPa)		0.32	1.19	0.62	1.34
Compressive strength enhancement/weakening (%) comparing with	M.P-10	–	38.96	55.38	57.71
	M.P-30	−28.04	–	11.82	13.49
	M.P-50	−35.64	−10.57	–	1.50
	M.P-70	−36.59	−11.89	−1.47	–

The porosity and other parameters obtained throughout the MIP analysis are summarized in Table 5 and shown in Figure 2.

Table 5. MIP results.

Parameter		Mixture Label M.P-10	M.P-30	M.P-50	M.P-70
Porosity (%)		22.34	18.99	16.90	15.80
Average pore diameter (μm)		0.07	0.05	0.06	0.05
Critical diameter (μm)		25.60	12.36	5.38	11.76
Bulk density at 14.50 psia (g/mL)		2.11	2.08	2.11	2.13
Apparent (skeletal) density (g/mL)		2.45	2.48	2.48	2.47
Pores typology (%) [14]	Mesopores (0.002–0.05 μm)	19.44	23.64	22.11	22.03
	Macropores (0.05–10 μm)	33.56	53.06	60.80	57.41
	Air voids/cracks (≥10 μm)	47.00	23.30	17.09	20.56

The analyses revealed that the mixture labels hold different porosity indices ranging from 15.80% to 22.34%. These results show that for the range of compaction pressure applied in the mortars' casting process of this study, higher values of static compaction pressure led to lower porosity indices and higher compressive strength results, as shown in Figure 2a. The M.P-30, M.P-50, and M.P-70 mixtures presented similar indices of mesopores, macropores, and air voids/cracks, while M.P-10 exhibited a much higher index of air voids/cracks, a lower index of macropores, and a similar index of mesopores than the rest (Figure 2c). Thus, such behaviour indicates that the compaction pressure applied in the mortars' casting process affects the pore typology up to a certain degree and acts especially on the transition of air voids/cracks to macropores and scarcely affects the increment of the mesopores index. Regarding the critical diameter (∅c), the results show that M.P-10 presented a much higher value than the other mixtures (M.P-10 = 25.6 μm; M.P-30 = 12.4 μm; M.P-50 = 5.4 μm; M.P-70 = 11.8 μm), where such behaviour seems to be related to the lower static compaction pressure applied in the casting process. Interestingly,

M.P-50 exhibited a critical diameter about two times lower than the one observed in M.P-70. Such behaviour may be attributed to the BFA pores since a peak in this diameter size is also present in other mixtures. The biomass fly ash was used as it was received, so differences in the pore structure of the raw materials may be found. Therefore, MIP results demonstrated that the static compaction pressure applied during the casting process plays an important role in the mortar's porosity index and in the characteristics of its pores which, in turn, affect the compressive strength results.

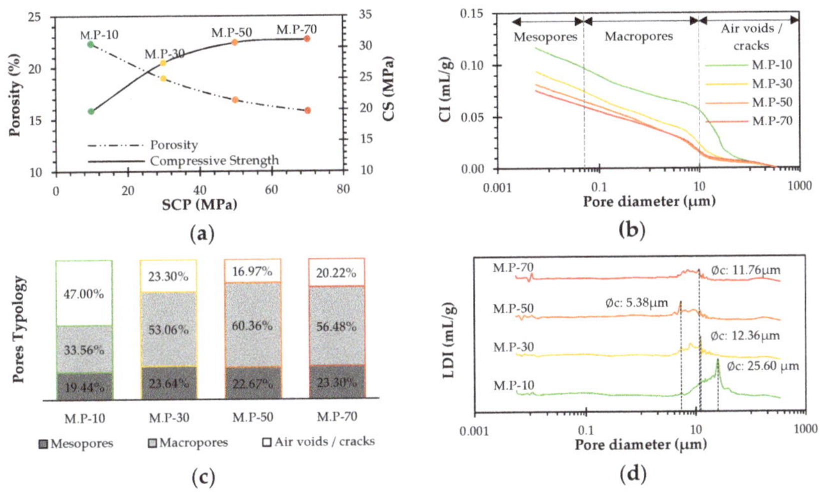

Figure 2. (**a**) Static compaction pressure (SCP) vs. compressive strength (CS) and porosity; (**b**) Cumulative intrusion (CI) vs. pore diameter; (**c**) Pores typology (%); (**d**) Log differential intrusion (LDI) vs. pore diameter.

The TG-DTG analyses (Figure 3) show that the mass loss of the devised materials was gradual, reaching the highest mass loss rate at a temperature range between 300 and 550 °C (Table 6). In this temperature range, the M.P-10 mixture had the highest index of mass loss (~15%), while the M.P-30, M.P-50, and M.P-70 lost ~12, 10, and 9% of their initial masses, respectively. Interestingly, the mass loss for this temperature interval does not follow the typical behaviour of compressive strength development in CRMC-based materials, since the mixture labels with lower compressive strength results had higher mass losses in the region of ~300–550 °C, thus indicating that more HMCs and carbonates were formed in these mixtures, probably due to the higher porosity index in the fresh moulded specimen thanks to the lower static pressure applied in their casting process. The DTG curves (Figure 3a) exhibit two initial peaks, which are attributed to the free water evaporation and dehydration of water bound to the HMCs [15–17]. The region between 300 and 550 °C presents a higher peak at ~420 °C, which refers to the overlapping curves of the dehydroxylation of Brucite [15,18,19] and Dypingite [15,18], as well as the decarbonisation of Hydromagnesite [20] and Nesquehonite [15,18,21], whereas the smaller peak at ~460 °C and the mass loss in the range of 450 to 600 °C correspond to the decarbonisation of Magnesite [19,22] and/or of undefined HMCs [4]. In turn, the last peak may be caused by the decarbonisation of Calcite and/or amorphous carbonates [23–25]. Finally, the TG-DTG analyses indicate that the four designed mortars were carbonated to a certain degree, in which the magnesia hydration and carbonation seem to be favourably affected by using a lower static compaction pressure in the casting process since the mass loss in the temperature range of 300−550 °C is higher as the applied static compaction pressure is lowered. Such a trend may occur due to a lower porosity index which negatively influences the CO_2 diffusion into the core of the designed materials, thus hindering the formation of HMCs.

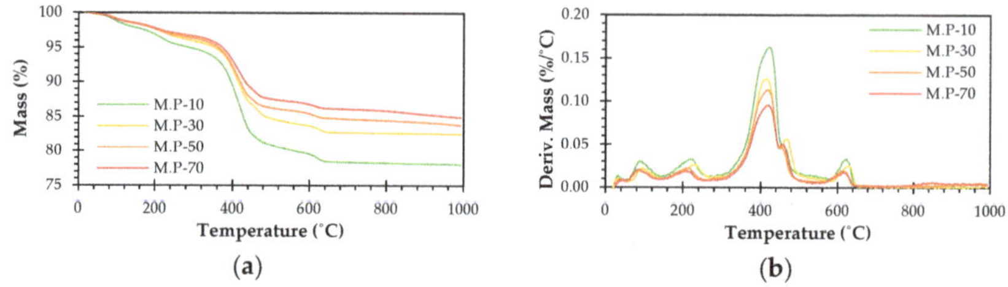

Figure 3. (**a**) TG curves; (**b**) DTG curves.

Table 6. Mass loss (%) in TG-DTG.

Mixture Label	Mass Loss (%) in TG-DTG by Temperature Range			
	25–300 °C	300–550 °C	550–1000 °C	Σ
M.P-10	5.19	14.63	2.09	21.92
M.P-30	4.07	11.77	1.63	17.47
M.P-50	3.66	10.41	2.10	16.17
M.P-70	3.34	9.43	2.28	15.05

The recorded FTIR spectra data of the four mixtures are exhibited in Figure 4. Despite the low intensity of the observed bands that may be attributed to the low MgO content (10 vol. %), some considerations may be made as follows: the absorbance bands located at ~680, 855, 880, 1420, and 1485 cm^{-1} indicate the presence of HMCs such as Nesquehonite, Hydromagnesite, and Dypingite [26–29], while the bands at ~720 and 1460 cm^{-1} could be attributed to the presence of Lansfordite [28]. Therefore, these absorbance bands may indicate that part of the r-MgO in the mixture was carbonated to a certain degree. Furthermore, the unidentified absorbance bands located between 800 and 1200 cm^{-1} may be due to the presence of MSH gels [30], which, along with the HMCs, tend to fill the cementitious matrix voids contributing to the enhancement in compressive strength [26].

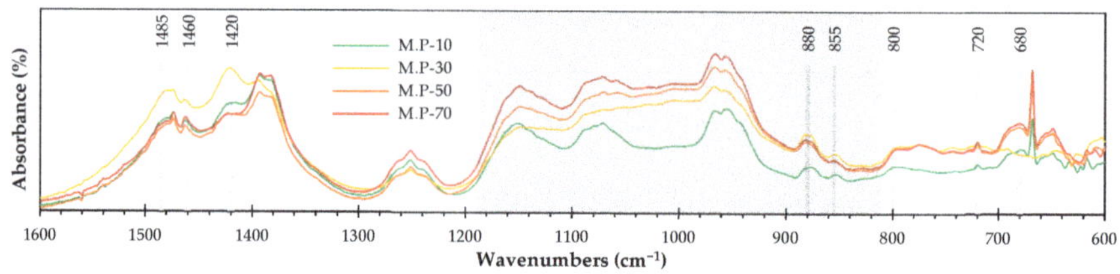

Figure 4. FTIR curves.

4. Conclusions

This work presented how CRMC-based mortars behave when changes in static compaction pressure on their casting process are made. For this purpose, the compressive strength of the developed mortars was evaluated, and MIP, TG-DTG, and FTIR analyses were carried out. Therefore, the main contributions of this study are as follows: higher values of static compaction pressure used during the casting process resulted in higher compressive strengths for the mortars developed and lower formation of HMCs. However, as the static compression pressure increases, the strength gain decreases, thus making it almost ineffective when applied at pressures greater than 50 MPa (i.e., M.P-70). MIP results showed that the static compaction pressure applied during the casting process of the devised mortars plays an important role in the porosity index and characteristics of the pores, thus affecting the results of the compressive strength obtained. TG-DTG and FTIR analyses indicated that the devised mortars were carbonated to a certain degree. More studies related to the microstructure, such as X-ray diffraction (XRD), and Scanning Electron Microscopy (SEM) with Energy Dispersive X-Ray Analysis (EDX) on the devised mortars analyses, should be carried out to better understand these materials.

Finally, the casting and the accelerated carbonation curing processes established and used in this study limit the suitability of the created mortars to the production of pre-cast building materials only.

Author Contributions: E.G.S.: Conceptualization, Methodology, Investigation, Writing—original draft, Writing—review & editing. J.C.-G.: Writing—review & editing, Supervision, Project administration, Funding acquisition. M.M.: Writing—review & editing. All authors have read and agreed to the published version of the manuscript.

Funding: Portuguese national funds partially financed this work through FCT, Foundation for Science and Technology, IP, within the research unit C-MADE, Centre of Materials and Building Technologies (CIVE-Central Covilhã-4082), University of Beira Interior, Portugal. This research was also partially funded by FCT grant number 2022.09813.BD.

Institutional Review Board Statement: Not applicable.

Informed Consent Statement: Not applicable.

Data Availability Statement: Not applicable.

Acknowledgments: The authors acknowledge Engineer Ana Paula Gomes from the Optical Centre of UBI (COUBI) for her contribution in carrying out the EDX analyses, and Engineer Carlos Alegria and Engineer Paulo Godinho from Magestop, Central de Biomassa do Fundão, for providing the biomass ashes used in this research. The publication cost of this paper was covered with funds from the Polish National Agency for Academic Exchange (NAWA): "MATBUD'2023 - Developing international scientific cooperation in the field of building materials engineering" BPI/WTP/2021/1/00002, MATBUD'2023.

Conflicts of Interest: The authors declare no conflict of interest.

References

1. Shi, C.; Qu, B.; Provis, J.L. Recent progress in low-carbon binders. *Cem. Concr. Res.* **2019**, *122*, 227–250. [CrossRef]
2. Sonat, C.; Lim, C.H.; Liska, M.; Unluer, C. Recycling and reuse of reactive MgO cements–A feasibility study. *Constr. Build. Mater.* **2017**, *157*, 172–181. [CrossRef]
3. Unluer, C.; Al-Tabbaa, A. Enhancing the carbonation of MgO cement porous blocks through improved curing conditions. *Cem. Concr. Res.* **2014**, *59*, 55–65. [CrossRef]
4. Dung, N.T.; Lesimple, A.; Hay, R.; Celik, K.; Unluer, C. Formation of carbonate phases and their effect on the performance of reactive MgO cement formulations. *Cem. Concr. Res.* **2019**, *125*, 105894. [CrossRef]
5. Liska, M.; Al-Tabbaa, A. Ultra-green construction: Reactive magnesia masonry products. *Proc. Inst. Civ. Eng. Waste Resour. Manag.* **2009**, *162*, 185–196. [CrossRef]
6. Grünhäuser Soares, E.; Castro-Gomes, J. Carbonation curing influencing factors of Carbonated Reactive Magnesia Cements (CRMC)–A review. *J. Clean. Prod.* **2021**, *305*, 17. [CrossRef]
7. Humbert, P.S.; Castro-Gomes, J.P.; Savastano, H. Clinker-free CO_2 cured steel slag based binder: Optimal conditions and potential applications. *Constr Build. Mater.* **2019**, *210*, 413–421. [CrossRef]
8. Grünhäuser Soares, E.; Castro-Gomes, J.; Sitarz, M.; Zdeb, T.; Hager, I. The Immobilisation of Heavy Metals from Sewage Sludge Ash in CO_2-Cured Mortars. *Sustainability* **2021**, *13*, 12893. [CrossRef]
9. Grünhäuser Soares, E.; Castro-Gomes, J.; Sitarz, M.; Zdeb, T.; Hager, I.; Hassan, K.; Al-Kuwari, M.S. Feasibility for co-utilisation of Carbonated Reactive Magnesia Cement (CRMC) and industrial wastes in circular economy and CO_2 mineralisation. *Constr. Build. Mater.* **2022**, *323*, 13. [CrossRef]
10. Grünhäuser Soares, E.; Castro-Gomes, J. The role of biomass bottom ash in Carbonated Reactive Magnesia Cement (CRMC) for CO_2 mineralisation. *J. Clean Prod.* **2022**, *380*, 135092. [CrossRef]
11. Shand, M.A. *The Chemistry and Technology of Magnesia*; John Wiley & Sons, Inc.: Hoboken, NJ, USA, 2006. [CrossRef]
12. Jin, F.; Al-Tabbaa, A. Characterisation of different commercial reactive magnesia. *Adv. Cem. Res.* **2014**, *26*, 101–113. [CrossRef]
13. Ma, H. Mercury intrusion porosimetry in concrete technology: Tips in measurement, pore structure parameter acquisition and application. *J. Porous Mater.* **2014**, *21*, 207–215. [CrossRef]

14. Everett, D.H. Manual of Symbols and Terminology for Physicochemical Quantities and Units, Appendix II: Definitions, Terminology and Symbols in Colloid and Surface Chemistry. *Pure Appl. Chem.* **1972**, *31*, 577–638. [CrossRef]
15. Dung, N.T.; Unluer, C. Carbonated MgO concrete with improved performance: The influence of temperature and hydration agent on hydration, carbonation and strength gain. *Cem. Concr. Compos.* **2017**, *82*, 152–164. [CrossRef]
16. Wu, H.; Zhang, D.; Ellis, B.R.; Li, V.C. Mechanical behavior of carbonated MgO-based Engineered Cementitious Composite (ECC) after high temperatures exposure. *Cem. Concr. Compos.* **2021**, *124*, 104255. [CrossRef]
17. Hay, R.; Otchere, C.; Kashwani, G.; Celik, K. Recycling carbonated reactive magnesium cement (RMC) as a building material. *J. Clean Prod.* **2021**, *320*, 128838. [CrossRef]
18. Dung, N.T.; Unluer, C. Improving the performance of reactive MgO cement-based concrete mixes. *Constr. Build. Mater.* **2016**, *126*, 747–758. [CrossRef]
19. Unluer, C.; Al-Tabbaa, A. The role of brucite, ground granulated blastfurnace slag, and magnesium silicates in the carbonation and performance of MgO cements. *Constr. Build. Mater.* **2015**, *94*, 629–643. [CrossRef]
20. Bhattacharjya, D.; Selvamani, T.; Mukhopadhyay, I. Thermal decomposition of hydromagnesite. *J. Therm. Anal. Calorim.* **2012**, *107*, 439–445. [CrossRef]
21. Jauffret, G.; Morrison, J.; Glasser, F.P. On the thermal decomposition of nesquehonite. *J. Therm. Anal. Calorim.* **2015**, *122*, 601–609. [CrossRef]
22. Tian, L.; Tahmasebi, A.; Yu, J. An experimental study on thermal decomposition behavior of magnesite. *J. Therm. Anal. Calorim.* **2014**, *118*, 1577–1584. [CrossRef]
23. Thiery, M.; Villain, G.; Dangla, P.; Platret, G. Investigation of the carbonation front shape on cementitious materials: Effects of the chemical kinetics. *Cem. Concr. Res.* **2007**, *37*, 1047–1058. [CrossRef]
24. Tu, Z.; Guo, M.Z.; Poon, C.S.; Shi, C. Effects of limestone powder on $CaCO_3$ precipitation in CO_2 cured cement pastes. *Cem. Concr. Compos.* **2016**, *72*, 9–16. [CrossRef]
25. Qin, L.; Gao, X. Properties of coal gangue-Portland cement mixture with carbonation. *Fuel* **2019**, *245*, 1–12. [CrossRef]
26. Bhagath Singh, G.V.P.; Sonat, C.; Yang, E.H.; Unluer, C. Performance of MgO and MgO–SiO_2 systems containing seeds under different curing conditions. *Cem. Concr. Compos.* **2020**, *108*, 103543. [CrossRef]
27. Khalil, A.; Sohn, S.; Celik, K. Thermal properties and stability of reactive magnesia cement. *Constr. Build. Mater.* **2021**, *308*, 125102. [CrossRef]
28. Coleyshaw, E.E.; Crump, G.; Griffith, W. Vibrational spectra of the hydrated carbonate minerals ikaite, monohydrocalcite, lansfordite and nesquehonite. *Spectrochim. Acta Part A* **2003**, *59*, 2231–2239. [CrossRef]
29. Kloprogge, J.T.; Martens, W.N.; Nothdurft, L.; Duong, L.V.; Webb, G.E. Low temperature synthesis and characterization of nesquehonite. *J. Mater. Sci. Lett.* **2003**, *22*, 825–829. [CrossRef]
30. Nied, D.; Enemark-Rasmussen, K.; L'Hopital, E.; Skibsted, J.; Lothenbach, B. Properties of magnesium silicate hydrates (M-S-H). *Cem. Concr. Res.* **2016**, *79*, 323–332. [CrossRef]

materials proceedings

Proceeding Paper

The Influence of the Quality of Brick Firing on Their Calcium Diffusion Capacity and Biodegradation Potential—A Preliminary Study †

Klára Kobetičová [1,*], Jana Náb̈elková [1], Martin Keppert [1], Igor Medved' [2], Zbigniew Suchorab [1,3] and Robert Černý [1]

1 Faculty of Civil Engineering, Czech Technical University in Prague, Thákurova 7, 16629 Prague, Czech Republic
2 Faculty of Civil Engineering, Slovak University of Technology, Radlinského 11, 810 05 Bratislava, Slovakia
3 Faculty of Environmental Engineering, Lublin University of Technology, Nadbystrzycka 40B, 20-618 Lublin, Poland
* Correspondence: klara.kobeticova@fsv.cvut.cz

† Presented at the 10th MATBUD'2023 Scientific-Technical Conference "Building Materials Engineering and Innovative Sustainable Materials", Cracow, Poland, 19–21 April 2023.

Abstract: The diffusion of calcium ions Ca^{2+} in aquatic solutions (10 g/l) was measured for two brick samples from a region in Bohemia (Vysoké Mýto, Holešov-Žopy). The experiment was performed under laboratory conditions at the ambient temperature of 20 ± 2 °C for a period of 240 h. The bricks were cut into three depth layers. The calcium concentrations were analyzed chelatometrically. The biodegradation potential of the individual layers was also studied. The results indicated that the depth and quality of firing are of importance regarding the transport of calcium, and they affect the success of bio-colonization.

Keywords: calcium; diffusion; brick; biodegradation; biofilm; porosity

Citation: Kobetičová, K.; Náb̈elková, J.; Keppert, M.; Medved', I.; Suchorab, Z.; Černý, R. The Influence of the Quality of Brick Firing on Their Calcium Diffusion Capacity and Biodegradation Potential—A Preliminary Study. *Mater. Proc.* **2023**, *13*, 6. https://doi.org/10.3390/materproc2023013006

Academic Editors: Katarzyna Mróz, Tomasz Tracz, Tomasz Zdeb and Izabela Hager

Published: 13 February 2023

1. Introduction

Fired bricks have been used by people since 5000 BC in Mesopotamia and many types of this building material are still produced. Many historic buildings are built from classic, load-bearing, non-perforated and non-relieved bricks, which can be expected to be subjected to relatively large diffusion processes or easier biodegradation than lightweight perforated bricks. It can be assumed that the quality of brick firing is of great importance. The middle part of fired bricks may contain a black or grey reduction core, which distinguishes it from the red color of the rim or surface of the fired clay body. Such a black core can be found in many commercial bricks [1], its presence indicating insufficient firing and incomplete burn-out of organic components present in the used clay.

Diffusion plays a fundamental role in the distribution and behavior of substances in porous building materials. This is well described, especially for Cl^- anions [2–5]. Cations have not been studied as much from this point of view [6]. The exception is radionuclides, but in this case the research is focused on the disposal of hazardous waste, in which less dangerous model elements [7–9] were used. At the same time, cations also play an important role in salinization of building materials, of which calcium is an example. This element occurs in free or bound form as Ca^{2+}. It is among the elements that are important in the metabolic processes of humans and other organisms. Calcium cations are naturally present in water and soil, but they can also enter the environment excessively during liming of fields [10] and, to a lesser extent, during the salting of roads ($CaCl_2$) as a less dangerous salt for the environment than NaCl or KCl [11]. The Ca^{2+} concentration is easy to assess via complexometric titration [12].

Biodegradation of fired bricks has been described in many studies performed in situ and also under laboratory conditions. The presence of algae, cyanobacteria and fungi [13–15] was studied. Their colonization depends on many factors, such as humidity, the amount of light, nutrients and also the structure of the surface of building materials on which organisms can create and maintain biofilm growth, e.g., Ref. [16]. So far, the influence of the degree of the firing of bricks on their ability to be bio-colonized has never been investigated.

In this work, the focus was placed on the influence of the quality of the firing of bricks from the surface to the depth on: (1) the ability of calcium diffusion; and (2) the effect of the firing on the success of the biological colonization of individual layers.

2. Materials and Methods

The model samples were fired bricks from brickyards situated in the towns of: (a) Vysoké Mýto and (b) Holešov-Žopy (Czech Republic) (Figure 1). Dry brick samples in the shape of discs (d = 5 cm, h = 1 cm) represented three layers from the surface to the depth of the brick (0–1 cm, 1–2 cm and 2–3 cm). The composition of brick samples is described in previous literature [17,18]. Pore size distribution, bulk density, total porosity and the specific surface area, determined by Mercury Intrusion Porosimetry, are described in Figures 2 and 3 and in Table 1.

Figure 1. Cutouts of bricks: Vysoké Mýto (left) and Holešov-Žopy (right).

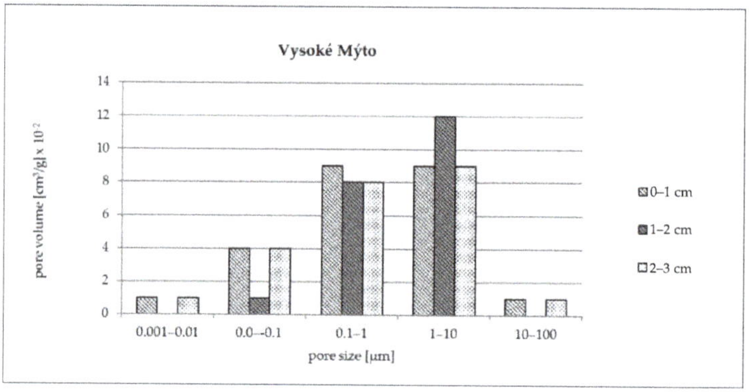

Figure 2. Pore size distribution of individual layer of the brick: Vysoké Mýto (0–1 cm, 1–2 cm, 2–3 cm from the cover).

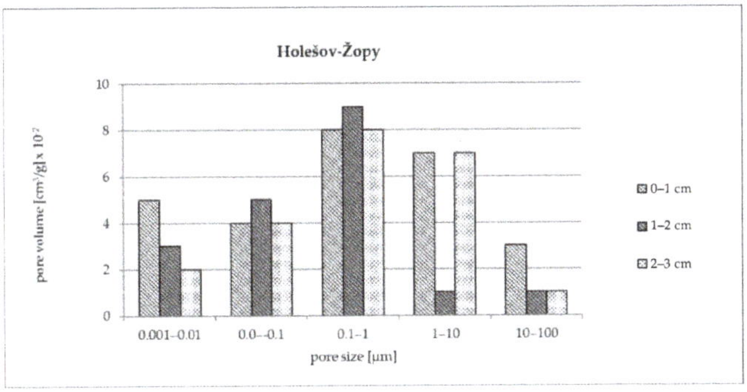

Figure 3. Pore size distribution of individual layer of the brick Holešov-Žopy (0–1 cm, 1–2 cm, 2–3 cm from the cover).

Table 1. Selected properties of the tested bricks—specific surface area (m/g), total cumulative volume (cc/g), total porosity (%) and bulk density (g/cm^3). The mean values from three replicates are used in the table.

Sample	Vysoké Mýto			Holešov-Žopy		
	Layer			Layer		
	0–1 cm	1–2 cm	2–3 cm	0–1 cm	1–2 cm	2–3 cm
Specific surface area (m/g)	14.805	12.695	8.742	9.417	8.184	6.638
Total cumulative volume (cc/g)	0.239	0.216	0.230	0.228	0.242	0.194
Total porosity (%)	39.861	43.273	43.023	41.692	48.870	37.246
Bulk density (g/cm^3)	1.665	2.006	1.868	1.829	2.021	1.917

$CaCl_2$ (99% purity) from Lach-Ner Ltd. (Prague, Czech Republic) was used in the experiment. The concentration of the $CaCl_2$ solution was chosen to be 10 g/L based on previous laboratory experiments. The solution was prepared by dissolving appropriate amounts of $CaCl_2$ powder in distilled water.

A two-compartment box made of plexiglass was used (see Figure 4). Initially, dry brick discs were sealed with silicone in a circular aperture cut in the plexiglass panel, splitting the box into two chambers. One of the chambers (chamber 1) was filled with 500 mL of distilled water, while the other chamber (chamber 2) was filled with calcium solution of the desired concentration (500 mL).

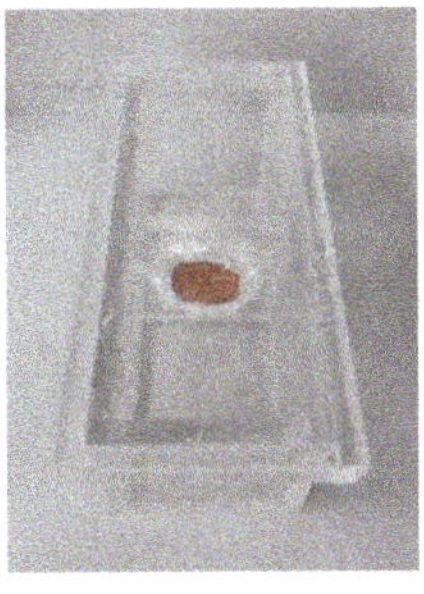

Figure 4. One of the model plexiglass boxes for experiments with calcium diffusion.

The box was closed with a plexiglass lid. The box was placed on a table at the ambient temperature of 20 ± 2 °C for a period of 240 h. The amount of calcium diffused through the sample was measured in chamber 1 by titration in selected time periods (0–240 h). When the accumulation started, the distilled water was replaced by the same new water and these replacements were conducted after each measuring of calcium concentrations.

A 0.05 M solution of chelaton III was used to measure the calcium concentration. Namely 500 µL of a 5 M solution of KOH was added to a specific amount of the monitored aquatic sample from chamber 1 and colored using a murexide indicator. Chelaton III was then dripped into the sample by an automatic burette [12]. A specific amount of titration reagent was used to calculate the calcium concentration c(Ca) in mg/L, according to the following Equation (1):

$$c(Ca)\left(\frac{mg}{L}\right) = \frac{V(ch).M(ch).\,M(Ca).1000}{V(sample)} \quad (1)$$

where:

V(ch) is the volume of used chelaton (mL);
M(ch) = 0.05 is its molarity;
M(Ca) = 40 g/mol is the calcium molar mass;
V(sample) is the volume of the analyzed aquatic sample (mL).

The disks from all of the studied brick layers (0–1 cm, 1–2 cm and 2–3 cm) from the two bricks were used. Every disc was separately placed in distilled water (V = 100 mL) in a glass vessel and left at a temperature of 20 ± 2 °C and under the illumination of 5000 LUX intensity in a light–dark period of 12:12. No organisms were added to the water. The samples were left in order for natural bio-colonization to occur on their surface. The samples were immersed in water for 30 days. They were then pulled out of the solutions and photographed. The biofilm scraped from the samples was then examined under the light microscope Olympus BX43 (Olympus, Prague, Czech Republic) with a CMOS camera (magnification 400×). The brick disks of the individual layers were then put into the glass test vessels into a 100 mL volume of distilled water. The test vessels were left at a temperature of 20 ± 2 °C and under the illumination of 5000 LUX intensity in a light-dark period of 12:12 for a time period of 168 h. The biomass of the biofilm was measured using a VIS spectrometer (under 680 nm of wavelength) (Thermo Fisher Scientific, Prague, Czech Republic).

3. Results

Table 2 shows the different rate of calcium diffusion through the samples. The intensity of calcium diffusion was higher for the Vysoké Mýto sample and the diffusion of calcium through this sample began during the first 24 h of the experiment. Calcium diffusion through the Holešov-Žopy sample began one day later and the diffusion was less intensive. The diffusion took place faster on the surface (0–1 cm) than in the core of the bricks (2–3 cm).

Table 2. Diffusion experiment with the Vysoké Mýto and Holešov-Žopy bricks. The results are expressed as a cumulation of calcium in distilled water over the time period.

Sample	Vysoké Mýto			Holešov-Žopy		
	Calcium Concentration in the Distilled Water (mg/L)					
	Layer			Layer		
	0–1 cm	1–2 cm	2–3 cm	0–1 cm	1–2 cm	2–3 cm
Time						
24	40	20	30	0	0	0
48	180	135	148	80	80	80
72	298	220	266	150	153	164
168	802	726	711	498	466	414
192	1160	1011	912	558	516	449
216	1260	1228	1198	678	616	509
240	1546	1416	1201	1078	912	709

The photographs from the biological experiment (Figures 5–8) confirmed that both types of bricks and all their investigated layers were very intensively covered by microorganisms The composition of biofilm was observed under a microscope (Figures 6 and 8). Both samples were covered by a mix of algae and cyanobacterial species. The density of the biofilm was expressed as an absorbance (see Table 3). The absorbance values were the highest for the surface layer of bricks (0–1 cm). The highest difference was found for the 2–3 cm layer where the two kinds of brick should be considered.

Figure 5. Samples of the Vysoké Mýto bricks covered by a biofilm layer of 0–1 cm, 1–2 cm and 2–3 cm (from left to right).

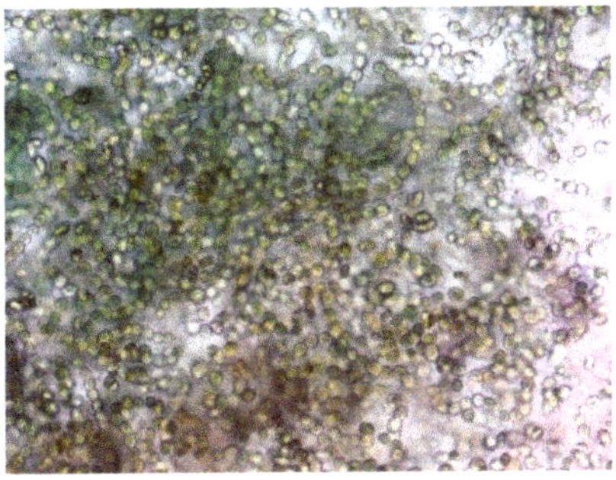

Figure 6. Biofilm on the surface of the Vysoké Mýto bricks. A photo was taken under the light microscope Olympus BX43 with a CMOS camera (magnification 400×).

Figure 7. Samples of the Holešov-Žopy bricks covered by a biofilm layer of 0–1 cm, 1–2 cm and 2–3 cm (from left to right).

Figure 8. Biofilm on the surface of the Holešov-Žopy bricks. The photo was taken under the light microscope Olympus BX43 with a CMOS camera (magnification 400×).

Table 3. Biological experiment with the Vysoké Mýto and Holešov-Žopy bricks. Biofilm density on the surface of samples expressed by absorbance values.

Sample	Vysoké Mýto			Holešov-Žopy		
	Layer			Layer		
	0–1 cm	1–2 cm	2–3 cm	0–1 cm	1–2 cm	2–3 cm
Replicate	Absorbance [A]					
1.	0.247	0.112	0.143	0.262	0.105	0.060
2.	0.244	0.112	0.141	0.263	0.113	0.066
3.	0.244	0.113	0.147	0.269	0.116	0.069
mean	0.246	0.113	0.145	0.266	0.110	0.066
SD	0.002	0.006	0.004	0.004	0.006	0.005

4. Discussion

Calcium diffusion through bricks has never been studied, with the exception of a study where calcium was used to detect the age of bricks [19]. For this reason, we are not able to compare our results with the data from the literature. However, calcium diffusion was studied for sandstone [6,20]. Calcium diffusion for the Holešov-Žopy sample was relatively similar to calcium diffusion through the Mšené sandstone, but the Vysoké Mýto sample caused higher diffusion than the sandstone by about one-fifth during the same time period. In the case of sandstone, in the sample from the Hořice locality, calcium diffusion was about one order of magnitude lower. These discrepancies are affected by the composition of the studied building materials and their physical–chemical properties. Moreover, the sandstones by themselves are a highly variable group of rocks with a wide range of physical properties.

The microstructure of brick samples may be described by several metrices, which are obviously interrelated. The highest and positive correlation between the increasing intensity of diffusion and an increasing specific surface area was observed (Table 1). It may indicate that the diffusion of chlorides is taking place by a surface diffusion mechanism where chloride ions are adsorbed on the brick surface. The Vysoké Mýto brick has lower differences in percentage composition between individual layers than the Holešov-Žopy brick; the total porosity of the Vysoké Mýto sample was from 40 to 43% of all the studied layers while the Holešov-Žopy sample had a total porosity of 42% (0–1 cm), 49% (1–2 cm) and 37% (2–3 cm). This fact could have significance not only for diffusion but also for biofilm density. The Holešov-Žopy brick also has a greater amount of the highest pores (10–100 μm).

Bio-colonization of buildings is affected by moisture and nutrient availability, favorable pH, essential and trace metal availability and suitable solar radiation, e.g., Ref. [20]. Biodegradation of bricks was studied many times [21] but species diversity and abundance have never been studied in depth among the individual layers in the bricks. In the present study, the bricks' layers were completely immersed in water to accelerate the growth of the biofilm and it was clearly visible both on the surface of the bricks and in the surrounding water. The biofilm coatings were photographed, and their analysis shows the presence of cyanobacteria and green algae. At least two species were observed microscopically in both types of bricks and their layers (Figures 6 and 8). These conclusions seem to be logical, and it likely cannot be assumed that more species of these organisms would be present in the depths.

The samples contained unicellular and multicellular organisms and it is therefore not possible to clearly determine the number of cells in the solution. The density of the organisms was indirectly expressed spectrometrically, but we did not recalculate the absorbance values on cell density for any volume unit. The density of the biofilm was once again higher in the samples with a higher visual porosity (0–1 cm layers) and for the Vysoké Mýto sample than for the Holešov-Žopy sample. Such broken surfaces and crevices probably allow algae to colonize the surface of building materials more intensively. The intensity of bio-colonization is more- or less-increasing with the rate of ions diffusion in the material as well as with the specific surface area. It again indicates that colonies are partially controlled by the available surface of the material.

5. Conclusions

Two fired bricks (Vysoké Mýto sample and Holešov-Žopy sample) were cut and their layers were studied in a diffusion experiment with calcium or in a biodegradation experiment. The results indicated that the structure of bricks' layers and especially their specific surface area affected the rate of calcium diffusion through the brick layers and probably also the density of colonization by biofilm (algae, cyanobacteria).

Author Contributions: Conceptualization, K.K and I.M. and R.Č.; methodology, K.K. and J.N. and M.K. and I.M..; investigation, K.K.; writing—original draft preparation, K.K. and M.K. and Z.S. and R.Č.; writing—review and editing, K.K. and Z.S. and R.Č.; supervision, R.Č. All authors have read and agreed to the published version of the manuscript.

Funding: This research was funded by Czech Technical University, grants No. SGS19/143/OHK1/3T/11, SGS22/137/OHK1/3T/11, by Technological Agency of Czech Republic, grant No. FW03010422 and by the Ministry of Education, Science, Research and Sport of the Slovak Republic, project VEGA No. 1/0682/19.

Institutional Review Board Statement: Not applicable.

Informed Consent Statement: Not applicable.

Data Availability Statement: The data supporting the reported results can be found with the authors.

Acknowledgments: The authors would like to thank Magdaléna Doleželová for measuring of bricks' properties, native speaker Marek Procházka for correction of English and two anonymous reviewers for their valuable advice. Publication cost of this paper was covered with founds of the Polish National Agency for Academic Exchange (NAWA): "MATBUD'2023—Developing international scientific cooperation in the field of building materials engineering" BPI/WTP/2021/1/00002, MATBUD'2023.

Conflicts of Interest: The authors declare no conflict of interest.

References

1. Gredmaier, L.; Banks, C.J.; Pearce, R.B. Calcium and Sulphur Distribution in Fired Clay Brick in the Presence of a Black Reduction Core Using Micro X-ray Fluorescence Mapping. Available online: https://core.ac.uk/download/pdf/1493405.pdf (accessed on 5 December 2022).
2. Pel, L.; Pishkari, R.; Casti, M. A simplified model for the combined wicking and evaporation of a NaCl solution in limestone. *Mater. Struct.* **2018**, *51*, 66. [CrossRef] [PubMed]
3. Ahl, J. Salt diffusion in brick structures—Part II—The effect of temperature, concentration and salt. *J. Mater. Sci.* **2004**, *39*, 4247–4254. [CrossRef]
4. Petkovic, J.; Huinink, H.P.; Pel, L.; Kopinga, K.; van Hees, R.P.J. Moisture and salt transport in three-layer plaster/substrate systems. *Constr. Build. Mater.* **2010**, *24*, 118–127. [CrossRef]

5. Pavlík, Z.; Michálek, P.; Pavlíková, M.; Kopecká, I.; Maxová, I.; Černý, R. Water and salt transport and storage properties of Mšené sandstone. *Constr. Build. Mater.* **2008**, *22*, 1736–1748. [CrossRef]
6. Kobetičová, K.; Keppert, M.; Medved', I.; Nábělková, J.; Černý, R. Effect of sample prewetting on the diffusion of calcium in sandstone. In Proceedings of the 6th International Conference on Chemical Technology (ICCT), Mikulov, Czech Republic, 16–18 April 2018.
7. Medved', I.; Kalvoda, L.; Vejmelková, E.; Vratislav, S.; Černý, R. Transport of gadolinium in a cement composite. In Proceedings of the 4th Central European Symposium on Building Physics (CESBP 2019), Prague, Czech Republic, 2–5 September 2019.
8. Medved', I.; Černý, R. Modeling of radionuclide transport in porous media: A review of recent studies. *J. Nucl. Mater.* **2019**, *526*, 151765. [CrossRef]
9. Medved', I.; Černý, R. Concentration Dependence of Diffusion Coefficients of Na-22(+) and Cs-134(+) In Opalinus Clay Rocks. In Proceedings of the International Conference on Numerical Analysis and Applied Mathematics (ICNAAM-2018), Rhodes, Greece, 13–18 September 2018.
10. Hlušek, J. Zásady Vápnění půd. Available online: http://web2.mendelu.cz/af_221_multitext/vyziva_rostlin/html/hnojiva/mineralni/cazasady.htm (accessed on 5 December 2022).
11. Coldsnow, K.D.; Relyea, R.A. The combined effects of macrophytes and three road salts on aquatic communities in outdoor mesocosms. *Environ. Pollut.* **2021**, *287*, 117652. [CrossRef]
12. *ČSN ISO 6058 (757416)*; Jakost vod. Stanovení vápníku. Odměrná metoda s EDTA. Český normalizační institut (ČNI): Prague, Czech republic, 1996.
13. Gaylarde, C.C.; Gaylarde, P.M. A comparative study of the major microbial biomass of biofilms on exteriors of buildings in Europe and Latin America. *Int. Biodeterior. Biodegrad.* **2005**, *55*, 131–139. [CrossRef]
14. Quagliarini, E.; Gianangeli, A.; D'Orazio, M.; Gregorini, B.; Osimani, A.; Aquilanti, L.; Clementi, F. Effect of temperature and relative humidity on algae biofouling on different fired brick surfaces. *Constr. Build. Mater.* **2019**, *199*, 396–405. [CrossRef]
15. Gutarowska, B.; Celikkol-Aydin, S.; Bonifay, V.; Otlewska, A.; Aydin, E.; Oldham, A.L.; Brauer, J.I.; Duncan, K.E.; Adamiak, J.; Sunner, J.A. Metabolomic and high-throughput sequencing analysis-modern approach for the assessment of biodeterioration of materials from historic buildings. *Front. Microbiol.* **2015**, *6*, 979. [CrossRef]
16. Prohlášení o vlastnostech. Available online: https://cihelna.hrabcuk.cz/cs/dokumenty-ke-stazeni (accessed on 5 January 2023).
17. Prohlášení o vlastnostech. Available online: http://www.cihelny-zlinsko.cz/files/modules/products//1/files/certifikat-vyrobku-zopy_516257b59d9dc.pdf (accessed on 5 January 2023).
18. Waddell, C.; Fountain, J.C. Calcium diffusion: A new dating method for archeological materials. *Geology* **1984**, *12*, 24–26. [CrossRef]
19. Kobetičová, K.; Medved', I.; Keppert, M.; Černý, R. Measurement and Modelling of Calcium Diffusion in a Sandstone. In Proceedings of the Thermophysics 2018, Smolenice, Slovakia, 7–9 October 2018.
20. Munyai, T.; Songishe, T.; Gumbo, J. Algae colonisation of brick pavement at the University of Venda: A potential slippery hazard. *Jamba J. Disaster Risk Stud.* **2019**, *11*, 1–7. [CrossRef] [PubMed]
21. Wilimzig, M. Biodeterioration of building materials like brick and mortar. *Int. Biodeterior. Biodegrad.* **1996**, *133*, 177–183.

materials proceedings

Proceeding Paper

Fire Resistance of Geopolymer Materials—A Change in Physical and Mechanical Properties †

Tarreck Mahaman Manssour Issa [1], Katarzyna Mróz [2,*], Mateusz Sitarz [2], Izabela Hager [2] and Ewa Słupska [2]

1 Buildings and Public Works, School of Engineering, École d'ingénieurs, bachelor et Mastère Spécialisé® (CESI), 76800 Saint-Étienne-du-Rouvray, France
2 Chair of Building Materials Engineering, Faculty of Civil Engineering, Cracow University of Technology, 31-155 Cracow, Poland
* Correspondence: katarzyna.mroz@pk.edu.pl; Tel.: +48-12-628-23-71
† Presented at the 10th MATBUD'2023 Scientific-Technical Conference "Building Materials Engineering and Innovative Sustainable Materials", Cracow, Poland, 19–21 April 2023.

Abstract: Geopolymers are new environmentally friendly cementitious materials that may offer an alternative to Ordinary Portland Cement. Not only do they have excellent mechanical properties, but they also have high temperature resistance. This research focuses on the high temperature exposure effect on geopolymer mortars. Two types of geopolymer mortars were prepared and exposed to different temperatures (20, 200, 400, 800 °C). The main goal of this work is to assess the changes in the mechanical and physical properties of geopolymer mortars after being exposed to a high temperature. Two types of mixtures were tested: one containing 100% fly ash (M0-K) and the other one containing 50% fly ash and ground Granulated blast furnace slag (M50-K). The paper presents an evaluation of compressive and tensile strength, density, porosity, and ultrasonic pulse velocity for both M50-K and M0-K after exposure to high temperatures.

Keywords: geopolymers; fly ash; ground blast furnace slag; high temperature; compressive and tensile strengths; ultrasound pulse velocity; density; porosity

check for updates

Citation: Issa, T.M.M.; Mróz, K.; Sitarz, M.; Hager, I.; Słupska, E. Fire Resistance of Geopolymer Materials—A Change in Physical and Mechanical Properties. *Mater. Proc.* **2023**, *13*, 7. https://doi.org/10.3390/materproc2023013007

Academic Editors: Tomasz Tracz and Tomasz Zdeb

Published: 14 February 2023

1. Introduction

Geopolymers are aluminosilicate materials of the Alkali-activated materials family. The geopolymerization reaction is not yet fully understood; many studies have shown that the activation of an aluminosilicate source by a highly basic alkaline solution results in the formation of amorphous materials, showing similar compressive strength to that of a traditional binder made with Ordinary Portland Cement (OPC) [1]. OPC has many disadvantages, such as excessive consumption of energy from non-renewable sources, intolerable volumes of CO_2 emissions, and questionable durability, among others. Given these facts, current environmental requirements aim to reduce the CO_2 impact of the construction industry, which is generally caused by the production of cement, one of the materials responsible for greenhouse gas emissions. Therefore, finding an alternative binder with the same strength but containing no cement is a major challenge that we are facing today. To avoid the effects of OPC on the environment, some studies have shown that geopolymers can indeed be used as binders in construction materials instead of OPC.

Fire is a frequent event that causes great damage to buildings. The degradation of physical and mechanical properties due to high temperature exposure is regarded as the primary cause of structural damage caused by fire [2]. Protecting these structures from fire is of extreme importance; therefore, the refractory materials for construction are particularly important.

This paper is focused on the (ambient and high) temperature resistance test, which is an engineering technique intended to evaluate the stability of a material under severe or unusual operating conditions. It was performed on geopolymer mortars manufactured in the building materials research laboratory of Cracow, University of Technology. The main goal of this test was to observe the behavior of those mortars in terms of their mechanical and physical properties after their exposure to ambient and high temperature.

Poland has extensive hard coal and brown coal resources. Therefore, the power industry is mostly based on these two original energy carriers. The power plants producing heat and electric energy create combustion byproducts. These products include carbon dioxide, but mainly slags and fly ash [3].

2. Materials and Samples

Two types of mixtures were tested: one containing 100% fly ash (FA), designated as M0-K, and the second one containing 50% fly ash and ground granulated blast furnace (GGBSF) slag, designated as M50-K. In addition to the two raw materials, sand with a grain size of 0/2 mm was used in the two mixtures. The oxide composition of the raw materials is presented in Table 1.

Table 1. Oxide compositions for the Raw materials used for the research.

Oxide	FA	GGBFS
SiO_2	52.3	39.91
Al_2O_3	28.05	7.61
Fe_2O_3	6.32	1.49
CaO	3.05	43.9
MgO	1.71	4.15
SO_3	0.28	0.51
K_2O	2.51	0.356
Na_2O	0.76	0.468

Potassium Silicate (K-Sil) was used as an Alkali source, which is a commercially available liquid, and a specific amount of water was added into the mixture. The compositions of the tested mixtures are presented in Table 2.

Table 2. Composition of the two mixtures in kg per m^3.

Mixtures	M0-K	M50-K
K-Sil + Water	330.7	347.5
FA	734.9	386.1
GGBFS	0.0	386.1
Sand 0/2 mm	1102.4	1158.3

A procedure involving the preparation of the binder was applied to the manufacture of geopolymer mortars. First, an alkaline-activating solution was prepared. The amount of liquid silicate was measured, then water was added into this liquid to obtain the assumed water/binder ratio of 0.3. Subsequently, FA was mixed with an Alkaline-activating solution for 10 min. Then, GGBFS was added, and mixing continued for 5 min. Next, the rotary mixer was switched off for 1 min while the binder mixing continued to remove the solids stuck onto the container. Finally, at a low mixing speed, sand was gradually added to the mixture for 3 min. The total mixing time was approximately 20 min. The manner of preparation was selected based on reports in the literature and optimized following the team's previous experience in the propagation of geopolymer blends.

The geopolymer mortars were cast into prismatic 40 × 40 × 160 mm specimens, as shown in Figure 1a. The samples were compacted on a shaking table and covered with plastic lids. Twenty-four hours after casting, the specimens were removed from the molds and stored at room ambient temperature with a relative humidity of about 75%. Plastic was used to protect them from water evaporation and to limit moisture exchange with the environment.

(**a**) (**b**)

Figure 1. (**a**) Samples after being heated and cooled down, (**b**) Electric Furnace NABETHERM.

3. Testing Procedures

3.1. Heating Procedures

After 21 days of curing, the samples were heated in an electrical furnace with a heating rate of 1 °C/min at a set temperature of 200, 400, 600, 800 and 1000 °C, Figure 1b. The heating rate followed RILEM recommendations [4] to minimize stress due to the temperature differences between hotter surfaces and the cooler inner parts of the sample. Concrete and mineral materials such as geopolymers are suitable for these heating conditions for evaluating the impact of temperature. Maintaining the temperature for 1 h until homogenization in the entire cross-section of the sample is required when the set temperature in the range of 200–1000 °C has been reached. After temperature homogenization, the samples were cooled down and their physical and mechanical properties were measured.

3.2. Physical and Mechanical Properties

The following properties of mortars were evaluated after temperature exposure.

3.2.1. Mechanical Properties

The compressive strength and tensile strength, tested in bending, were measured for the undamaged materials and for the materials after high-temperature exposure.

Firstly, the mortar samples were tested in a three-point bending test using the control testing machine. The loading rate applied was 50 N/s, as this is recommended in PN-EN 196-1 for testing cement mortars. Apart from the bending test, the compressive strength was evaluated according to the remaining prisms (40 × 40 × approx. 80 mm) produced by bending or three-point tests. The load was applied at a loading rate of 2400 N/s for the compressive strength, according to PN-EN 196-1.

The specimens after compressive and bending tests are presented in Figure 2.

(a) (b)

Figure 2. Samples after mechanical tests: (**a**) after bending tests, (**b**) after compressive tests.

3.2.2. Physical Properties

➢ *Specific and apparent density*

The material's specific density and its drop after thermal exposure were determined using a helium pycnometer ULTRAPYCNOMETR 1000, Quantachrome, while the apparent density was determined by the geometrical method, using an electronic caliper (measuring accuracy up to 0.01 mm) and a RADWAG analytical balance, with a measurement accuracy of 0.001/0.01 g.

➢ *Porosity*

Based on the specific and apparent densities, porosity can be evaluated by the formula 1:

$$P = 1 - \frac{\rho_0}{\rho} \tag{1}$$

where ρ_0 *and* ρ are, respectively, apparent and specific density.

➢ *Ultrasonic Pulse Velocity (UPV)*

Ultrasonic pulse velocity (UPV) testing was used to determine the integrity and quality of the mortar, structure concrete, or stone (up to 6 feet thick) by measuring the speed and attenuation of an ultrasonic wave passing through the specimen being tested. The method consists specifically of measuring the time of travel of an ultrasonic pulse passing through the sample being evaluated as per EN 12504-4:2021 [5]. Comparatively a higher velocity is obtained when material quality has sufficient density, uniformity, homogeneity, etc.

Determining the time as well as the distance the wave was sent through the sample in formula 2 allows us to determine the value of UPV used.

$$V_p = \frac{d}{t}\ \mathrm{m/s} \tag{2}$$

where t and d are, respectively, the travel time and the distance. The UPV (Vp) is expressed in m/s.

4. Results and Discussion

4.1. Mechanical Properties

In general, a decrease in mechanical properties was observed (except for an increase observed at 200 °C for M0-K due to the geopolymerisation progress). For compressive strength, the reference value (at 20 °C) for M0-K equaled 11.16 MPa, and after exposure to 800 °C was 6.77 MPa, while for M50-K, the reference value was 74.4 MPa and at 800 °C was 11.50 MPa. A decrease of 39.33% and 84.54%, respectively, in compressive strength for the M0-K and M50-K was noted at 800 °C.

Above 400 °C, it is reported that the mechanical properties of mortar significantly decrease compared to undamaged samples. This is ascribed to the movement of physically bonded water and OH groups, phase transformation of geopolymers, the development of anhydrous products, and the sintering process [6,7]. At elevated temperatures, non-uniform recrystallization occurred, which resulted in the formation of large cracks that consequently reduced the strength [8,9].

Considering the mechanical properties of the two tested mortars, it can be highlighted that replacing 50% of FA with GGBSF resulted in an increase of over 700% in compressive strength and over 350% in tensile strength. Nevertheless, the drop in compressive strength in the M50-K was more significant at 800 °C, reaching the same level as the M0-K. Similarly, the tensile strength of the M50-K dropped suddenly after being exposed to 200 °C and continued to decrease, reaching the same value as M0-K at 800 °C. It seems that the geopolymer bond created by GGBSF is very sensitive to high temperature exposure and is susceptible to damage at temperatures over 200 °C.

The changes in compressive strength and tensile strength for the M0-K and M50-K are presented in Figure 3.

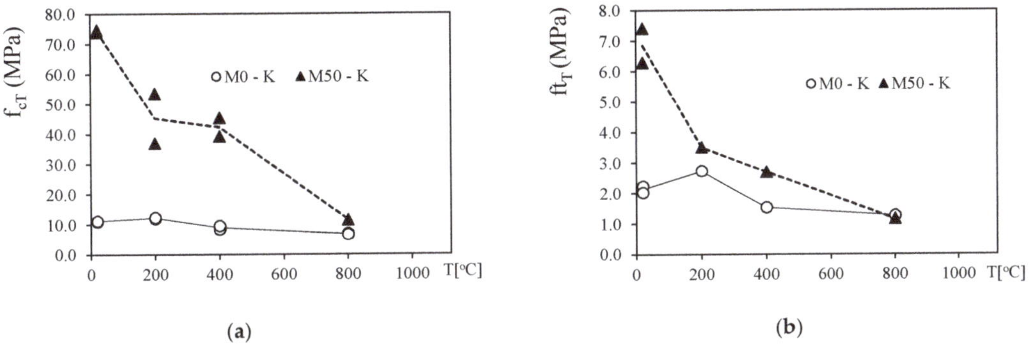

Figure 3. The changes of compressive and tensile strength with temperature (T) for M0-K and M50-K: (**a**) compressive strength f_{cT}, (**b**) tensile strength, f_{tT}.

4.2. Physical Properties

The initial (20 °C) apparent densities were 1825 kg/m^3 for M0-K and 2070 kg/m^3 for M50-K, and their true (specific) densities evaluated with a helium pycnometer were, respectively, 2490 kg/m^3 and 2475 kg/m^3. Samples containing 50% (M50-K) of FA and Metakaolin had a higher apparent density compared to those containing 100% of FA (M0-K), as shown in Figure 4. The evaluated total porosities reflected this observation. The porosity of the M0-K was 40% and that of M50-K was 27% (Figure 5). The lower apparent density of M0-K was attributed to a more significant number of pores in the material and their larger size. This also affects the differences in the initial mechanical properties.

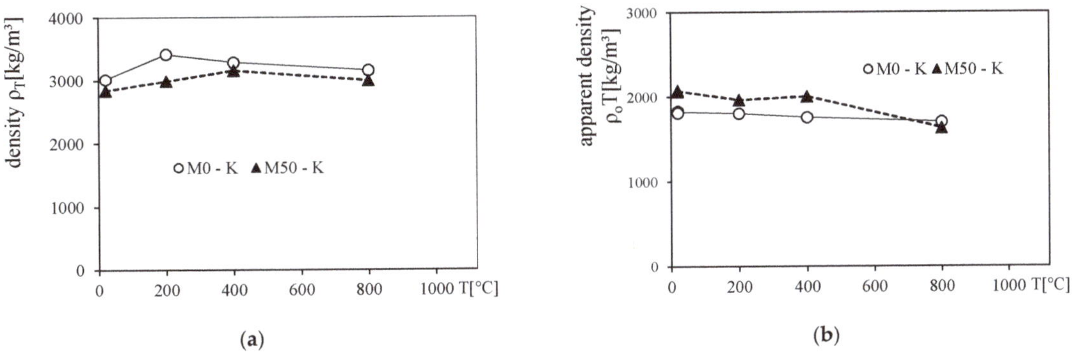

Figure 4. (**a**) Specific density ρ_T and (**b**) apparent density ρ_{oT} in kg/m^3 of heated samples.

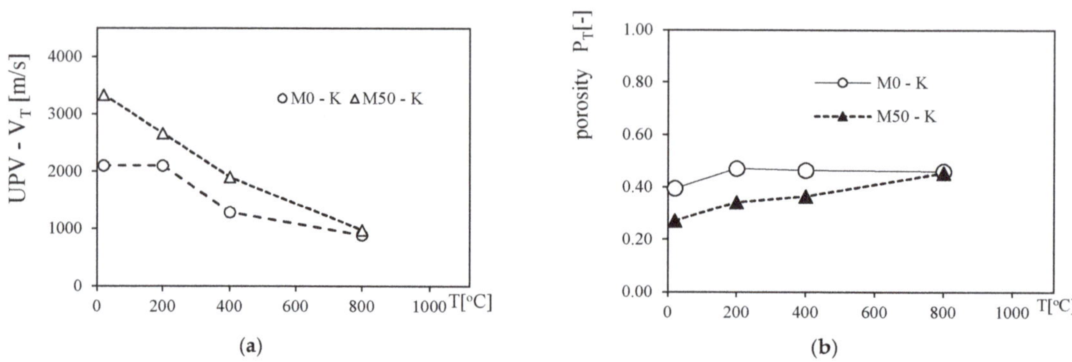

Figure 5. (**a**) Ultrasound Pulse Velocity (UPV—V_T) and (**b**) porosity (P) with temperature.

After high temperature exposure, the apparent density of the samples slightly decreased due to the drying process and changes in the matrix. Once exposed to high temperatures, the components of the samples, such as aggregate, became dry. The moisture contained in the material was gradually removed through heating. Furthermore, heating did not significantly affect the change in density. The physically bound water was entirely evaporated, and the hydroxyl groups were removed at high temperatures. Some sources showed that the dihydroxylation process begins at 250 °C and continues up to 600 °C [10], which causes the geopolymer binder to shrink.

For the M0-K, an increase in density after exposure to 200 °C was observed, while for the M50-K, density increased to 400 °C. This is related to the continuous geopolymerization process under elevated temperature; the mortar behaves similarly to an autoclaving mortar, with binder connections developing and strengthening. This was confirmed by the increase in tensile strength and applies more to a matrix without GGBSF. Another insignificant drop in density over 400 °C was related to the cracking of the matrix.

The density (ρ) and the apparent density (ρ_0) values enabled the total porosity (P) to be determined. The variation in total porosity P with increasing temperature is presented in Figure 5. The M0-K presents an increase in porosity between 20 °C and 200 °C. In this temperature range, the porosity remained stable (47%) from 200 °C up to 800 °C. On the other hand, for the M50-K, a slight increase in porosity from 20 °C (27%) to 800 °C (45%) was observed. The stable porosity level in the entire range of tested temperatures ensured constant thermal insulation parameters with increasing temperature, which is extremely important in cases of fire protection or fire-resistant material. Therefore, the M0-K, independently from its lower initial mechanical properties, can be developed in the direction of fire protection applications.

The UPV values decreased in the entire temperature range (20–800 °C), Figure 5a. However, in M0-K, a stable velocity was measured between 20 °C and 200 °C. The stable UPV up to 200 °C confirms the development of the M0-K matrix at elevated temperatures, as the UPC reflects the amount of air pores in the material. Moreover, the final UPV after exposure to 800 °C reached similar values for both the M0-K and M50-K. The same quality was confirmed for both the mechanical and physical properties of the tested materials when exposed to 800 °C.

5. Conclusions

The effect of elevated temperature ranging from 200 °C to 800 °C on samples with 100% fly ash and 50% fly ash, and ground granulated blast furnace slag blended geopolymer mortar was investigated in the presented research by evaluating compressive strength, tensile strength, apparent density, specific density, porosity, and the ultrasound pulse velocity. Based on the results, the following conclusions are drawn:

- The initial mechanical properties (compressive and tensile strengths) of a geopolymer mortar containing 50% fly ash and ground-granulated blast furnace Slag (M50-K) are greater than those of a geopolymer mortar containing 100% fly ash (M0-K). This is probably due to the presence of two precursors (fly ash and metakaolin) in the mixture. Previous research has shown that the geopolymer materials based on FA can be improved with the addition of other wastes or precursors.
- After exposure to a high temperature, the M50-K specimens continue to present higher values of compressive and tensile strengths through the temperature range of 200–600 °C, while the rate of decrease with temperature is more significant than for the M0-K.
- The M0-K, despite having lower mechanical properties than the M50-K, presents a developing bond at elevated temperatures (up to 200 °C), remaining more stable until 800 °C than the M50-K. Therefore, the M0-K can be considered a valuable solution to develop in the direction of fire protection applications.
- The physical properties have also been evaluated. The high temperature exposure contributes to the damage of the material's cross section, resulting in a decrease in density, apparent density, and Ultrasound Pulse Velocity. The M50-K is slightly higher than the M0-K in terms of apparent density, but we had the opposite principle in terms of true densities after exposure to high temperatures. The M0-K porosity remains quasi-stable across the entire temperature range, whereas the M50-K porosity increases until 800 °C.
- It should be noted that the tested samples exhibited a high temperature a relatively short time after production. In the case of the geopolymer mortars with a fly ash precursor, curing in ambient conditions, the matrix properties developed much slower. Previous team research results show that after 3 months, the FA-based geopolymer matrix is much more resistant to high temperatures.

According to the obtained results, both materials can be used as the basis for developing fire insulation materials or thermal barriers. They remain stable in terms of integrity under high-temperature conditions, presenting porosity at an affordable level. Despite the drop in mechanical properties, compressive strength, even after exposure to 800 °C, remained at a satisfactory level in terms of the structural application.

Author Contributions: Conceptualization, T.M.M.I., I.H., M.S., K.M., E.S.; methodology, I.H, T.M.M.I. and K.M.; software, M.S., K.M. and T.M.M.I.; validation, I.H., K.M. and M.S.; formal analysis, T.M.M.I., K.M. and M.S.; investigation T.M.M.I., K.M., E.S. and M.S.; data curation, T.M.M.I., K.M. and M.S.; writing—original draft preparation, I.H., K.M. and M.S.; writing—review and editing, T.M.M.I., I.H., M.S., K.M., E.S.; visualization, K.M. and M.S.; supervision, I.H.; project administration, I.H.; funding acquisition, I.H., K.M. and M.S. All authors have read and agreed to the published version of the manuscript.

Funding: This research was funded by Cracow University of Technology and the Polish National Agency for foreign students in the framework of the project NAWA Spinaker, POWR.03.03.00-00-PN16/18. This work has been supported by the National Centre for Research and Development in Poland in the framework of the project SMART-G Smart geopolymers (ERA-MIN2-3/SMART-G/1/2022).

Institutional Review Board Statement: Not applicable.

Informed Consent Statement: Not applicable.

Data Availability Statement: Not applicable.

Acknowledgments: This research was part of the work provided by two interdisciplinary research group of Building Materials Engineering, Faculty of Civil Engineering (Cracow University of Technology, 31-155 Cracow, Poland) & Buildings and Public Works (School of Engineering—Cesi, 76800 Saint Etienne du Rouvray, France). Publication cost of this paper was covered with founds of the Polish National Agency for Academic Exchange (NAWA): "MATBUD'2023—Developing international scientific cooperation in the field of building materials engineering" BPI/WTP/2021/1/00002, MATBUD'2023.

Conflicts of Interest: The authors declare no conflict of interest.

References

1. Pouhet, R. Formulation and Durability of Metakaolin Based Geopolymers. Ph.D. Thesis, Université Paul Sabatier Toulouse III, Toulouse, France, 2015. (In English).
2. Davidovits, J. Properties of geopolymer cements. In Proceedings of the First International Conference onn Alkaline Cements and Concretes, Scientific Research Institute on Binders and Materials, Kiev State Technical University, Kiev, Ukraine, 11–14 October 1994; pp. 131–149.
3. Zhou, C.; Pan, J.; Zhang, Z.; Zhu, Y. Comparative study on the tensile mechanical behavior of GFRP bars under and after high temperature exposure. *Case Stud. Constr. Mat.* **2022**, *16*, e00905. [CrossRef]
4. RILEM Technical Committee 200-HTC. Recommendation of RILEM TC 200-HTC: Mechanical concrete properties at high temperatures—Modelling and applications. *Mater. Struct.* **2007**, *40*, 841–853. [CrossRef]
5. *EN 12504-4:2021-12*; Testing Concrete in Structures-Part 4: Determination of Ultrasonic Pulse Velocity. CEN: Brussels, Belgium, 2021.
6. Yang, Z.; Mocadlo, R.; Zhao, M.; Sisson, R.D., Jr.; Tao, M.; Liang, J. Preparation of a geopolymer from red mud slurry and class F fly ash and its behavior at elevated temperatures. *Constr. Build. Mater.* **2019**, *221*, 308–317. [CrossRef]
7. Behera, P.; Baheti, V.; Militky, J.; Louda, P. Elevated temperature properties of basalt microfibril filled geopolymer composites. *Constr. Build. Mater.* **2018**, *163*, 850–860. [CrossRef]
8. Hager, I.; Sitarz, M.; Mróz, K. Fly Ash based geopolymer mortar for high-temperature application—Effect of slag addition. *J. Cleaner. Prod.* **2021**, *316*, 128168. [CrossRef]
9. Omer, S.A.; Demirboga, R.; Khushefati, W.H. Relationship between compressive strength and UPV of GGBFS based geopolymer mortars exposed to elevated temperatures. *Constr. Build. Mater.* **2015**, *94*, 189–195. [CrossRef]
10. Sitarz, M.; Figiela, B.; Łach, M.; Korniejenko, K.; Mróz, K.; Castro-Gomes, J.; Hager, I. Mechanical Response of Geopolymer Foams to Heating—Managing Coal Gangue in Fire-Resistant Materials Technology. *Energies* **2022**, *15*, 3363. [CrossRef]

materials proceedings

Proceeding Paper

Influence of Spent Fluid Catalytic Cracking Catalyst on the Properties of the New Binder Based on Fly Ash and Portland Cement †

Jelena Rakić and Zvezdana Baščarević *

Institute for Multidisciplinary Research, Belgrade University, Kneza Višeslava 1, 11000 Belgrade, Serbia
* Correspondence: zvezdana@imsi.bg.ac.rs
† Presented at the 10th MATBUD'2023 Scientific-Technical Conference "Building Materials Engineering and Innovative Sustainable Materials", Cracow, Poland, 19–21 April 2023.

Abstract: One of the measures to reduce the carbon footprint of the Portland cement (PC) manufacturing process is through a wider use of supplementary cementitious and waste materials. The main objective of this work was to produce a new binder using two different waste materials: fly ash (FA) from thermal power plants and spent fluid catalytic cracking catalyst (sFCCC) from petrol refineries. In order to improve their reactivity, both FA and sFCCC were mechanically activated prior to the preparation of the binder. The new binder consisted mostly of the waste materials (70 mass %), with PC as a minor component (30 mass %). It was found that using sFCCC as the binder component accelerated cement hydration and the pozzolanic reaction. The new binder had a shorter setting time and a higher early strength than the binder prepared without sFCCC.

Keywords: spent catalyst; fly ash; catalytic cracking; Portland cement; pozzolanic reaction

check for updates

Citation: Rakić, J.; Baščarević, Z. Influence of Spent Fluid Catalytic Cracking Catalyst on the Properties of the New Binder Based on Fly Ash and Portland Cement. *Mater. Proc.* **2023**, *13*, 8. https://doi.org/10.3390/materproc2023013008

Academic Editors: Katarzyna Mróz, Tomasz Tracz, Tomasz Zdeb and Izabela Hager

Published: 14 February 2023

1. Introduction

Fly ash (FA) is waste material generated by coal combustion in power plants. It is a fine aluminosilicate powder and owing to these properties, FA has been used for decades as cement and/or a concrete component. However, when FA is used as the major component in high volume FA–Portland cement (PC) blends, it causes prolonged setting time and delayed strength development of a binder [1]. Although the reactivity of FA and early age properties of high volume FA binders can be improved by using chemical activators [1,2], it often results in decreased strength of a binder at later ages.

Spent fluid catalytic cracking catalyst (sFCCC) is a waste material from petrol refineries that consists mostly of zeolite and an amorphous aluminosilicate matrix. Due to relatively small quantities of sFCCC produced (200,000–400,000 tons per year globally [3]), landfilling is often considered as the most economical option for its disposal. This waste material can be used as a cement additive, although its amount in binder is limited by the fact that it can cause short setting time and increased water demand [2,4,5].

Previous studies demonstrated that using sFCCC as a component of blended cements (70–80% of cement in the binder) had a positive impact on the binder properties [6–8]. However, little work has been done so far on investigating the possibilities of using sFCCC as part of a binder in which higher amounts of cement were replaced by waste materials, such as FA. Calorimetric and thermal analyses of very high volume FA binders (20% of cement in the binder) showed that using sFCCC as a FA replacement in the binder accelerated both cement hydration and the pozzolanic reaction [4].

The aim of this work was to further investigate the effects of replacing part of FA in the high volume FA binder (70 mass %) with sFCCC. It was expected that the use of sFCCC as the binder component would shorten setting times and increase early compressive strength of the binder. In order to evaluate the effects of using sFCCC as the binder component, setting time, compressive strength, mineral composition and microstructure of the new binders were analyzed.

2. Materials and Methods

FA used in this work was from "Nikola Tesla B" power plant (Serbia). The sFCCC sample was obtained from a local petrol refinery. Portland cement CEM I 52.5 N (CEM, Našice cement, Nexe d.d., Našice, Croatia) was used as a minor component of the new binder.

As-received FA and sFCCC were mechanically activated in a planetary ball mill (Fritsch Pulverisette 05 102, Fritsch GmbH, Idar-Oberstein, Germany). The mechanical activation of FA was conducted in stainless steel bowls (500 cm^3 in volume) using 13 mm diameter steel balls, with a 1:3 FA to balls ratio, at 380 rpm and lasting for 15 min. The sFCCC sample was ground for 20 min at 200 rpm in corundum grinding bowls (500 cm^3), and the sFCCC to corundum grinding balls (5 mm in diameter) mass ratio was 1:3.

Particle size distribution (PSD) of the materials was analyzed using Mastersizer 2000 (Malvern Panalytical, Malvern, UK).

The chemical composition of the binder components was determined by energy dispersive X-ray fluorescence spectrometer ED2000 (Oxford Instruments, Abingdon, UK).

X-ray diffraction analysis (XRD, Rigaku Smart Lab, Rigaku, Tokyo, Japan), with Cu anticathode operating at 40 kV and 30 mA, was used to determine mineral composition of the starting materials and the new binder pastes. The XRD analyses were conducted in the 5–55 °2θ range, with 0.01° step and 2°/min recording speed.

As-received and grinded FA and sFCCC, as well as resulting binder samples, were examined using a scanning electron microscope (SEM, VEGA TS 5130 MM, Tescan, Brno, Czech Republic) equipped with a backscattered detector (BSE, Tescan). Before the analyses, the samples were Au-coated.

The new binders were prepared by dry mixing the waste materials and CEM in 70:30 mass ratios for 5 min. Paste samples were made by mixing the binder with water (Table 1), provided that the standard consistency was achieved [9]. The setting time of the pastes was determined following the recommendations given in EN 196-3 [9].

Table 1. Binder samples denotation and composition.

Binder	Composition (mass %)			Water/Binder Ratio
	FA	sFCCC	CEM	
CEM	0	0	100	0.29
FCCC70	0	70	30	0.47
FCCC35	35	35	30	0.46
FCCC21	49	21	30	0.45
FA70	70	0	30	0.43

Binder mortars were prepared by mixing the binder with water and standard sand [10]. Water to binder (w/b) ratios of the new binder mortars were aimed at providing workability similar to the workability of the CEM mortar prepared with w/b = 0.50 [10,11]. Therefore, the w/b ratio of the mortar based only on FA (binder FA70) was 0.57 and the w/b ratio of the mortar prepared with both of the waste materials (FCCC21) was 0.58.

Paste and mortar samples were cured in a humid chamber (relative humidity ~90%, temperature 20 ± 2 °C) until testing.

The compressive strength of the mortars was determined using Matest testing machine E161 (Matest, Treviolo, Italy) [9].

3. Results and Discussion

3.1. Characterization of the Binder Components

3.1.1. Effects of the Mechanical Activation on Morphology and Particle Size Distribution of FA and sFCCC

The as-received FA sample consisted mainly of large particles (~100 μm), irregular in shape (Figure 1a). The grinding of the FA sample in the planetary ball mill for only 15 min provided a material with fine particles, mostly smaller than 10 μm (Figure 1b).

(**a**) (**b**)

Figure 1. SEM micrographs: (**a**) as-received FA; (**b**) ground FA.

The sFCCC sample obtained from the local petrol refinery contained spherical particles, ~20 μm in diameter, and smaller irregular particles (Figure 2a). After the mechanical activation of the as-received sFCCC sample, most of the spherical particles were broken. However, SEM analysis of the ground sFCCC sample showed the presence of agglomerates (Figure 2b).

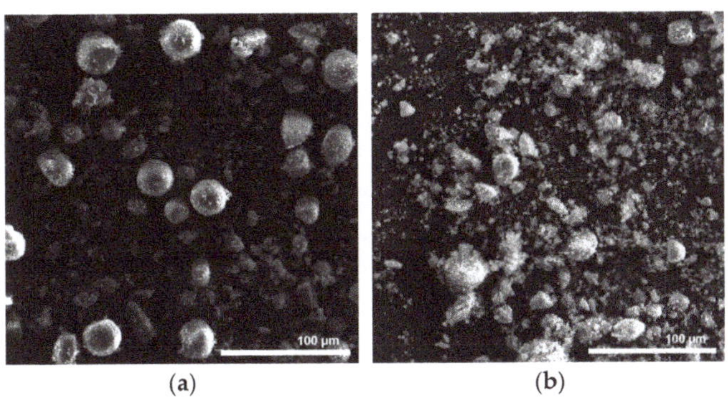

(**a**) (**b**)

Figure 2. SEM micrographs: (**a**) as-received sFCCC; (**b**) ground sFCCC.

Particle size distribution analyses of the as-received and ground FA and sFCCC demonstrated that mechanical activation, for only 15 min, resulted in a substantial decrease in particle sizes of both of the materials (Figure 3a,b). It has already been demonstrated that mechanical activation increases the reactivity of these waste materials [12–15]. The fact that the PSD of the mechanically activated FA and sFCCC was almost similar to the PSD of CEM (Figure 3c) could also have a positive impact on the mixing of the powders when preparing the binder [16].

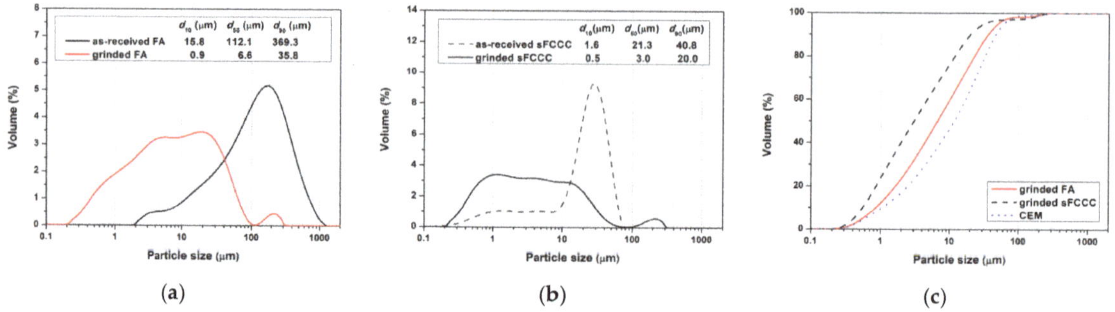

Figure 3. Particle size distribution: (**a**) as-received and ground FA; (**b**) as-received and ground sFCCC; (**c**) binder components (cumulative curve).

3.1.2. Chemical and Mineral Composition of the Binder Components

Table 2 shows the chemical composition of the materials used for the preparation of the new binder (the CEM sample and the ground FA and sFCCC samples, hereinafter referred to as FA and sFCCC). The FA produced by Serbian power plants originates from burning lignite and usually belongs to class F [17,18]. The FA sample used in this work consisted mostly of SiO_2 and Al_2O_3, with a relatively high content of CaO (Table 2). The sFCCC sample was rich in Al_2O_3 and contained approximately equal amounts of SiO_2 and Al_2O_3 (~42 mass %), while the chemical composition of the CEM sample was typical for the material.

Table 2. Chemical composition of the starting materials.

Composition (mass %)	FA	sFCCC	CEM
LOI (1000 °C)	2.69	5.55	2.54
SiO_2	61.51	41.97	20.17
Al_2O_3	15.91	42.86	5.36
Fe_2O_3	7.79	1.86	2.86
CaO	8.45	4.79	62.04
MgO	0.35	0.43	2.17
SO_3	0.19	0.10	3.38
Na_2O	0.35	0.01	0.45
K_2O	1.01	0.07	0.75
P_2O_5	0.06	0.47	-
TiO_2	1.46	1.61	-
SUM	99.77	99.72	99.72

XRD analyses also showed that the mineral composition of the FA, sFCCC and CEM samples was typical for these materials (Figure 4). Clinker minerals and anhydrite were detected in the CEM sample, while quartz peaks were the most prominent in the XRD graph of the FA sample. Crystalline zeolite faujasite was detected in the sFCCC sample. The presence of the amorphous phases in the both of the waste materials was indicated by the shape of the baselines in the 15–35 °2θ range (Figure 4).

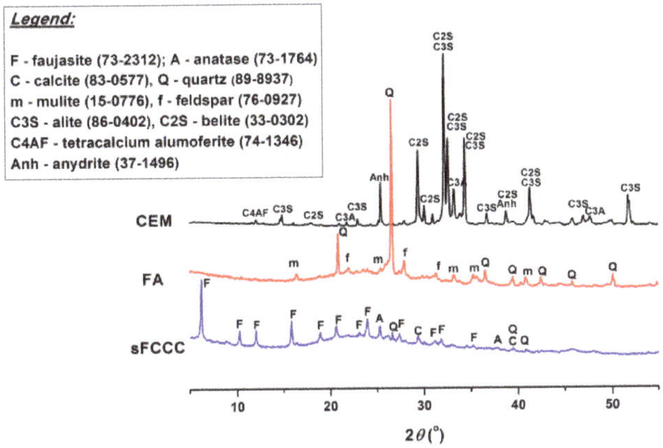

Figure 4. Mineral composition of the binder components.

3.2. *Influence of sFCCC on Setting Times of the New Binder*

Figure 5 shows the setting times of the prepared binder pastes. The paste made only with the sFCCC and Portland cement (FCCC70) showed a flash setting (initial setting was observed already after 15 min), while the setting times of the paste prepared with FA and CEM (FA70) were significantly longer than the setting times of the paste prepared with pure cement (CEM). Decreasing percentages of the FA were substituted with the sFCCC in the binder mixes (starting from 50 mass %, Figure 5, binder FCCC35) in order to obtain the new binder with setting times shorter than the setting times of the FA70 paste, but comparable to setting time of the cement paste. Finally, the binder prepared with 21 mass % of the sFCCC (FCCC21) showed the envisaged properties regarding setting times, and was used for further studies. The goal of this approach was to use sFCCC as an activating component of the binder. The setting times of the new binder should be shorter than the setting times of the binder based only on FA. The obtained results confirmed that sFCCC is a very active pozzolanic material [4]. The accelerating effect of the sFCCC in the blended cement pastes has also been attributed to the adsorption of water on the sFCCC particles and to decreased sulfate content in the blended cement [5].

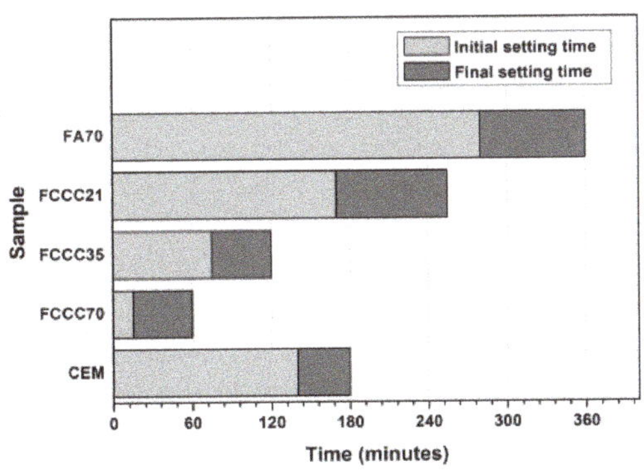

Figure 5. Setting times of the new binder and CEM pastes.

3.3. *Compressive Strength of the New Binder*

Analyses of the compressive strength of the new binder mortars showed that using sFCCC as the binder component had a positive impact on early (2 days) compressive strength (Table 3). Afterwards, compressive strength of the FA70 binder was higher than the strength of the FCCC21 binder. Relatively high compressive strengths of the FA70 sample were probably achieved due to the fact that mechanically activated FA was used in the preparation of the new binder. Very fine particles of the ground FA acted as nucleation sites and contributed to cement hydration at the early stages [6].

Table 3. Compressive strength of the new binder and CEM mortars.

Compressive Strength (MPa)	CEM	FA70	FCCC21
2 days	31.9 ± 0.9	6.0 ± 0.2	7.0 ± 0.2
7 days	48.8 ± 0.6	15.0 ± 0.5	14.8 ± 0.3
28 days	57.8 ± 2.1	32.9 ± 0.7	26.9 ± 0.9

3.4. Mineral Composition and Microstructure of the New Binder

Mineral compositions of the binder pastes based on FA and the combination of FA and sFCCC after 28 days of curing were similar (Figure 6). The main hydration products of the new binders were calcium aluminate hydrate (CAH) and ettringite (E). The shape of the peak at 29.3 °2θ indicated that calcium silicate hydrate (CSH) was probably formed in the new binder pastes too. As seen from Figure 6, portlandite (calcium hydroxide, CH) could not be detected in the XRD graphs of the binders FA70 and FCCC21, cured for 28 days, which could be explained by the low content of cement in the binders and portlandite consumption by the pozzolanic reactions of FA and sFCCC [4]. The XRD analysis of the CEM paste showed a mineral composition typical for the material.

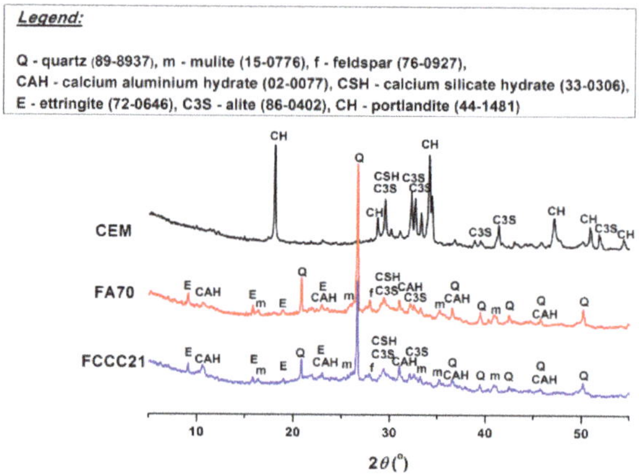

Figure 6. Mineral composition of the new binder and CEM pastes cured for 28 days.

SEM analyses of the FA70 and FCCC21 binder pastes cured for 28 days showed that the microstructure of the new binders was heterogeneous and relatively porous, as seen from the BSE micrographs (Figures 7a and 8a). It seems that the FCCC21 paste was slightly more porous than the FA70 paste, which corresponds with the observed differences in compressive strength (Table 3), and could partly be explained by a higher water/binder ratio of the FCCC21 paste (Table 1). SEM micrographs taken at the higher magnifications confirmed the presence of the hydration products identified by the XRD analyses, probably CSH (Figure 7b) and ettringite (Figure 8b).

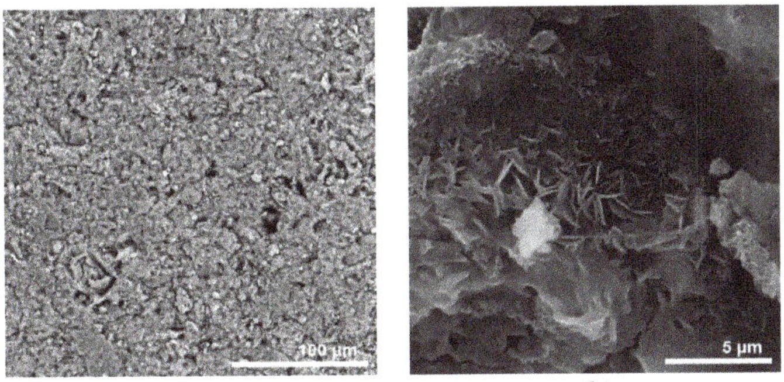

(a) (b)

Figure 7. SEM analyses of the FA70 paste after 28 days of curing: (**a**) microstructure; (**b**) reaction products.

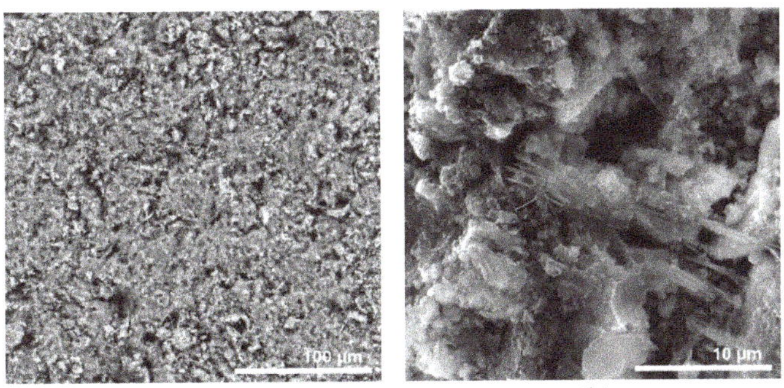

(a) (b)

Figure 8. SEM analyses of the FCCC21 paste after 28 days of curing: (**a**) microstructure; (**b**) reaction products.

The structural characterization of the new binder pastes did not provide an explanation for the observed differences in the strength development of the binder mortars after the first 2 days of curing (Table 3). It is possible that portlandite was consumed in the FCCC21 more quickly than in the FA70 binder due to the high reactivity of sFCCC [4]. The lower strength of the FCCC21 binder after 28 days of curing, compared to the FA70, could also be due to the very fast hydration of the binder containing sFCCC at the beginning of the reaction, and the attachment of the reaction products on the surface of the un-reacted particles [19,20].

4. Conclusions

The spent catalyst from the catalytic cracking in the petrol refinery was used as a component of the high volume fly ash binder. Both of the waste materials were mechanically activated prior to the preparation of the new binder and the ground materials had a PSD comparable to the PSD of commercial cement. It was found that the binder that contained 21 mass % of the sFCCC had setting times comparable to the setting times of cement pastes. The use of the sFCCC as the binder component also had a positive effect on the early (2 days) compressive strength of the binder. However, the compressive strength of the binder prepared without sFCCC was higher at later ages. Structural analyses revealed that the main reaction products of the new binders were calcium aluminate hydrate, calcium silicate hydrate and ettringite, while the microstructure of the binders was heterogeneous and relatively porous. Differences in the observed properties of the binders prepared with and without sFCCC were attributed to the high reactivity of the mechanically activated waste materials.

Author Contributions: Conceptualization, Z.B.; methodology, Z.B. and J.R.; validation, J.R. and Z.B.; formal analysis, J.R.; investigation, J.R.; resources, Z.B. and J.R.; data curation, Z.B. and J.R.; writing—original draft preparation, J.R. and Z.B.; writing—review and editing, Z.B.; supervision, Z.B.; project administration, Z.B.; funding acquisition, Z.B. All authors have read and agreed to the published version of the manuscript.

Funding: This research work was funded by the Ministry of Education, Science and Technological Development, Republic of Serbia, contract number 451-03-68/2022-14/200053 and by the funding of the EUREKA project E! 9980 INBYCON.

Data Availability Statement: The data presented in this study are available on request from the corresponding author. The data are not publicly available due to legal restrictions.

Acknowledgments: The authors are thankful to Sabina Kovač for performing XRD analyses. The publication cost of this paper was covered with funds from the Polish National Agency for Academic Exchange (NAWA): "MATBUD'2023-Developing international scientific cooperation in the field of building materials engineering" BPI/WTP/2021/1/00002, MATBUD'2023.

Conflicts of Interest: The authors declare no conflict of interest.

References

1. Alahrache, S.; Winnefeld, F.; Champenois, J.; Hesselbarth, F.; Lothenbach, B. Chemical activation of hybrid binders based on siliceous fly ash and Portland cement. *Cem. Concr. Compos.* **2016**, *66*, 10–23. [CrossRef]
2. Wilinska, I.; Pacewska, B. Influence of selected activating methods on hydration processes of mixtures containing high and very high amount of fly ash. *J. Therm. Anal. Calorim.* **2018**, *133*, 823–843. [CrossRef]
3. Alonso-Fariñas, B.; Rodríguez-Galán, M.; Arenas, C.; Arroyo Torralvo, F.; Leiva, C. Sustainable management of spent fluid catalytic cracking catalyst from a circular economy approach. *Waste. Manag.* **2020**, *110*, 10–19. [CrossRef] [PubMed]
4. Wilińska, I.; Pacewska, B. Calorimetric and thermal analysis studies on the influence of waste aluminosilicate catalyst on the hydration of fly ash–cement paste. *J. Therm. Anal. Calorim.* **2014**, *116*, 689–697. [CrossRef]
5. Payá, J.; Monzó, J.; Borrachero, M.V. Physical, chemical and mechanical properties of fluid catalytic cracking catalyst residue (FC3R) blended cements. *Cem. Concr. Res.* **2001**, *31*, 57–61. [CrossRef]
6. Soriano, L.; Payá, J.; Monzó, J.; Borrachero, M.V.; Tashima, M.M. High strength mortars using ordinary Portland cement–fly ash–fluid catalytic cracking catalyst residue ternary system (OPC/FA/FCC). *Constr. Build. Mater.* **2016**, *106*, 228–235. [CrossRef]
7. Da, Y.; He, T.; Wang, M.; Shi, C.; Xu, R.; Yang, R. The effect of spent petroleum catalyst powders on the multiple properties in blended cement. *Constr. Build. Mater.* **2020**, *231*, 117203. [CrossRef]
8. Payá, J.; Monzó, J.; Borrachero, M.V. Fluid catalytic cracking catalyst residue (FC3R). An excellent mineral by-product for improving early strength development of cement mixtures. *Cem. Concr. Res.* **1999**, *29*, 1773–1779. [CrossRef]
9. *SRPS EN 196-3:2017*; Methods of Testing Cement–Part 3: Determination of Setting Times and Soundness. Institute for Standardization of Serbia: Belgrade, Serbia, 2017.
10. *SRPS EN 196-1:2017*; Methods of Testing Cement–Part 1: Determination of Strength. Institute for Standardization of Serbia: Belgrade, Serbia, 2017.
11. *SRPS EN 1015-3:2008*; Methods of Test for Mortar for Masonry–Part 3: Determination of Consistence of Fresh Mortar (by Flow Table). Institute for Standardization of Serbia: Belgrade, Serbia, 2008.
12. Marjanović, N.; Komljenović, M.; Baščarević, Z.; Nikolić, V. Improving reactivity of fly ash and properties of ensuing geopolymers through mechanical activation. *Constr. Build. Mater.* **2014**, *57*, 151–162. [CrossRef]
13. Marjanović, N.; Komljenović, M.; Baščarević, Z.; Nikolić, V. Comparison of two alkali-activated systems: Mechanically activated fly ash and fly ash-blast furnace slag blends. 7th Scientific-Technical Conference Material Problems in Civil Engineering (MATBUD'2015). *Procedia Eng.* **2015**, *108*, 231–238. [CrossRef]
14. Pacewska, B.; Wilińska, I.; Bukowska, M.; Blonkowski, G.; Nocuń-Wczelik, W. An attempt to improve the pozzolanic activity of waste aluminosilicate catalyst. *J. Therm. Anal. Calorim.* **2004**, *77*, 133–142. [CrossRef]
15. Baščarević, Z.; Rakić, J.; Petrović, R. Possibility to use spent catalyst from fluid catalytic cracking process for geopolymer synthesis. In Proceedings of the (Tagungsband 1) Internationale Baustofftagung, IBAUSIL, Weimar, Germany, 12–14 September 2018.
16. Musha, H.; Chandratilleke, G.R.; Chan, S.L.I.; Bridgwater, J.; Yu, A.B. Effects of Size and Density Differences on Mixing of Binary Mixtures of Particles. *Powders Grains 2013 AIP Conf. Proc.* **2013**, *1542*, 739–742. [CrossRef]
17. *ASTM C618-22*; Standard Specification for Coal Fly Ash and Raw or Calcined Natural Pozzolan for Use in Concrete. ASTM International: West Conshohochen, PA, USA, 2022.
18. Komljenović, M.; Baščarević, Z.; Bradić, V. Mechanical and microstructural properties of alkali-activated fly ash geopolymers. *J. Hazard Mater.* **2010**, *181*, 35–42. [CrossRef] [PubMed]

19. Lam, L.; Wong, Y.L.; Poon, C.S. Degree of hydration and gel/space ratio of high-volume fly ash/cement systems. *Cem. Concr. Res.* **2000**, *30*, 747–756. [CrossRef]
20. He, T.; Da, Y.; Xu, R.; Yang, R. Effect of multiple chemical activators on mechanical property of high replacement high calcium fly ash blended system. *Constr. Build. Mater.* **2019**, *198*, 537–545. [CrossRef]

materials proceedings

MDPI

Proceeding Paper

Study on the Blending Characteristics of Ternary Cementless Materials †

Yi-Hua Chang [1], Lukáš Fiala [2], Martina Záleská [2], Dana Koňáková [2], Wei-Ting Lin [1,*] and An Cheng [1]

1 Department of Civil Engineering, National Ilan University, Ilan 26047, Taiwan
2 Department of Materials Engineering and Chemistry, Faculty of Civil Engineering, Czech Technical University in Prague, Thákurova 7/2077, 166 29 Prague 6, Czech Republic
* Correspondence: wtlin@niu.edu.tw; Tel.: +886-3-9317567; Fax: +886-3-9360125
† Presented at the 10th MATBUD'2023 Scientific-Technical Conference "Building Materials Engineering and Innovative Sustainable Materials", 19–21 April 2023.

Abstract: In this study, three industrial by-products (ultrafine fly ash, ground granulated blast-furnace slag (ggbs) and circulating fluidized bed co-fired fly ash) were used to produce ternary cementless composites without using alkali activators. The finenesses of ultrafine fly ash, ggbs and co-fired fly ash were 33,800, 5830 and 5130 cm^2/g, respectively. The composite material was developed by mixing supplementary cementing materials of different particle sizes and exploiting the high-alkaline properties of the co-fired fly ash to develop a substantial hardening property like cement. The test specimens were made in the form of pastes and the water-to-cementitious-material ratio for the test was fixed at 0.55. The test results show that the flowability of the six different mixtures could be up to 120% and the setting time could be controlled within 24 h. At 60% of the ggbs proportion, the setting time could be held for 8 h. The compressive strength of each proportion reached 7 MPa at 7 days and 14 MPa at 28 days. The water-cured specimens exhibited better strength behavior than the air-cured specimens. Scanning electron microscopy found the main components of strength growth of the specimens to be hydrated reactants of C-A-S-H or ettringite. The results of the XRF analysis show that the specimens responded to higher compressive strengths as the Ca/Si and Ca/Al ratios increased.

Keywords: cementless composites; microscopic analysis; reactants; net-zero carbon emissions

check for updates

Citation: Chang, Y.-H.; Fiala, L.; Záleská, M.; Koňáková, D.; Lin, W.-T.; Cheng, A. Study on the Blending Characteristics of Ternary Cementless Materials. *Mater. Proc.* **2023**, *13*, 9. https://doi.org/10.3390/materproc2023013009

Academic Editors: Katarzyna Mróz, Tomasz Tracz, Tomasz Zdeb and Izabela Hager

Published: 14 February 2023

1. Introduction

Cement production has become very energy-intensive, consuming more than 5% of the world's total energy demand, and a large portion of CO_2 emissions is associated with the cement industry [1]. However, cement concrete is widely used as a construction material due to its low cost and long service life. Although this material has high compressive strength and durability, its structural use is limited by the low tensile strength of the material and its susceptibility to crack expansion. In addition, the cement production process has a significant influence on global warming [2,3]. Presently, global production of Portland cement is approximately 4.6 billion tons per year. It is expected to reach a capacity of more than 6 billion tons by 2050 [4]. For countries around the world to achieve the goal of net-zero CO_2 emissions by 2050, there is an urgent need to find suitable substitutes for cement materials and significantly reduce cement usage [5], or to use alternative fuels, such as natural gas, biomass and agriculture-related secondary wastes (e.g., tires, sewage sludge, and municipal solid waste). This could significantly reduce indirect carbon emissions from burning fossil fuels or coal mines in heating rotary kilns [6]. Other strategies aim to use blended cement (such as ground granulated blast-furnace slag (ggbs), fly ash and other supplementary cementitious materials to replace cement) through the addition of cementitious materials; other chemical additives or aggregate types and grading optimization strategies may change the characteristics of concrete proportions and their environmental impact [5,7]. These strategies represent a significant research area focused on developing methods to achieve the same material design parameters as those found in concrete structures using pure cement concrete but with a lower demand for Portland cement. The most effective approach is to completely replace cement with appropriate industrial wastes to achieve net-zero carbon emissions. This would significantly contribute to reducing emissions.

In this study, three industrial by-products were blended to completely replace cementitious materials without adding alkali activators to produce ternary cementless composites. After demolding, the test specimens were tested for cross-comparison of the compressive strength between air and water curing. The ternary cementless composites were tested using the flowability, setting time, X-ray fluorescence (XRF), and scanning electron microscopy (SEM) to verify their applicability and feasibility.

2. Experimental Details

2.1. Materials and Mix Proportions

This study used cementless materials made from ultrafine fly ash from thermal power plants, ggbs from continuously operating steel plants, and co-fired fly ash from circulating fluidized bed boilers. The finenesses of ultrafine fly ash, ggbs and co-fired fly ash were 33,800, 5830 and 5130 cm^2/g, respectively. The specific gravity and chemical composition of the three raw materials were analyzed as described in Table 1, where ultrafine fly ash and ggbs are the amorphous-oriented materials. The tests were conducted on paste specimens, and the proportions are shown in Table 2, which shows that there were 6 mixtures of various proportions for a fixed water-to-cementitious-material ratio (w/c) of 0.55. After casting, the molds were removed once the specimens had hardened at room temperature for 24 h. After demolding, each group of mixtures was kept in water and exposed to air, respectively. Relevant tests were conducted when the test age was reached.

Table 1. The specific gravity and chemical compositions of the three raw materials.

Raw Materials	Specific Gravity	Chemical Compositions						
		CaO	SiO_2	Al_2O_3	Fe_2O_3	SO_3	MgO	Others
ultrafine fly ash	2.21	8.44	53.21	21.66	9.50	0.18	0.32	6.69
ggbs	2.88	40.24	33.68	14.37	0.29	0.66	7.83	2.93
co-fired fly ash	2.73	35.54	29.47	19.27	3.49	7.36	1.82	3.05

Table 2. Mix proportions (unit: g).

Mix No.	Ultrafine Fly Ash	ggbs	Co-Fired Fly Ash	Water
G1	0	500	500	550
G2	50	500	450	550
G3	0	600	400	550
G4	50	600	350	550
G5	0	400	600	550
G6	50	400	550	550

2.2. Test Procedures

The setting time tests were conducted using the Vicat Needle method following ASTM C191. The flow tests were conducted in accordance with ASTM C1437. Compressive strength tests were performed according to ASTM C109 for 7 and 28 days. The average value was measured after each of the three tests. The SEM observations were made on fragments left after the compression tests, and all the tests were carried out according to the procedures specified in ASTM C1723. XRF analysis was performed on the powder (about 5 g) produced in the compression test, and the test procedure was based on ASTM C1365.

3. Results and Discussion

3.1. Flowability and Setting Time

The test results are summarized as shown in Table 3. The test results showed that the flowability of the specimen with a w/c of 0.55 exceeded the standard flowability (110%) for all groups. ggbs had better flowability when the ratio of ggbs to mixed fly ash reached 6:4. Using ultrafine fly ash instead of partially co-fired fly ash resulted in slightly lower flow values due to the ultra-fine particles. Only the final setting time was observed as a test indicator for cementless materials. Six groups were used to control the final setting time within 24 h. The final setting time can be significantly shortened to less than 8 h for specimens with a ggbs dosage up to 60% of the cementless blend. These test results were similar to those of the alkali-activated specimens with ggbs added. The setting time was accelerated when the ggbs exceeded 50%, meaning the setting times were dramatically reduced as the ggbs content increased [8,9].

Table 3. Results of flowability and setting time.

Mix No.	Flowability	Final Setting Time
G1	115%	21 h
G2	115%	24 h
G3	135%	8 h
G4	125%	7 h
G5	130%	21 h
G6	125%	20 h

3.2. Compressive Strength

The results of the compressive strength tests at 7 and 28 days are shown in Table 4, including air and water curing. The test results showed that the strength of the water-cured specimens was better than that of the air-cured specimens, which demonstrated that this type of cementless material should be water-hardened. The hardening behavior of the ternary cementless materials was assumed to be due to the Ca-Si-Al colloidal system, and the strength development was due to calcium silicate hydrate (C-S-H) or calcium aluminum silicate hydrate (C-A-S-H) produced by the hydration process, which also supported the hypothesis that cementless materials could be made without alkali activators. The results were consistent with the results of previous tests [10,11]. The strength increased with hydration time, and the 28-day compressive strength of the specimens cured in water reached between 15 and 18 MPa. The cementless material had a specific compressive strength, which benefited a large number of industrial by-products, reduced the cement consumption, and reduced the carbon emissions.

Table 4. Results of compressive strength.

Mix No.	Compressive Strength (MPa)			
	Air Curing		Water Curing	
	7 Days	28 Days	7 Days	28 Days
G1	11.7	12.8	7.3	13.3
G2	6.4	13.7	8.6	15.6
G3	6.4	11.5	9.2	15.1
G4	6.1	11.0	12.1	15.0
G5	9.7	10.2	10.5	17.5
G6	10.2	11.1	9.8	16.4

3.3. XRF Analysis

The results of the XRF analysis are shown in Table 5; the main chemical components of the cementless materials were found to be CaO, SiO_2 and Al_2O_3. The proportions of the mentioned oxides were similar among the groups of specimens, which also indicated that the hydrated products should be C-S-H or C-A-S-H. It was noticed that the specimens all had a high content of SO_3, which quickly formed well-crystallized calcium-alumina crystals or ettringite in the voids of the specimens and helped to improve the hydrated bulk densities of the microstructures [12]. It may also be the leading cause of the higher compressive strength of G5 and G6 specimens (60% and 55% of co-fired fly ash). Moreover, the ratios of Ca/Si and Ca/Al in the air-cured and water-cured specimens were similar. The higher Ca/Si and Ca/Al specimens corresponded to better compressive strengths and represented more crystallization and gelation.

Table 5. XRF analysis.

Mix No.	Curing Type	Chemical Composition (%)				
		CaO	SiO_2	Al_2O_3	Fe_2O_3	SO_3
G1	Water	47.13	24.91	11.80	1.73	8.41
	Air	47.13	24.57	11.43	1.76	8.54
G2	Water	45.01	25.76	12.38	1.94	7.47
	Air	44.38	26.09	12.42	1.98	7.82
G3	Water	48.31	25.08	11.72	1.54	7.30
	Air	47.50	25.08	11.84	1.54	7.53
G4	Water	44.13	27.40	13.09	1.66	6.36
	Air	43.42	27.69	13.19	1.68	6.71
G5	Water	46.98	23.83	11.30	2.07	9.79
	Air	46.91	23.60	11.37	2.05	9.91
G6	Water	42.22	26.49	12.82	2.03	8.86
	Air	43.96	25.54	12.17	2.10	8.93

3.4. SEM Observation

The SEM photographs of each group of specimens are shown in Figure 1a–e. Many needle-like products (C-A-S-H or ettringite) could be seen on the surface of the microstructure of the specimen maintained in water. The needle-like hydration was more pronounced on the surface of the G1 specimen with lower strength, while hexagonal calcium hydroxide crystals were also observed on the surface of the G1 and G5 air-cured specimen. It was also indirectly stated that water curing was favorable for the continuous hydration of calcium hydroxide to produce C-S-H or C-A-S-H colloids. Figure 1e is an SEM photograph of the specimen with ultrafine fly ash, and transparent unreacted fly ash particles can be observed on its surface. It was clear that the reaction of ultrafine fly ash was favorable to the generation of C-A-S-H and ettringite. Figure 1f shows the experimental spectrum of SEM combined with energy dispersive X-ray (EDX) analysis for G5 water-cured specimen. It was found that elements with higher percentages were Si, Ca, Al, etc. It was evident that needle-like hydrates would contain C-S-H, C-A-S-H colloid, and calcium alumina, which was also the source of strength in the ternary cementless specimens.

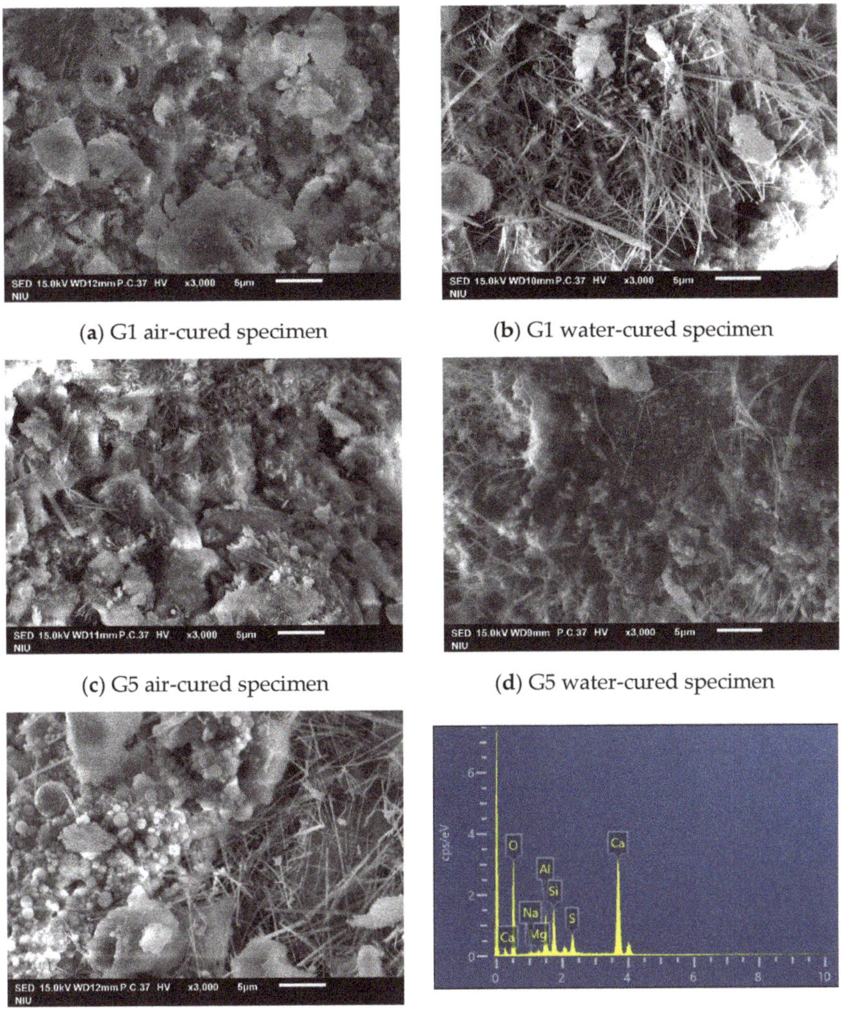

(**a**) G1 air-cured specimen (**b**) G1 water-cured specimen

(**c**) G5 air-cured specimen (**d**) G5 water-cured specimen

(**e**) G2 water-cured specimen (**f**) EDX for G5 water-cured specimen

Figure 1. SEM photos (x3000).

4. Conclusions

This study found that ternary cementless composites can be formed using ggbs, co-fired fly ash and ultrafine fly ash. The highest compressive strength was 17.5 MPa for two raw materials and 16.4 MPa for three raw materials. The average 28-day compressive strength ranged from 13 to 18 MPa, making it an innovative cementless material that meets net-zero carbon emissions. The cementless materials had better hydration and crystallization reactions when they were maintained in water, and the main products were C-S-H, C-A-S-H colloids and calcium alumina. These crystalline reactants can be verified using XRF, SEM and EDX analysis. The results also suggest that the novel ternary cementless composites developed in this study are worthy of further investigation into their engineering properties for future applications in construction materials.

Author Contributions: Conceptualization, W.-T.L. and Y.-H.C.; methodology, Y.-H.C.; validation, W.-T.L., Y.-H.C., M.Z. and D.K.; investigation, L.F.; resources, A.C.; data curation, M.Z. and D.K.; writing—original draft preparation, W.-T.L. and Y.-H.C.; writing—review and editing, W.-T.L. and L.F.; visualization, A.C.; supervision, W.-T.L. and L.F.; project administration, W.-T.L. and L.F.; funding acquisition, A.C. All authors have read and agreed to the published version of the manuscript.

Funding: The National Science and Technology Council (NSTC) in Taiwan supported this research with grant NSTC 111-2923-E-197-001-MY3 and the Czech Science Foundation in Czech Republic with project GAČR 22-00987J.

Institutional Review Board Statement: Not applicable.

Informed Consent Statement: Not applicable.

Data Availability Statement: Not applicable.

Acknowledgments: Publication cost of this paper was covered with founds of the Polish National Agency for Academic Exchange (NAWA): "MATBUD'2023-Developing international scientific cooperation in the field of building materials engineering" BPI/WTP/2021/1/00002, MATBUD'2023.

Conflicts of Interest: The authors declare no conflict of interest.

References

1. Le Quéré, C.; Andrew, R.M.; Friedlingstein, P.; Sitch, S.; Pongratz, J.; Manning, A.C.; Jackson, R.B. Global carbon budget 2017. *Earth Syst. Sci. Data.* **2017**, *10*, 405–448. [CrossRef]
2. Kawashima, A.B.; Martins, L.D.; Rafee, S.A.A.; Rudke, A.P.; de Morais, M.V.; Martins, J.A. Development of a spatialized atmospheric emission inventory for the main industrial sources in Brazil. *Environ. Sci. Pollut. Res.* **2020**, *27*, 35941–35951. [CrossRef] [PubMed]
3. Gil, A. Current insights into lignocellulose related waste valorization. *Chem. Eng. J. Adv.* **2021**, *8*, 100186. [CrossRef]
4. Andrew, R.M. Global CO_2 emissions from cement production, 1928–2018. *Earth Syst. Sci. Data.* **2019**, *11*, 1675–1710. [CrossRef]
5. Miller, S.A.; John, V.M.; Pacca, S.A.; Horvath, A. Carbon dioxide reduction potential in the global cement industry by 2050. *Cement Concr. Res.* **2018**, *114*, 115–124. [CrossRef]
6. Chatziaras, N.; Psomopoulos, C.S.; Themelis, N.J. Use of waste derived fuels in cement industry: A review. *Manag. Environ. Qual. Int. J.* **2016**, *27*, 178–193. [CrossRef]
7. Amran, M.; Al-Fakih, A.; Chu, S.H.; Fediuk, R.; Haruna, S.; Azevedo, A.; Vatin, N. Long-term durability properties of geopolymer concrete: An in-depth review. *Case Stud. Constr. Mater.* **2021**, *15*, e00661. [CrossRef]
8. Xie, J.; Wang, J.; Rao, R.; Wang, C.; Fang, C. Effects of combined usage of GGBS and fly ash on workability and mechanical properties of alkali activated geopolymer concrete with recycled aggregate. *Compos. Part B* **2018**, *164*, 179–190. [CrossRef]
9. Nath, P.; Sarker, P.K. Effect of GGBFS on setting, workability and early strength properties of fly ash geopolymer concrete cured in ambient condition. *Constr. Build. Mater.* **2014**, *66*, 163–171. [CrossRef]
10. Lin, K.L.; Lin, W.T.; Korniejenko, K.; Hsu, H.M. Application of ternary cementless hybrid binders for pervious concrete. *Constr. Build. Mater.* **2022**, *346*, 128497. [CrossRef]
11. Li, C.; Zhang, N.; Zhang, J.; Song, S.; Zhang, Y. C-A-S-H gel and pore structure characteristics of alkali-activated red mud–iron tailings cementitious mortar. *Materials* **2022**, *15*, 112. [CrossRef] [PubMed]
12. du Toit, G.; van der Merwe, E.M.; Kruger, R.A.; McDonald, J.M.; Kearsley, E.P. Characterisation of the hydration products of a chemically and mechanically activated high coal fly ash hybrid cement. *Minerals* **2022**, *12*, 157. [CrossRef]

materials proceedings

MDPI

Proceeding Paper

The Impact of Waste Fluid Catalytic Cracking Catalyst Addition on the Selected Properties of Cement Pastes †

Paweł Niewiadomski [1,*] and Michał Cisiński [2]

1 Department of Materials Engineering and Construction Processes, Wroclaw University of Science and Technology, Wybrzeże Wyspiańskiego 27, 50-370 Wroclaw, Poland
2 Department of Advanced Material Technologies, Wroclaw University of Science and Technology, Wybrzeże Wyspiańskiego 27, 50-370 Wroclaw, Poland
* Correspondence: pawel.niewiadomski@pwr.edu.pl; Tel.: +48-508-859-290

† Presented at the 10th MATBUD'2023 Scientific-Technical Conference "Building Materials Engineering and Innovative Sustainable Materials", Cracow, Poland, 19–21 April 2023.

Abstract: The significant reduction in CO_2 emissions arising from the cementitious composites industry is one of the highest priorities for the construction sector's movement towards climate neutrality and sustainable development. One of the approaches to cope with this issue is to partially substitute cement with supplementary cementitious materials. Recently, various oil refinery wastes (ORW) have attracted researchers' attention in terms of being investigated for such an application. As such, the present paper shows the preliminary results of investigations conducted on cement pastes with the addition of a spent fluid catalytic cracking catalyst derived from a Polish oil refinery company. It is worth mentioning that the incorporation of ORW in cementitious composites might enable the production of more environmentally friendly construction materials without sacrificing quality, whilst, simultaneously providing an opportunity for recycling petrochemical wastes.

Keywords: oil refinery wastes; spent FCC catalyst; cementitious composites; cement pastes; supplementary cementitious materials; sustainable development

check for updates

Citation: Niewiadomski, P.; Cisiński, M. The Impact of Waste Fluid Catalytic Cracking Catalyst Addition on the Selected Properties of Cement Pastes. *Mater. Proc.* **2023**, *13*, 10. https://doi.org/10.3390/materproc2023013010

Academic Editors: Katarzyna Mróz, Tomasz Tracz, Tomasz Zdeb and Izabela Hager

Published: 14 February 2023

1. Introduction

1.1. Supplementary Cementitious Materials towards Construction Sustainability

Global urbanization and economic growth have increased society's demand for new buildings and infrastructure and hence for concrete, the scale of usage of which makes it currently the most commonly consumed material, after water, worldwide [1] (pp. 35–42). However, the production process of cement, with its crucial constituent used as a binder, is based on quarrying, and requires high temperatures and considerable energy consumption, contributing to natural resource depletion and significant CO_2 emission; it accounts for 5–8% of global emissions of this greenhouse gas [2] (pp. 169–174). Therefore, for the sake of natural environment, the construction sector is obligated to drastically reduce its CO_2 emissions to reach the goal of sustainable development whilst simultaneously filling the increasing market demand for cementitious composites. One of the approaches to cope with this issue is to partially substitute cement with supplementary cementitious materials (SCMs) with pozzolanic properties [3]. Despite the fact that a huge number of various waste materials and by-products derived from agricultural and industrial activities were investigated regarding such applicability, it is mainly fly ash, ground-granulated blast-furnace slag and silica fume that have already been implemented on industrial scale [4] (p. 3136).

1.2. Oil Refinery Waste Recycling

Recently, various waste materials derived from petrochemical industry, i.e., oil refinery wastes (ORW) have caught researchers' attention in terms of being investigated as a potential green resource for the construction sector. One considerable type of solid ORW is spent catalysts; they are generally used to promote the conversion processes of crude oil into petrochemical products. These materials become deactivated after a period of time being used in the refining processes, due to the mechanical degradation and most notably contamination with various pollutants from processing resources which, while adsorbed on their structure, reduce their specific surface area and thus their catalytic activity. Consequently, spent catalysts are withdrawn from petrochemical installations of single-complex oil refinery companies in an estimated annual amount of 5515 tonnes [5]. It should also be pointed out that almost 10% of that quantity is represented itself by spent catalysts withdrawn from fluidized catalytic cracking units, which makes them the largest group of catalytic wastes in the worldwide petrochemical industry [5]. Unfortunately, the majority of these waste materials are consequently sent to landfills, thereby providing economical losses for companies, and serious health risks related to the release of chemical pollutants into the natural environment [2] (pp. 169–174). Apart from organic compounds, spent FCC catalysts might also contain heavy metals, such Cr, Pb, and Ni [6] (pp. 111–118). Therefore, finding a newer, safer application for these waste materials is necessary. A reasonable research direction for this purpose seems to be the use of ORW as partial cement substitutes in cementitious composites. Such a procedure, apart from reducing cement usage, might potentially allow the binding of chemical pollutants in their hardened state to the composite structure, thereby minimizing the risk of their release to the environment [6] (pp. 111–118). Moreover, as waste recycling is the most preferable option for managing waste materials and moving towards a circular economy, the suitability of such direction is clearly underlined.

1.3. Recent State of the Art Knowledge

In the literature, spent FCC catalysts are predominantly investigated in ORW material, regarding their applicability in construction sector. Accordingly, the substitution of up to 25–30% of cement, by mass, with spent FCC catalysts derived from, e.g., Lithuanian [7] (pp. 103–108), Chinese [8] (pp. 1773–1783), Spanish [9] (pp. 12–17) Portuguese [10] (pp. 109–121) and Oman [11] (pp. 77–81) oil refineries contributed to the mechanical strength enhancement of hardened state cementitious composites. However, even though the incorporation of this waste material typically contributes to rheological properties' deterioration of cementitious composites mixes, which has been confirmed in many works such as [8–10], this negative effect might be reduced by using plasticizing admixtures. Nonetheless, it should be pointed out that the oil-refining catalysts are manufactured individually for each oil refinery company to obtain the most favorable properties, i.e., activity and selectivity, tailored to the specific conditions in the petrochemical installations and depending on many different factors such as type of processing crude oil resources and oil refinery complexity. Therefore, the specific properties, chemical composition and potential pollutants of spent FCC catalysts, as well as of other ORW, might be significantly variable between the oil refineries from which they are withdrawn. This might be considered a key limitation in terms of freely using these materials in construction practice. As such, an individual approach is required for assessing the feasibility of recycling ORW in cementitious composites [2] (pp. 169–174).

1.4. Aim of the Research

The aim of the conducted research was to assess the potential of spent FCC catalysts derived from a Polish oil refinery company, to be recycled as a partial cement substitute for the production of sustainable cementitious composites. As such, the work consists of preliminary research carried out on basic cementitious composites, i.e., cement pastes, with the addition of a spent FCC catalyst used as a partial cement substitute within the range of 0 to 20%, by mass. The cement pastes have been prepared with water to a binding (cement + spent catalyst) ratio of 0.3. The influence of spent catalysts on their selected properties was investigated through the measurements of slump flow, compressive strength, flexural strength, water absorptivity and microstructure. For the purpose of future investigations, we plan to expand the scope of the current research by investigating cement pastes prepared with different w/b ratios (e.g., 0.35, 0.40). Moreover, tests will also be conducted on cement mortars. Apart from spent FCC catalysts, we plan also to investigate a few other types of ORW.

2. Materials and Methods

2.1. Materials Characteristic

The spent FCC catalyst used in this research was supplied by a Polish oil refinery company. The role of FCC catalysts in the petrochemical industry is to promote the cracking reaction of high-boiling oil fractions into lower-boiling hydrocarbons for blending commercial fuels. From the physicochemical viewpoint, this waste material might be characterized as a white-grey powder. In the research, it was used as a substitute of cement (CEM I 52.5R). The microstructures of these materials have been presented in Figure 1 in the form of SEM images at ×100 magnification. Their chemical compositions, obtained using X-ray fluorescence spectrometer analysis (XRF) and from [12] for the spent catalyst and cement, respectively, have been presented in Table 1. Accordingly, the mean particle size of cement grains is finer than in case of the spent catalyst which, as a typical pozzolanic material, is mainly composed of SiO_2 and Al_2O_3.

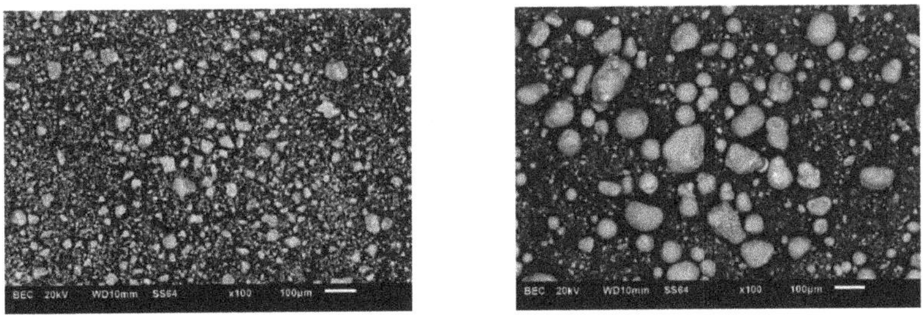

(a) (b)

Figure 1. SEM images of microstructure of the materials: (**a**) cement; (**b**) spent FCC catalyst.

Table 1. Chemical composition of the materials, wt.%.

Material\Oxide	SiO_2	Al_2O_3	CaO	Fe_2O_3	MgO	SO_3	Other
Cement (CEM I 52.5R)	20.09	4.84	64.02	3.87	1.15	2.83	3.20
Spent FCC catalyst	44.48	38.11			3.43	3.88	10.10

2.2. Formulations of Cement Paste Mixes

The cement paste mixes were prepared by mixing of ingredients in the laboratory bowl. The spent FCC catalyst was used as a substitute of 0%, 5%, 10%, 15% and 20% of cement. The water to binder (w/b) ratio of 0.3 was kept constant for all of the formulations of mixes. Accordingly, the reference (plain cement) mix is described as 0.30 ref, while subsequent substituted mixes are described as 0.30 5%, 0.30 10%, 0.30 15% and 0.30 20%, corresponding to the cement substitution level with the spent catalyst, i.e., 5%, 10%, 15% and 20%, by mass, respectively. The complete formulations of paste mixes have been listed in Table 2.

Table 2. Formulations of 1 dm^3 of cement paste mixes.

Material\Mix (1 dm^3)	0.30 ref	0.30 5%	0.30 10%	0.30 15%	0.30 20%
Cement [kg]	1.593	1.508	1.423	1.340	1.257
Spent FCC catalyst [kg]	0.000	0.079	0.158	0.236	0.314
Substitution level [%]	0	5	10	15	20
Water [kg]	0.478	0.476	0.474	0.473	0.471

2.3. Evaluation of the Flowability of Cement Paste Mixes

The flowability of cement paste mixes was measured through the flow table test, according to the PN-EN 1015–3:2000. The truncated cone-shaped mold was placed on the spreader disc, and the cement paste mix was applied in two layers, thickening each for at least 10 strokes of the compactor. Then, the excess paste was scraped off, and after 15 s, the mold was slowly removed by lifting it vertically. After shaking the disc 15 times, two diameters of the mix oriented perpendicularly to each other were measured, and their average value was assumed to be the final one.

2.4. Evaluation of Mechanical Properties of Cement Paste Samples

The mechanical properties of the cement paste samples were determined through measurements of flexural and compressive strength. The tests were carried out on beam (40 × 40 × 160 mm) and cubic cement paste samples (40 × 40 × 40 mm), respectively, both at 7th and 28th curing day with the use of an electromagnetic universal testing machine with a load cell capacity of 300 kN.

2.5. Evaluation of Water Absorptivity of Cement Paste Samples

The water absorptivity of the cement paste samples was evaluated as follows: the mass 'A' of cement pastes was measured after removing all uncombined water through drying the samples to a constant weight. Subsequently, the samples were placed in the vessel with water. Immediately after the samples were removed from the vessel, their surfaces were carefully dried with a cloth. After that, the mass 'B', corresponding to the maximum saturated cement paste samples, was measured. The water absorptivity was calculated as the difference between B and A in relation to A, and is expressed in percentages.

2.6. Evaluation of Microstructure of Cement Paste Samples

The microstructure of the surface of the cement paste samples was observed using scanning electron microscopy (SEM JEOL model JSM-6610A) with secondary electron mode.

3. Results

3.1. Rheological Properties of Cement Paste Mixes

The results of slump flow values obtained for the respective cement paste mixes have been presented in Figure 2. It may be seen that the slump flow of the investigated mixes generally tends to decrease with the increase in cement substitution level. However, the 0.30 5% mix reveals the highest slump flow of all investigated mixes, i.e., 182.5 mm.

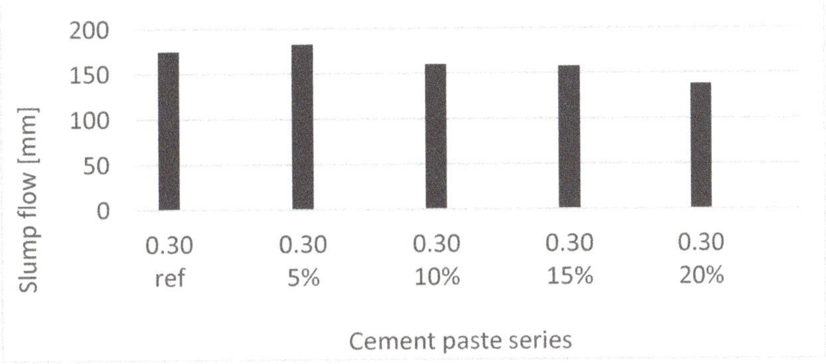

Figure 2. Effects of cement substitution with spent FCC catalyst on the slump flow of cement paste mixes.

3.2. Mechanical Properties of Cement Paste Samples

3.2.1. Compressive Strength

The results of 7- and 28-day compressive strength (CS) values obtained for the cement paste samples have been presented in Figure 3. Accordingly, the CS of each cement paste with the addition of the spent catalyst was found to be higher in comparison with the 0.30 ref sample, both at the 7th and 28th day of hydration. The highest 7- and 28-day CS values were measured for 0.30 15% and 0.30 10% sample, respectively, exceeding the CS of 0.30 ref for 36% and 17%, respectively. As expected, the CS enhances with the curing age.

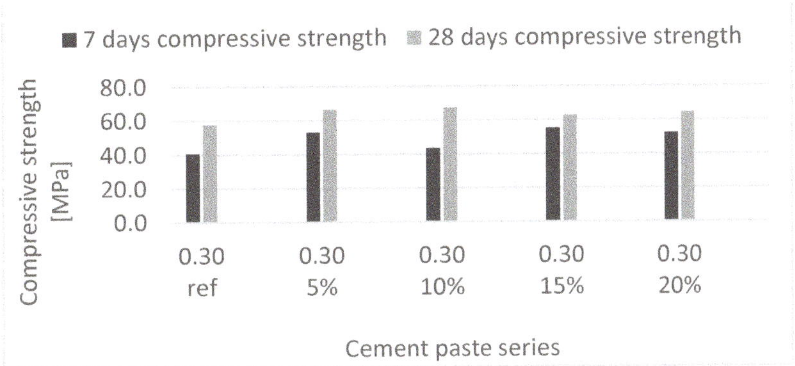

Figure 3. Effects of cement substitution with a spent FCC catalyst on the 7- and 28-day compressive strength of cement paste samples.

3.2.2. Flexural Strength

The results of 7- and 28-day flexural strength (FS) values obtained for the respective cement paste samples have been presented in Figure 4. Accordingly, the FS tends to decrease with the increase in cement substitution level, mentioning that the FS of the respective samples is generally similar between the 7th and 28th day of hydration.

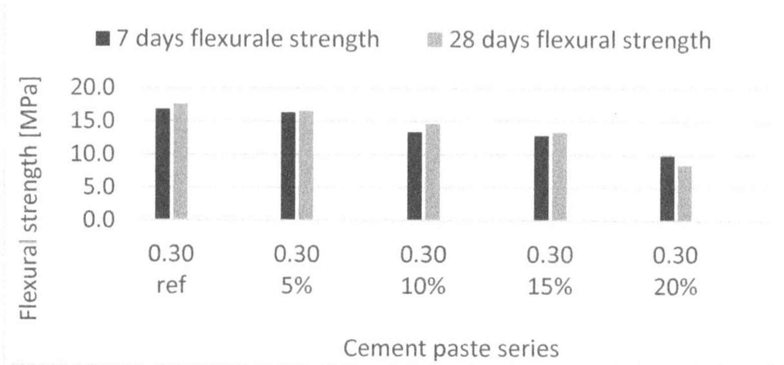

Figure 4. Effects of cement substitution with a spent FCC catalyst on the 7- and 28-day flexural strength of cement paste samples.

3.3. Water Absorptivity of Cement Paste Samples

The results of water absorptivity values obtained for the respective cement paste samples have been presented in Figure 5. Considering the results, it might be stated that the water absorptivity of 0.30 5% is similar to 0.30 ref. However, as 10% and more of cement is substituted with spent catalyst, the water absorptivity increases. The values obtained for 0.30 10%, 0.30 15% and 0.30 20% are also similar to each other, nonetheless, exceeding 0.30 ref for 17%, 14% and 17%, respectively.

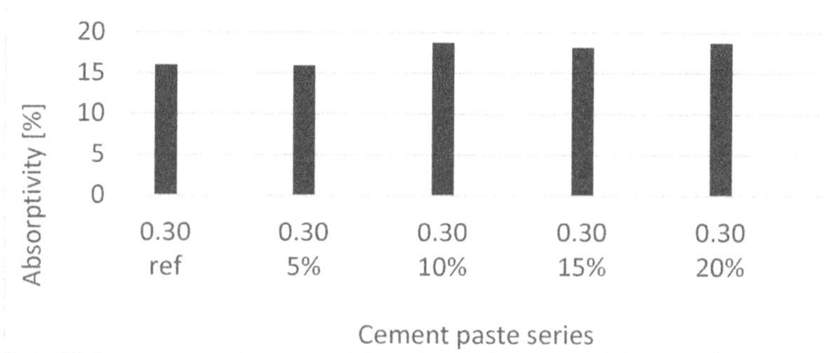

Figure 5. Effects of cement substitution with spent FCC catalyst on the water absorptivity of cement paste samples.

3.4. Microstructure of Cement Paste Samples

The microstructures of the cement paste samples have been presented in Figure 6 in the form of SEM images at ×200 magnification. On the basis of the greyscale of the images, the components and pores were identified. The light grey shapes correspond to the unhydrated cement grains, whilst the coarser spherical shapes in darker color are attributed to unreacted spent catalyst particles. Accordingly, the pores, which might be seen as a black pixels, are mainly identified in the interphase transition zone (ITZ) between spent catalyst and CSH gel. Therefore, their content was found to increase with the increasing cement substitution level. As such, the matrix of the 0.30 ref was found to the be more homogenous and uncracked when compared to those matrices of the samples with spent catalyst addition.

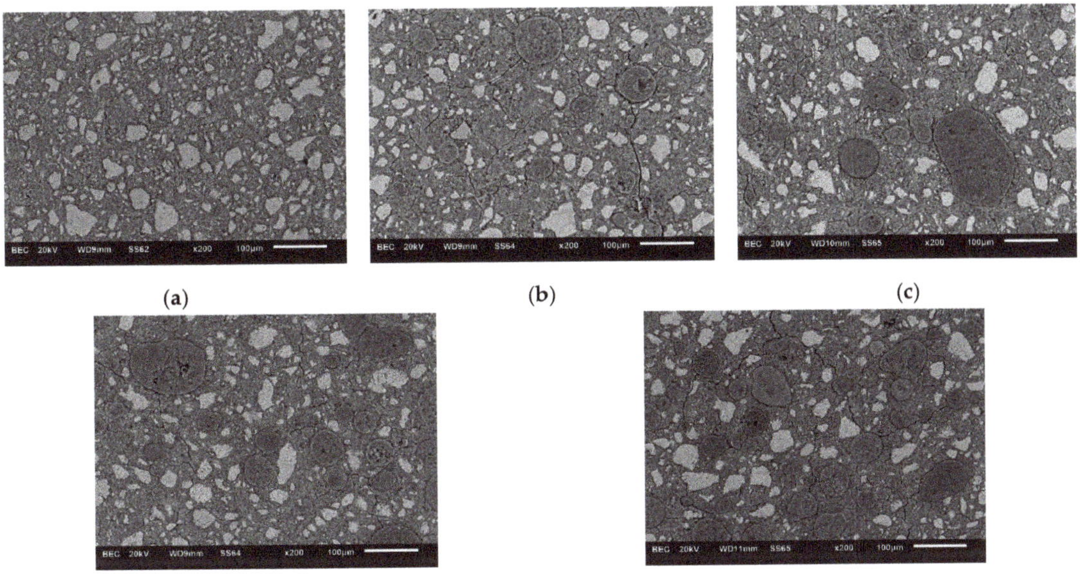

(**a**) (**b**) (**c**) (**d**) (**e**)

Figure 6. SEM images of microstructure of the respective cement paste samples: (**a**) 0.30 ref; (**b**) 0.30 5%; (**c**) 0.30 10%; (**d**) 0.30 15%; (**e**) 0.30 20%.

4. Discussion

The negative effect of the addition of spent FCC catalysts on the slump flow and rheological properties of cement paste mixes was found to be in accordance with the literature results [8–10]. This finding might be generally attributed to the very high specific surface area of spent catalysts, which is provided by highly porous zeolite structure and identified in [8] (pp. 1773–1783). Thereby, spent FCC catalysts are a water-demanding material which contribute to the reduction of available water in the system and hence the workability of fresh mixes. The slump flow value of the 0.30 5% mix was found to be in the scope of standard deviation thresholds of 0.30 ref, and thus the slight increase in slump flow between those mixes was assumed to be the result of a measurement statistic.

The development of the compressive strength of the cement paste samples with the addition of spent FCC catalysts has been strengthened by the number of works, such as [7] (pp. 103–108). This phenomenon might be attributed to the vital pozzolanic activity of spent catalysts, identified in [13]. Pozzolanic activity is the ability of a material to react with $Ca(OH)_2$ in the presence of moisture to form additional CSH gel in the system, thus contributing to compressive strength gain. Such behavior is promoted by the typical pozzolanic properties of spent FCC catalysts, i.e., aluminosilicate composition and high specific surface area.

The deterioration of the flexural strength of cement paste samples in relation to the increasing spent catalyst addition seems to be caused by the porous structure of the respective samples, as has been investigated through SEM analysis. Accordingly, the interphase transition zone (ITZ) reveals a lack of adhesion between the spent catalyst particles and CSH gel. This might be associated with the poor binding provided by the significant consumption of water by spent catalyst.

The results of water absorptivity of cement pastes indicated that spent FCC catalysts, while substituting 10% and more of cement, generally promote the water absorptivity of the respective paste samples. This finding is in accordance with [6] (pp. 111–118), and might be attributed to the fact that those samples tend to dry up, again as the catalyst provides high water absorption and thus also prompts water content reduction in the system. Moreover, such behavior seems to be strengthen by the low w/b ratio used to prepare the mixes.

Overall, based on the preliminary exploration of the possibility of using spent FCC catalysts as a pozzolanic additive to cementitious composites, this waste material was found to be a promising substitute for up to 20% of cement in primary cementitious composites, i.e., cement pastes, with the slight decrease in workability and increase in absorptivity acceptable from a construction materials quality viewpoint. It is vital that their addition results in higher compressive strength of hardened samples, whilst simultaneously decreasing the cost and carbon footprint arising from the cement industry. Therefore, the obtained results, combined with the conducted research into the applicability of spent FCC catalysts in cement mortars described in [14], allow us to conclude on the rightness of further research into using spent FCC catalysts from the Polish petrochemical industry in cementitious composites. The authors have to perform planned future studies in terms of investigating the impact of the addition of spent FCC catalysts on the properties of cement mortars. The research will include purification of spent catalysts from pollutants, and toxicity analyses of the resulting materials. The pozzolanic activity of spent catalysts and the setting time of related binders will be also measured. Moreover, other spent catalysts and petrochemical wastes, i.e., molecular sieves, will be investigated regarding such applicability.

5. Conclusions

On the basis of the conducted research, the following main conclusions may be drawn:

- Spent FCC catalysts derived from the Polish petrochemical industry reveal the potential to be recycled as a sustainable resource material with pozzolanic properties for cement substitution in cementitious composites. However, in terms of assessing their impact on the properties of cementitious composites more extensively, further research activity is planned, to be carried out on cement mortars by the authors.
- According to the research results, this material, while substituting 0–20% of cement, contributed to enhancement of the compressive strength of cement paste samples by up to 36% and 17%, respectively, on the 7th and 28th day of curing. Despite the fact that the workability, flexural strength and water absorptivity of the resulting composites are slightly deteriorated, they are still acceptable from a construction materials quality viewpoint. Moreover, these factors may potentially be minimized by using appropriate plasticizing admixtures.
- The recycling of ORW in the construction sector, if achieved, will contribute to numerous benefits such as decreased CO_2 emissions, natural resource protection and cost savings, whilst solving ORW utilization issues. Considering this, the correctness of further research regarding such an application of ORW is clearly underlined.

Author Contributions: Conceptualization, P.N. and M.C.; methodology, P.N. and M.C.; software, P.N.; validation, P.N.; formal analysis, P.N. and M.C.; investigation, P.N.; resources, P.N. and M.C.; data curation, P.N. and M.C.; writing—original draft preparation, M.C.; writing—review and editing, P.N.; visualization, M.C.; supervision, P.N.; project administration, P.N.; funding acquisition, P.N. All authors have read and agreed to the published version of the manuscript.

Funding: This research was funded by National Science Centre (NCN), grant number MINIATURA-5 2021/05/X/ST8/00483. The APC was funded by the Department of Materials Engineering and Construction Processes from Wroclaw University of Science and Technology.

Institutional Review Board Statement: Not applicable.

Informed Consent Statement: Not applicable.

Data Availability Statement: Not applicable.

Acknowledgments: The study is part of research project entitled MINIATURA-5 2021/05/X/ST8/00483 supported by the National Science Centre (NCN). Publication cost of this paper was covered with funds from the Polish National Agency for Academic Exchange (NAWA): "MATBUD'2023—Developing international scientific cooperation in the field of building materials engineering" BPI/WTP/2021/1/00002, MATBUD'2023.

Conflicts of Interest: The authors declare no conflict of interest.

References

1. Castellanos, N.T.; Agredo, J.T. Using spent fluid catalytic cracking (FCC) catalyst as pozzolanic addition-a review. *Ingeniería e investigación* **2010**, *30*, 35–42. [CrossRef]
2. Niewiadomski, P.; Nieswiec, M.; Cisinski, M.; Sadowski, L. Short review on the feasibility assessment of the recycling of oil refinery wastes in cementitious composites. *Chem. Eng. Trans.* **2022**, *94*, 169–174.
3. Ahmad, W.; Ahmad, A.; Ostrowski, K.A.; Aslam, F.; Joyklad, P. A scientometric review of waste material utilization in concrete for sustainable construction. *Case Stud. Constr. Mater.* **2021**, *15*, e00683. [CrossRef]
4. Costa, C.; Marques, J.C. Feasibility of eco-friendly binary and ternary blended binders made of fly-ash and oil-refinery spent catalyst in ready-mixed concrete production. *Sustainability* **2018**, *10*, 3136. [CrossRef]
5. Odpowiedzialność środowiskowa. Available online: https://raportzintegrowany2015.orlen.pl/pl/odpowiedzialna-firma/odpowiedzialnosc-srodowiskowa.html (accessed on 20 November 2022).
6. Su, N.; Chen, Z.H.; Fang, H.Y. Reuse of spent catalyst as fine aggregate in cement mortar. *Cem. Concr. Compos.* **2001**, *23*, 111–118. [CrossRef]
7. Vaičiukynienė, D.; Grinys, A.; Vaitkevičius, V.; Kantautas, A. Purified waste FCC catalyst as a cement replacement material. *Ceramics–Silikáty* **2015**, *59*, 103–108.
8. Su, N.; Fang, H.Y.; Chen, Z.H.; Liu, F.S. Reuse of waste catalysts from petrochemical industries for cement substitution. *Cem. Concr. Res.* **2000**, *30*, 1773–1783. [CrossRef]
9. Payá, J.; Monzó, J.; Borrachero, M.V.; Velázquez, S. Cement equivalence factor evaluations for fluid catalytic cracking catalyst residue. *Cem. Concr. Compos.* **2013**, *39*, 12–17. [CrossRef]
10. Nunes, S.; Costa, C. Numerical optimization of self-compacting mortar mixture containing spent equilibrium catalyst from oil refinery. *J. Clean. Prod.* **2017**, *158*, 109–121. [CrossRef]
11. Al-Jabri, K.; Baawain, M.; Taha, R.; Al-Kamyani, Z.S.; Al-Shamsi, K.; Ishtieh, A. Potential use of FCC spent catalyst as partial replacement of cement or sand in cement mortars. *Constr. Build. Mater.* **2013**, *39*, 77–81.
12. Cement Foreign Trade. Available online: http://www.musayapi.com.tr/index.asp (accessed on 30 November 2022).
13. Matos, A.M.; Nunes, S.; Costa, C.; Barroso-Aguiar, J.L. Spent equilibrium catalyst as internal curing agent in UHPFRC. *Cem. Concr. Compos.* **2019**, *104*, 103362. [CrossRef]
14. Abdolpour, H.; Niewiadomski, P.; Sadowski, Ł.; Chowaniec, A. Self-compacting ultra-high performance mortars produced with waste catalysts from petrochemical industry: Rheological, mechanical and microstructural properties. *J. Clean. Prod.* **2022**, *369*, 133225. [CrossRef]

materials proceedings

MDPI

Proceeding Paper

Properties of the Cement, Slag and Fly Ash Mixture Composition Corresponding to CEM II/C-M and CEM VI †

Ewelina Tkaczewska and Grzegorz Malata *

Department of Building Materials Technology, Faculty of Materials Science and Ceramics, AGH University of Science and Technology, 30 Mickiewicza Avenue, 30-059 Krakow, Poland

* Correspondence: gmalata@agh.edu.pl; Tel.: +48-617-4552

† Presented at the 10th MATBUD'2023 Scientific-Technical Conference "Building Materials Engineering and Innovative Sustainable Materials", Cracow, Poland, 19–21 April 2023.

Abstract: In the study, cement mixtures containing granulated blast furnace slag (GBFS) and siliceous fly ash (SFA) were tested, including those corresponding to special cements according to the PN-B-19707: 2013 standard. Measurements included the period of development of standard strength (up to 28 days) and concerned the compressive strength, linear changes and phase composition of cement mixtures. Furthermore, an evaluation of the microstructure of cement mortar was carried out by SEM. The mixture of composition CEM II/C-M (S-V) satisfies the requirements of the 32.5R or 32.5N strength class, whereas that of CEM VI (S-V) is of the 32.5N strength class, as opposed to stronger mixtures richer in Portland clinker.

Keywords: cement mixture; granulated blast furnace slag; siliceous fly ash; compressive strength; linear changes; phase composition; microstructure

check for updates

Citation: Tkaczewska, E.; Malata, G. Properties of the Cement, Slag and Fly Ash Mixture Composition Corresponding to CEM II/C-M and CEM VI. *Mater. Proc.* **2023**, *13*, 11. https://doi.org/10.3390/materproc2023013011

Academic Editors: Katarzyna Mróz, Tomasz Tracz, Tomasz Zdeb and Izabela Hager

Published: 14 February 2023

1. Introduction

In 2021, worldwide emissions from the cement industry reached almost 2.9 billion tons of CO_2, which is more than 7% of the total global CO_2 emissions [1]. For the production of 1 Mg of Portland cement clinker, about 1.7 Mg of natural resources are used, mainly carbonate raw materials, such as limestone and marl [2]. About 60% of the total CO_2 emissions released by a cement plant come from the calcination of carbonates in the raw material bulk. The remaining 40% of all CO_2 emissions in a cement plant are derived from the burning of fossil fuels. Therefore, the world cement industry has to meet the constantly growing environmental requirements, which mainly concern the reduction in CO_2 emissions [3].

One way to reduce CO_2 emissions is the production of multicomponent cements CEM II-CEM V according to the PN-EN 197-1 standard [4] using significant amounts of main ingredients other than Portland clinker, mainly granulated blast furnace slag or siliceous fly ashes. Cements containing significant amounts of these components are characterized by low hydration heat, higher compressive strength, longer curing periods, and higher resistance to chemical aggression [5–10].

The PN-EN 197-5 standard [11] defines the framework conditions for a significant reduction in the clinker content of cements. It extends the range of Portland multicomponent cements (the possibility of using several main components in the composition of cement) by a group of Portland multicomponent cements CEM II/C-M with a minimum content of Portland clinker of 50% and a newly created group of multicomponent cements CEM VI, in which the share of non-clinker components can reach 65% at a maximum.

This paper presents the results of research into the properties of the CEM II/C-M cements with a mixture of granulated blast furnace slag and siliceous fly ashes at 20 and 15% or 30 and 15%, respectively, and the CEM VI cement containing 40% granulated blast furnace slag and 15% siliceous fly ashes. Cement tests were performed to analyze the compressive strength, linear changes, phase composition, and microstructure after a specified period of time.

2. Materials and Methods

2.1. Materials

The materials used were the ordinary Portland cement type of CEM I 42.5R (PC) according to the PN-EN 197-1 standard [4], granulated blast furnace slag (GBFS), siliceous fly ash (SFA), and gypsum (G). The chemical composition and physical characteristics of these materials are presented in Table 1.

Table 1. Chemical composition and physical characteristics of the materials.

Parameter		PC	GBFS	SFA	G
Chemical composition (%)	SiO_2,	21.62	36.25	49.67	3.70
	Al_2O_3	2.37	8.68	27.77	1.29
	Fe_2O_3	1.51	0.23	6.96	0.64
	CaO	62.12	42.70	2.78	35.61
	MgO	1.57	8.22	1.37	0.40
	SO_3	2.03	1.51	0.93	39.20
	Na_2O	0.42	0.57	1.48	0.12
	K_2O	0.84	0.39	2.51	0.22
	TiO_2	0.15	0.41	1.60	0.07
	P_2O_5	0.18	0.03	0.82	0.03
	Residue	3.42	0.92	0.48	0.28
	LOI	3.77	0.10	3.66	18.45
	Cl^-	0.097	0.014	0.013	0.036
	Reactive SiO_2	—	—	38.87	—
Density (g/cm^3)		3.04	2.89	2.11	1.49
Blaine surface area (cm^2/g)		3810	4120	2740	5250
Particle size distribution (%) (1)	D_{10}, μm	4.4	0.7	5.8	1.0
	D_{50}, μm	17.4	13.2	52.5	5.8
	D_{90}, μm	45.7	34.7	239.9	45.7

(1) D_{10}, D_{50}, D_{90}: the portion of particles with diameters smaller than this value is 10, 50, and 90%, respectively; D_{50} is also known as the median diameter.

The XRD analysis of PC, GBFS and SFA (at 10–65°2θ range) is presented in Figure 1. The GBFS is composed of a glassy phase (a background signal at 24–38°2θ) with XRD peaks from the crystalline phases, such as akermanite and merwinite. For SFA, the identified phases were the glassy phase (a broad background signal at 16–38°2θ), mullite, quartz and hematite.

The GBFS and SFA used meet the requirements according to the PN-EN 197-1 standard [4]. The chemical modulus of GBFS (CaO + MgO + SiO_2) and ((CaO + MgO)/SiO_2) were 87.2% and 1.4%, respectively. For SFA, the reactive SiO_2 content was 38.87%, while the contents of MgO, SO_3, and Na_2O_e ($Na_2O_e = Na_2O + 0.658K_2O$) were 8.22, 1.51, and 3.13%, respectively. SFA showed 0.013% chloride ion composition.

2.2. Composition of Cement Mixtures and Measured Properties

In the study, three cement mixtures were prepared, and their compositions are shown in Table 2. The proportions between the components were selected to achieve the compositions of the cement types CEM II/C-M (S-V) and CEM VI (S-V) included in the PN-EN 197-5 standard [11]. Measurements included the period of development of standard strength (up to 28 days) and concerned the compressive strength, linear change, and phase composition of the cement mixture specimens. Evaluation of the microstructure of the cement mortars was carried out by SEM.

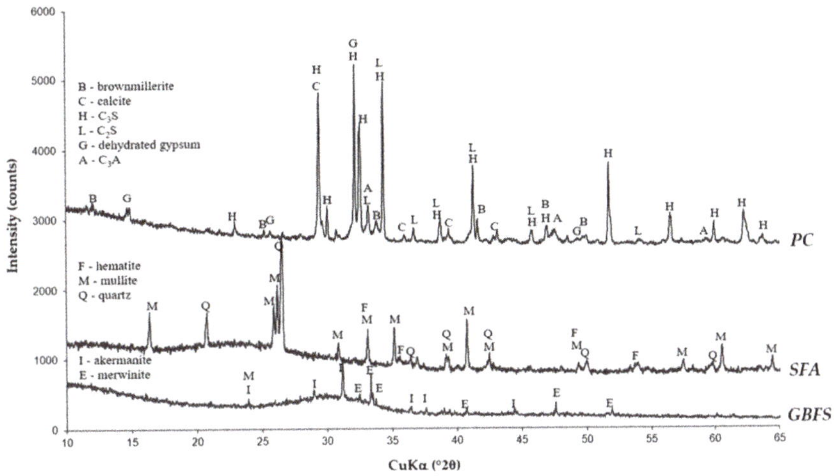

Figure 1. XRD patterns of PC, GBFS and SFA.

Table 2. Compositions of the cement mixtures studied.

Symbol	PC (%)	GBFS (%)	SFA (%)	G (%)	Cement Type According to the PN-EN 197-5 Standard [11]
60PC-20GBFS-15SFA	60	20	15	5	CEM II/C-M (S-V)
50PC-30GBFS-15SFA	50	30	15	5	CEM II/C-M (S-V)
40PC-40GBFS-15SFA	40	40	15	5	CEM VI (S-V)

Compressive strength tests of the cement mortars were carried out on prismatic samples of 40 mm height, 40 mm width, and 160 mm length after 2, 7, and 28 days according to the procedure described in the PN-EN 196-1 standard [12].

Linear change measurements were performed on cement paste samples prepared with a water-to-cement ratio of 0.33. Cement pastes were molded into prismatic samples of 25 mm height, 25 mm width, and 100 mm length. Molded samples were stored for 24 h in a high-moisture atmosphere and at a temperature of 20 °C. Samples were then removed from the molds and stored under water at a temperature of 20 °C until testing. Linear changes in cement paste prisms were investigated in the Grauf–Kaufman apparatus every day for up to 7 days.

The identification of the phase composition of the cement paste samples prepared with a water-to-cement ratio of 0.33 was determined using the X-ray diffraction (XRD) technique. A Philips X'Pert Pro MD diffractometer (Cu Kα1 line monochromatized with a Ge(111) monochromator) was used. The standard Bragg–Brentano geometry with a θ-2θ setup was applied (0.008° step size and 5–90°2θ range). Studies were performed for cement pastes 60PC-20GBFS-15SFA and 40PC-40GBFS-15SFA after 2, 7, 14, and 28 days.

The microstructure of the cement paste was observed with the FEI Nova NanoSEM 200 scanning electron microscope equipped with an EDS microanalyzer. Polished cross-sections of pastes were covered with a thin layer of carbon to avoid charging. Studies were made for the cement paste 60PC-20GBFS-15SFA after 2 and 28 days.

3. Results and Discussion

3.1. Compressive Strength of the Cement Mortar Samples

The compressive strength measurements were performed for all cement mixtures studied. Results are given in Table 3.

Table 3. Compressive strength of the cement mortar samples.

Cement Mixture	Compressive Strength (MPa)		
	2 Days	7 Days	28 Days
60PC-20GBFS-15SFA	13.73	23.49	41.73
50PC-30GBFS-15SFA	9.85	19.75	40.13
40PC-40GBFS-15SFA	6.34	17.69	37.68

According to Table 3, after 2 days the strength of the 60PC-20GBFS-15SFA mixture was 13.73 MPa and achieved a minimum of 10 MPa as required by the PN-EN 197-1 standard [4]. After the same time period the 50PC-30GBFS-15SFA mixture represented a strength of 9.85 MPa, a drop of 28% compared to the 60PC-20GBFS-15SFA mixture. In the case of the 40PC-40GBFS-15SFA mixture, the strength was only 6.34 MPa and about two times lower compared to the 60PC-20GBFS-15SFA mixture. The factor determining the drop in the early strength of the 50PC-30GBFS-15SFA and 40PC-40GBFS-15SFA mixtures is lower content of CEM I 42.5R in the mixtures, forcing a lower content of alite (C_3S). The lower content of C_3S gives a lower increase in the calcium silicate hydrate phase (C-S-H) after 2 days of hydration, especially in the case of the 40PC-40GBFS-15SFA mixture.

From Table 3, after 7 days, the strength of the 50PC-30GBFS-15SFA and 40PC-40GBFS-15SFA mixtures was still lower than that of the 60PC-20GBFS-15SFA mixture, but the difference between the strength of these cement mixtures was less. After 7 days the strength of the 40PC-40GBFS-15SFA mortar was 23.49 MPa, while the strength of the 50PC-30GBFS-15SFA and 60PC-20GBFS-15SFA mixtures was 19.75 MPa and 17.69 MPa, respectively. The high increase in the strength of the 40PC-40GBFS-15SFA mortar in the period from 2 to 7 days, two and half times compared to the strength after 2 days, results from the highest content of granulated blast furnace slag of laten hydraulic properties in this mixture. After 7 days the strength of the 40PC-40GBFS-15SFA mortar met the required 16 MPa according to the PN-EN 197-1 standard [4].

As is shown in Table 3, after 28 days, the strength of the 60PC-20GBFS-15SFA, 50PC-30GBFS-15SFA, and 40PC-40GBFS-15SFA mortars was 41.73 MPa, 40.13 MPa, and 37.68 MPa, respectively, and achieved a minimum of 32.5 MPa as required by the PN-EN 197-1 standard [4]. After 28 days the drop in strength of the 40PC-40GBFS-15SFA mixture was only 4% compared to the 60PC-20GBFS-15SFA mixture.

According to requirements of the PN-EN 197-1 standard [4], the 60PC-20GBFS-15SFA and 50PC-30GBFS-15SFA mixtures can be classified as CEM II/C-M (S-V) cements the 32.5R and 32.5N strength classes, respectively, while the 40PC-40GBFS-15SFA mixture corresponds to CEM VI (S-V) cement of the 32.5N strength class.

3.2. Linear Changes in the Cement Paste Samples

The linear changes were defined as the change in the length of the prismatic cement paste samples. Length changes are given in Table 4.

Table 4. Length changes of the cement paste prismatic samples.

Cement Mixture	Length Changes (mm)						
	1 Day	2 Days	3 Days	4 Days	5 Days	6 Days	7 Days
60PC-20GBFS-15SFA	0.00	−0.05	−0.01	−0.01	−0.01	−0.01	0.00
50PC-30GBFS-15SFA	0.00	−0.09	−0.01	0.00	0.00	0.01	0.00
40PC-40GBFS-15SFA	0.00	−0.05	−0.06	−0.04	−0.06	−0.05	−0.04

All the cement pastes studied showed dimensional stability for up to 7 days, as shown in Table 4. No or little shrinkage was observed due to the hydration process. These slight linear changes in the initial curing period probably resulted from the disappearance of monocarboaluminates in favor of higher amounts of ettringite.

3.3. Phase Composition of the Cement Paste Samples

Phase composition analysis was performed for the 60PC-20GBFS-15SFA and 40PC-40GBFS-15SF cement pastes. The XRD patterns after 2, 7, 14, and 28 days of hydration are presented in Figure 2.

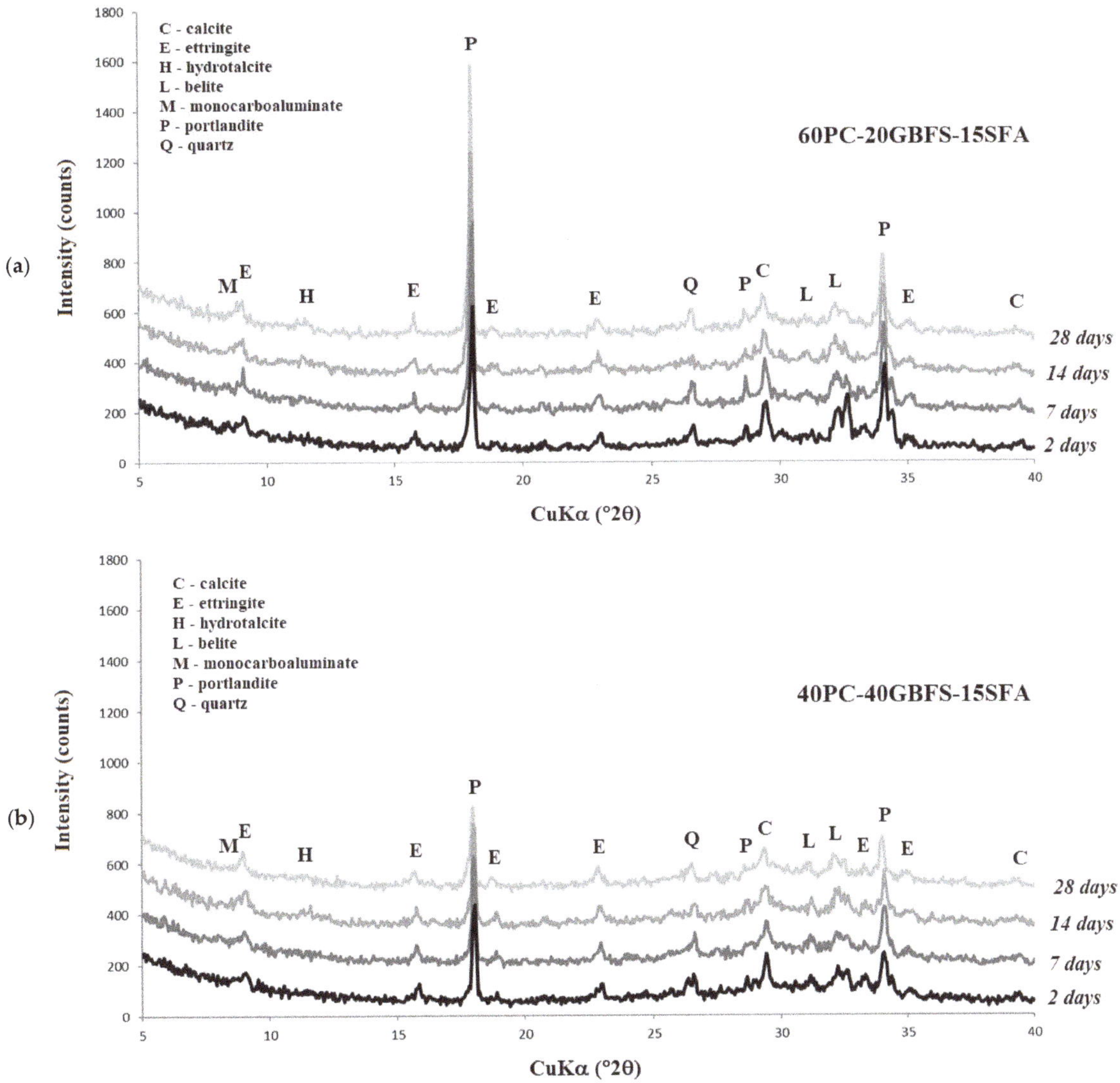

Figure 2. XRD patterns of the (**a**) 60PC-20GBFS-15SFA and (**b**) 40PC-40GBFS-15SFA cement pastes after 2, 7, 14, and 28 days.

For the cement paste samples with a higher proportion of Portland cement (60PC-20GBFS-15SFA cement paste), more hydration products were formed (Figure 2a); mainly portlandite. Ettringite dominated among the reaction products of the aluminate phase. After 7 and 14 days the monocarboaluminate was visible. After 28 days the hydrotalcite also appeared.

3.4. Microstructure of Cement Paste Samples

Figure 3a and b present the SEM images of the 60PC-20GBFS-15SFA cement paste with a magnification of 2000 times after 2 and 28 days of hydration, respectively.

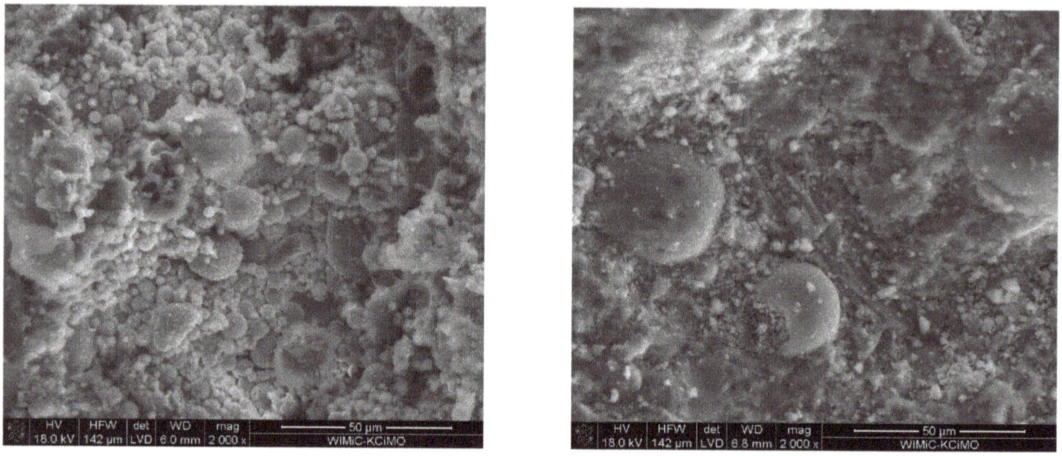

(a) (b)

Figure 3. SEM micrographs of the 60PC-20GBFS-15SFA cement paste after (**a**) 2 days and (**b**) 28 days.

As shown in Figure 3a, after 2 days of hydration, the microstructure of the 60PC-20GBFS-15SFA cement paste was porous, which, as is known from the literature [2], is associated with a slower hydration process of the 60PC-20GBFS-15SFA mixture in the early stages. This is due to the fact that the 60PC-20GBFS-15SFA mixture contains less CEM I 42.5R and, as a result, less C-S-H phase is formed after 2 days of hydration. As the hydration process is carried out, the microstructure of the 60PC-20GBFS-15SFA becomes denser, similar to previous results [13] (hydration products of Portland cement and granulated blast furnace slag, as shown in Figure 3b). However, after 28 days, the siliceous fly ashes were still unreacted; siliceous fly ashes do not represent hydraulic properties, but only pozzolanic properties, which become apparent over a longer period of time [2].

4. Conclusions

The following conclusions can be drawn based on the results presented above:

1. According to requirements of the PN-EN 197-1 standard the 60PC-20GBFS-15SFA and 50PC-30GBFS-15SFA mixtures can be classified as CEM II/C-M (S-V) cements of the 32.5R and 32.5N strength classes, respectively, while the 40PC-40GBFS-15SFA mixture corresponds to the CEM VI (S-V) cement of the 32.5N strength class.
2. Slight linear changes in the initial curing period of the cement pastes studied may be due to the disappearance of monocarboaluminates in favor of an increase in the amount of ettringite.
3. Changes in the dynamics of the hydration process and an increase in the strength of the mortar may result in sealing the cement mortar microstructure with a compact C-S-H.

Author Contributions: E.T. and G.M.: Conceptualization, methodology, formal analysis, investigation, resources, visualization; E.T.: writing—original draft preparation, writing—review and editing, supervision. All authors have read and agreed to the published version of the manuscript.

Funding: Publication cost of this paper was covered with founds of the Polish National Agency for Academic Exchange (NAWA): "MATBUD'2023–Developing international scientific cooperation in the field of building materials engineering" BPI/WTP/2021/1/00002, MATBUD'2023.

Institutional Review Board Statement: Not applicable.

Informed Consent Statement: Not applicable.

Data Availability Statement: Not applicable.

Acknowledgments: This research was supported from the subsidy of the Ministry of Education and Science for the AGH University of Science and Technology in Kraków (Project No 16.16.160.557).

Conflicts of Interest: The authors declare no conflict of interest.

References

1. Andrew, R.; Peters, G. *The Global Carbon Project's fossil CO_2 Emissions Dataset: 2021 Release*; CICERO Center for International Climate Research: Oslo, Norway, 2021; Available online: https://zenodo.org/record/5569235/files/GCP%20fossilCO2%202021v34.pdf (accessed on 20 November 2022).
2. Kurdowski, W. *Chemistry of Cement and Concrete*; Polish Scientific Publishers PWN: Warszawa, Poland, 2010.
3. Bjegovic, D.; Strimer, N.; Serdar, M. Ecological aspects of concrete production. In Proceedings of the Second International Conference on Sustainable Construction Materials and Technologies, Ancona, Italy, 28–30 June 2010; Zachar, J., Claisse, P., Naik, T.R., Ganjian, E., Eds.; UWM Center for By-Products Utlilization: Milwaukee, WI, USA, 2010; pp. 1483–1492.
4. *PN-EN 197-1*; Cement—Part 1: Composition, Specifications and Conformity Criteria for Common Cements. iTeh Standards: Warszawa, Poland, 2012.
5. Giergiczny, Z. Fly ash and slag. *Cem. Concr. Res.* **2019**, *124*, 105826. [CrossRef]
6. Król, A.; Giergiczny, Z.; Kuterasińska-Warwas, J. No TitleProperties of Concrete Made with Low-Emission Cements CEM II/C-M and CEM VI. *Materials* **2020**, *13*, 2257. [CrossRef] [PubMed]
7. Deschner, F.; Winnefeld, F.; Lothenbach, B.; Seufert, S.; Schwesig, P.; Dittrich, S.; Goetz-Neunhoeffer, F.; Neubauer, J. Hydration of Portland cement with high replacement by siliceous fly ash. *Cem. Concr. Res.* **2012**, *42*, 1389–1400. [CrossRef]
8. Sakai, E.; Miyahara, S.; Ohsawa, S.; Lee, S.H.; Daimon, M. Hydration of fly ash cement. *Cem. Concr. Res.* **2005**, *35*, 1135–1140. [CrossRef]
9. Schuldyakov, K.V.; Kramar, L.Y.; Trofimov, B.Y. The properties of slag cement and its influence on the structure of the hardened cement paste. *Procedia Eng.* **2016**, *150*, 1433–1439. [CrossRef]
10. Roy, D.M.; Malek, R.I. Hydration of slag cement. In *Progress in Cement and Concrete: Mineral Admixtures in Cement and Concrete*; ABI Books Pvt. Ltd.: Delhi, India, 1993; pp. 84–117.
11. *PN-EN 197-5*; Cement—Part 5: Composition, Specifications and Conformity Criteria for Portland Composite-Cement and Composite Cement. iTeh Standards: Warszawa, Poland, 2021.
12. *PN-EN 196-1:2016*; Methods of Testing Cement—Part 1: Determination of Strength. iTeh Standards: Warszawa, Poland, 2016.
13. Wieczorek, M.; Pichniarczyk, P. Wpływ zmiennego stosunku popiołu lotnego krzemionkowego do granulowanego żużla wielkopiecowego na właściwości cementu. In Proceedings of the XI Konferencja „Dni Betonu: Tradycja i nowoczesność", Wisła, Poland, 11–13 October 2021; Stowarzyszenie Producentów Cementu: Kraków, Poland, 2021; pp. 683–696.

materials proceedings

MDPI

Proceeding Paper

Numerical Modelling of Structural Behaviour of Curved Insulating Glass Units †

Marcin Kozłowski * and **Kinga Zemła**

Department of Structural Engineering, Silesian University of Technology, Akademicka 5, 44-100 Gliwice, Poland
* Correspondence: marcin.kozlowski@polsl.pl
† Presented at the 10th MATBUD'2023 Scientific-Technical Conference "Building Materials Engineering and Innovative Sustainable Materials", Cracow, Poland, 19–21 April 2023.

Abstract: The application of curved Insulating Glass Units (IGUs) in facades has emerged as a novel solution to meet aesthetic and energy performance objectives. Due to improved stiffness, curved IGUs cannot equalise internal and atmospheric pressure changes by pillowing, as flat IGUs do. Thus, the climatic loads in curved IGUs may be several times higher than those in flat units. This paper deals with experiments and numerical simulations of cylindrically curved IGUs. It presents the results of an experimental campaign designed to collect data that validate the numerical model of a curved IGU developed within this study. The model is sequentially used to analyse a case study that compares the resulting internal pressure, displacement, and stress in glass in a flat and curved IGU subjected to the characteristic climatic actions given by DIN18008-1.

Keywords: insulating glass units; curved glass; internal pressure; climatic load; experiments; numerical simulations

1. Introduction

The application of curved Insulating Glass Units (IGUs) in facades has emerged as a novel solution to meet aesthetic and energy performance objectives [1]. The increased stiffness resulting from curvature has been identified as a remarkable advantage over flat glass, decreasing support requirements and improving the aesthetics or increasing spans. However, there are many constraints regarding its design, production, and performance during operation [2]. In general, as the temperature of the gas entrapped in the IGU's gap changes its value, the gas volume inside varies relative to its original volume. The bending stiffness of the component panes limits their deformations (pillowing effect of the panes) and causes a change in the pressure of the gas inside the IGU gap. Changes in internal pressure result in unfavourable visual effects that can be observed as distorted images reflected in windows. Due to improved stiffness, curved IGUs cannot equalise internal and atmospheric pressure changes by pillowing, as flat IGUs do. Therefore, the climatic loads in curved IGUs can be several times higher compared to flat units, leading to glass fracture or failure of the silicone seal, particularly in combination with mechanical stresses [3].

Although research that focuses on the performance of flat IGUs under climatic loads has been presented in many studies, research on curved IGUs is still limited. Some results from numerical studies have been reported, but there are practically no experimental studies. Nizich et al. [4] proved numerically that the equalisation rate of the curved IGU is significantly reduced compared to a flat unit. Furthermore, it was found that the curvature and resulting increased internal pressure require increasing the horizontal bite of the seal several times. In the paper by Bao and Gregson [5], a thorough sensitivity study on the internal pressure of cylindrically curved IGUs was presented, considering a series of geometric variables. The main finding was that, given the same climatic action and for the same dimension, curved IGUs always generate significantly higher internal pressure than flat IGUs.

check for updates

Citation: Kozłowski, M.; Zemła, K. Numerical Modelling of Structural Behaviour of Curved Insulating Glass Units. *Mater. Proc.* **2023**, *13*, 12. https://doi.org/10.3390/materproc2023013012

Academic Editors: Katarzyna Mróz, Tomasz Tracz, Tomasz Zdeb and Izabela Hager

Published: 14 February 2023

This paper deals with experiments and numerical simulations of cylindrically curved IGUs. It presents the results of an experimental campaign designed to collect data that validate a numerical model of curved IGU developed within this study. The validated model is sequentially used to analyse a case study comparing the resulting internal pressure and other static quantities in flat and curved IGUs subjected to characteristic climatic actions given by DIN18008-1 [6].

2. Research Methodology

2.1. Experiments

In the experimental part of the work, a curved IGU specimen was produced, with dimensions of 500 × 1000 mm^2 (width × arch length) and a 1500 mm radius of curvature (Figure 1a). The specimen was composed of two 4 mm panes made of toughened glass, a 16 mm wide flexible silicone foam spacer, and a 6 mm wide hot-melt secondary sealant. Before testing, two holes were drilled in the spacer, and copper tubes, 6 mm in diameter, were glued into the holes (Figure 1b). Subsequently, silicone conduits were mounted at the ends of the tubes and connected to some laboratory equipment (a set of syringes with a maximum capacity of 300 mL and a pressure sensor). Insulating glass units are typically mounted in frames through flexible gaskets, which partly support the glass. The specimen was tested in a vertical position in a simply supported set-up that allowed for its unrestrained deformation during testing. This set-up was chosen to detect any deformation of the IGU specimen under loading.

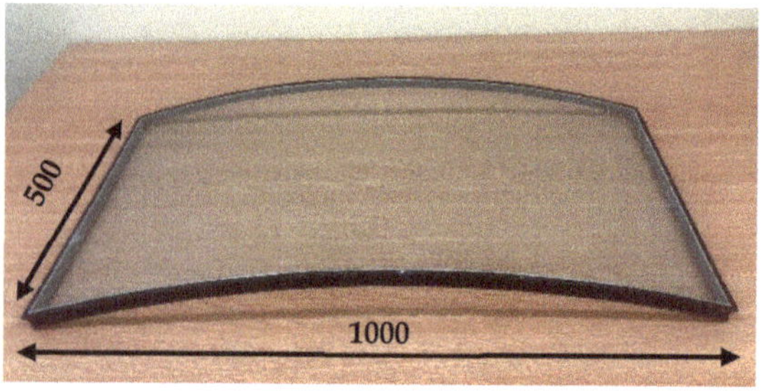

(**a**)

(**b**)

Figure 1. Experimental study: (**a**) Curved IGU sample; (**b**) Copper tube glued into a hole drilled in the spacer.

A single series of tests involved injecting or withdrawing a defined air volume into/from the IGU cavity [7]. After the procedure, pressure measurements were performed for 5 min, until the stabilisation of the pressure. The test campaign consisted of eight tests, including air injection and withdrawal into/from the cavity ranging from ±75 mL to ±300 mL with a 75 mL margin. More details on the experimental campaign can be found in [8].

2.2. Development of the Numerical Model of the Curved IGU

Numerical models were developed in the ABAQUS CAE finite element analysis programme to investigate the behaviour of IGUs in the experiments described in Section 2 [9]. A script in Python language was developed to improve the speed of the model generation process. The numerical models differ only by the volume of injected/withdrawn gas, whereas other parameters, such as the geometry of the sample, material properties, and boundary conditions, remain the same. Since creating eight numerical models in the ABAQUS GUI (Graphical User Interface) would consume significant time, a Python script was developed to accelerate the process, in which each syntax reflected the ABAQUS program commands. The variable parameter, in this case the volume of gas, was assigned as input data. The script was developed in the loop, taking another set of input data each time while passing through the code. Applying the Python script to the modelling process significantly reduces the time of the process. The numerical analysis time was limited to the execution of the Python code and computing of the data. As a result, a simple 3D model was created, with dimensions of 500 × 1000 mm^2 and a 1500 mm radius of curvature. The model consists of shell parts with square 4-node finite elements with full integration (S4 from the ABAQUS library [9]) and a fluid cavity. Two parallel shells represent the glass panes, while four perpendicular shells correspond to the foamed spacer bar and the secondary sealing (see Figure 2). The fluid cavity was implemented for gas simulation in the air cavity of the IGU. It was vital to create fluid exchange as an injected/withdrawn volume of gas similar to the test conditions.

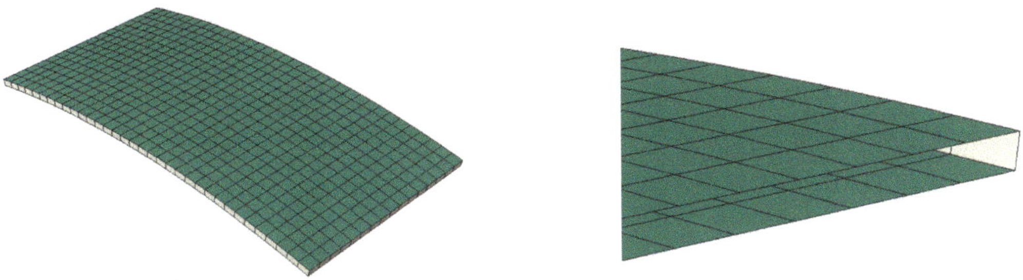

(**a**) (**b**)

Figure 2. Numerical model of the curved IGU: (**a**) Finite element mesh; (**b**) Detail of the glass–spacer interface. Note: glass panes are shown in green, while spacers are grey.

The following gas constants described the fluid cavity: the universal gas constant Ru = 8.314 J/(K·mol) [10] and the molecular weight of dry air: $M_{air} = 28.97 \times 10^{-3}$ kg/mol [11]. In addition, it was crucial to assign the appropriate physical parameters, such as the thermal expansion coefficient and density, to the air, which were necessary to perform the temperature load analysis. The thermal expansion coefficient was equal to 0.00343 1/K, and the density was 1.204 kg/m^3. The model was simply supported and free to deform. The material properties for the glass parts were taken from EN 572-1 (E = 70 GPa, μ = 0.23) [12], while for the spacer, E = 4.6 GPa and μ = 0.49 were used according to [13]. Kozłowski et al. [8] reported a more detailed description of the material properties and the parameters of the fluid cavity. In the same article, a mesh convergence study for maximal stress was performed for similar model parameters, and the model converged to an acceptable degree (change in stress not greater than 2% compared to the previous iteration step). Consequently, in the following research, the same mesh size was assumed, which was equal to 34 points of division along the longer edge of a glass component.

2.3. *Case Study*

Within the case study, the influence of the curvature of the component panes of an IGU on the internal pressure and static values (stress in glass and panes' deflection) under the same loading conditions was investigated. For this purpose, the validated numerical model and methodology described in the previous section were used.

Two geometries of an IGU were considered. The first case is a flat reference IGU with the dimensions of 500 × 1000 mm^2 (width × length) composed of two 4 mm flat panes and a 16 mm wide flexible silicone foam spacer. The build-up and spatial geometry of the second model are the same as for the reference IGU; however, the component panes are bent to a radius of curvature of 1500 mm.

The analysis was limited to two load cases, according to [6]. The standard provides two climatic design situations for IGUs that generate extreme internal pressure values in the cavity in summer and winter (Table 1). It can be seen that the climatic load in winter is almost a reverse version of summer. The load cases involve temperature differences in the gas in the cavity and changes in atmospheric pressure.

Table 1. Load actions by [6].

Load Case	Temperature Difference [K]	Change in Atmospheric Pressure [kN/m^2]
'Summer'	+20	−2.0
'Winter'	−25	+4.0

3. Results

3.1. Validation of the Curved IGU Model by Physical Experiments

Table 2 presents the quantitative results of the experiments and numerical simulations. In general, the numerical model underestimates the measured values by approximately 8%, which proves its correctness and the sound reproduction of the experimental results. Considering all the measured and simulated values pairs, the relationship shows excellent agreement (R2-value 0.99). The most significant difference was 10.8% for the withdrawal of 75 mL from the cavity. The difference between the measured and simulated values of the internal pressure is probably due to how the interface between the spacer and the panes is modelled. It exhibits a hinged connection, which does not transfer any bending moments between the components. In physical samples, the interface provides a certain degree of restraint that limits the deformation of the panes and results in increased pressure inside the space between the panes.

Table 2. Comparison of measured and simulated values of internal pressure in the IGU cavity.

Loading (Air Injection/Withdrawal to/from IGU Cavity)	Internal Pressure [Pa]		Difference (Simulated vs. Measured)
	Measured Experimentally	Simulated Numerically	
+75 mL	528.6	482.5	−8.7%
+150 mL	1066.8	960.1	−10.0%
+225 mL	1582.4	1435.7	−9.3%
+300 mL	2092.7	1906.7	−8.9%
−75 mL	−543.1	−484.4	−10.8%
−150 mL	−1064.3	−970.4	−8.8%
−225 mL	−1556.3	−1464.0	−5.9%
−300 mL	−2097.2	−1594.5	−6.8%

3.2. Validation of the Reference (Flat) IGU Model with Alternative Software

The modelling methodology and material parameters for the flat IGU were the same as those for the validated bent glass IGU model. Since laboratory tests did not include a flat IGU, additional model validation was carried out using the engineering programme SJ-Mepla, widely used for the structural design of IGUs in the glass industry [14]. Only the temperature difference load case included in Table 1 was selected for validation purposes.

Figures 3 and 4 present the validation results for the 'summer' case. The results agree regarding the deformation and maximum principal (tensile) stress. Differences are less than 11%. In terms of internal pressure, the numerical model overestimates the reference values (obtained from SJ-Mepla) by approximately 4.5% (701 Pa vs. 733 Pa). The same overestimation was achieved in the 'winter' load case. The difference between the models and consequent results is related to the fact that, contrary to the modelling approach developed within this study, the SJ-Mepla software never regards the bending rigidity, i.e., no bending stiffness of the spacer nor the sealing material is used [14]. The results obtained prove the correctness of the flat IGU model.

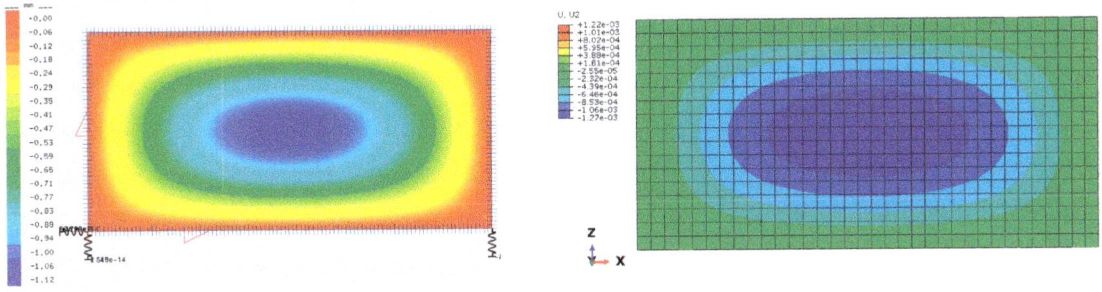

(a) (b)

Figure 3. Validation of the flat IGU model ('summer' load case): (**a**) deflections obtained from the SJ-Mepla software (values in mm); (**b**) deflections obtained from the ABAQUS software (values in m).

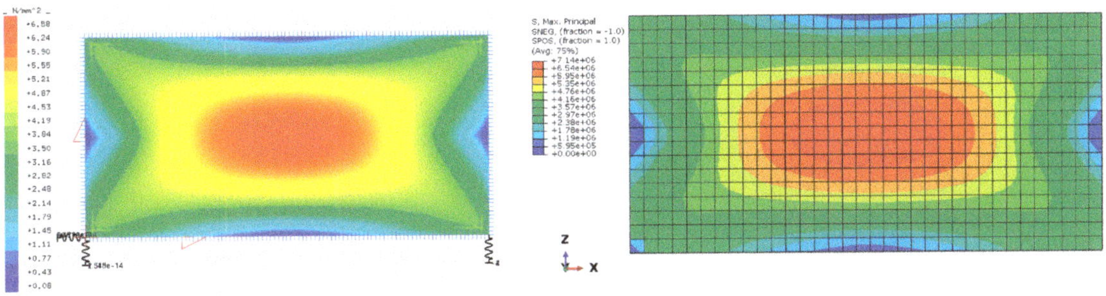

(a) (b)

Figure 4. Validation of the flat IGU model ('summer' load case): (**a**) maximum principal (tensile) stress obtained from the SJ-Mepla software (values in MPa); (**b**) maximum principal (tensile) stress obtained from the ABAQUS software (values in Pa).

3.3. Results of the Case Study

Figures 5 and 6 present qualitative results of flat and curved IGUs for a selected climatic load case. It should be noted that the graphical results are for the same 'summer' load case (Table 1), while quantitative results for both climatic load cases are shown in Table 3.

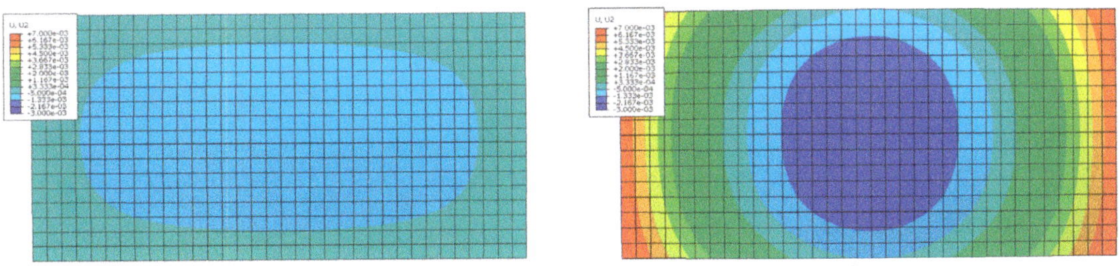

(a) (b)

Figure 5. Out-of-plane displacements ('summer' load case) for: (**a**) flat IGU; (**b**) curved IGU. Note: Bottom view; values in m; the same scale was used for the legend on both figures (max: 7.0 mm, min: −3.0 mm).

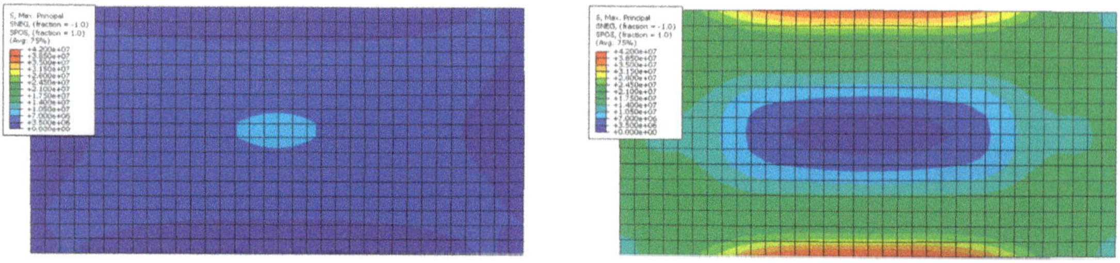

(**a**) (**b**)

Figure 6. Maximum principal tensile stress ('summer' load case) for: (**a**) flat IGU; (**b**) curved IGU. Note: Bottom view; values in MPa; the same scale was used for the legend on both figures (max: 42 MPa).

Table 3. Results of the case study.

Load Case		Flat IGU	Curved IGU	Difference
'Summer'	Internal pressure [Pa]	1190	5680	+477%
	Maximum principal tensile stress [MPa]	7.63	41.21	+540%
	Deflection in the Y direction [mm]	1.35	6.79	+503%
'Winter'	Internal pressure [Pa]	−2870	−6250	+218%
	Maximum principal tensile stress [MPa]	8.89	51.46	+579%
	Deflection in the Y direction [mm]	1.75	7.86	+449%

Figure 5 presents results for flat and cylindrically curved IGU in terms of out-of-plane displacements for the 'summer' load case. From the comparison, it can be observed that the curvature of the component pane has a significant influence on the deformation pattern. In the case of flat IGU, the panes bulge with the maximum displacement value in the centre of the pane. This behaviour is typical for rectangular, simply supported flat slabs at all edges and subjected to uniform pressure. Surprisingly, the curved IGU shows a completely different deformation behaviour. The increased pressure in the cavity due to gas expansion caused by the temperature increase and negative air pressure surrounding the IGU causes the model to 'flatten', being closer to a flat shape. This way, the short edges are lifted concerning the model's centre and achieve the maximum values at these locations.

Figure 6 shows maps of maximum principal (tensile) stress in glass, which is a critical factor in structural glazing design. For the flat IGU, the stress map follows the deformation pattern so that the location of the maximum stress is identical to the case of maximum deformation. As for deformation, the stress map is typical for a simply supported flat slab at all edges and subjected to uniform pressure. As seen in the figure, the location of the maximum stress in the curved IGU changes its location. The maximum stress for curved IGU is located at the curved edges of the component panes. This phenomenon is directly related to flattening panes, generating tension in the centre of the curved edges.

Table 3 presents the quantitative results of the case study. In terms of internal pressure, the radius of curvature of 1500 mm causes an average 477% increase in the value for the 'summer' load case, while for the 'winter' case, the increase is still significant; however, to a lower extent (218%). Regarding maximum deflection and stress, the curvature generates static values higher by, on average, 517% in curved IGU compared to the flat model.

4. Conclusions and Further Work

This paper presents the results of experiments and numerical simulations of a case study involving a flat and cylindrically curved IGU subjected to characteristic climatic actions given in the standards.

It was found that the curvature of the component panes significantly influences the internal pressure, deformation pattern, and stress in the glass of the IGU subjected to climatic loads. The flat IGU uniformly bulges (the pillowing effect), while the curved IGU tends to 'flatten' with lifting short edges. This phenomenon is also reflected in the stress distribution. Its maximum value changes location from the centre of the pane (flat IGU) to the centre of the long edge in the curved IGU. The value of principal tensile stress increases significantly by more than 500% in the curved IGU. This finding is critical for the structural design of the panes and the secondary sealant and therefore justifies undertaking the research topic.

It should be emphasised that the above conclusions are valid only for the assumptions used in the case study. The influence of the curvature of the component panes of an IGU on the resulting internal pressure is highly complex and requires further research. It includes investigations of the impact of spacer stiffness and the interface between the glass panes and the spacer.

Author Contributions: Conceptualisation, M.K. and K.Z.; methodology, M.K. and K.Z.; software, M.K. and K.Z.; validation, M.K. and K.Z.; formal analysis, M.K. and K.Z.; investigation, M.K. and K.Z.; resources, M.K. and K.Z.; data curation, M.K. and K.Z.; writing—original draft preparation, M.K.; writing—review and editing, M.K. and K.Z.; visualisation, M.K. and K.Z.; supervision, M.K.; project administration, M.K.; funding acquisition, M.K. All authors have read and agreed to the published version of the manuscript.

Funding: This research was funded by the research project "Analysis of internal pressure in 3D bent insulated glass units" (grant number 2021/05/X/ST8/00168) financed by The National Science Centre (NCN) within the MINIATURA 5 programme.

Institutional Review Board Statement: Not applicable.

Informed Consent Statement: Not applicable.

Data Availability Statement: Data will be provided upon request.

Acknowledgments: The authors thank their colleagues and students from the Faculty of Civil Engineering at the Silesian University of Technology, Gliwice, Poland, for their support during the experiments. Moreover, the insight and expertise provided by Zbigniew Respondek, PhD from the Faculty of Civil Engineering at the Czestochowa University of Technology, Częstochowa, Poland, is acknowledged. The publication cost of this paper was covered with funds from the Polish National Agency for Academic Exchange (NAWA): "MATBUD'2023—Developing international scientific cooperation in the field of building materials engineering" BPI/WTP/2021/1/00002, MATBUD'2023.

Conflicts of Interest: The authors declare no conflict of interest.

References

1. Kosić, T.; Svetelć, I.; Cekić, Z. Complexity of Curved Glass Structures. *IOP Conf. Ser. Mater. Sci. Eng.* **2017**, *262*, 012158. [CrossRef]
2. Malewski, A.; Kozłowski, M.; Sumelka, W.; Połedniok, M. Large Scale Architectural Glass Slumping Process-Challenges and Limitations. *Arch. Civ. Eng.* **2020**, *66*, 485–505. [CrossRef]
3. Kocer, C.; Aronen, A.; Collins, R.; Asano, O.; Ogiso, Y. Measurement of heat flow through vacuum insulating glass part 2: measurement area separated from glass sheets with buffer plates. *Glass Struct. Eng.* **2022**. [CrossRef]
4. Nizich, A.; Baer, S.; Prandelli, S.; Burkhart, I.K.; Moore, W.P. *Enhancing the Light-Curved Insulated Glass Unit Design Study Keywords*; GPD Glass Performance Days: Tampere, Finland, 2019.
5. Bao, M.; Gregson, S. Sensitivity study on climate induced internal pressure within cylindrical curved IGUs. *Glass Struct. Eng.* **2019**, *4*, 29–44. [CrossRef]
6. *DIN 18008-1*; Glass in Building—Design and Construction Rules—Part 1: Terms and General Bases. Deutsches Institut fur Normung E.V. (DIN): Berlin, Germany, 2010.
7. Kozłowski, M.; Respondek, Z.; Wiśniowski, M.; Cornik, D.; Zemła, K. Experimental and Numerical Simulations of Climatic Loads in Insulating Glass Units by Controlled Change of Pressure in the Gap. *Appl. Sci.* **2023**, *13*, 1269. [CrossRef]

8. Kozłowski, M.; Respondek, Z.; Cornik, D.; Wiśniowski, M.; Zemła, K. Influence of Curvature and Geometrical Parameters on Internal Pressure in Cylindrical Insulating Glass Units. *Thin-Walled Struct.* 2023, *in review*.
9. SIMULIA. *Abaqus/CAE User's Guide*; SIMULIA: Johnston, RI, USA, 2014.
10. Jensen, W. The Universal Gas Constant R. *J. Chem. Educ.* **2003**, *60*, 7. [CrossRef]
11. Rajput, R.K. *Engineering Thermodynamics: A Computer Approach*; Transatlantic Publishers: London, UK, 2003.
12. *EN 572-1:2012*; Glass in Building Basic Soda Lime Silicate Glass Products—Definitions and General Physical and Mechanical Properties. CEN: Brussels, Belgium, 2012.
13. Minasyan, M. *Randverbundbeanspruchung Gebogener Zweischeiben-Isoliergläser mit Silikonschaum-Abstandhalter*; Hamburg University: Hamburg, Germany, 2014.
14. SJ Software GmbH. *SJ MEPLA Software and User's Manual*; SJ Software GmbH: Aachen, Germany, 2021.

materials proceedings

MDPI

Proceeding Paper

Effect of Potassium Formate on Alkali–Silica Reaction in Aggregates with Different Categories of Reactivity †

Aneta Antolik *, Daria Jóźwiak-Niedźwiedzka, Kinga Dziedzic and Paweł Lisowski

Institute of Fundamental Technological Research, Polish Academy of Sciences, Adolfa Pawińskiego 5B, 02-106 Warsaw, Poland

* Correspondence: aantolik@ippt.pan.pl; Tel.: +48-826-12-81 (ext. 207)

† Presented at the 10th MATBUD'2023 Scientific-Technical Conference "Building Materials Engineering and Innovative Sustainable Materials", Cracow, Poland, 19–21 April 2023.

Abstract: During the wintertime, concrete pavements experience harsh exposure conditions due to the presence of both the freezing–thawing and wetting–drying cycles. Airport concrete pavements are commonly de-iced using chloride-free organic salts such as potassium formate or potassium acetate. However, these materials contain alkali ions which can have harmful effects on both the cement matrix and the aggregate. Specifically, there is an increased risk of occurrence of the alkali–silica reaction (ASR). The goal of this research was to estimate the influence of potassium formate on the potential of causing alkali–silica reaction in aggregates with different categories of reactivity (R0, R1, R2). The accelerated mortar bar test and its modification (which involves replacing sodium hydroxide solution with a potassium formate solution) were used. Detailed SEM-EDS examinations were performed to confirm the presence of alkali–silica reaction and to analyze the influence of potassium formate on the microstructure of mortar.

Keywords: alkali–silica reaction; potassium formate; de-icing agent; reactive aggregate

check for updates

Citation: Antolik, A.; Jóźwiak-Niedźwiedzka, D.; Dziedzic, K.; Lisowski, P. Effect of Potassium Formate on Alkali–Silica Reaction in Aggregates with Different Categories of Reactivity. *Mater. Proc.* **2023**, *13*, 13. https://doi.org/10.3390/materproc2023013013

Academic Editors: Katarzyna Mróz, Tomasz Tracz, Tomasz Zdeb and Izabela Hager

Published: 14 February 2023

1. Introduction

The alkali–silica reaction (ASR) is a chemical reaction between the alkali and hydroxyl ions present in the concrete pore solution and certain reactive forms of silica present in the aggregate [1]. This phenomenon is characterized by formation of an alkali–silica gel that absorbs water and expands, ultimately causing cracks in the aggregate and then in the cement matrix [2]. Although the predominant source of alkalis in concrete is Portland cement, other sources of alkalis, such as de-icing salts or agents, have been shown to contribute to ASR in concrete, [3–5].

Chloride based de-icers (NaCl, $MgCl_2$ and $CaCl_2$) are the most common de-icing salts used during winter months to clean the concrete road pavement and enable a normal flow of traffic [6]. However, they may have detrimental effects on concrete durability, which is often manifested by expansion, mass change, reduction in the dynamic modulus of elasticity [1] and, most of all, reduction in strength [5]. At the same time, due to the risk of aircraft corrosion, their use on airfield pavements is not allowed. Alternative agents used for de-icing airport pavements include sodium and potassium formates or acetates [7].

Studies on the effects of formate- and acetate-based de-icing agents have already been published, [3,4,8]. Wang et al. [3] analyzed the influence of de-icing agents on concrete degradation due to ASR. They used potassium acetate, which has a similar effect to potassium formate. The authors did not observe significant damage in the macro- or micro-scale and they did not find ASR products, despite using a high concentration of potassium acetate (54.4 wt. %). On the other hand, Rangaraju et al. [8] found that the two common airfield de-icing chemicals—potassium acetate and sodium acetate-based de-icer solutions—have significant potential to cause ASR in test specimens containing reactive aggregates. The research conducted by Giebson et al. [4] came to similar conclusions. They revealed that deleterious ASR can be initiated and accelerated in concretes with reactive aggregates exposed to alkali-containing de-icers, especially ones based on acetates and formates. They showed that even low-alkali cement was not able to prevent the ASR.

Real-life experience shows that even in concrete made of nominally non-reactive aggregates, ASR can sometimes occur under adverse conditions of additional external alkali supply, humid environment, or heavy traffic and load [9]. The goal of the current research was to determine the impact of potassium formate on the potential for the occurrence of ASR in aggregates with different levels of reactivity.

2. Materials

Three aggregates characterized by different categories of reactivity were selected for testing. ASR reactivity was tested according to PB/1/18 [10], which is based on expansion measurements. Two granite aggregates (G1, G2) and one quartzite aggregate (Q) met the initial requirements of the research regarding the ASR reactivity category; they were R0—non-reactive (G2), R1—moderately reactive (G1), and R2—highly reactive (Q). The category of reactivity of aggregate according to PB/1/18 is presented in Table 1.

Table 1. Category of ASR reactivity according to PB/1/18 [10].

Category of ASR Reactivity	Reactivity	14-Day Expansion [%]
R0	Non-reactive	≤0.10
R1	Moderately reactive	>0.10; ≤0.30
R2	Highly reactive	>0.30; ≤0.45
R3	Very highly reactive	>0.45

Ordinary Portland cement CEM I 42.5 R was used for mortar specimens. The alkali content in cement was 0.86%.

3. Methods

Aggregate for petrographic analysis was prepared according to Technical Guidelines [11]. The identification of reactive minerals presented in tested aggregates was conducted on thin sections of 30 ± 1 μm thickness. Polarizing microscope Olympus BX51 equipped with a digital colour camera and an automatic stage Prior ES11BX/B was used for petrographic observations. Transmitted light with parallel (PPL) and crossed (XPL) polarizers was used.

The expansion tests of mortars were performed according to the accelerated mortar bar test described in PB/1/18 [9]. The above method is similar to the one given in the ASTM C 1260 [12]. The specimens (25 × 25 × 285 mm) were exposed to 1 M NaOH and 80 °C for 28 days. The method was modified using de-icing agent (50% concentration of potassium formate (HCOOK)) as a soak solution. The Federal Aviation Administration Advisory Circular: Standard Specifications for Construction of Airports concerning cement concrete pavement [13] stipulates that alkali–silica reactivity of the aggregates used in Portland cement concrete pavements shall be tested and evaluated using the modified ASTM C1260 [12], among other guidelines. The modification of the testing method involves extending the length of the test from 14 to 28 days.

After the accelerated mortar bar test, the specimens were cut (40 × 25 × 10 mm) and prepared for post-mortem microscopic observations. The specimens were dried, impregnated with epoxy resin, ground on diamond discs (125, 75, 54, 18 and 9 μm), and polished on polishing pads using diamond pastes (6, 3, 1, 0.25 μm). Then, specimens were coated with a thin carbon layer. Microstructure analysis was performed on a JEOL JSM-6460LV scanning electron microscope (SEM) equipped with an X-ray energy dispersion (EDS) detector. The analysis was performed at high vacuum in the backscattered electron mode. The SEM was operated with acceleration voltage set to 20 kV and a working distance of 10 mm.

4. Results and Discussion

In the first step, the petrographic analysis was performed on thin sections. The aggregate was analyzed for the presence of reactive minerals (Figure 1). A significant presence of microcrystalline quartz was found in granite aggregate G1 and quartzite aggregate Q. Microcrystalline quartz was not found in the granite G2 sample, but strained quartz was found.

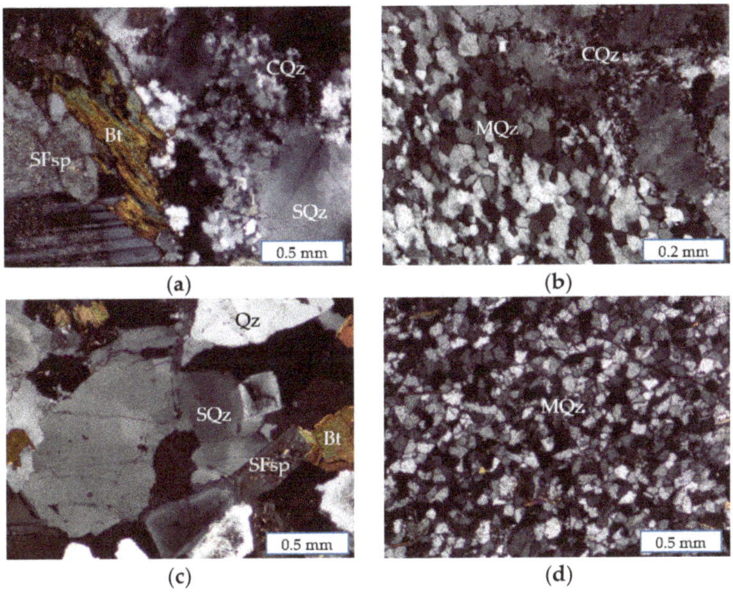

Figure 1. The microphotographs of the analyzed aggregates in cross-polarized light (XPL) (**a**,**b**) granite G1 aggregate, (**c**) granite G2 aggregate, (**d**) quartzite Q, aggregate; recognized minerals: Qz—quartz, SQz—strained quartz, MQz—microcrystalline quartz, CQz—cryptocrystalline quartz, SFsp—feldspar alteration to sericite, Bt—biotite.

Expansion of mortar bars over a 28-day period is presented on Figure 2. The expansion of mortar bars exposed to 1 M NaOH after 14 days confirmed the aggregate reactivity categories. Granite aggregate G1 characterized by 0.07% expansion was classified as the R0 category, granite G2 with 0.16% expansion as the R1 category, and quartzite with 0.31% expansion as the R2 category. Extending the PB/1/18 [10] to 28 days did not significantly slow down the rate of expansion.

Mortar bars stored in potassium formate solution expanded significantly more than those stored in sodium hydroxide solution. In the case of granite aggregates, expansion was found to increase at a constant rate for 28 days, without a significant slowdown. However, a slowdown in the expansion of mortar bars with quartzite aggregate was observed after 17 days of storage in a potassium formate solution. After 28 days of storage in 50% HCOOK, the following expansion values were found: G1—0.97%, G2—1.23%, Q—0.99%. The expansion of mortars with G1 and Q aggregates was similar after 28 days of storage in HCOOK, although in the sodium hydroxide solution, the mortar with the G1 aggregate had four times lower expansion than the mortar with the Q aggregate. The slowdown of the reaction in the case of quartzite aggregate may be related to the presence of reactive minerals in the form of micro- and cryptocrystalline quartz, which are fast-reacting minerals. In the first few days of the test, the highest increase in expansion among all the tested aggregates is visible. However, as a result of the exhaustion of reactive components, the rate of further increase in expansion decreases. However, in specimens with granite aggregate (G1, G2), apart from micro- and cryptocrystalline quartz, we also discovered strained quartz, which is classified as a slowly reacting mineral [14–16]. The reaction was not slowed down due to the presence of both fast and slow reacting minerals.

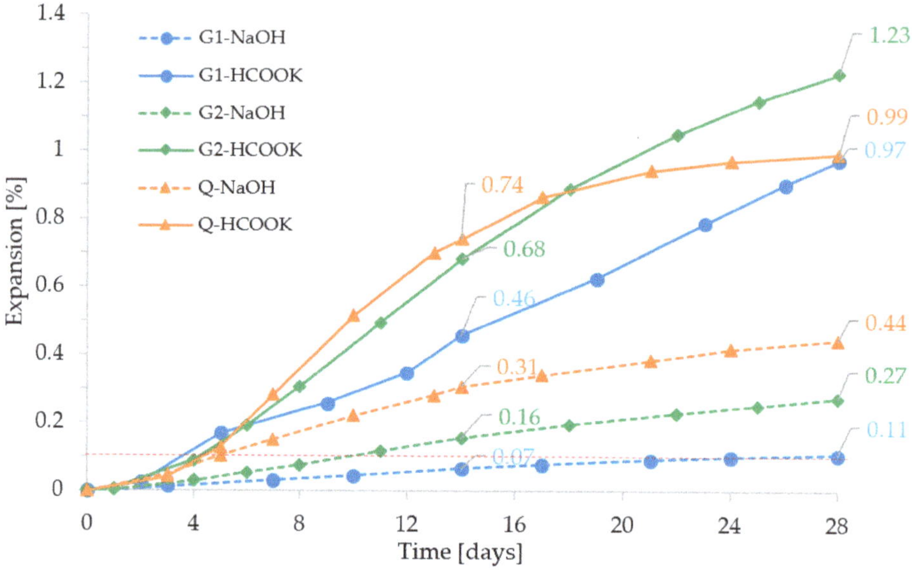

Figure 2. Expansion of mortar bars with aggregate with different ASR reactivity (R0, R1, R2) exposed to 1 M NaOH or 50% HCOOK solution.

Cracks were observed on the surface of all mortar specimens. Larger cracks were seen in mortar bars stored in potassium formate solution (Figure 3).

(a) (b)

Figure 3. Cracks on the surface of mortar bars with quartzite Q aggregate immersed in (**a**) 1 M NaOH, (**b**) 50% HCOOK.

For a more detailed analysis, microscopic examinations were carried out on polished specimens using SEM. Results are presented in Figures 4–6.

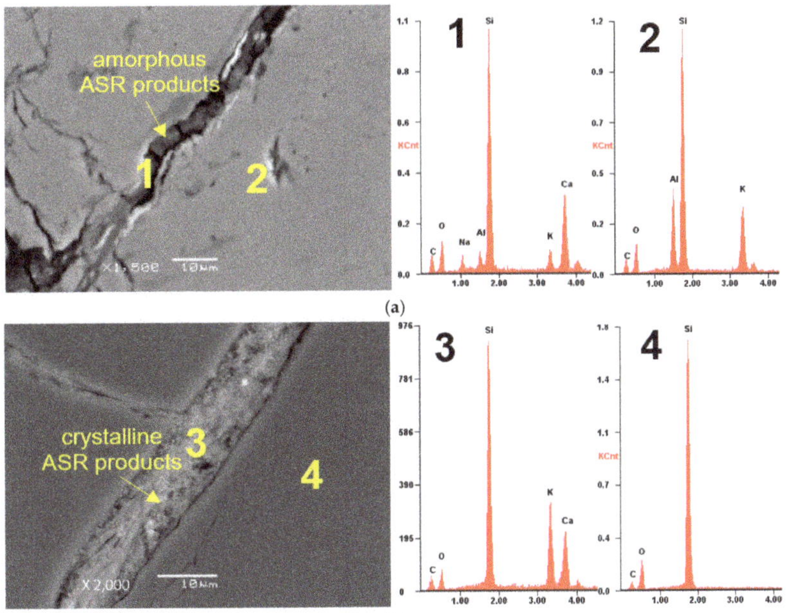

Figure 4. Microstructure of the mortar with aggregate G1 after 28 days of storage at 80 °C in (**a**) 1 M NaOH, crack in the aggregate grain filled with amorphous ASR products (1), potassium feldspar (2); (**b**) 50% HCOOK, crack in the aggregate grain filled with crystalline ASR products (3), quartz (4).

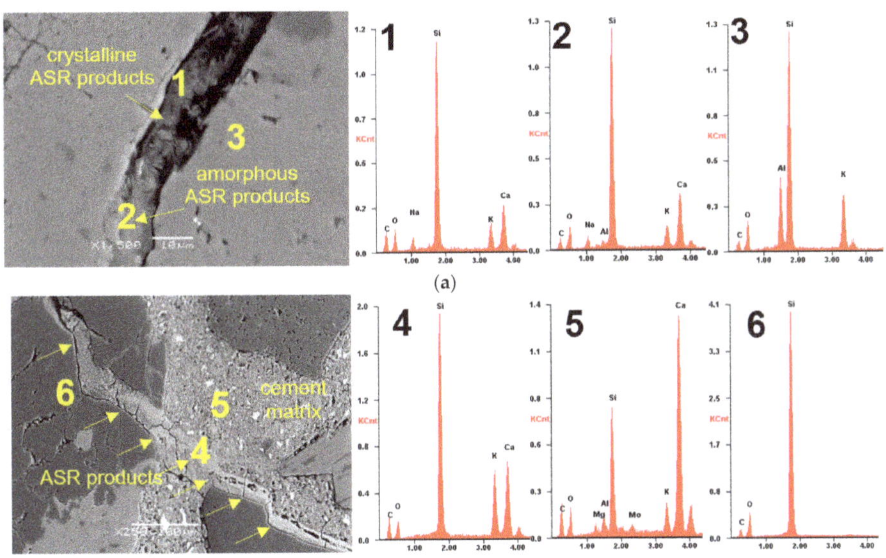

Figure 5. Microstructure of the mortar with aggregate G2 after 28 days of storage at 80 °C in (**a**) 1 M NaOH, crack in the aggregate grain filled with crystalline (1) and amorphous (2) ASR products, (3) potassium feldspar; (**b**) 50% HCOOK, crack in the aggregate grain filled with ASR products that penetrates the cement matric (5), C-S-H (5), quartz (6).

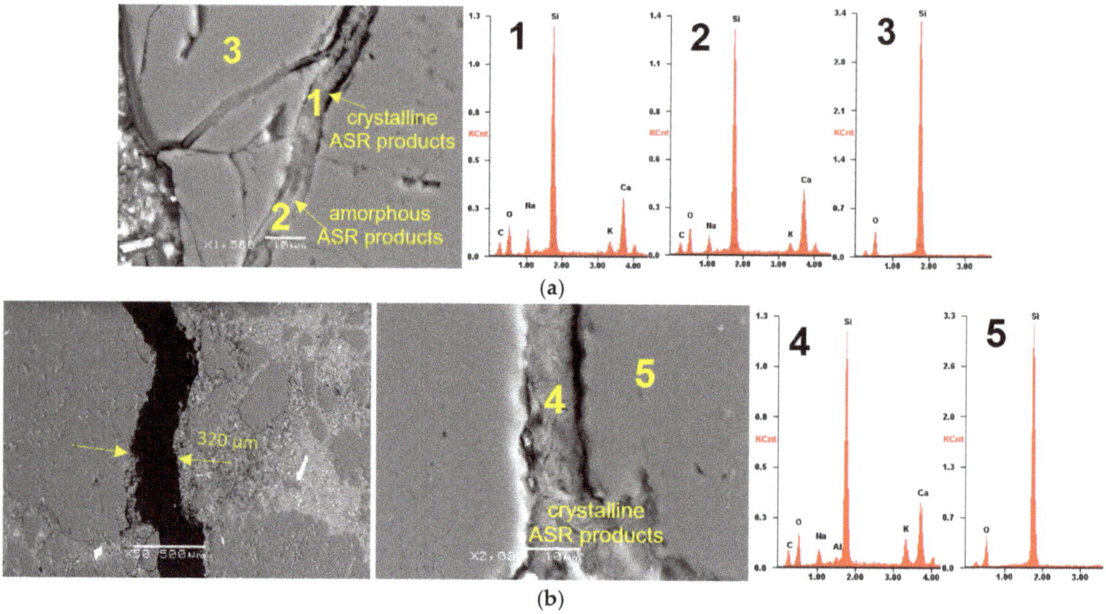

Figure 6. Microstructure of the mortar with aggregate Q after 28 days of storage at 80 °C in (**a**) 1 M NaOH, cracks in the aggregate grain filled with crystalline (1) and amorphous (2) ASR products; (**b**) 50% HCOOK, crack of 320 µm and crack in the aggregate grain filled with crystalline ASR products (4), quartz (5).

The presence of ASR products was found in all analyzed mortar bars. The identified reaction products were found in both crystalline and amorphous forms (distinguished by morphology). The ASR products were mainly found in the cracks of the aggregate grains, but also in the cement matrix (Figure 5b). Large macro-cracks were observed in the specimens stored in potassium formate solution. Their width was between 100 and 500 µm. The chemical composition of the ASR products identified in the aggregate varied depending on the storage solution (NaOH, HCOOH). There was no significant influence of the type of aggregate (granite, quartzite) on chemical composition in the reaction products. In the ASR gel found in mortar bars after storage in NaOH solution, the values of the Ca/Si and (Na+K)/Si were 0.25–0.35 and 0.20–0.25, respectively. However, the ASR products in the mortar bars stored in the HCOOK solution were characterized by Ca/Si and (Na+K)/Si of 0.3–0.45 and 0.25–0.35, respectively (alkali was present only in the form of potassium).

The obtained test results are in line with expectations. Sodium salts of acetate and formate have a significant potential to exert detrimental effects on concretes, causing deleterious ASR in mortar and concrete specimens containing reactive aggregates [17]. However, the deleterious ASR symptoms were not seen in specimens containing non-reactive aggregate [18]. The conducted research showed the effect of potassium formate on the potential for ASR occurrence in aggregate of various reactivity categories. Leeman et al. [19] analyzed crystalline and amorphous ASR products. The chemical composition of the ASR gel was similar to the results obtained in the conducted research, but only for the specimens tested in 1 M NaOH solution. For amorphous and crystalline ASR products, Ca/Si 0.21–0.23 and 0.22–0.26, respectively, (Na+K)/Si: 0.21–0.24 and 0.22–0.34, respectively.

The next stage of research will concern the impact of de-icing chemicals based on the formates on real concrete mixes used in airport concrete pavements as well as more accurate characterization of ASR products using Transmission Electron Microscopy, Raman spectroscopy and nanoindentation techniques.

5. Conclusions

Based on the conducted research, the following conclusions can be drawn:

- The presence of reactive minerals: micro- and cryptocrystalline quartz was found in the quartzite and granite aggregate. In addition, strained quartz was present in one granite aggregate.
- The accelerated mortar bar test confirmed different categories of reactivity of the tested aggregate.
- Regardless of the aggregate reactivity category, the mortar bars stored in the 50% potassium formate solution reached very high expansion values, including granite G1 with the R0 reactivity category.
- Reduction in the expansion rate of quartzite aggregate in potassium formate solution was probably associated with the depletion of fast-reacting minerals (micro- and cryptocrystalline quartz). Such phenomena were not observed in mortars with granite aggregate, in which strained quartz (a slow-reacting mineral) was the main reactive mineral.
- The presence of ASR products was found in all analyzed mortar bar specimens.
- The chemical composition of the gel varied depending on the mortar storage solution (NaOH, HCOOK). The gel identified in the cracks of aggregate grains in the mortar in the potassium formate solution was characterized by a higher alkali content, specifically potassium.

Author Contributions: Conceptualization, A.A.; methodology, A.A. and D.J.-N.; validation, A.A., D.J.-N., K.D. and P.L.; investigation, A.A.; resources, A.A.; writing—original draft preparation, A.A.; writing—review and editing, A.A., D.J.-N., K.D. and P.L.; supervision, D.J.-N. All authors have read and agreed to the published version of the manuscript.

Funding: This research was funded by Polish National Science Centre as a part of Preludium Project "Influence of de-icing agents on the properties of alkali-silica reaction products in cement-matrix composites" (2021/41/N/ST8/03799).

Institutional Review Board Statement: Not applicable.

Informed Consent Statement: Not applicable.

Data Availability Statement: Not applicable.

Acknowledgments: The publication cost of this paper was covered with funds from the Polish National Agency for Academic Exchange (NAWA): "MATBUD'2023-Developing international scientific cooperation in the field of building materials engineering" BPI/WTP/2021/1/00002, MATBUD'2023.

Conflicts of Interest: The authors declare no conflict of interest.

References

1. Stanton, T.E. Expansion of concrete through reactionn between cement and aggregate. *Proc. Am. Soc. Civ. Eng.* **1940**, *66*, 1781–1811.
2. Fournier, B.; Bérubé, M.A. Alkali-aggregate reaction in concrete: A review of basic concepts and engineering implications. *Can. J. Civ. Eng.* **2000**, *27*, 167–191. [CrossRef]
3. Wang, K.; Nelsen, D.E.; Nixon, W.A. Damaging effects of deicing chemicals on concrete materials. *Cem. Concr. Compos.* **2006**, *28*, 173–188. [CrossRef]
4. Giebson, C.; Seyfarth, K.; Stark, J. Influence of acetate and formate-based deicers on ASR in airfield concrete pavements. *Cem. Concr. Res.* **2010**, *40*, 537–545. [CrossRef]
5. Jain, J.; Olek, J.; Janusz, A.; Jóźwiak-Niedźwiedzka, D. Effects of Deicing Salt Solutions on Physical Properties of Pavement Concretes. *Transp. Res. Rec.* **2012**, *2290*, 69–75. [CrossRef]
6. Broekmans, M.A.T.M. Deleterious Reactions of Aggregate With Alkalis in Concrete. *Rev. Mineral. Geochem.* **2012**, *74*, 279–364. [CrossRef]
7. Kowalska, D.; Pietruszewski, P. Wpływ środków odladzających na beton przeznaczony na nawierzchnie lotniskowe. In *Dni Betonu, Monografie Technologii Betonu*; Stowarzyszenie Producentów Cementu: Krakow, Poland, 2016; Volume 2, pp. 111–129.
8. Rangaraju, P.R.; Sompura, K.R.; Olek, J. Investgation into Potential of Alkali-Acetate-Based Deicers to Cause Alkali-Silica Reaction in Concrete. *Transp. Res. Rec.* **1979**, *1*, 69–78.

9. Fernandes, I.; Noronha, F.; Teles, M. Microscopic analysis of alkali-aggregate reaction products in a 50-year-old concrete. *Mater. Charact.* **2004**, *53*, 295–306. [CrossRef]
10. Procedura Badawcza GDDKiA PB/1/18, Instrukcja badania reaktywności kruszyw metodą przyspieszoną w 1M roztworze NaOH w temperaturze 80 °C. 2018. Available online: https://www.gov.pl/web/gddkia/reaktywnosc-kruszyw (accessed on 15 February 2022).
11. Garbacik, A.; Glinicki, M.A.; Jóźwiak-Niedźwiedzka, D.; Adamski, G.; Gibas, K. Technical Guidelines for the Classification of Domestic Aggregates and Prevention of the Alkali-Aggregate Reaction in Concrete Used in Road Pavements and Road Engineering Facilities, The Institute of Ceramics and Building Materials, Institute of Fundamental Technological Research Polish Academy of Sciences. 2019. Available online: https://www.gddkia.gov.pl/pl/1118/dokumenty-techniczne (accessed on 15 February 2022). (In Polish)
12. *ASTM Standard C 1260*; Standard Test Method for Potential Alkali Reactivity of Aggregates (Mortar-Bar Method). ASTM International: West Conshohocken, PA, USA, 2016.
13. *AC 150/5370-10H*; FAA Advisory Circular: Standard Specifications for Construction of Airports (Item P-501 Cement Concrete Pavement). Mead & Hunt: Watertown, SD, USA, 2018.
14. Shayan, A. Alkali reactivity of deformed granitic rocks: A case study. *Cem. Concr. Res.* **1993**, *23*, 1229–1236. [CrossRef]
15. Antolik, A.; Jóźwiak-Niedźwiedzka, D. Assessment of the alkali-silica reactivity potential in granitic rocks. *Constr. Build. Mater.* **2021**, *295*, 123690. [CrossRef]
16. Velasco-Torres, A.; Alaejos, P.; Soriano, J. Comparative study of the alkali-silica reaction (ASR) in granitic aggregates. *Estud. Geol.* **2010**, *66*, 105–114. [CrossRef]
17. Desai, J. Investigation into Mitigation of Alkali-Silica Reaction Using Selected SCMs in Presence of Potassium Acetate Deicer. Ph.D. Thesis, Clemson University, Clemson, SC, USA, 2007; Volume 139, p. 307.
18. Rangaraju, P.R. *Mitigation of ASR in Presence of Pavement Deicing Chemicals*; IPRF 01-G-002-04-8; Innovative Pavement Research Foundation: Skokie, IL, USA, 2007; p. 99.
19. Leemann, A.; Shi, Z.; Lindgård, J. Characterization of amorphous and crystalline ASR products formed in concrete aggregates. *Cem. Concr. Res.* **2020**, *137*, 106190. [CrossRef]

materials proceedings

Proceeding Paper

Low Carbon Concrete Possibilities: EPD and Regulations in Northern Periphery and Arctic †

Iveta Novakova [1,*], Priyadharshini Perumal [2], Andrzej Cwirzen [3] and Olafur Haralds Wallevik [4]

1 Department of Building, Energy and Material Technology, The Arctic University of Norway, Lodve Langesgate 2, 8514 Narvik, Norway
2 Fiber and Particle Engineering Research Unit, University of Oulu, Erkki Koiso-Kanttilankatu, Koehalli/Ovi V6, 90570 Oulu, Finland
3 Department of Civil, Environmental and Natural Resources Engineering, Luleå University of Technology, 97187 Luleå, Sweden
4 Department of Applied Engineering, Reykjavik University, Menntavegi 1, 102 Reykjavík, Iceland
* Correspondence: iveta.novakova@uit.no; Tel.: +47-76-96-61-20

† Presented at the 10th MATBUD'2023 Scientific-Technical Conference "Building Materials Engineering and Innovative Sustainable Materials", Cracow, Poland, 19–21 April 2023.

Abstract: Concrete is one of the most common building materials in the Northern Periphery and Arctic, and therefore we should pay attention to its quality while reducing its carbon footprint. The concrete industry has established many measures to limit greenhouse gas emissions from concrete, as stated in the environmental product declaration (EPD). The most significant contributor is cement (common dose between 250 and 600 kg per 1 m^3 of concrete) in a concrete binder. Aside from the use of alternative fuels for cement production, new alternative materials for cement replacement are being sought. Those materials are called supplementary cementitious materials and mainly originate from industrial waste streams. Some of the materials are already standard and limited by the maximum allowed replacement, and some are new and still under investigation. The benefits and limitations of low-carbon concrete regulations in Norway, Sweden, Iceland, and Finland are demonstrated on three different concrete mixes in this article. The sorting of a reference mix and two low-carbon concrete mixes according to 4 different systems showed the informative character of the Icelandic system and the underestimation of possibilities for the carbon footprint of concrete in the Swedish classification system.

Keywords: concrete; low carbon concrete (LCC); environmental product declaration (EPD); carbon footprint; Northern Periphery and Arctic

Citation: Novakova, I.; Perumal, P.; Cwirzen, A.; Wallevik, O.H. Low Carbon Concrete Possibilities: EPD and Regulations in Northern Periphery and Arctic. *Mater. Proc.* **2023**, *13*, 14. https://doi.org/10.3390/materproc2023013014

Academic Editors: Katarzyna Mróz, Tomasz Tracz, Tomasz Zdeb and Izabela Hager

Published: 14 February 2023

1. Introduction to Low-Carbon Concrete

There is currently a strong effort underway to reduce greenhouse gas emissions and mitigate global warming. To achieve this goal, most industries and human activities must be revised, and a change from the linear approach/model (from cradle to grave) to a circular approach/model (from cradle to cradle) has to take place.

Concrete is probably the most common building material or man-made product, and it was almost entirely produced from virgin materials until recently. Nowadays, the trend is changing, and a more circular approach is being implemented. New types of concrete, such as Low Carbon Concrete (LCC), substitute various materials, primarily waste streams from other industries, for cement. Cement's production has a significant carbon footprint, which also translates to concrete's carbon footprint. The materials that can replace cement are called supplemental cementitious materials (SCM), and the mechanism behind their influence on final concrete properties is described in Section 2.1.

2. Low Carbon Concrete

There are several terms referring to concrete with a lower content of clinker. In this study, we will keep to the term "low carbon concrete." As indicated, the design of LCC is based on direct clinker reduction (linked with CO_2), resulting in binder volume reduction, and therefore the mix design process deviates from standard concrete design practices. Mix design is more demanding, and its optimization is required. The main compounds of concrete are aggregates (fine and coarse) and binder (cement, supplementary cementitious materials, water, and eventually superplasticizer and air), and their proportion defines the fresh and hardened properties of final concrete. New trends connected to various aggregate types such as crushed sand, recycled aggregates, or artificial aggregates are becoming popular as deposits of natural sand and good-quality natural aggregates are shrinking. In this study, the focus will be given mainly to binder composition and its contribution to concrete´s carbon footprint as expressed by the environmental product declaration (EPD) value [1]. Supplementary cementitious materials have a significant impact on both fresh and hardened properties, and therefore it is important to have good knowledge of their characteristics.

Throughout the whole publication, a demonstration series of three mixes will be used as an example for mix design, characterization, and environmental assessment. The series consists of a reference mix (REF) and two different low-carbon concretes (LCC1 and LCC2). See Table 1.

Table 1. Concrete proportions in volume (%).

	C45/55 REF	C45/55 LCC1	C45/55 LCC2
Binder	16.0	9.9	11.0
Aggregates	60.9	69.1	71.0
Water	17.7	14.6	13.7
Air	5.0	6.0	4.0
Admixtures	0.4	0.4	0.3
Total	100	100	100

All designed mixes belong to the same strength class C45/55 according to European standard EN206 [2], with a minimum cube strength of 55 MPa after 28 days (see Table 2). Concrete mixes were designed to be air-entrained for exposure classes X0, XC1-4, FX1, XD1-3, XA1-4, and XS1-3. The testing of fresh concrete consisted of workability evaluation by a slump test and air content determination by a pressure test.

Table 2. Concrete characteristics.

	C45/55 REF	C45/55 LCC1	C45/55 LCC2
Slump [mm]	240	210	210
Slump class [-]	S5	S4	S4
Compressive strength 3 days [MPa]	38.4	29.2	20.1
Compressive strength 28 days [MPa]	65.0	61.5	58.6
Compressive strength 56 days [MPa]	73.1	74.3	69.1

2.1. Binder Composition

Supplementary cementitious materials are powder additions used as a replacement for clinker (in some cases referred to as Portland cement) with the aim of reducing the carbon footprint of cement while still preserving sufficient cement or concrete properties. The SCM can be blended directly into cement during its production or later during concrete production. The origin of SCM is mostly from various industries' waste streams, which, with or without treatment, are changed into secondary raw materials that are beneficially used in cement or concrete production. The most widely used SCMs are fly ash (FA), ground-granulated blast furnace slag (SL), and silica fume (SF). More alternative types are calcined clay, ground glass, pumice, or rice husk ash. Nowadays, the trend is highly focused on existing and new SCMs because of the emphasis on the reduction of CO_2 emissions, the use of environmentally friendly materials, the circularity of materials through the reuse of waste from various industries, and the reduction of landfilling.

The most important property of SCM is the content of the amorphous (glassy) phase, which is reactive under certain conditions and contributes to the strength and other properties of the binder. Other properties influencing the reactivity of SCMs are fines, the shape of particles, chemical composition, replacement level, and the water/cementitious materials ratio. Another reason why it is essential to keep testing the potential of new materials that could become SCMs is due to the closure of thermal power plants that produce the by-product FA, which is presently the most common SCM on the market.

Reactivity is a factor that sorts SCMs into inert, latent hydraulic, or pozzolanic types based on their chemical reaction principle. For example, the reactivity of SCMs could be improved by the addition of a sulphate constituent (gypsum) and a carbonate constituent (lime). The reactivity of individual SCMs is possible to analyze using various test methods. There are tests specifically for FA and SL described in ASTM and EN, but they are based only on the comparison of strength properties to the reference mix. Other methods, such as the Chapelle test, the Frattini test, the strength activity index test, and the SCM dissolution test, have been introduced. Those methods are useful, but they are not able to distinguish whether the SCM is pozzolanic or latently hydraulically reactive (see Figure 1. Therefore, isothermal calorimetry and thermogravimetry analysis, or "rapid, relevant, and reliable" test methods, were introduced [3].

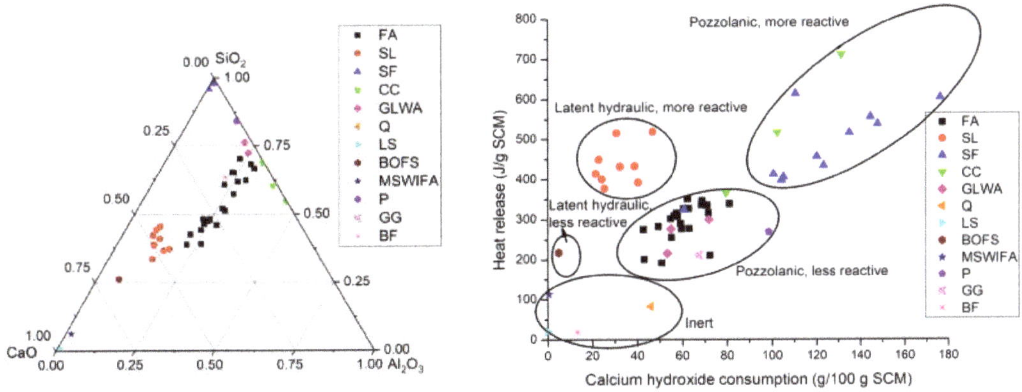

(a) (b)

Figure 1. Characteristic of various SCMs: (**a**) ternary phase diagram showing SiO2, Al2O3, and CaO amounts for various SCMs [4]; (**b**) results from isothermal calorimetry and thermogravimetry analysis on various types of SCMs and their classification. Key to SCM types: fly ash (FA), ground granulated blast furnace slag (SL), silica fume (SF), calcinated clay (CC), ground lightweight aggregates (GLWA), quartz (Q), ground limestone (LS), basic oxygen furnace slag (BOFS), municipal solid waste incineration fly ash (MSWIFA), ground pumice (P), ground glass (GG), and basalt fines (BF) [4].

There are two main standards for the utilization of SCMs: directly in cement, later during concrete production, or both to be followed in Europe. The amount of SCMs used as a replacement for clinker is regulated by standard EN 197-1, "Cement—Part 1: Composition, specifications, and conformity criteria for common cements" [5]; and standard EN 197-5:2021, "Cement Part 5: Portland-composite cement CEM II/C-M and composite cement CEM VI" [6]. Only EN197-1 [5] is mentioned in the standard for concrete, EN206, "Concrete—Specification, Performance, Production, and Conformity" [2], and therefore, the maximum replacement of clinker in concrete is approximately 35% (this can vary for individual countries by the national annex). If the SCMs are added during concrete production, EN206 [2] shall be followed, where in the section "5.2.5 Use of Additives", it is specified the maximal proportions of FA, SF, and SL to cement. The k-value concept is regulative, and if the amount of SCM in question exceeds the limit value, an excessive amount is considered an additive of type I (mineral filler).

All these regulations must be followed, so it is challenging for industry to abide by them and still produce concrete with a low carbon footprint. The demonstration series of mixes contains fly ash, silica fume, and limestone powder in various proportions to cement (see Table 3).

Table 3. Binder composition in kg and % per 1 m^3 of concrete.

	C45/55 REF		C45/55 LCC1		C45/55 LCC2	
	[kg]	[%]	[kg]	[%]	[kg]	[%]
CEM II/B-M 42,5R	455	96	190	68	210	70
Fly ash	0	0	0	0	48	16
Silica fume	20	4	30	11	42	14
Limestone powder	0	0	60	21	0	0
Total binder	475	100	280	100	300	100

The most common method for evaluating CO_2 connected to concrete produced in a given concrete plant is the EPD. This certificate allows concrete producers around Europe to compare their concrete in terms of carbon footprint. A detailed description of EPD is described in the following Section 3.

3. Environmental Product Declaration for Low Carbon Concrete

The environmental impact in terms of CO_2 associated with the production of low-carbon concrete is calculated in an EPD consisting of four major sections: CO_2 related to raw materials (A1) and their transportation (A2), concrete production (A3), and delivery of the final product (A4), concrete, to the construction site where it is used (A4). Stages A1, A2, and A3 belong to the product stage, and stage A4 is part of the construction installation stage [1]. The use of SCMs can help reduce CO_2, which is predominantly related to Stage A1. In general, limestone powder and SCMs such as fly ash or microsilica have a significantly lower CO_2 footprint connected to their production than Portland cement. The EPDs for the demonstration series were created in two different concrete plants in Northern Norway, Nordland Betong (NB) and HGB Betong AS (HGB). The difference in carbon footprint connected to individual mixes is related to the A2 stage, as all input materials are the same besides aggregates; see Table 4.

Table 4. EPD calculations for demonstration series in two different concrete productions.

	C45/55 REF		C45/55 LCC1		C45/55 LCC2	
Production	NB	HGB	NB	HGB	NB	HGB
EPD [kg CO_2—eqv. per 1 m^3 of concrete]	334	292	173	191	180	143

4. Areas for Utilisation of LCC in Northern Periphery and Arctics

Different countries in the Northern Periphery and the Arctic have different national regulations and recommendations for LCC, but all of them are aligned with European standards, particularly EN197-1 [5] and EN206 [2]. Individual concrete types are regulated by a classification system that sorts concretes in low-carbon concrete classes defined by kg CO_2—eqv. per 1 m^3 of concrete. LCC can be designed for all concrete strength classes, and the emitted CO_2 varies for each class. The exposure classes defined in EN 206 [2] further describe the properties of concrete, and if the LCC fulfills those, it can be used for any application. Nevertheless, based on extensive research, we know that LCC suffer in harsh conditions, and their use is even more limited than in countries with a warmer climate.

Recommendations and regulations for LCC classification for Norway, Sweden, Iceland, and Finland follow.

4.1. Norway

For Norway, the system is described in the Norwegian Concrete Association's publication NB 37, "Lavkarbonbetong" [7]. In this publication, it is clearly explained that the production of LCC is more feasible in the southern part of Norway than in the northern part, mainly due to input material transportation. Northern Norway has only one realistic supplier of cement: Norcem in Kjøpsvik. The nearest supplier apart from this is Norcem Brevik (near Oslo). Therefore, CO_2 footprint that concrete producers from Northern Norwegian use is 616 kg CO_2/ton compared to southern producers, which are using Norcem Brevik with 573 kg CO_2/ton [8]. Classification system in NB 37 sorts of concrete in 4 LCC classes and compare them to industrially produced concrete (Industry reference [7]). The following Table 5 presents a system for the classification of LCC in Norway. There are already discussions that The Norwegian Public Roads Administration will impose requirements for low carbon at B in the concrete structures.

Table 5. NB's No. 37 overview of greenhouse gas emissions to the low-carbon classes compiled with the industry reference [7].

Strength Class/Low Carbon Class	B20	B25	B30	B35	B45	B55	B65
	Maximum Permitted Greenhouse Gas Emissions [kg CO_2—eqv. per 1 m^3 of Concrete]						
Industry reference	240	260	280	330	360	370	380
Low carbon B	190	210	230	280	290	300	310
Low carbon A	170	180	200	210	220	230	240
Low carbon plus [2)]	-	-	150	160	170	180	190
Low carbon Extrem [2)]	-	-	110	120	130	140	150

[2)] Possible level for some projects, but with several restrictions in the standard work, and limited availability. Feasibility must be clarified in each individual project.

4.2. Sweden

The Swedish national climate target goal aims for LCC by 2045. The industry is working on strategies to reach these goals and thereby attain a lower climate impact. Now, some LCC are currently available, but generally, there is a lack of LCC on the Swedish market.

A very good overview of Swedish progress is given in a master's thesis by Strombom (2021) on a survey of climate-improved concrete [9]. The purpose of the work is to investigate how the concrete industry works towards climate neutrality and to see how far the actors have come in that work. The survey and review of the EPDs showed that some LCC already exists today, but that it is not used to such a large extent. Just over half of the concrete producers do not have any LCC on the market now, but most are working on developing it, and most expect to have it on the market within one to three years. Most concrete producers are confident that all of their concrete will be climate neutral by 2045. Because the Swedish classification system is quite extensive, the only category into which we classified our test series is shown in Table 6.

Table 6. Swedish concrete industry reference [10].

	Exposure Class	Industry Reference			Climate Improved Step 1		Climate Improved Step 2		Climate Improved Step 3	
		w/c Ratio	Cement	Max kg CO_2/m^3	Max kg CO_2/m^3	Improve from ref. %	Max kg CO_2/m^3	Improve from ref. %	Max kg CO_2/m^3	Improve from ref. %
5. House Outdoors, salt										
For example, car park, external stairs, loft passage, plinth, outer wall near coast.	XD3	0.40	470 kg CEM II **	340	305	10	255	25	205	40

** In the industry references for houses, a mix of 50% cement type CEM II/A-V and 50% cement type CEM II/A-LL has been used.

4.3. Iceland

Iceland has had its own system for low-carbon concrete qualification since 2009. At the ICI Rheocenter at IBRI and Reykjavík University, special carbon classes have been defined for concrete, taking into account their EPDs [11].

Such a classification can be useful for evaluating the environmental friendliness of concrete, but conventional, vibrated concrete was then estimated to release 400 kg CO_2/m^3. However, this value has decreased somewhat as the carbon footprint of cement has continuously decreased. According to the Icelandic classification, the original low carbon concrete classes (LCCC) are given in Table 7. It is now a relatively simple system, as when one specifies the concrete, the LCCC is just added after the strength and durability classification, ex. C25/30 FX2 LCC_{200}, meaning the carbon footprint of this concrete should not exceed 200 kg CO_2/m^3.

Table 7. Icelandic classification system for LCC [11].

Class	Marking	Limits
Industry reference	IR	400 kg CO_2/m^3
LKS [1] 300	Semi-LCC	≤300 kg CO_2/m^3
LKS 250	LCC_{250}	≤250 kg CO_2/m^3
LKS 200	LCC_{200}	≤200 kg CO_2/m^3
LKS 150	LCC_{150}	≤150 kg CO_2/m^3
EcoCrete	-	≤125 kg CO_2/m^3
EcoCrete-Xtreme	-	≤105 kg CO_2/m^3

[1] LKS is a seldom-used acronym for low-carbon footprint concrete; LCC is usually applied.

4.4. Finland

In Finland, concrete types are assigned to classes depending on the associated CO_2 emissions. The classification system covers a total of 16 different concrete types and a total of 5 different low-carbon classes [12]. The classification is assigned separately for each concrete mix and concrete plant. Concrete manufacturers can assign the concrete mix recipes of their choice to low-carbon classes, provided the emission value of the mix meets the classification requirements.

Low-carbon classes are indicated as GWP.NN, where GWP stands for global warming potential and NN indicates the emission level in comparison with the reference level. The reference level is given for each type of concrete as the average emission level of Finnish concrete manufacturers in 2021. Table 8 shows selected section of the Concrete Association of Finland (BY) low-carbon classes. The GWPtotal values include modules A1–A3. The values are given in kg (GWPtotal)/m^3 of concrete. As the table is rather extensive, sections relevant for the demonstration series are present in Table 8.

Table 8. Concrete types included in the BY low-carbon classification and limit values for low-carbon classes in Finland [12].

Concrete Quality	GWP.REF	GWP.85	GWP.70	GWP.55	GWP.40
C30/37—Air-entrained	290	245	205	160	115
C35/45—Air-entrained	330	280	230	180	130
C45/55—Air-entrained	375	320	265	205	150
C50/60—Air-entrained	395	335	275	215	160

5. Application of Classification Systems to Demonstration Series

The demonstration series of one reference mix and two LCC mixes was classified according to systems from all four countries: Norway (NO: NB37), Sweden (SE: SCI), Iceland (IS: LKS), and Finland (FI: LC class), see Table 9. Several benefits and limitations for all systems were identified. As the carbon footprint for the demonstration series was evaluated in two different locations, 6 EPDs were classified.

Table 9. Classification of demonstration series according to LCC guidelines of NO, SE, IS, and FI.

	C45/55 REF		C45/55 LCC1		C45/55 LCC2	
Production	NB	HGB	NB	HGB	NB	HGB
EPD [kg CO_2—eqv. per 1 m^3 of concrete]	334	292	173	191	180	143
NO:NB37	IR		A		A	Pluss
SE:SCI	IR	STEP1	STEP3		STEP3	
IS:LKS	IR	LKS 300	LKS 200		LKS 200	LKS 150
FI:LC class	GWP.REF	GWP.85	GWP.55		GWP.55	GWP.40

Prepared concrete mixes were air-entrained with a minimum cube strength of 55 MPa after 28 days. According to the individual systems, the demonstration series was classified as B45 in NO: NB37; in SE:SCI classified as 5. House Outside, salt (For example. Car park, external stairs, loft passage, plinth, an outer wall near the coast). The Icelandic system does not distinguish any strength or exposure classes and is only informative.According to Finish system, FI:LC, was demonstration series classified as C45/55—Air-entrained. Demonstration LCC improved in most cases by 2–3 classes, which shows very good environmental potential for newly designed concrete types.

In all systems besides NO:NB37, the REF mix was classified in different classes due to the 12.6% (42 kg CO_2—eqv. per 1 m^3 of concrete) difference based on the production plant (NB or HGB). According to NO:NB37, REF belongs to the "industry reference" class, as the limit for a subsequent class B is 290 kg CO_2—eqv. per 1 m^3 of concrete.

The C45/55 LCC1 produced in both plants fits in the same class in all systems. The difference between producers was 18 kg CO_2—eqv. per 1 m^3 of concrete, and the limits for all countries were between 220 and 200 kg CO_2—eqv. per 1 m^3 of concrete. The lower EPD for LCC1 produced in NB is mainly due to the transport method and distance for limestone powder.

Classification of C45/55 LCC2 was more variable as HGB managed to have EPD value below 150 kg CO_2—eqv. per 1 m^3 of concrete, which is a common level for IS: LKS and FI: LC class. In addition, in NO:NB37 system, LCC2 fit in lower class "Lavkarbon Pluss" with a limit 170 kg CO_2—eqv. per 1 m^3 of concrete. The Swedish system´s highest class has a limit of 205 kg CO_2—eqv. per 1 m^3 of concrete, with which both LCCs complied regardless of producer. Transportation method and distance are the main reasons for the low EPD value of LCC2 produced in HGB, as cement and fly ash are transported only 95 km by trucks compared to 200 km to NB by coastal barge.

6. Conclusions

The overall design of LCC in the demonstration series was successful, as both mixes reached the requested strength class despite lacking 9.8 and 5.3% of the REF mix. But the second intention to design concrete for freeze-thaw exposure was not successful, as the mixes exceeded the scaling limit level despite showing high chloride migration resistivity. Example: LCC mixes had a higher aggregate share, which allowed for cement reduction and only partial replacement by SCM. EPD is for two particular concrete productions located in Norway, and even with that, we can see differences in EPD values. Therefore, the same mix design produced in a different location will have a different EPD value/carbon footprint per 1 m^3 of concrete. The difference between the REF mix produced in two different concrete plants was 12.6%, according to the demonstration series used in this article. The LCC1 carbon footprint was reduced by 48.2 and 34.6%, and the carbon footprint of LCC2 was reduced by 46.1 and 51.0% in NB and HGB.

In all classification systems from Norway, Sweden, Iceland, and Finland, the REF mix complied with the criteria for industry reference, and when produced in HGB, it was even classified as the first level of LCC according to SE:SCI, IS:LCC_{xxx}, and FI:LC class. The LCC mixes moved by 2 to 3 classes, which points out that systems are comparable; only the Norwegian is more ambitious, with an extreme class of 130 kg CO_2—eqv. per 1 m^3 of concrete. In addition, the Icelandic system has classes down to a value <105 kg CO_2 per m^3 of "EcoCrete-Xtreme" and is mainly informative without direct comparison to industry reference. On the contrary, the Swedish system is well developed and clearly describes classes based on exposure classes, but the highest recognized class, "climate improved class 3", is rather conservative, and other classes with a lower carbon footprint might be essential.

Author Contributions: I.N.: conceptualization, validation, writing—original draft preparation, funding acquisition. P.P.: methodology, writing—review and editing. A.C.: methodology, writing—review and editing. O.H.W.: methodology, writing—review and editing. All authors have read and agreed to the published version of the manuscript.

Funding: This research was funded by Northern Periphery and Arctic programme 2021-2027, Project No. NPA0100039.

Data Availability Statement: No new data were created or analyzed in this study. Data sharing is not applicable to this article.

Acknowledgments: I would like to acknowledge the Interreg bridging call project ArCorD, within which the first draft of the study took place. The publication cost of this paper was covered by the funds of the Polish National Agency for Academic Exchange (NAWA): "MATBUD'2023—Developing international scientific cooperation in the field of building materials engineering" (BPI/WTP/2021/1/00002, MATBUD'2023).

Conflicts of Interest: The authors declare no conflict of interest.

References

1. Environmental Product Declaration. Available online: https://www.environdec.com (accessed on 5 July 2022).
2. *NS-EN 206:2013+A2:2021+NA:2022*; Concrete—Specification, Performance, Production and Conformity. Standard Norge: Oslo, Norway, 2022.
3. Avet, F.; Snellings, R.; Alujas, A.; Scrivener, K. Development of a new rapid, relevant and reliable (R^3) testing method to evaluate the pozzolanic reactivity of calcined clays. In *Calcined Clays for Sustainable Concrete: Proceedings of the 1st International Conference on Calcined Clays for Sustainable Concrete*; Springer: Dordrecht, The Netherlands, 2015; pp. 539–544. [CrossRef]
4. Suraneni, P.; Hajibabaee, A.; Ramanathan, S.; Wang, Y.; Weiss, J. New insights from reactivity testing of supplementary cementitious materials. *Cem. Concr. Comp.* **2019**, *103*, 331–338. [CrossRef]
5. *NS-EN 197-1:2011*; Cement—Part 1: Composition, Specifications and Conformity Criteria for Common Cements. Standard Norge: Oslo, Norway, 2011.
6. *NS-EN 197-5:2021*; Cement Part 5: Portland-Composite Cement CEM II/C-M and Composite Cement CEM VI. Standard Norge: Oslo, Norway, 2021.

7. 37 PDF Lavkarbonbetong (2020)—Norsk Betongforening. Available online: https://betong.net/nettbutikk/nb-publikasjoner/37-pdf-lavkarbonbetong-2015-gratis-nedlasting-klikk-les/ (accessed on 2 October 2022).
8. Norcem Standardsement FA | Norcem. Available online: https://www.norcem.no/en/StandardsementFA_eng (accessed on 3 December 2022).
9. BORÅS BYGGINGENJÖR HANNA STRÖMBOM, Högskoleingenjörsutbildningen. Kartläggning av klimatförbättrad betong [online]. 2021. Available online: http://urn.kb.se/resolve?urn=urn:nbn:se:hb:diva-26057 (accessed on 3 October 2022).
10. Klimatförbättrad Betong. Available online: https://xn--klimatfrbttradbetong-jzb70b.guide/ (accessed on 5 October 2022).
11. Wallevik, O.H.; Mansour, W.I.; Yazbeck, F.H.; Kristjansson, T.I. Ecocrete-Xtreme: Extreme Performance Of A Sustainable Concrete. In Proceedings of the International Symposium on Environmentally Friendly Concrete Ecocrete, Reykjavik, Iceland, 13–15 August 2014; pp. 3–10.
12. PUNKKI, Jouni a Mirva VUORI. Low-Carbon Classification of Concrete in Finland. Available online: www.vahahiilinenbetoni.fi (accessed on 10 October 2022).

materials proceedings

MDPI

Proceeding Paper

Effect of Low-Quality Calcined Clay on the Suppression of the Alkali–Silica Reaction †

Daria Jóźwiak-Niedźwiedzka [1,*], Roman Jaskulski [2], Kinga Dziedzic [1] and Aneta Antolik [1]

1 Institute of Fundamental Technological Research, Polish Academy of Sciences, A. Pawińskiego 5b, 02-106 Warsaw, Poland

2 Department of Building, Wrocław University of Environmental and Life Sciences, C. K. Norwida 25, 50-375 Wrocław, Poland

* Correspondence: djozwiak@ippt.pan.pl; Tel.: +48-826-12-81 (ext. 217)

† Presented at the 10th MATBUD'2023 Scientific-Technical Conference "Building Materials Engineering and Innovative Sustainable Materials", Cracow, Poland, 19–21 April 2023.

Abstract: This article presents the results of an experimental investigation into the mitigation of the alkali–silica reaction (ASR) resulting from using low-grade clay calcined at 850 °C. The clay used in the experiment was domestic clay with an Al_2O_3 content equal to 26% and a SiO_2 content of 58%. The performance of calcined clay in ASR mitigation was evaluated according to ASTM C1567 using reactive aggregates. The control mortar mixture consisted of 100% Portland cement (Na_2O_{eq} = 1.12%) binder and reactive aggregate. The test mixtures used the same reactive aggregate and binders, in which part of the cement was replaced with either 10%, 20% or 30% calcined clay. The microstructure of specimens was examined on the polished sections using a scanning electron microscope (SEM) operated in the backscattered mode (BSE). The results of expansion obtained from the mortar bars made with the reactive aggregate showed that replacing cement by calcined clay reduced their expansion, with the level of expansion decreasing with the increase in the level of cement replacement.

Keywords: calcined clay; alkali–silica reaction; expansion; mitigation; microstructure

check for updates

Citation: Jóźwiak-Niedźwiedzka, D.; Jaskulski, R.; Dziedzic, K.; Antolik, A. Effect of Low-Quality Calcined Clay on the Suppression of the Alkali–Silica Reaction. *Mater. Proc.* **2023**, *13*, 15. https://doi.org/10.3390/materproc2023013015

Academic Editors: Katarzyna Mróz, Tomasz Tracz, Tomasz Zdeb and Izabela Hager

Published: 14 February 2023

1. Introduction

Nowadays, the continuous development of the cement-based materials industry, fueled by unflagging demand, is a cause for concern due to the large amount of CO_2 and NO_x emitted during the production process. One of the most effective ways to reduce the negative environmental impact of the cement industry is the use of so-called supplementary cementitious materials (SCMs). The most commonly used SCMs are fly ash and ground-granulated blast furnace slag. However, their availability will continue to decline, thus increasing the need for alternative sources of SCMs. One of the most promising alternative sources are calcined clays, which are abundant and widely available [1–3]. Among the most well-known material to be used as a calcined clay in cement and concrete technology is metakaolin. Metakaolin is a pozzolanic material obtained by calcination of kaolinitic clay, and it has been shown to considerably enhance the mechanical and chemical properties of concrete [1]. Calcined clays other than metakaolin are rarely used as SCMs due to the complex composition of clay minerals, such as the calcite impurities in clays containing kaolinite and insufficient knowledge of the basic reaction mechanisms [2–4]. Additionally, although there is an increasing interest in demonstrating the effectiveness of various treatments that can lead to improvements in the reactivity of clays with respect to their usage as an SCM [3], issues such as the required quality control procedures and the acceptance criteria also need to be taken into account [1].

One of the possible applications of calcined clays, making use of their composition and pozzolanic properties, is as an additive to cement or concrete with the objective of reducing the expansion induced by an alkali silica reaction [5,6]. However, previous studies have mostly focused on the application of metakaolin [7–9], thus overlooking the fact that the efficacy of calcined clays in mitigating ASR can vary significantly due to differences in quality and sources [5]. As a result, the information on the efficacy of ASR mitigation of calcined clays obtained from various sources and of different quality is still limited [6].

The research presented in this paper is focused on the effectiveness of locally available low-quality calcined clay with respect to its ability to control the expansion due to the alkali–silica reaction. The scope of experimental work includes a characterization of the raw and calcined clay, an assessment of the effectiveness of the calcined clay in mitigation according to ASTM C1567 [10] and a microstructural analysis. Although a different clay was tested than that used in [6], similar assumptions were made. One calcination temperature (850 °C) was used because the highest pozzolanic reactivity was reached when the clay was calcined at 800 °C [11,12], and the strength performance was maximized for raw clays calcined at temperatures up to 850 °C [3]. In the conducted research, the main factor, which is the replacement level, was taken into consideration.

2. Materials

The material used in the research was a domestic clay calcined at 850 °C with the loss on ignition equal to 9.20%. The material was classified as a low-quality clay due to the fact that it only contains 26% Al_2O_3 [2]. The main associated mineral, except for kaolinite and muscovite, found in the raw clay by XRD was quartz, Figure 1. A previously analyzed aggregate from igneous rocks with medium reactivity (aggregate reactivity class R1 according to [13,14]) was selected for the reference aggregate. Ordinary Portland cement (CEM I 42.5R) with a high alkali content (Na_2O_{eq} = 1.12%) was used in this research. A detailed chemical composition of the raw clay and cement determined by XRF is shown in Table 1.

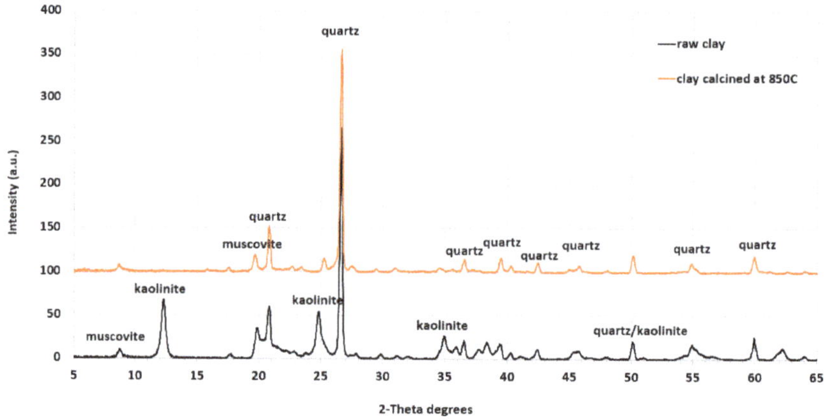

Figure 1. Mineral composition of the raw and calcined clay.

Table 1. Chemical composition of raw clay and Portland cement CEM I [15] by XRF (wt%).

Constituent	Raw Clay	CEM I
	Content %	
SiO_2	58.23	19.43
TiO_2	2.04	0.27
Al_2O_3	26.18	4.84
Fe_2O_3	1.53	3.18
Mn_2O_3	0.03	0.06
MgO	0.27	2.56
CaO	0.22	61.81
Na_2O	0.13	0.41
K_2O	1.86	1.08
P_2O_5	0.06	0.11
SO_3	<0.01	3.93
Cl	0.01	0.03
F	0.03	-

3. Methods

Clay calcination was carried out in a laboratory chamber furnace with a heating capacity of 8.8 kW. Prior to loading into the furnace, the clay was dried to a constant weight at 110 °C, then ground and sieved through a 0.125 mm mesh. The heating ramp was 4 °C/min until a temperature of about 150 °C was reached, and 12 °C/min after this threshold was exceeded. After a temperature 850 °C was reached, it was maintained for 60 min, after which the material was allowed to cool in the oven to a temperature below 50 °C.

X-ray diffraction was used to estimate the main minerals in the investigated clays. XRD diffractograms were obtained with a diffractometer (Cu Kα radiation source) equipped with a Göbel mirror and a GADDS 2D detector system. The operation parameters of the equipment were 40 kV and 40 mA. Diffraction patterns were collected over a 2θ range from 5° to 65° with a 1°/min step using flat plane geometry.

Evaluation of the calcined clay efficiency in ASR mitigation was performed following the ASTM C1576 standard specification [10]. The test involved exposing standard mortar bars to a 1 M NaOH solution at a temperature of 80 °C. The test was carried out on reference mortar and on mortars with 10%, 20% and 30% of cement replacement by calcined clay. One set of test specimens consisted of three mortar bars of 25 × 25 × 285 mm. After 24 h, the specimens were disassembled and stored in distilled water at 80 ± 2 °C for the next 24 h. Then, the initial length reading was taken for each of the bars and then specimens were placed in 1 M NaOH at 80 ± 2 °C. Systematic measurements of the expansion of the mortar bars were carried out at least 4 times within 14 days. An expansion of less than 0.10% after 14 days of testing in 1 M NaOH at 80 °C indicated acceptable ASR performance behavior; however, the test was extended to 28 days for expansion curve analysis.

A post-mortem evaluation of the microstructure of the mortars was conducted using a scanning electron microscope (SEM) operated in the backscattered electrons (BES) mode and equipped with an energy dispersion X-ray (EDX) detector. The beam specimens, previously used for monitoring the expansion, were cut to the dimensions of 45 × 30 × 15 mm, dried at 50 °C for 3 days, vacuum-impregnated with a low-viscosity epoxy resin and lapped and polished using a special procedure for SEM specimens. The specimens were coated with a carbon layer (~20 nm) and a strip of conductive tape was attached to each specimen to improve the conduction properties. Each of the specimens was examined using a JEOL JSM-6380 LA SEM-EDX with an acceleration voltage of 20 kV and a working distance of 8–10 mm. More than 50 EDX point analyses were collected to assess the composition of the ASR products in the mortar during post-mortem analysis.

4. Results and Discussion

The accelerated mortar bar test results are presented in Figure 2. For the first 14 days, for all specimens the mortar bar expansion increased with the test duration. The fastest and the most extensive expansion development was found for the reference mortar and for the mortar containing 10% calcined clay. The expansion of specimens with 10% calcined clay exceeded the 0.1% limit after just 9 days of exposure and reached values higher than 0.15% after 14 days. Mortar bars with 20% and 30% calcined clay showed significantly lower expansions. The elongations were similar (below 0.06%) after 14 days of testing. It is worth noting that even after 28 days of testing, specimens with 20% and 30% calcined clay showed an expansion of less than 0.1%. Moreover, the slowing of the expansion is clearly visible in the expansion curve. The slowing phenomenon is more pronounced for 20% calcined clay specimens when compared to the specimens with 30% clay replacement. It is also worth noting that this effect is weaker for the 30% calcined clay content, suggesting that a 20% replacement rate is a slightly better solution.

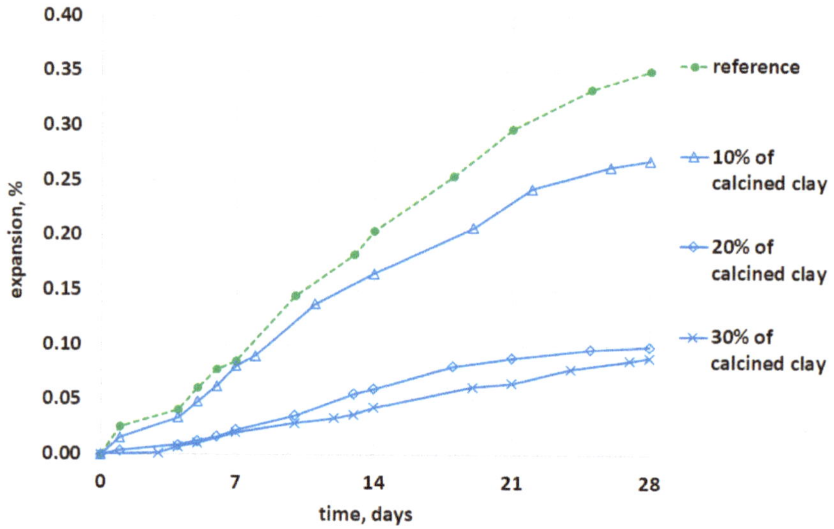

Figure 2. Expansion of mortar specimens with various content of calcined clay as a function of exposure time to 1 M of NaOH solution and a temperature of 80 °C.

A microscopic analysis confirmed the results of the mortar bar expansion test according to ASTM C1567 [10]. The observed cracks in the aggregate and in the cement matrix were caused by an alkali–silica reaction, Figures 3 and 4. The degree of damage and the size and content of cracks in the aggregate decreased with an increase in the content of calcined clay. The same trend was also observed in the amount of ASR of the gel in the air voids, Figure 4. The ASR gel layer in the air voids in the reference mortar was approx. 35 μm, while in the mortar with 20% calcined clay it was half of that. There were also differences in the morphology and composition of the ASR gel. Based on the SEM observations, it was found that in the reference specimens, the ASR gel was largely amorphous and was characterized by a homogeneous composition of Si-Ca-Na-K (Figure 4a), typical for this kind of product [16]. Regardless of the location, the contents of Si, Ca and alkali ions in the gel were similar, while the ASR gel visible in the specimens containing calcined clay seemed to consist of two layers, outer and inner, Figure 4b. The inner (thinner) part of the ASR gel was composed of Si, Ca, Na and K, i.e., a typical composition. However, the outer part of the ASR gel, apart from being significantly cracked, also contained Mg and Al in its composition, Figure 4b. Additionally, the content of K ions was lower compared to the composition of the gel in the inner part.

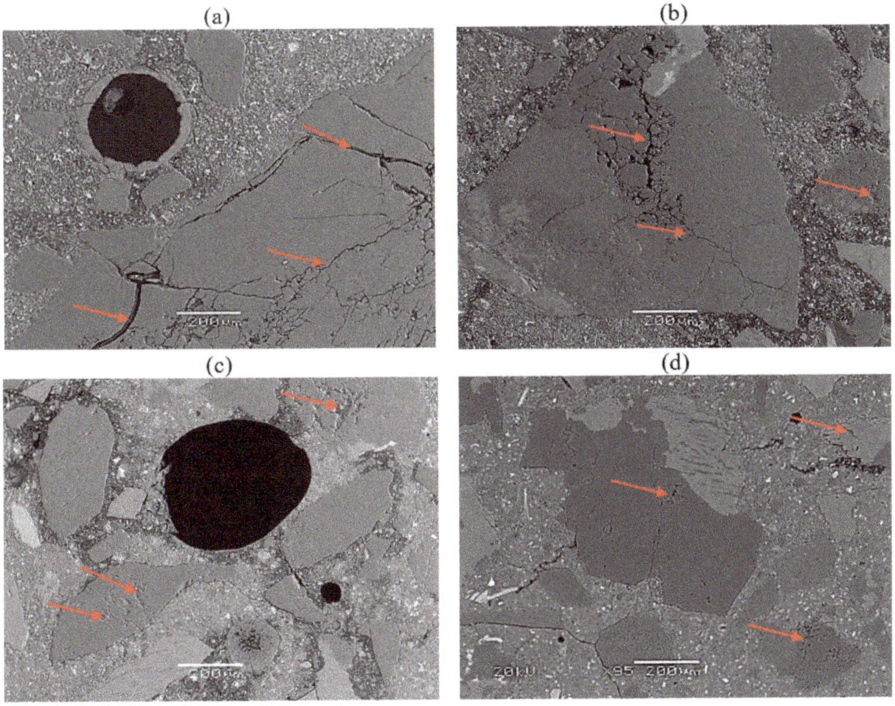

Figure 3. Dependency of the degree of cracking of the aggregate on the replacement level of cement with calcined clay: (**a**) 0%, (**b**) 10%, (**c**) 20% and (**d**) 30%; SEM-BES, scale bar = 200 μm.

The dependence of the expansion of the mortar bars on the content of calcined clay as a substitute for mass cement is presented in Figure 5. The obtained results are in line with expectations, assuming that the calcined clay exhibits pozzolanic properties as a supplementary cementitious material [2,4]. A similar dependence was demonstrated when using high-quality calcined clay, i.e., metakaolin [5,6].

The significantly different composition of the ASR gel in the mortar containing calcined clay is a new finding. In previous studies [6], no significant differences were found in the composition of the ASR gel between the two mixtures, with or without calcined clay. In the conducted test, a relationship between the Si/Al ratio in the ASR gel localized in air voids and the content of calcined clay, as well as a relationship between the mortar bars expansion and the Si/Al ratio in the ASR gel, were found, Figures 5 and 6.

An analysis was performed on the ASR gel in the air voids to determine the effect of the calcined clay and avoid the effect of aggregates. The differences in the ASR gel may be due to the presence of aluminum in calcined clay, which can influence the alkali binding ability. Although it is not presented in the article, the effectiveness of suppressing the alkali–silica reaction by low-quality calcined clay may be influenced by its pozzolanic reactivity, which may primarily affect the alkali binding capacity.

The next step of the research will be focused on the detailed characterization of domestic calcined clay and the properties of the obtained ASR gel.

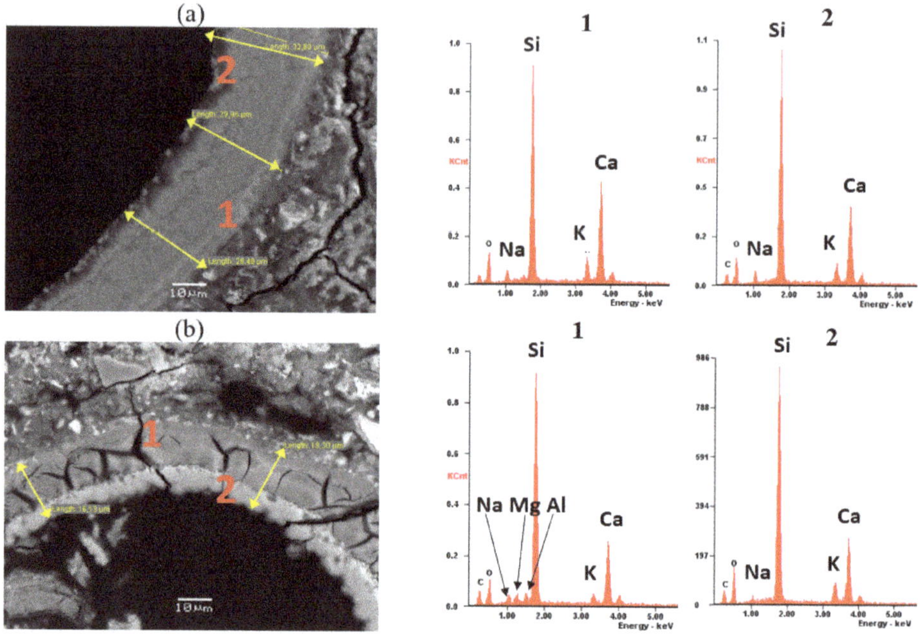

Figure 4. Microstructure and chemical composition of the ASR gel in air voids in specimens: (**a**) reference, 0% calcined clay and (**b**) 20% calcined clay, SEM-EDS, scale bar = 10 μm.

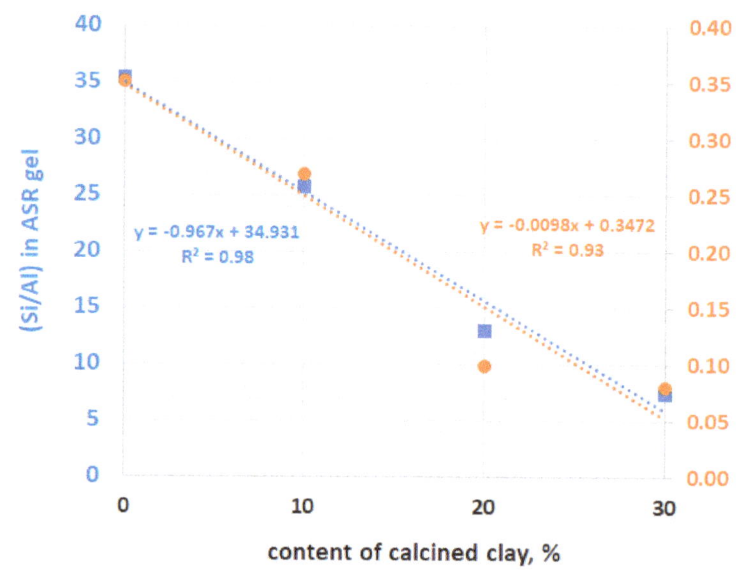

Figure 5. The relationship between the ASTM C1567 expansion and the Si/Al ratio in the ASR gel in relation to the content of the calcined clay cement substitute.

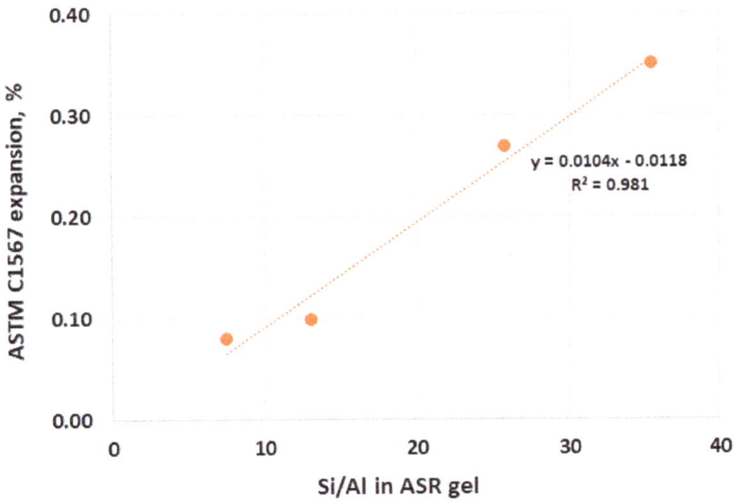

Figure 6. The relationship between the ASTM C1567 expansion and the Si/Al ratio in the ASR gel.

5. Conclusions

Based on the experimental analyses performed on mortars containing reactive aggregates and various contents of low-quality calcined clay, the following conclusions can be drawn:

- The effect of low-grade calcined clay on the ASR expansion was found.
- The results indicated a reduced alkali silica expansion in mortar containing low-grade calcined clay regardless of the percentage substitution with cement.
- At a lower percentage (10%) of calcined clay, the expansion results did not satisfy ASTM standard limits.
- Replacing 20% of the cement with calcined clay appeared to be sufficient to mitigate the alkali–silica reaction.
- The morphology and chemical composition, as well as the width of the ASR gel in air voids, was different depending on the calcined clay content.

The substitution method used in ASR-resistant composites results in less CO_2 being released into the atmosphere, making the material more sustainable. Future research will therefore focus on the study of eco-friendly concretes that use locally available aggregates as well as locally available, low-quality calcined clay as a partial substitute for Portland cement.

Author Contributions: Conceptualization, D.J.-N. and R.J.; methodology, D.J.-N., K.D. and A.A.; validation, D.J.-N., R.J., K.D. and A.A.; investigation, D.J.-N., K.D. and A.A.; resources, D.J.-N. and R.J.; writing—original draft preparation, D.J.-N. and R.J.; writing—review and editing, D.J.-N., R.J., K.D. and A.A.; All authors have read and agreed to the published version of the manuscript.

Funding: This research received no external funding.

Institutional Review Board Statement: Not applicable.

Informed Consent Statement: Not applicable.

Data Availability Statement: Not applicable.

Acknowledgments: The publication cost of this paper was covered with funds from the Polish National Agency for Academic Exchange (NAWA): "MATBUD'2023—Developing international scientific cooperation in the field of building materials engineering" BPI/WTP/2021/1/00002, MATBUD'2023.

Conflicts of Interest: The authors declare no conflict of interest.

References

1. Okashah, A.M.; Zainal, F.F.; Hayazi, N.F.; Nordin, M.N.; Abdullah, A. Pozzolanic properties of calcined clay in geopolymer concrete: A review. *AIP Conf. Proc.* **2021**, *2339*, 020234. [CrossRef]
2. Jaskulski, R.; Jóźwiak-Niedźwiedzka, D.; Yakymechko, Y. Calcined clay as supplementary cementitious material. *Materials* **2020**, *13*, 4734. [CrossRef] [PubMed]
3. Yanguatin, H.; Tobón, J.; Ramírez, J. Pozzolanic reactivity of kaolin clays, a review. *Rev. Ing. Constr.* **2017**, *32*, 13–24. [CrossRef]
4. Zunino, F.; Boehm-Courjault, E.; Scrivener, K. The impact of calcite impurities in clays containing kaolinite on their reactivity in cement after calcination. *Mater. Struct.* **2020**, *53*, 44. [CrossRef]
5. Li, S.C.; Ideker, J.H.; Drimalas, T.; Adams, M.P. Review on using calcined clays to mitigate alkali-silica reaction. In Proceedings of the 15th International Conference on Alkali-Aggregate Reaction in Concrete (ICAAR), São Paulo, Brazil, 3–7 July 2016.
6. Li, C.; Ideker, J.H.; Drimalas, T. The Efficacy of Calcined Clays on Mitigating Alakli-Silica Reaction (ASR) in Mortar and Its Influence on Microstructure. In *Calcined Clays for Sustainable Concrete*; Scrivener, K., Favier, A., Eds.; RILEM Bookseries; Springer: Dordrecht, The Netherlands, 2015; Volume 10. [CrossRef]
7. Walters, G.V.; Jones, T.R. Effect of metakaolin on alkali-silica reaction (ASR) in concrete manufactured with reactive aggregate, durability of concrete. In Proceedings of the 2nd International Conference, Montreal, QC, Canada, 4–9 August 1991; Volume II, pp. 941–953.
8. Ramlochan, T.; Thomas, M.; Gruber, K.A. The effect of metakaolin on alkali-silica reaction in concrete. *Cem. Concr. Res.* **2000**, *30*, 339–344. [CrossRef]
9. Gruber, K.; Ramlochan, T.; Boddy, A.; Hooton, R.; Thomas, M. Increasing concrete durability with high-reactivity metakaolin. *Cem. Concr. Compos.* **2001**, *23*, 479–484. [CrossRef]
10. *ASTM C1567-21*; Standard Test Method for Determining the Potential Alkali-Silica Reactivity of Combinations of Cementitious Materials and Aggregate (Accelerated Mortar-Bar Method). Book of Standards. ASTM International: West Conshohocken, PA, USA, 2021; Volume 04.02, p. 6. [CrossRef]
11. Alujas, A.; Fernandez, R.; Quintana, R.; Scrivener, K.; Martirena, F. Pozzolanic reactivity of low grade kaolinitic clays: Influence of calcination temperature and impact of calcination products on OPC hydration. *Appl. Clay Sci.* **2015**, *108*, 94–101. [CrossRef]
12. Ferreiro, S.; Canut, M.M.C.; Lund, J.; Herfort, D. Influence of fineness of raw clay and calcination temperature on the performance of calcined clay-limestone blended cements. *Appl. Clay Sci.* **2019**, *169*, 81–90. [CrossRef]
13. *AASHTO R80-17*; Standard Practice for Determining the Reactivity of Concrete Aggregates and Selecting Appropriate Measures for Preventing Deleterious Expansion in New Concrete Construction. American Association of State and Highway Transportation Officials: Washington, DC, USA, 2017; p. 25.
14. Garbacik, A.; Glinicki, M.A.; Jóźwiak-Niedźwiedzka, D.; Adamski, G.; Gibas, K. Technical Guidelines for the Classification of Domestic Aggregates and Prevention of the Alkali-Aggregate Reaction in Concrete Used in Road Pavements and Road Engineering Facilities, ICiMB and IPPT PAN, Kraków-Warszawa, 2019. Available online: https://www.gddkia.gov.pl/pl/1118/dokumenty-techniczne (accessed on 12 September 2022). (In Polish)
15. Glinicki, M.A.; Jóźwiak-Niedźwiedzka, D.; Antolik, A.; Dziedzic, K.; Gibas, K. Susceptibility of selected aggregates from sedimentary rocks to alkali-aggregate reaction. *Roads Bridges Drog. Mosty* **2019**, *18*, 5–24. [CrossRef]
16. Antolik, A.; Jóźwiak-Niedźwiedzka, D. Assessment of the alkali-silica reactivity potential in granitic rocks. *Constr. Build. Mater.* **2021**, *295*, 123690. [CrossRef]

materials proceedings

Proceeding Paper

High-Strength Concrete Using Ash and Slag Cements †

Leonid Dvorkin *, Vadim Zhitkovsky, Vitaliy Marchuk and Ruslan Makarenko

Institute of Civil Engineering and Architecture, National University of Water and Environmental Engineering, 33028 Rivne, Ukraine
* Correspondence: l.i.dvorkin@nuwm.edu.ua; Tel.: +380-68-353-33-38

† Presented at the 10th MATBUD'2023 Scientific-Technical Conference "Building Materials Engineering and Innovative Sustainable Materials", Cracow, Poland, 19–21 April 2023.

Abstract: The article discusses the possibility of improving the properties of concrete made with composite ash-containing cement (CC) by introducing complex chemical additives. The pozzolanic activity of CC and its degree of hydration increases with the increasing dispersion of cement resulting from the introduction of a complex additive in the form of a polyfunctional modifier (PFM) during grinding, including a grinding intensifier-propylene glycol and a superplasticizer. The analysis of mathematical models of water demand and strength of concrete made with the PFM additive showed that a significant reduction in water demand and an increase in the strength of concrete based on the ash-containing CC is possible with the introduction of the PFM additive. For concrete based on CC with the addition of PFM, it has been demonstrated that it is possible to design concrete mixes on the basis of the obtained models. Relevant examples are provided.

Keywords: high-strength concrete; composite cement; fly ash; slag; multifunctional modifier; planning of experiments

Citation: Dvorkin, L.; Zhitkovsky, V.; Marchuk, V.; Makarenko, R. High-Strength Concrete Using Ash and Slag Cements. *Mater. Proc.* **2023**, *13*, 16. https://doi.org/10.3390/materproc2023013016

Academic Editors: Katarzyna Mróz, Tomasz Tracz, Tomasz Zdeb and Izabela Hager

Published: 14 February 2023

1. Introduction

As is known, high-strength concrete currently includes concrete with a compressive strength of 60–100 MPa and higher at the age of 28 days [1,2]. With the use of classic technology, obtaining high-strength concrete is possible by ensuring the high quality of raw materials, low values of the water-cement ratio, and sufficient compaction of the concrete mixture. In practice, these requirements are achieved by reducing the water content of the concrete mixture when using stiff mixtures, or by introducing plasticizing additives, increasing the activity and specific surface of cement, introducing hardening accelerators, reducing water consumption, and optimizing the grain composition of aggregates with high physical and mechanical properties [1–6].

The technology of high-strength concrete of the new generation involves the introduction of superplasticizer additives with a high (20–40%) water-reducing effect and dispersed active mineral fillers into the composition of concrete mixtures. These components provide both important individual effects, namely achieving extremely low values of the water-cement ratio while maintaining ease of installation, an increase in the volume of hydration products, the degree of cement hydration, and a significant synergistic effect. With the simultaneous introduction of superplasticizer additives and a dispersed active filler, the rheological potential of the plasticizing additive is fully realized on the one hand, whereas on the other hand, in the conditions of a limited water-cement ratio, the positive effect of the dispersed active filler on the structure formation of cement stone and concrete is manifested [7–10].

In hydration systems, the physical interaction of cement particles and the filler is significantly affected by the so-called "compressed conditions" [11], which are characterized by a sharp increase in the concentration of the solid phase and the transition of part of the volume water into film water. At the same time, the change in free surface energy between the solid and liquid phases contributes to the intensive formation of crystal nuclei of hydration products, which is in accordance with the Gibbs–Volmer theory [12]. With the optimal concentration and dispersion of the active additive, a fine-grained structure of the binder is formed, which has a positive effect on the technical properties of the artificial stone.

The structuring effect of dispersed additives in concrete increases when the physico-chemical activity of their surface increases. Methods of mechanical activation of additives include grinding them with cement in the presence of plasticizers and, if necessary, other chemical admixtures. This method of activation is the basis for obtaining composite cement with low water consumption [13].

An important indicator of the quality of concrete as a structural material is the specific consumption of cement per unit of concrete strength. In traditional concretes, this indicator is on average close to 10 kg/MPa, in new generation concretes it decreases two or more times [7–9].

The best active fillers for high-strength concrete, as demonstrated by many studies and by practice, are microsilica additives [14,15]. At the same time, the use of microsilica is complicated due to the need for its granulation or briquetting for transportation and dosing, as well as to significant fluctuations in its composition and properties.

As regards active fillers for high-strength concrete, the use of dispersed siliceous and aluminosilicous materials based on local raw materials and industrial waste is a practical interest [14,16]. Among such materials, the most common is coal fly ash and granulated blast furnace slag. The main task of our work was to figure out the technological parameters for obtaining high-strength concrete using composite ash and slag-containing cement and to research their complex structure and technical properties.

2. Materials and Methods

The main studies were performed using (EN 197-1):

Composite CEM V/A with a fly ash content of 27.4%, blast furnace slag—18.3%, clinker—45.7%, and gypsum—8.6%;

Portland cement clinker with mineralogical composition was used in the research: C_3S—63.95%; C_2S—15.15%; C_3A—7.42%, C_4AF—12.48%.

Fly ash and blast furnace granulated slag have the following chemical composition:

- fly ash: (SiO_2 + Al_2O_3 + Fe_2O_3)—85.8%, SO_3—2.3%, CaO_{free}—2.8%, MgO—2%, Na_2O + K_2O—1.2%, LOI—5.1%;
- blast furnace slag: SiO_2—39.5%, Al_2O_3—6.4%, Fe_2O_3—0.2%, CaO—47.2%, MgO—3.1%, MnO—1.1%, SO_3—1.7%.

The Samples of composite cement for research were obtained by joint grinding of the components in the laboratory ball mill. To regulate the properties of the composite cement and the concretes made with the use thereof, the influence of naphthalene formaldehyde (SP-1) and polycarboxylate (Sika VS 225) superplasticizers [17] and grinding intensifier-propylene glycol (PG) [18] was used.

For mineral additives, the pozzolanic activity was determined by the CaO uptake method [19,20]. Additives (ash, slag, and their mixture) before the test were milled to the required specific surface area (Sssa) (Table 1).

Table 1. Pozzolanic activity of fly ash and ash of slag composition.

Material	Sssa, m^2/kg	Absorption of CaO mg/g		
		7 Days	28 Days	60 Days
Fly ash	350	15	52	78
Fly ash	450	18	65	97
Fly ash	550	25	90	135
Fly ash + slag (1:1)	350	20	70	83
Fly ash + slag (2:1)	350	17	63	81
Fly ash + slag (1:1)	450	21	83	105
Fly ash + slag (2:1)	450	19	73	101

For the obtained composite cements, the grain composition (Table 2) was determined by the sedimentation method [21]. Also, the normal consistency, the change in the degree of hydration of the cement paste [22], and the standard strength of cement-sand samples over time were determined (Table 3) [23].

Table 2. Composite cements grain composition.

No.	Additives	Content of Fractions, %				
		<10 μm	10–20 μm	20–40 μm	40–60 μm	>60 μm
1	PG—0.04%, Sika VC 225—0.5%;	35.5	33.1	15.5	12.4	3.3
2	PG—0.04%	31.2	36.4	14.2	14.6	3.6
3	PG—0.02%, Sika VC 225—0.5%;	28.2	36.1	16.7	15.1	3.9
4	PG—0.02 %	26.5	33.7	18.4	17.2	4.2
5	PG—0.04%, SP-1—0.5%	29.8	35.5	14.3	15.8	4.6
6	SP-1—0.5%	22.8	35.1	19.6	17.3	5.4
7	Sika VC 225—0.5%	17.4	36.6	21.8	17.5	6.7
8	Without additives	15.6	35.5	22.5	18.2	8.2

Table 3. The main properties of composite cement (CC).

No.	Specific Surface Area S_{ssa}, m^2/kg	Additive PFM, %	Normal Consistency, %	Compressive/Bending Strength, MPa in Age, Days			
				1	3	7	28
1	350	without additives	27.8	15.5/2.5	22.4/3.2	31.6/4.1	41.5/5.8
2	450	-//-	28.3	19.3/3.1	25.5/3.6	39.7/4.8	52.3/6.2
3	450	PFM_1 (PG—0.04%, Sika VC 225—0.5%)	18.5	24.7/4.1	32.3/4.8	45.8/5.6	61.5/6.7
4	550	-//-	19.7	30.3/4.3	39.6/4.9	50.6/6.1	71.8/7.5
5	450	PFM_2 (PG—0.04%, SP-1-0.5%)	21.5	21.8/3.6	28.3/4.1	37.4/4.6	57.8/6.8
6	550	-//-	22.8	25.2/3.9	31.3/4.2	41.2/5.6	61.3/7.2

To analyze the influence of the composition of binders and concrete, as well as modifier additives under normal conditions of hardening and when subjected to heat treatment, experimental-statistical models were obtained using the method of mathematical planning [24–26]. When studying concrete mixtures and concretes, the three-level B4 plan was implemented [26]. According to plan B4, 24 series of experiments were performed. In each series, the water demand was determined until the specified slump was reached, and 6 concrete cubes 100 × 100 × 100 mm were made to determine the compressive strength [27]. To obtain concrete mixtures, quartz sand with fineness modulus M_f = 1.95

and crushed stone with D_{max} = 20 mm were used. Samples were tested after 1 and 28 days. The conditions for planning the experiments are given in Table 4. After the implementation and statistical processing of experiments, mathematical models of the water consumption and the concrete compressive strength were obtained.

Table 4. Conditions for planning experiments when obtaining models of the concrete mixture water consumption and the concrete at the CC strength.

No.	Factors		Levels of Variation			Interval
	Natural	Coded	−1	0	+1	
1	The content of PFM_1 additive in CC, %	X_1	0.4	0.7	1.0	0.3
2	Specific surface area of CC, S_{ssa}, m^2/kg	X_2	350	450	550	100
3	Water-cement ratio, *W/C*	X_3	0.25	0.35	0.45	0.1
4	Slump, Sl, cm	X_4	2	13	24	11

3. Results and Discussion

Experiments on laboratory ground cement showed that the pozzolanic activity of the ash-slag composition in the cement most significantly depends on the fineness of the grind, characterized by the specific surface area (S_{ssa}), and, to a lesser extent, on the ash:slag ratio (Table 1). However, decreasing the ratio of ash to slag significantly affects the kinetics of CaO bonding, increasing the amount of bound CaO, especially in the period from 7 to 28 days.

Achieving an increased composite cement's specific surface area while minimizing energy consumption is possible with the use of grinding intensifier additives [28,29]. The propylene glycol (PG) has become widespread as such an additive. Along with the propylene glycol, the addition of superplasticizers and their compositions with PG have a certain influence on the grinding kinetics and grain composition of cement. The experiment results are shown in Figure 1 and Table 2.

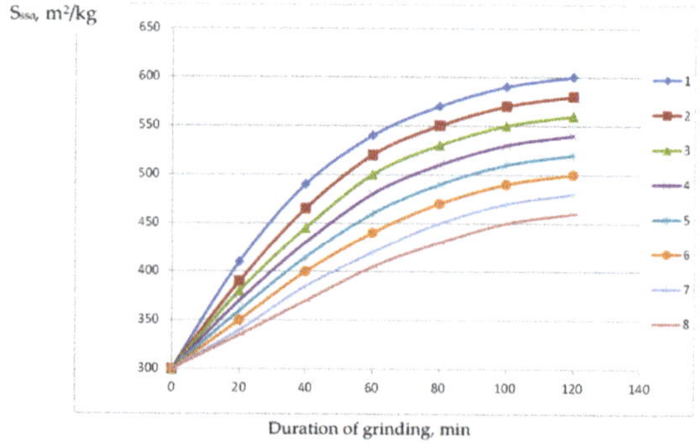

Figure 1. The influence of additives on the specific surface area of CC at different durations of grinding. 1—PG—0.04%, Sika VC 225—0.5%; 2—PG—0.04%; 3—PG—0.02%, Sika VC 225—0.5%; 4—PG—0.02%; 5—SP-1—0.5%, PG—0.04%; 6—SP-1—0.5%; 7—Sika VC 225—0.5%; 8—without additives.

The addition of PG practically doubled the content of the cement's finest fraction. The composite additive PG + Sika VC 225 provides a grain composition of cement that is slightly different from the grain composition of cement with a single PG additive. The naphthalene-formaldehyde superplasticizer introduction into the cement also to some extent intensifies the grinding of the cement. This effect is significantly lower than the propylene glycol effect.

Enrichment of composite cement during grinding with the finest fractions significantly accelerates its hydration. This is evidenced by the results of experiments on determining the content of hydrated water in samples of cement stone with $W/C = 0.3$ at different durations of their hardening (Figure 2).

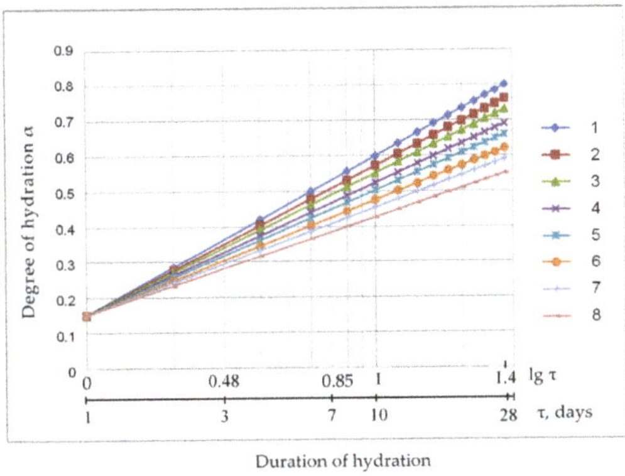

Figure 2. The influence of the hydration duration on the degree of hydration of composite cements (α) (line numbering is given for cements according to Figure 1).

The hydration kinetics of composite cements ground with PG additives, superplasticizers SP-1 and Sika VC 225 can be approximated by a general equation [18]:

$$\alpha = k\lg\tau + B \tag{1}$$

where τ—hydration duration; k i B—coefficients depending on the additives type and grain composition of cements.

Complex additives, including grinding intensifier (PG) and superplasticizers SP-1 and Sika VC 225 can be considered polyfunctional cement modifiers (PFM):

- PFM_1—PG + Sika VC 225;
- PFM_2—PG + SP-1.

The composite cements' main properties values, which were obtained by grinding in a laboratory ball mill, are given in Table 3. It follows from them that the strength of composite cement with a clinker content of 50% when adding PFM_1 and $S_{ssa} = 450\ m^2/kg$ reaches 60 at the age of 28 days, $S_{ssa} = 550\ m^2/kg$—70. When PFM_2 is introduced, it is 55 and 60 MPa, respectively. At the same time, at the 3 days, the strength of cement with PFM_1 reaches 50% of the 28-day strength.

The study of the concrete mixtures and concrete properties based on modified composite cement with the addition of PFM_1 using mathematical planning of experiments (Table 4) made it possible to obtain mathematical models of the concrete mixture water demand (W) and the concrete compressive strength at the 1 ($f^1{}_{cm}$) and 28 days ($f^{28}{}_{cm}$):

$$W = 142.1 - 22X_1 + 4.17X_2 - 10.1X_3 + 17.1X_4 + 6X_1^2 + 1.5X_2^2 + 6.6X_3^2 - 1.5X_4^2 + 0.6X_1X_2 - 0.6X_1X_3 + \\ +1.3X_2X_3 - 0.3X_2X_4 + 0.4X_3X_4 \tag{2}$$

$$f'_{cm} = 33.1 + 1.3X_1 + 8.9X_2 - 7.5X_3 - 1.2X_1^2 - 4.1X_2^2 + 2.8X_3^2 + 0.3X_1X_2 - 0.3X_1X_3 - 2.5X_2X_3 \tag{3}$$

$$f_{cm}^{28} = 70.9 - 0.3X_1 + 9.8X_2 - 13.8X_3 - 2.5X_1^2 - 4.28X_2^2 + 3.78X_3^2 + 3.9X_2X_3 \tag{4}$$

For the concrete compressive strength models, the influence of concrete mixture slump, provided that other varied factors are constant, was statistically insignificant. Factors can be arranged in the following sequence according to the decreasing influence on the studied properties:

$$W : X_1 > X_4 > X_3 > X_2$$
$$f'_{cm} : X_2 > X_3 > X_1$$
$$f_{cm}^{28} : X_3 > X_2 > X_1$$

When analyzing models of water consumption of a concrete mixture (Figures 3 and 4), attention is drawn to the practically identical values of water consumption of a concrete mixture in the interval $W/C = 0.45 - 0.35$ and its significant increase at $W/C < 0.35$. this confirms the well-known [1] rule of water consumption constancy at $W/C < W/C_{critical}$. At the same time, the results of the study show that the polycarboxylate superplasticizers addition makes it possible to shift the critical value $W/C_{critical}$ towards lower values. For concrete without superplasticizers, it is usually in the range of 0.43–0.4 [28]. The water demand model also reflects the increase in the concrete mixture water demand as the specific surface area of the cement increases.

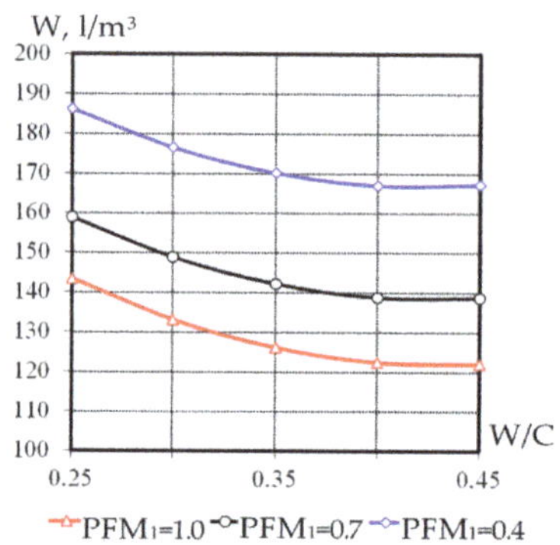

Figure 3. The influence of *W/C* and PFM_1 additives on the concrete mixture water consumption (S_{ssa} = 450 m^2/kg, Sl = 13 cm).

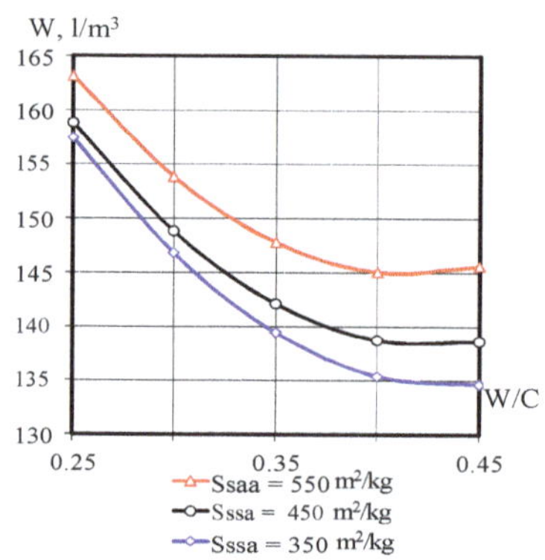

Figure 4. The influence of *W/C* and the specific surface area of the CC on the concrete mixture water consumption (PFM_1 = 0.7%, Sl = 13 cm).

Analysis of the strength mathematical models (Figures 5–7) shows the unconditional leading value of the factors W/C and Sssa. The influence of the specific surface area of the CC increases significantly for the one-day strength of the concrete. With increased values of

Spyt, the effect of increasing W/C on the early strength of concrete becomes less significant (Figure 5).

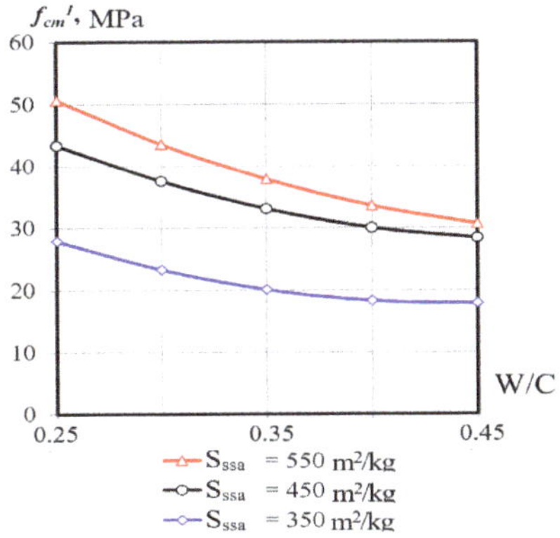

Figure 5. The influence of *W/C* and specific surface area on the concrete compressive strength at the 1 day (PFM_1 = 0.7%).

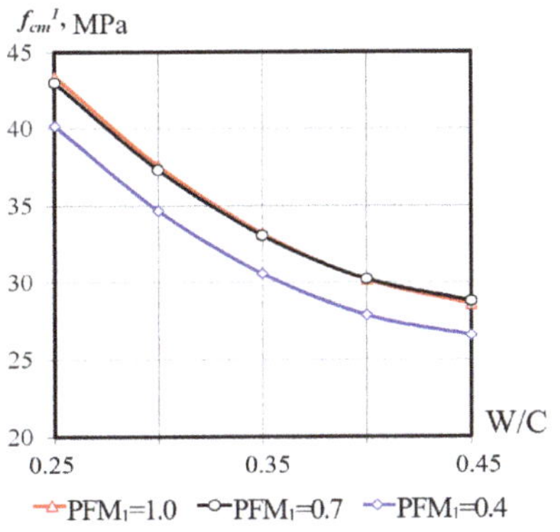

Figure 6. The influence of *W/C* and PFM_1 content on the concrete compressive strength at the 1 day (S_{ssa} = 450 m^2/kg).

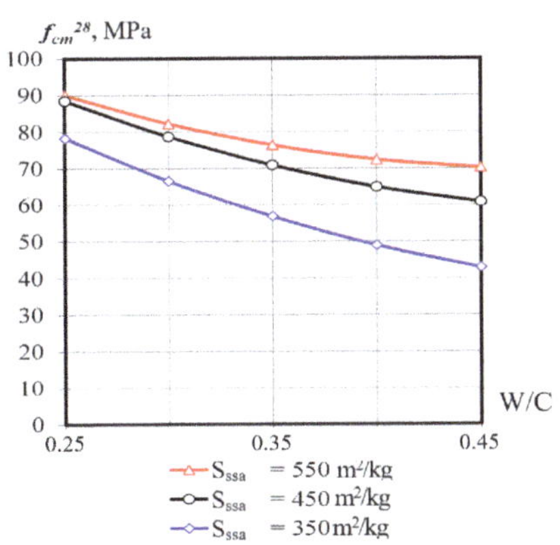

Figure 7. The influence of *W/C* and specific surface area on the concrete compressive strength at 28 days (PFM_1 = 0.7).

The models make it possible to design compositions of the concrete mixture on the CC with the PFM, specified workability, and strength.

Numerical Example

It is necessary to design the concrete class C 40/50 composition with compressive strength at a 1 day of not less than 30% 28 of the daily strength with the slump Sl = 13 cm when using composite cement with S_{ssa} = 350 m^2/kg and PFM_1 0.7% of the mass of cement. Materials: granite crushed stone fractions 5–20 mm with true density (ρ_{cs}) −2720 kg/m^3 and bulk density (ρ_{bcs}) −1350 kg/m^3, sand with true density (ρ_s) −2650 kg/m^3.

1. Let us find the required average strength of concrete class C 40/50, determined on sample cubes with a variation coefficient of 13.5% [1]. f_c^{28} = 50: 0.778 = 64.3 MPa
2. With the consumption of the additive PFM_1 = 0.7% (X_1 = 0) and the specific surface area of the CC S_{ssa} = 350 m^2/kg ($X_2 = -1$), calculate the necessary *W/C* from model (4). *W/C* = 0.311 With *W/C* = 0.311 ($X_3 = -0.39$), S_{ssa} = 350 m^2/kg ($X_2 = -1$), as well as the slump Sl = 13 cm (X_4 = 0), the concrete at one-day strength from model (3) is 22.4 MPa, which exceeds 30% f_c^{28}. Thus, for further calculations, we accept *W/C* = 0.311.
3. By model (2), when $X_1 = 0$, $X_2 = -1$, $X_3 = -0.39$, $X_4 = 0$, we will find the concrete mixture water demand: *W* = 145 L/m^3
4. Cement Consumption: $C = \frac{W}{W/C} = \frac{145}{0.311} = 466$ kg/m^3
5. Let us find the crushed stone and sand consumption, applying known formulas [1,2].

$$Cs = \frac{1000}{\alpha\frac{v_{cs}^{II}}{\rho_{bcs}} + \frac{1}{\rho_{cs}}} = \frac{1000}{1.375\frac{0.5}{1.35} + \frac{1}{2.72}} = 1140 \text{ kg/m}^3$$

where α—coefficient of displacement of coarse aggregate grains,
$V_{cs}{}^{p}$—the crushed stone intergranular voids.

$$S = \left(1000 - \left(\frac{C}{\rho_c} + W + \frac{Cs}{\rho_{cs}}\right)\right)\rho_s = \left(1000 - \left(\frac{466}{3.1} + 145 + \frac{1140}{2.72}\right)\right) \times 2.65 = 757 \text{ kg/m}^3.$$

Calculated composition of concrete, kg/m^3: *C* = 466 kg/m^3, *W* = 145 kg/m^3, *Cs* = 1140 kg/m^3, *S* = 757 kg/m^3, PFM_1 = 466 × 0.007 = 3.26 kg/m^3, (PG = 0.0004*C* = 0.18 kg/m^3, SP = 0.0066*C* = 3.07 kg/m^3).

4. Conclusions

1. The rational ratio of fly ash and blast-furnace granulated slag in the composition of the composite additive provides its increased pozzolanic activity, which increases significantly with an increase in the cement-specific surface area.
2. The joint addition of propylene glycol and superplasticizers into the composite cement during its grinding ensures that its specific surface is achieved without a significant increase in the duration of grinding and, as a result, a significant increase in the degree of hydration and strength, especially in the early stages of hardening.
3. Using mathematical planning, experimentally obtained experimental-statistical models of water demand, 1 daily and 28 daily strength of concrete on composite cement containing a polyfunctional modifier, including a polycarboxylate superplasticizer and a grinding intensifier. The models made it possible to quantify the influence of the main factors of concrete compositions and their interaction, as well as to design compositions of high-strength concretes with given values of workability and strength.

Author Contributions: Conceptualization, L.D. and V.Z.; methodology, R.M.; software, V.Z.; validation, V.M. and R.M.; formal analysis, L.D.; investigation, R.M.; resources, R.M.; data curation, V.Z.; writing—original draft preparation, V.M.; writing—review and editing, L.D.; visualization, V.Z.; supervision, L.D.; project administration, R.M.; funding acquisition, R.M. All authors have read and agreed to the published version of the manuscript.

Funding: This research received no external funding.

Institutional Review Board Statement: Not applicable.

Informed Consent Statement: Not applicable.

Data Availability Statement: Not applicable.

Acknowledgments: In this section, you can acknowledge any support given which is not covered by the author contribution or funding sections. This may include administrative and technical support, or donations in kind (e.g., materials used for experiments). The publication cost of this paper was covered with funds from the Polish National Agency for Academic Exchange (NAWA): "MATBUD'2023-Developing international scientific cooperation in the field of building materials engineering" BPI/WTP/2021/1/00002, MATBUD'2023.

Conflicts of Interest: The authors declare no conflict of interest.

References

1. Dvorkin, L.I. *Concrete Science (In 2 Volumes)*; Infra-Engineering: Moscow, Russia, 2021; 1300p. (In Russian)
2. Amin, M.; Abu El-Hassan, K. Effect of using different types of nano materials on mechanical properties of high strength concrete. *Constr. Build. Mater.* **2015**, *80*, 116–124. [CrossRef]
3. Pu, X. *Super-High-Strength High Performance Concrete*; CRC Press: Boca Raton, FL, USA, 2012; 276p.
4. Kodeboyina, G.B. *High Performance Self Consolidating Cementitious Composites*; CRC Press: Boca Raton, FL, USA, 2018; 433p.
5. Caldarone, M.A. *High-Strength Concrete: A Practical Guide*; CRC Press: Boca Raton, FL, USA, 2008; 272p.
6. Papayianni, I.; Anastasiou, E. Production of high-strength concrete using high volume of industrial by-products. *Constr. Build. Mater.* **2010**, *24*, 1412–1417. [CrossRef]
7. Bazhenov, Y. *Modified High Quality Concrete*; Association of Civil Engineering Education: Moscow, Russia, 2006; 368p. (In Russian)
8. Kalashnikov, V.; Tarakanov, O.; Kusnetsov, Y.; Volodin, V.; Belyakov, E. Next generation concrete on the basis of fine grained dry powder mixes. *Mag. Civ. Eng.* **2012**, *8*, 47–53. [CrossRef]
9. Kalashnikov, V.I.; Valiyev, D.M.; Gutseva, Y.V.; Volodin, V.M. Vysokoprochnyye poroshkovo-aktivirovannyye proparivayemyye betony novogo pokoleniya. *Iz-vo Vysshikh Uchebnykh Zavedeniy* **2011**, *5*, 14–19. (In Russian)
10. Dvorkin, L.; Zhitkovsky, V.; Sitarz, M.; Hager, I. Cement with Fly Ash and Metakaolin Blend—Drive towards a More Sustainable Construction. *Energies* **2022**, *15*, 3556. [CrossRef]
11. Svatovskaya, L.B.; Sychev, M.M. *Activated Hardening of Cements [Aktivirovannoye Tverdeniye Tsementov]*; Stroyizdat: Moscow, Russia, 1983; 160p. (In Russian)
12. Gibbs, D. *Thermodynamic Works [Termodinamicheskiye Raboty]*; Gostekhizdat: Moscow, Russia, 1950; 492p. (In Russian)
13. Siddique, R. Utilization of silica fume in concrete: Review of hardened properties. *Resour. Conserv. Recycl.* **2011**, *55*, 921–932. [CrossRef]
14. Khan, M.; Rehman, A.; Ali, M. Efficiency of silica-fume content in plain and natural fiber reinforced concrete for concrete road. *Constr. Build. Mater.* **2020**, *244*, 118382. [CrossRef]
15. Gonen, T.; Vazicioglu, S. The influence of mineral admixtures on the short and long-term performance of Concrete. *Build. Environ.* **2007**, *42*, 3080–3085. [CrossRef]
16. Dvorkin, L.I.; Dvorkin, O.L.; Ribakov, Y. *Construction Materials Based on Industrial Waste Products*; Nova Science Publishers: Hauppauge, NY, USA, 2016; 242p.
17. Batrakov, V.G. *Modified Concretes. Theory and Practice. [Modifitsirovannyye Betony. Teoriya i Praktika]*; Stroyizdat: Moscow, Russia, 1998; 768p. (In Russian)
18. Dvorkin, L.I.; Dvorkin, O.L.; Garnitskiy, Y.u.V.; Chorna, I.V.; Marchuk, V.V. *High-Strength Concretes on Low-Water-Consumption Cements Using Dusty Industrial Waste. [Visokomitsni betoni na tsementakh nizkoï vodopotrebi z vikoristannyam pilovidnikh vidkhodiv promislovosti]*; Budivelni Materiali, Virobi ta Sanitarna Tekhnika: Kiïv, Ukraine, 2012; pp. 73–81. (In Ukrainian)
19. Donatello, S.; Tyrer, M.; Cheeseman, C. Comparison of test methods to assess pozzolanic activity. *Cem. Concr. Compos.* **2010**, *32*, 121–127. [CrossRef]
20. *EN 196–5:2011*; Methods of Testing Cement. Pozzolanicity Test for Pozzolanic Cement. NSAI: Nashua, NH, USA, 2011.
21. *ISO 13317-1:2001*; Determination of Particle Size Distribution by Gravitational Liquid Sedimentation Methods. ISO: Geneva, Switzerland, 2001; 17p.
22. Liao, W.; Sun, X.; Kumar, A.; Sun, H.; Ma, H. Hydration of Binary Portland Cement Blends Containing Silica Fume: A Decoupling Method to Estimate Degrees of Hydration and Pozzolanic Reaction. *Front. Mater.* **2019**, *6*, 78. [CrossRef]
23. *EN 196-1*; Methods of Testing Cement–Part 1: Determination of Strength. CEN: Brussels, Belgium, 2005; 12p.
24. Montgomery, D.C. *Design and Analysis of Experiments*, 5th ed.; Wiley: Hoboken, NJ, USA, 2000; 688p.
25. Box, G.E.P.; Hunter, J.S.; Hunter, W.G. *Statistics for Experimenters: Design, Discovery, and Innovation*, 2nd ed.; Wiley: Hoboken, NY, USA, 2005; 672p.

26. Dvorkin, L.; Dvorkin, O.; Ribakov, Y. *Mathematical Experiments Planning in Concrete Technology*; Nova Science Publishers: Hauppauge, NY, USA, 2012; 175p.
27. *EN 12390-1:2021*; Testing Hardened Concrete-Part 1: Shape, Dimensions and Other Requirements for Specimens and Moulds. CEN: Brussels, Belgium, 2021; 12p.
28. Spiratos, N.; Page, M.; Mailvaganam, N.; Malhotra, V.; Jolicoeur, C. *Superplasticizers for Concrete: Fundamentals, Technology and Practice*; Supplementary Cementing Materials for Sustainable Development Inc.: Amsterdam, The Netherlands, 2017; 322p.
29. Locher, F.W. *Cement: Principle of Production and Use*; Verlag Bau + Technic.: Nordrhein-Westfalen, Germany, 2006; 534p.

materials proceedings

Proceeding Paper

Geopolymers Based on Fly Ash from the Bełchatów Power Plant †

Agnieszka Bąk *, Patrycja Bazan, Kinga Pławecka and Michał Łach

Faculty of Material Engineering and Physics, Cracow University of Technology, Jana Pawła II 37, 31-864 Cracow, Poland
* Correspondence: agnieszka.bak@pk.edu.pl

† Presented at the 10th MATBUD'2023 Scientific-Technical Conference "Building Materials Engineering and Innovative Sustainable Materials", Cracow, Poland, 19–21 April 2023.

Abstract: Geopolymer materials are attracting increasing interest from scientists due to their specific properties and the possibility of using waste materials from the power industry for their production. The most common raw material for their production is fly ash from the combustion of hard coal in pulverized coal boilers, but it is also possible to use ash from the combustion of lignite for this purpose. This article presents the results of a study of geopolymers produced with the use of ashes from lignite combustion at the Bełchatów Power Plant. It includes characterization of the ashes (laser particle size analysis, SEM/EDS morphology, XRD phase composition) and the strength properties of geopolymers made from these ashes and activated with 10 M and 14 M aqueous sodium hydroxide solution with water glass. As a result of this study, it was found that it is possible to obtain geopolymers characterized by flexural strength of almost 3 MPa and compressive strength of 30 MPa. A comparison of activators with concentrations of 10 M and 14 M made it possible to conclude that, due to the lack of significant differences in the properties of the obtained geopolymers, from an economic point of view, it is more advantageous in this case to use activators with a lower concentration.

Keywords: geopolymers; fly ash; fly ash from lignite; Bełchatów; alkaline activation; flexural strength; compressive strength

Citation: Bąk, A.; Bazan, P.; Pławecka, K.; Łach, M. Geopolymers Based on Fly Ash from the Bełchatów Power Plant. *Mater. Proc.* **2023**, *13*, 17. https://doi.org/10.3390/materproc2023013017

Academic Editors: Katarzyna Mróz, Tomasz Tracz, Tomasz Zdeb and Izabela Hager

Published: 14 February 2023

1. Introduction

Fly ash is a waste material that settles in electrostatic precipitators as a residual after the combustion of coal dust in the furnaces of a power plant or thermal power plant. Fly ashes vary in chemical and phase composition. The best-known fly ashes are those with a high silica or calcium content. In this paper, the latter will be presented. Fly ash from the Bełchatów mine is ash with a high calcium oxide content. This material is a by-product of lignite coal combustion. This ash can come from conventional furnaces or be generated by dry flue gas desulfurization (from conventional or fluidized bed furnaces). The concentration of reactive calcium oxide required as a pozzolan and hydraulic component in the whole mixture is more than 10% for lignite fly ash [1–3]. According to EN 197-1 [4], only the fly ash from boiler furnaces can be used. However, EN 206 [5] states that lignite fly ash cannot be used as a Type II active additive.

In Poland, lignite fly ash is a by-product of combustion mainly in the boilers of the Bełchatów Power Plant and Combined Heat and Power Plant, as well as Pątnów, Adamowa and Konin (the PAK region). Table 1 shows the chemical composition of the lignite fly ash from the Bełchatów Power Plant and, for comparison, Table 2 shows the coal fly ash from the Skawina Power Plant. Tables 3 and 4 show the phase composition of the fly ash from the Bełchatów Power Plant and, for comparison, the phase composition of fly ash from hard coal. The data presented in Tables 1–4 highlight the great diversity in domestic fly ashes. The fly ash from the Bełchatów Power Plant has the characteristic of calcium aluminosilicates and is characterized by a high content of silicon, aluminum and calcium oxide and a small amount of other oxides [6,7].

Table 1. Oxide composition of the Belchatów lignite fly ash.

Precursor	Oxide Composition (wt%)									
	LOI *	SiO_2	CaO	Al_2O_3	Fe_2O_3	SO_3	CaO Free	MgO	K_2O	Na_2O
F.A. Bełchatów **	2.1	42.8	22.5	20.5	4.4	4.3	4.1	0.9	0.2	0.1

* LOI—Loss on ignition; ** F.A. Bełchatów—Fly ash from Bełchatów.

Table 2. Oxide composition of the Skawina charcoal fly ash.

Precursor	Oxide Composition (wt%)							
	SiO_2	Al_2O_3	Fe_2O_3	K_2O	CaO	MgO	TiO_2	Na_2O
F.A. Skawina ***	55.9	23.49	5.92	3.55	2.72	2.61	1.09	0.59

*** **F.A. Skawina**—Fly ash from Skawina.

Table 3. Phase composition of the Belchatów lignite fly ash.

Fly Ash from Bełchatów
Mineral components listed in the decreasing order of content
Quartz Gehlenite Anhydrite Hematite AnorthiteLarnite Ye'elemit $C_4A_3\hat{S}$ $C_{12}A_7$ C_3A Free lime Mullite
Amorphous phase
Calcium aluminosilicate glass

Table 4. Phase composition of the Skawina charcoal fly ash.

Fly Ash from Skawina
Mineral components listed in the decreasing order of content
Mullite Quartz Hematite Magnetite
Amorphous phase
Aluminosilicate glass

Fly ash from Bełchatów lignite is characterized by a more complex mineral composition compared to fly ash from Skawina coal. As shown in Tables 3 and 4, this observation applies to both the glass phase and minerals. The main minerals in lignite fly ash are quartz, gelenite, anorthite, anhydrite and calcium oxide. Typical cement clinker phases, i.e., $C_{12}A_7$, C_3A and $C_4A_3\hat{S}$, have also been identified [6]. These phases determine the hydraulic properties of fly ash. The pozzolanic and hydraulic properties of lignite fly ash are also related to the concentration of the amorphous phase. In the case of lignite fly ash, aluminosilicate glass is identified as the amorphous phase [8]. Fly ash from lignite Bełchatów is characterized by a relatively high variability in chemical and phase composition. This is also true for other characteristics that affect their functional properties [9].

The purpose of this paper is to study lignite fly ash from the Bełchatów Power Plant and Combined Heat and Power Plant as a replacement for hard coal fly ash from the Skawina Power Plant and Combined Heat and Power Plant. In this paper, the base material was tested and then specimens were made for bending and compression tests, so the mechanical strength of the finished structural material was shown. For the alkaline activation of the raw material, the same activator with different molar concentrations of 10 and 14 M NaOH was used.

2. Materials and Methods

The test samples were made from lignite fly ash, which came from the Bełchatów Power Plant (Bełchatów, Poland), with waste code 10 01 02, and river sand from a sand plant in Świetochłowice (Świetochłowice, Poland). The test material was made in a 50/50 wt.% ratio. The base material used for the study, fly ash, was tested for phase analysis and particle size distribution. Table 5 show the XRD phase analysis, while Table 6 shows the particle size distribution analysis. The reaction activator was a 10 M and 14 M NaOH solution. Technical sodium hydroxide flakes and an aqueous solution of sodium silicate R-145 with a molar modulus of 2.5 and a density of about 1.45 g/cm^3 were used. The ratio of the two components was 1:2.5. Distilled water was not used—the supplementary water added was "mains" water. To prepare the alkaline solution, solid sodium hydroxide was poured over an aqueous solution of sodium silicate and water. The solution was mixed thoroughly and allowed to equilibrate until it reached a constant concentration and temperature. The compositions of the geopolymer mixture are shown in Table 7. The samples were cured in an SLW 750 laboratory dryer (POLEKO, Wodzisław Śląski, Poland), at 60 °C for 24 h. After 28 days of sample preparations, the geopolymers were tested for flexural and compressive strength.

Table 5. Phase analysis of the Belchatów lignite fly ash.

Fly Ash from Bełchatów			
Identified Phase	**Chemical Formula**	**Percentage Share [wt.%]**	**Data Sheet Number**
Quartz	SiO_2	1.7	01-074-1811
Gehlenite	$Ca_2Al_2SiO_7$	31.3	04-015-3030
Anhydrite	$CaSO_4$	15.9	00-006-0226
Hematite	Fe_2O_3	9.8	04-006-2616
Anorthite	$CaAl_2Si_2O_8$	15.1	00-041-1486
Ye'elimite	$Ca_4Al_6(SO_4)$	5.9	04-009-7268
Chlormayenite ($C_{12}A_7$)	$C_{12}A_{14}O_{33}$	3.4	00-048-1882
Lime	CaO	3.2	04-005-4757
Mullite	$Al_6Si_2O_{13}$	13.7	00-015-0776

Table 6. Particle size analysis of the Belchatów lignite fly ash.

Material	D_{10} [μm]	D_{50} [μm]	D_{90} [μm]	Mean Size [μm]
Fly ash from Bełchatów	3.29	20.74	37.24	21.46
	3.35	20.80	37.06	21.45
	3.46	20.83	36.81	21.43
	3.87	21.88	37.30	22.21
	3.88	21.42	37.23	21.91

Table 7. Compositions of geopolymer mixture of the Belchatów lignite fly ash.

Index	Base Materials (S) [Weight Ratio]		Alkaline Activator (L)	Liquid/Solid Ratio [Weight Ratio]
	Fly Ash	Sand		
R10	1	1	10 M NaOH + sodium water glass (weight ratio: 1:2.5)	1:0.30
R14	1	1	14 M NaOH + sodium water glass (weight ratio: 1:2.5)	1:0.30

3. Results

3.1. Tests of Mechanical Properties—Flexural Strength Tests

Flexural strength tests were carried out on an MTS Criterion 43 testing machine with TestSuites 1.0 software (MTS System Corp., Eden Prairie, MN, USA) with a measuring range of up to 30 kN. The method for determining the flexural strength of cement mortar specimens is specified by EN 196-1:2016-07 (Cement test methods—Part 1: Determination of strength—Section 9.1) [10]. Test specimens in the form of cuboids are subjected to a bending moment by applying a uniform load, induced by the lower and upper rollers of the testing machine. The test continues until the maximum load is reached, which causes the failure of the component, and the bending strength is calculated based on this parameter. The bending strength is determined by the formula for the three-point method:

$$R_f = \frac{1.5 * F_f * l}{b^3}\ [\text{MPa}] \quad (1)$$

where:

R_f—flexural strength (MPa)

b—lateral length of the section (mm)

F_f—maximum load (N)

l—length between supports (mm).

For the tests, six specimens activated with 10 M NaOH and six specimens activated with 14 M NaOH with dimensions of 40 × 40 × 160 mm were prepared. Based on the results obtained, a graph (Figure 1) was made showing the results of flexural strength tests.

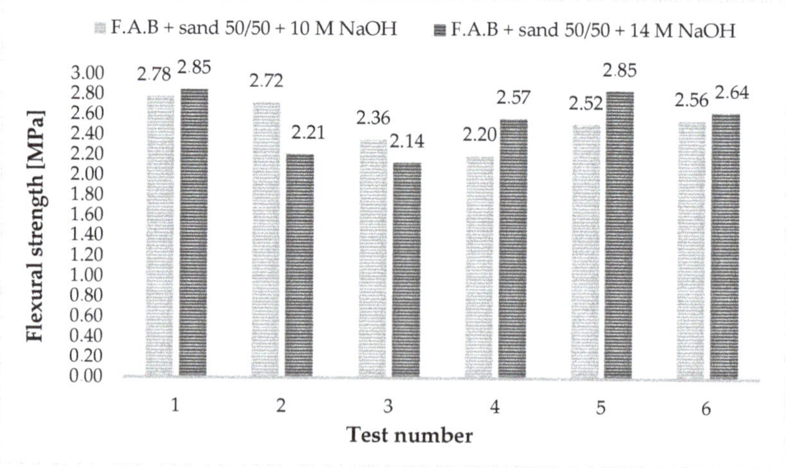

Figure 1. Flexural strength of the Belchatów lignite fly ash samples.

The above image shows that the test results for samples activated with 10 M solution and 14 M solution are comparable. The values for all samples are in a range of 2–3 MPa. The average strength value for both activators is 2.5 MPa. However, for economic reasons, a better choice would be solutions with a lower molar concentration.

3.2. *Tests of Mechanical Properties—Compressive Strength Tests*

Compressive strength tests were carried out on an MATEST 3000 kN testing machine (Matest, Treviolo, Italy). In the building and construction sector, the document that regulates the method for determining the compressive strength of cement mortar specimens is PN-EN 196-1:2016-07 (Methods for testing cement—Part 1: Determination of strength—Section 9.2) [10]. Compression testing involves loading specimens until a critical value is reached that will cause the material to fail. The maximum load is the basis for calculating the compressive strength of the concrete material according to the following formula:

$$R_c = \frac{F_c}{1600}\ [\text{MPa}] \tag{2}$$

where:

R_c—compressive strength (MPa)

1600—surface of tiles (or auxiliary tiles) (mm^2)

F_c—maximum load (N).

Specimens formed after flexural strength tests—12 specimens activated with 10 M NaOH and 12 specimens activated with 14 M NaOH with dimensions of 40 × 40 × 40 mm—were used for the tests. Based on the results obtained, a graph (Figure 2) was made showing the results of compressive strength tests.

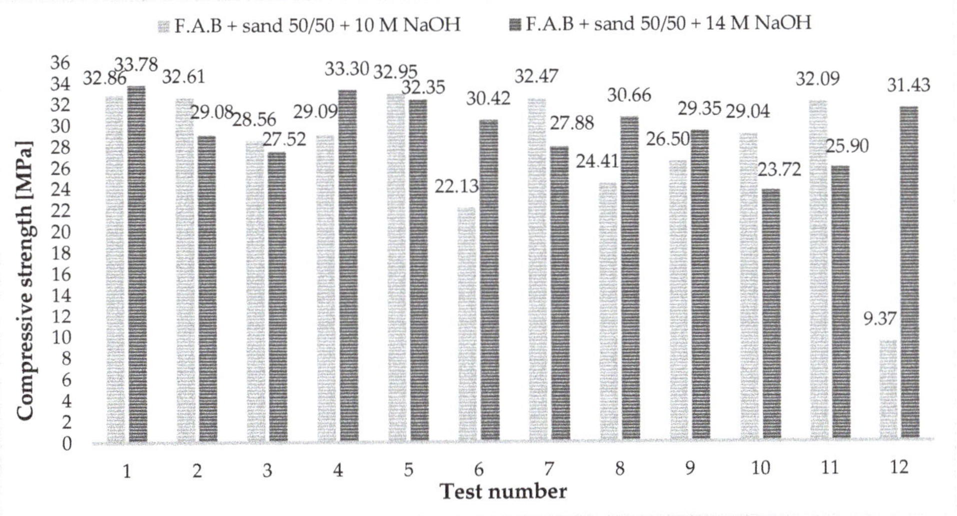

Figure 2. Compressive strength of the Belchatów lignite fly ash samples.

The above picture shows that the test results for samples activated with 10 M solution and 14 M solution are comparable, as is the case for the testing of crushing strength. The values for all samples are in a range of 20–35 MPa (the last result for the 10 M activator is only a deviation from this norm). The average strength value for both activators is 28 MPa. Again, for economic reasons, a solution with a lower molar concentration would be a better choice.

3.3. Evaluation of the Microstructure and Analysis of the Oxide Chemical Composition of the Resulting Geopolymers

A JEOL IT200 SEM scanning microscope (JEOL, Warszawa, Poland) was used to perform oxide analysis of the resulting samples, as shown in Figure 3.

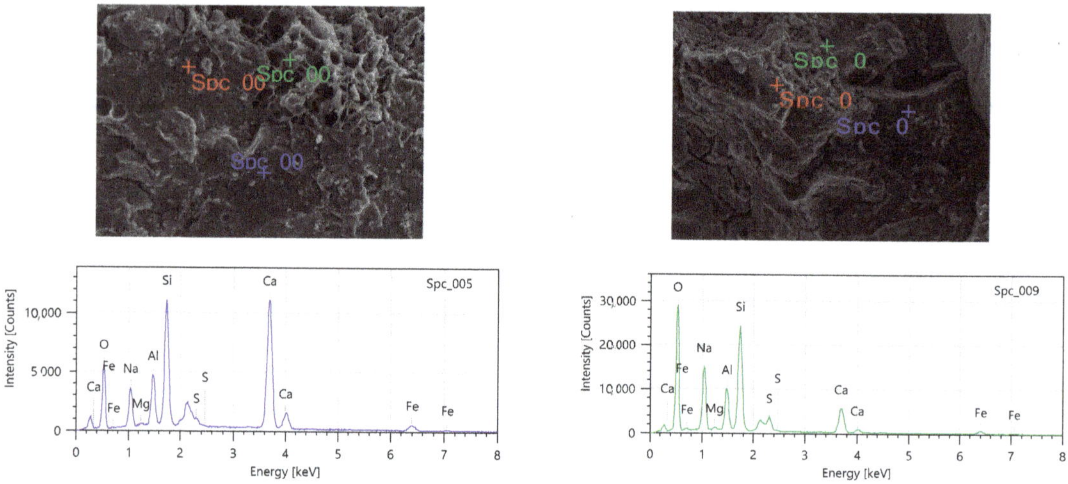

Chemical formula	Line	Mass%	Mol%	Cations
O	K			
Na2O	K	10.29±0.13	10.28±0.13	3.29
MgO	K	0.51±0.04	0.78±0.06	0.12
Al2O3	K	11.69±0.14	7.10±0.08	2.27
SiO2	K	33.69±0.24	34.72±0.25	5.55
SO3	K	0.77±0.05	0.60±0.04	0.10
CaO	K	38.91±0.22	42.96±0.25	6.87
FeO	K	4.13±0.11	3.56±0.09	0.57
Total		100.00	100.00	
Spc_005				Fitting ratio 0.0501

(a)

Chemical formula	Line	Mass%	Mol%	Cations
O	K			
Na2O	K	21.85±0.13	22.88±0.14	6.29
MgO	K	0.90±0.03	1.45±0.06	0.20
Al2O3	K	14.27±0.11	9.09±0.07	2.50
SiO2	K	42.43±0.20	45.83±0.22	6.30
SO3	K	6.47±0.09	5.24±0.07	0.72
CaO	K	11.02±0.09	12.75±0.10	1.75
FeO	K	3.06±0.07	2.76±0.06	0.38
Total		100.00	100.00	
Spc_009				Fitting ratio 0.0327

(b)

Figure 3. Analysis of the oxide composition of the Belchatów lignite fly ash—(**a**) 10 M NaOH, (**b**) 14 M NaOH.

The above photo shows the structure of a geopolymer based on limestone fly ash Belchatów, activated with solutions of different molar concentrations. There are no visible differences in the structure and the oxide analysis performed as well [11,12].

4. Short Discussion

The analysis of the phase composition performed within the scope of the article is consistent and comparable with the analysis presented in the literature [6]. Only larnite and one phase of C_3A cement clinker were not identified during the study.

Oxide chemical composition analysis was carried out using a scanning microscope with an EDS system. A comparison of the literature condition with the test results (Figure 3) showed that the same oxides are mostly present in the base material [6].

The presented tests of mechanical properties, i.e., bending and compressive strength, showed that there is no major difference between the activator 10 M and 14 M. The obtained results are comparable with the results of other researchers [13]. Fly ash from lignite combustion has a less amorphous phase (glassy particles), so it is less reactive compared to hard coal ash from pulverized coal boilers. The use of an activator with a higher concentration does not bring results, because the phases that can be solubilized and activated become active at a concentration of 10 M. For economic reasons, i.e., ever-increasing inflation and current fuel prices, a better activator is one with a lower concentration.

5. Conclusions

Based on the above discussion of the research results, several conclusions can be drawn to summarize the research work:

- Limestone fly ash contains quartz, gelenite, anorthite, hematite, anhydrite, mullite and calcium oxide, as well as typical cement clinker phases, i.e., $C_{12}A_7$, C_3A and $C_4A_3\hat{S}$.
- The following oxides can be identified in the base material: SiO_2, CaO, Al_2O_3, Fe_2O_3, SO_3, CaO free, MgO, K_2O and Na_2O.
- The flexural strength for samples activated with the two solutions is at a similar level and is no more than 3 MPa. Compressive strength for both activators is similar, with results not exceeding 35 MPa.

Author Contributions: Conceptualization, M.Ł. and A.B.; methodology, M.Ł. and A.B.; investigation, A.B., K.P. and P.B.; resources, A.B.; writing—original draft preparation, A.B.; writing—review and editing, A.B., K.P., P.B. and M.Ł.; supervision, A.B. and M.Ł.; funding acquisition, M.Ł. All authors have read and agreed to the published version of the manuscript.

Funding: This work was funded by the National Center for Research and Development in Poland under grant: M-ERA.NET3/2021/70/GEOSUMAT/2022 "Materials for Circular Economy—Industrial Waste Based Geopolymers Composites with Hybrid Reinforcement" under M-ERA.NET 3 Call 2021.

Institutional Review Board Statement: Not applicable.

Informed Consent Statement: Not applicable.

Data Availability Statement: Not applicable.

Acknowledgments: Special thanks go to Joanna Marczyk, who performed the phase analysis studies of the base material. Publication cost of this paper was covered with funds from the Polish National Agency for Academic Exchange (NAWA): "MATBUD'2023—Developing international scientific cooperation in the field of building materials engineering" BPI/WTP/2021/1/00002, MATBUD'2023.

Conflicts of Interest: The authors declare no conflict of interest. The funders had no role in the design of the study; in the collection, analyses, or interpretation of data; in the writing of the manuscript, or in the decision to publish the results.

References

1. Baran, T.; Drożdż, W. Evaluation of properties of domestic calcareous fly ash and its processing methods. *Roads Bridges* **2013**, *12*, 5–15.
2. Gibas, K.; Glinicki, M.A.; Nowowiejski, G. Evaluation of impermeability of concrete containing calcareous fly ash in respect to environmental media. *Roads Bridges* **2013**, *12*, 159–171.
3. Giergiczny, Z.; Garbacik, A.; Ostrowski, M. Pozzolanic and hydraulic activity of calcareous fly ash. *Roads Bridges* **2013**, *12*, 71–81.
4. *PN-EN 197-1:2012*; Cement—Część 1: Skład, Wymagania i Kryteria Zgodności Dotyczące Cementów Powszechnego Użytku. Polish Version. Available online: https://sklep.pkn.pl/pn-en-197-1-2012p.html (accessed on 5 December 2022).
5. *PN-EN 206:2014-04*; Beton—Wymagania, Właściwości, Produkcja i Zgodność. Polish Version. Available online: https://sklep.pkn.pl/pn-en-206-2014-04p.html (accessed on 5 December 2022).
6. Giergiczny, Z. The role of calcium and silica fly ashes in shaping the properties of modern construction binders and cementitious plastics. *Bull. Tadeusz Kosciuszko Crac. Univ. Technol.* **2006**, 1–50.
7. Figiela, B.; Brudny, K.; Lin, W.T.; Korniejenko, K. Investigation of mechanical properties and microstructure of construction—And demolition—Waste—Based geopolymers. *J. Compos. Sci.* **2022**, *6*, 191. [CrossRef]
8. Kurdowski, W. *Cement and Concrete Chemistry*; Polish Cement Publishing, PWN Scientific Publishers: Warsaw, Poland, 2010.
9. Pachowski, J. Development of technologies for the formation of power plant by-products and their characteristics and potential applications in road construction technologies. *Roads Bridges* **2002**, *1*, 59–99.
10. *PN-EN 196-1:2016-07*; Metody Badania Cementu—Część 1: Oznaczanie Wytrzymałości. Polish Version. Available online: https://sklep.pkn.pl/pn-en-196-1-2016-07p.html (accessed on 5 December 2022).
11. Nath, S.K.; Maitra, S.; Mukherjee, S.; Kumar, S. Microstructural and morphological evolution of fly ash based geopolymers. *Constr. Build. Mater.* **2016**, *111*, 758–765. [CrossRef]

12. Kumar, S.; Mucsi, G.; Kristály, F.; Pekker, P. Mechanical activation of fly ash and its influence on micro and nano-structural behaviour of resulting geopolymers. *Adv. Powder Technol.* **2017**, *28*, 805–813. [CrossRef]
13. Baran, T.; Garbacik, A.; Synowiec, K.; Żak, A. Use of calcareous fly ash as an active constituent of hydraulic road binders. *Roads Bridges* **2013**, *12*, 17–29.

materials proceedings

MDPI

Proceeding Paper

Three-Dimensional, Printable Paving Stone: A Preliminary Study †

Volkan Arslan * and **Zekeriya Dogan**

Department of Civil Engineering, Faculty of Engineering, Zonguldak Bulent Ecevit University, 67100 Zonguldak, Turkey
* Correspondence: volkanarslan@beun.edu.tr; Tel.: +90-3722911940
† Presented at the 10th MATBUD'2023 Scientific-Technical Conference "Building Materials Engineering and Innovative Sustainable Materials", Cracow, Poland, 19–21 April 2023.

Abstract: Three-dimensional (3D) printing applications have emerged as a new production method in the construction industry. The materials that are to be used in 3D production process play an important role for a sustainable built environment. The main objective of this study is to design a suitable mixture to produce 3D printed concrete paving stones. In this respect, a unique 3D printer was also developed. The results show that the setting time of cement-based mortars was shortened by increasing the ratio of the added accelerator admixture. However, the optimum mixture proportions for 3D printed concrete paving stones were not reached. The results of the study are expected to develop a sustainable method of paving stone production.

Keywords: 3D printer; construction technology; construction material; paving stone

Citation: Arslan, V.; Dogan, Z. Three-Dimensional, Printable Paving Stone: A Preliminary Study. *Mater. Proc.* **2023**, *13*, 18. https://doi.org/10.3390/materproc2023013018

Academic Editors: Katarzyna Mróz, Tomasz Tracz, Tomasz Zdeb and Izabela Hager

Published: 14 February 2023

1. Introduction

Developments and research in technology have resulted in new design, fabrication, and construction techniques in the construction industry [1]. Three-dimensional printing as one of these techniques stands out as it requires minimum human intervention and the minimum pre-processing of raw materials [2]. The 3D printing technique draws attention in the construction industry due to its advantages over traditional manufacturing methods. The major advantages of the 3D printing of concrete are its higher precision, safer working conditions, faster construction, and lower cost of construction, owing to the decrease in costs of formwork and labor [3]. It is also more expeditious than conventional construction methods are, optimizes the site works, enhances the constructability, and reduces the amount of material, and decreases the occurrence of labor-related risks [2].

Three-dimensional printing literature encompasses studies in different industries. However, research on the construction industry on 3D printing technology is relatively scarce. Regarding the current literature, studies examining concrete mixture designs and the mechanical properties of 3D printed concrete exist. There are also studies on the production of bricks/briquettes from clay and ceramic materials. Moreover, researchers have investigated the technical properties of these materials, such as strength and thermal insulation [3–6]. In fact, cementitious materials are more popular than clay-based ones are. Despite the fact that concrete paving stones are one of the most frequently used construction materials, there are no studies in the literature on the mechanical or design properties of 3D printed concrete paving stones. The aim of this study was to design a suitable mixture to produce 3D printed concrete paving stones. To do this, a series of experiments in the construction materials laboratory were conducted.

2. Past Studies

The constructability and mechanical properties of 3D printed concrete can be seen as the most significant issues at the moment. Therefore, many researchers have concentrated on mixture designs or the strength of 3D printed concrete. Paul et al. [7] introduced a mixing ratio for 3D printing using cement, and the 28 day compressive strength of the produced concrete was measured, 36–57 MPa, and the bending strength was 10 MPa. Le et al. [8] designed a high-strength fiber-reinforced concrete with a compressive strength of 92 MPa and a flexural strength of 11 MPa. Ma et al. [5] utilized copper waste to develop a printable cementitious mix with good workability and 50 MPa compressive strength. Marais et al. [9] measured the thermal performance of 3D printed lightweight foam concrete and high-performance concrete elements. Ting et al. [10] analyzed the effects of the glass-to-binder ratio, fineness modulus, and nano-clay content on the extrudability and constructability of concrete. Gomaa et al. [11] and Alqenaee and Memari [3] developed a printable clay-based cob mix design in which the components have sufficient strength for construction. Hojati et al. [12] investigated ways to replace cement in the mixtures with other cementitious alternatives and to design sustainable mixtures suitable for 3D printing.

Most of the current literature in this field is focused on printing cement-based concrete. They aimed to analyze concrete mixture designs by utilizing different materials. In addition, most of the 3D printing research has been carried out at a laboratory scale. However, there are no studies in the literature on the mechanical or design properties of the 3D printed concrete paving stones. Concrete paving stones are popular around the world [13]. They are widely used in sidewalks, urban roads, etc. The most common grade of concrete paving stones is interlocking pacing stones [14]. Compared with concrete and asphalt pavements, interlocking paving stones offer numerous advantages such as minimal maintenance and economic benefits. Therefore, it can be stated that concrete paving stones deserve enough attention to be produced by utilizing advanced technology. The aim of this study was to develop a new method to produce 3D printed concrete paving stones. As a result, to the best of our knowledge, the present study is the first attempt to design a mixture for 3D printed concrete paving stones.

3. Materials and Methods

3.1. Materials

In this study, cement-based mortar was developed for 3D printed paving stones. The cement-based mortar comprises cement, water, and a set-accelerating admixture. In this process, the pumpability and the stability of the extrusion were considered as the major properties of the cement-based mortar.

The main material used to produce paving stones is CEM I 42.5 N Portland cement, which is in accordance with the standard TS EN 197–1. The water-to-cement ratio was set as 0.4, and Polisan Antiton 100 was used as set accelerating admixture to prevent collapses during the printing process.

3.2. Methods

In this stage, 3D printed concrete paving stones were produced in accordance with TS 2824 EN 1338 [15]. To observe the pumpability and printability characteristics of a cement-based mortar, experiments were carried out in the Construction Materials Laboratory of the Faculty of Engineering of Zonguldak Bulent Ecevit University. The initial and final setting times of the fresh mixture were measured through the Vicat Needle test.

In this study, a new 3D printer was designed and produced (Figure 1). The printer chassis was formed with a 30 × 30 aluminum sigma profile. Plastic parts produced from FDM (fused deposition modeling), stainless steel bolts, and nuts were used. Flexible plastic elevations were attached to the printer's feet to prevent mechanical vibration. The movement in the axes was provided by three Nema 17 step motors utilizing the delta arm design. A double extruder was used to mix cement-based mortar and accelerator admixture during the printing process. Flow settings in each extruder were made using mechanical and software calibrations. Cement-based mortar and accelerating admixture were extruded using a stepper motor and archimedean screws. In the calibration phase, the open source computer supported Repetier Host program was used. Stages such as step settings of the axes, speed, and acceleration were performed through the Repetier Host program. The stl (standard triangle language) file format of the paving stones was sliced with the open source G-code (program language of CNC machines) generator, and then was saved to the SD card. Production was started via the control panel by inserting the SD card into the printer.

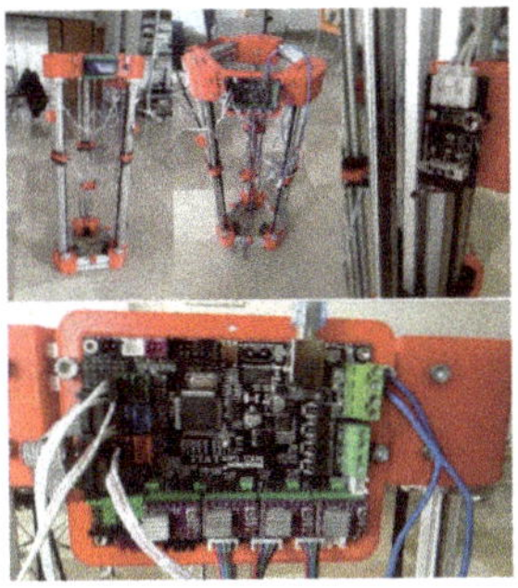

Figure 1. Three-dimensional printer designed for this study.

Three-dimensional printed paving stone samples were cured using a standard water curing protocol. The compressive strengths of the hardened concrete were measured to elucidate the development of its strength over time in accordance with TS EN 12390-3 [16]. After that, they were compared with those for reference paving stones as per the standard TS 2824 EN 1338.

The machines and equipment required to carry out the above mentioned tests are available in the Construction Materials Laboratory of the Faculty of Engineering of Zonguldak Bulent Ecevit University.

4. Results and Discussion

4.1. Cement-Based Mortar Design

Figure 2 gives the results of the setting times of cement-based mortar samples. There are seven samples given in the figure. They were designed through adding accelerator admixture in certain ratios ranging from 0‰ to 3‰ (0:CH00; 5‰:CH05; 10‰:CH10; 15‰:CH15; 20‰:CH; 25‰:CH25; 30‰:CH30). The results show that the setting time of cement-based mortars was shortened by increasing the ratio of the added accelerator admixture. Designing the optimum mixture for the 3D printed concrete paving stones was the aim. In fact, recent findings have some defects and design proportions that should be improved.

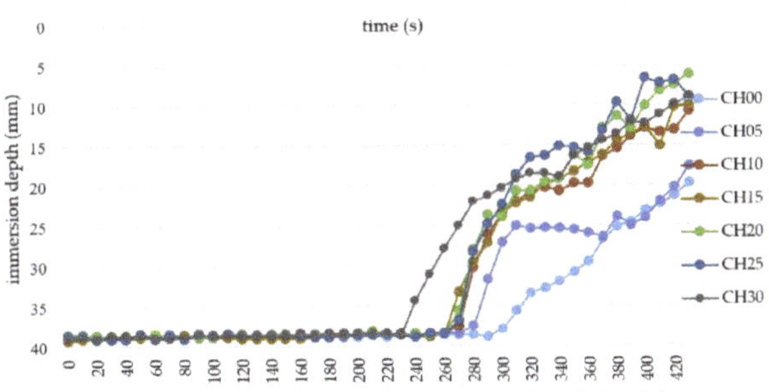

Figure 2. Cement-based mortar setting times.

The print time, setting time, and layer interval time in 3D printing should be considered as the loading increases on newly printed layers [17]. In general, the initial setting time of cement-based mortars is usually longer than ten hours, and this is not acceptable for the 3D printing abilities of the materials [5]. The printable construction materials should exhibit good workability, an appropriate setting time, and high strength values. The setting time should be kept at 20–80 min, so that the material coagulates rapidly and exhibits high strength values and the desired styling ability within a short time [18]. According to Figure 2, the cement-based mortars designed were insufficient in terms of consistency and quality of printing. Therefore, in the context of the study, more experiments should be conducted to obtain the most suitable mortar design to produce 3D printed concrete paving stones.

4.2. Results of the Compressive Strength Test

The 3D printed paving stones are also required to meet specific mechanical strength parameters in accordance with TS EN 12390-3. The compressive strength of the 3D printed materials was measured 1, 7, and 28 day ages to monitor the strength development over time. To conduct a compressive strength test, the printed samples were placed in the moist cabinet for proper curing. After that, the mechanical properties of hardened 3D printed paving stones were measured according to TS EN 12390-3.

Although the primary findings of the study show promising results to obtain suitable cement-based mortar for 3D printing, the 3D printing process of concrete paving stones was successfully completed. Therefore, the immature results of the compressive strength tests have not been presented in this paper. It should be noted that there is no study which revealed the mechanical properties of 3D printed concrete paving stones. Therefore, the results of the study will be compared to compressive strength of the paving stones produced with conventional construction methods.

5. Conclusions

The goal of this study was to produce 3D printed concrete paving stones. To do this, a unique 3D printer was developed, and a digitally controlled printing process which can build 3D printed paving stones without formwork was designed. As soon as the suitable cement-based mortar was obtained, the 3D printed paving stones were produced. Thereafter, the mechanical results of compressive strength tests were published. Three-dimensional printing could be used instead of a conventional production machines to achieve real, rapid manufacturing. Further research will be conducted to assess the structural behavior of 3D printed construction materials under the provided conditions to improve the mechanical and architectural properties of these materials.

Author Contributions: Conceptualization, V.A. and Z.D.; methodology, V.A.; software, Z.D. All authors have read and agreed to the published version of the manuscript.

Funding: This research was funded by Zonguldak Bulent Ecevit University Scientific Research Projects (BAP) Coordination Unit, grant number 2022-37891158-01.

Institutional Review Board Statement: Study did not require approval.

Informed Consent Statement: We did not need any consent in the study.

Data Availability Statement: The study did not report any data.

Acknowledgments: The authors thank the technical support provided by Zonguldak Bulent Ecevit University. The publication cost of this paper was covered by the funds of the Polish National Agency for Academic Exchange (NAWA): "MATBUD'2023—Developing international scientific cooperation in the field of building materials engineering" BPI/WTP/2021/1/00002, MATBUD'2023.

Conflicts of Interest: The authors declare no conflict of interest.

References

1. Ghanbari-Ghazijahani, T.; Kasebahadi, M.; Hassanli, R.; Classen, M. 3D printed honeycomb cellular beams made of composite materials (plastic and timber). *Constr. Build. Mater.* **2022**, *315*, 125541. [CrossRef]
2. Ulubeyli, S. Lunar shelter construction issues: The state-of-the-art towards 3D printing technologies. *Acta Astronaut.* **2022**, *195*, 318–343. [CrossRef]
3. Alqenaee, A.; Memari, A. Experimental study of 3D printable cob mixtures. *Constr. Build. Mater.* **2022**, *324*, 126574. [CrossRef]
4. Wolf, A.; Rosendahl, P.L.; Knaack, U. Additive manufacturing of clay and ceramic building components. *Autom. Constr.* **2022**, *133*, 103956. [CrossRef]
5. Ma, G.; Li, Z.; Wang, L. Printable properties of cementitious material containing copper tailings for extrusion based 3D printing. *Constr. Build. Mater.* **2018**, *162*, 613–627. [CrossRef]
6. Sangiorgio, V.; Parisi, F.; Fieni, F.; Parisi, N. The new boundaries of 3D-printed clay bricks Design: Printability of complex internal geometries. *Sustainability* **2022**, *14*, 598–613. [CrossRef]
7. Paul, S.C.; Tay, Y.W.D.; Panda, B.; Tan, M.J. Fresh and hardened properties of 3D printable cementitious materials for building and construction. *Arch. Civ. Mech. Eng.* **2018**, *18*, 311–319. [CrossRef]
8. Lim, S.; Buswell, R.A.; Le, T.T.; Austin, S.A.; Gibb, A.G.F.; Thorpe, T. Developments in construction-scale additive manufacturing processes. *Autom. Constr.* **2012**, *21*, 262–268. [CrossRef]
9. Marais, H.; Christen, H.; Cho, S.; De Villiers, W.; Van Zijl, G. Computational assessment of thermal performance of 3D printed concrete wall structures with cavities. *J. Build. Eng.* **2021**, *41*, 102431. [CrossRef]
10. Andrew Ting, G.H.; Noel Quah, T.K.; Lim, J.H.; Daniel Tay, Y.W.; Tan, M.J. Extrudable region parametrical study of 3D printable concrete using recycled glass concrete. *J. Build. Eng.* **2022**, *50*, 104091. [CrossRef]
11. Gomaa, M.; Jabi, W.; Veliz Reyes, A.; Soebarto, V. 3D printing system for earth-based construction: Case study of cob. *Autom. Constr.* **2021**, *124*, 103577. [CrossRef]
12. Hojati, M.; Li, Z.; Memari, A.M.; Zahabi, M.; Nazarian, S.; Duarte, J.P.; Radlińska, A. 3D-printable quaternary cementitious materials towards sustainable development: Mixture design and mechanical properties. *Results Eng.* **2022**, *13*, 100341. [CrossRef]
13. Gunatilake, D.; Mampearachchi, W.K. Finite element modelling approach to determine optimum dimensions for interlocking concrete blocks used for road paving. *Road Mater. Pavement Des.* **2019**, *20*, 280–296. [CrossRef]
14. Bakis, A. Interlocking paving stones made of limestone sand and volcanic ash aggregates. *Road Mater. Pavement Des.* **2022**, *23*, 1505–1522. [CrossRef]
15. *TS 2824 EN 1338*; Concrete Paving Blocks-Requirements and Test Methods. Turkish Standards Institution (TSE): Ankara, Turkey, 2005.
16. *TS EN 12390-3*; Testing Hardened Concrete Compressive Strength of Test Specimens. Turkish Standards Institution (TSE): Ankara, Turkey, 2010.
17. Bos, F.; Wolfs, R.; Ahmed, Z.; Salet, T. Additive manufacturing of concrete in construction: Potentials and challenges of 3D concrete printing. *Virtual Phys. Prototyp.* **2016**, *11*, 209–225. [CrossRef]
18. Sun, X.; Zhou, J.; Wang, Q.; Shi, J.; Wang, H. PVA fibre reinforced high-strength cementitious composite for 3D printing: Mechanical properties and durability. *Addit. Manuf.* **2022**, *49*, 102500. [CrossRef]

materials proceedings

MDPI

Proceeding Paper

Feasibility Study of Waste Gypsum as a Full Replacement for Fine Aggregates of Controlled Low-Strength Material †

Wei-Ting Lin [1,*], Kinga Korniejenko [2], Dariusz Mierzwiński [2], Michał Łach [2], An Cheng [1] and Kae-Long Lin [3]

1 Department of Civil Engineering, National Ilan University, Ilan 26047, Taiwan
2 Institute of Material Engineering, Faculty of Material Engineering and Physics, Cracow University of Technology, 31-864 Cracow, Poland
3 Department of Environmental Engineering, National Ilan University, Ilan 26047, Taiwan
* Correspondence: wtlin@niu.edu.tw; Tel.: +886-3-9317567, Fax: +886-3-9360125

† Presented at the 10th MATBUD'2023 Scientific-Technical Conference "Building Materials Engineering and Innovative Sustainable Materials", Cracow, Poland, 19–21 April 2023.

Abstract: The waste gypsum used in this study was a by-product of petroleum coke burning by petrochemical companies which has been treated with hydration. The waste gypsum has been stored in the atmosphere for over ten years and can be considered an inert filler. Its main chemical components were calcium sulfate dihydrate (47.90%), calcium hydroxide (21.64%) and calcium carbonate (14.80%). In this study, Portland cement and fly ash were used as cementitious materials, and waste gypsum of the particle size from 9.53 mm to 0.149 mm was selected as the fine aggregate to produce a controlled low-strength material (CLSM) and to verify the suitability of reusing waste gypsum. The water to binder ratio of 0.65 was used for the specimen. The test results showed that the CLSM specimen with a high amount of waste gypsum had air-hardening properties. The placement of the specimen in water caused abnormalities, such as cracking and disintegration of the specimens. The compressive strength of atmospherically maintained specimens increased with age, with 4.71 MPa and 6.08 MPa at 28 and 56 days, respectively. No significant changes in weight or volume were measured after the specimens had been left for 56 days and then immersed in seawater and water for 28 days. As specimens were immersed in seawater for up to 100 days, needle-shaped ettringite and C-S-H colloids filled the interface between the pores and the colloids. In accordance with the concept of eco-engineering, special consideration should be given to avoid long-term contact with water and to ensure the safety and durability of waste gypsum reuse through the design of multiple protective layers.

Keywords: air-hardening properties; microscopic analysis; waste reuse

Citation: Lin, W.-T.; Korniejenko, K.; Mierzwiński, D.; Łach, M.; Cheng, A.; Lin, K.-L. Feasibility Study of Waste Gypsum as a Full Replacement for Fine Aggregates of Controlled Low-Strength Material. *Mater. Proc.* **2023**, *13*, 19. https://doi.org/10.3390/materproc2023013019

Academic Editors: Katarzyna Mróz, Tomasz Tracz, Tomasz Zdeb and Izabela Hager

Published: 14 February 2023

1. Introduction

Climate change disasters have made reducing greenhouse gases more important than ever, and cutting CO_2 emissions from energy-intensive industries has become the main focus of net-zero carbon emissions. Among Taiwan's biggest energy consumers are thermal power generation, petrochemicals and cement manufacturing industries, including the second-largest petrochemical industrial zone in the world. Globally, it has been investing heavily in reducing this energy-intensive carbon emission problem in recent years, with the aim to achieve net zero carbon emissions eventually [1,2]. Petrochemical companies have used circular fluidized bed technology to reduce wastewater, reduce coal combustion, improve combustion efficiency and reduce carbon emissions. The derived wastes included fly ash (also known as mixed gypsum) and bottom ash (also known as by-product lime), which were then hydrated to produce hydrated by-product lime (also known as waste gypsum) for reuse to achieve the circular economy effect. The suitability of waste gypsum, which is an industrial by-product of secondary treatment, as a component material for replacing construction concrete was also a subject worthy of in-depth study and research. Using such materials as replacement for construction materials would be the most promising solution for waste resource recovery and disposal on a large scale [3].

In recent years, researchers worldwide have focused on recycling and disposing of the ash generated by circulating fluidized beds. These studies covered the basic material properties, hydration mechanisms, activation methods and applications of circulating fluidized bed combustion ashes [4]. They also provided an overview of the prospects of applying it in cementitious composites, alkali-activated materials or geopolymers [5]. The application of civil construction materials also included soil conditioners, cement substitutes, lightweight aggregates, road construction materials, controlled low-strength material (CLSM), roller compacted concrete and other technologies [6–9].

In this study, the waste gypsum provided by the plastic chemical manufacturer has been exposed to the air for more than 5 years in an interior. This has been carried out to stabilize it so it can be used as a complete replacement for fine aggregates in cement mortar. Ordinary Portland cement was used as the cementitious material, fly ash as the supplementary cementitious material and waste gypsum as the inert filler (fine aggregates). Testing for compressive strength, volume stability, water/sea immersion and scanning electron microscope (SEM) observations were used to determine the feasibility of the CLSM application.

2. Experimental Details

2.1. Materials and Mix Proportions

The test materials in this study were ordinary Portland cement, fly ash, waste gypsum and water in cement mortar specimens, in which waste gypsum of different particle-size scales was used to completely replace the fine aggregates. The specific gravity of Portland cement is 3.15 and the fineness is 3690 cm^2/g; the specific gravity of fly ash is 2.25 and the fineness is 9100 cm^2/g. The main oxidation components of fly ash contained 54.98% of SiO_2, 29.23% of Al_2O_3 and 3.38% of CaO. Waste gypsum was sieved into the particle size $\leqq$ 4.75 mm, and the particle size between 4.75 mm and 9.53 mm. The specific gravity of waste gypsum was 2.08 and the fineness modulus (less than 4.75 mm) was 3.58 in the oven-dried condition. The main chemical components of waste gypsum were calcium sulfate dihydrate (47.90%), calcium carbonate (14.80%), silicon dioxide (4.31%), calcium hydroxide (21.64%) and calcium oxide (5.82%). The external appearance of waste gypsum, as shown in Figure 1, reveals grayish-white multi-angular-shaped particles. The test mixture is presented in Table 1. The water/cementitious ratio was fixed at 0.65, the amount of fly ash instead of cement was 33%, and waste gypsum (fine aggregates) was 20% between 4.75 and 9.53 mm.

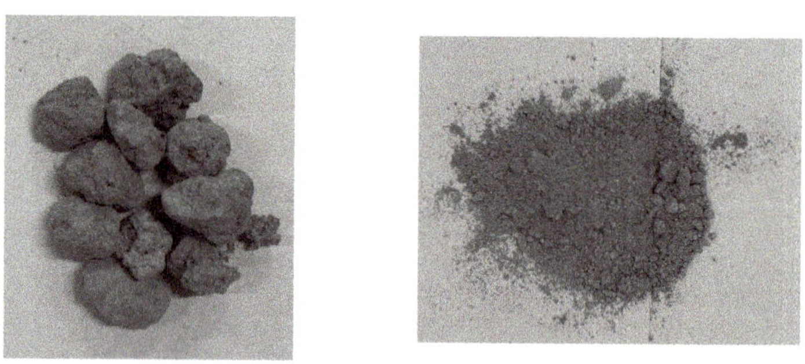

(**a**) Waste gypsum (large particles) (**b**) Waste gypsum (small particles)

Figure 1. Appearance of waste gypsum.

Table 1. Mix proportions (kg/m^3).

Mix No.	Water	Cement	Fly Ash	Waste Gypsum (0.149–4.75 mm)	Waste Gypsum (4.75–9.53 mm)
A	225	200	100	896	224

2.2. Test Procedures

Following the ASTM C109 standard procedure, 5 × 5 × 5 cm cubic specimens were used for compressive strength tests. The specimens were kept in the atmosphere and were water-cured. The compressive strength tests were conducted at different ages (7, 14, 28 days), and the average values of the four specimens were taken. The volume stability test specimens were made according to ASTM C596 standard as 2.5 × 2.5 × 28.5 cm prismatic specimens, and the length was taken as the average of 10 specimens. The test specimens for the immersion and exposure tests were 5 × 5 × 5 cm cubic specimens. The evaluation indices included specimen scale, weight and strength, and the evaluation was based on the average of five measurements. The exposure environment included natural seawater immersion, tap water immersion and atmospheric exposure. For SEM observations, the crushed specimens (about 1 × 1 × 3mm) after the compressive strength test were selected and tested according to the test procedure of ASTM C1723, and the age of the observed specimens was 56 days.

3. Results and Discussion

3.1. Compressive Strength

After demolding, the waste gypsum mortar specimens were first maintained in water and air for comparison. When the age of curing reached 7 days, the appearance of the specimens in water curing was irregularly cracked (as shown in Figure 2), which revealed that during the hydration of the specimens, the calcium sulfate dihydrate, calcium hydroxide and calcium oxide of waste gypsum might still react with water to produce expansion, resulting in cracking of the specimen after water immersion. On the contrary, there was no significant difference in the appearance of the air-cured specimens. Obviously, the specimens with waste gypsum completely replacing the fine aggregates should be placed in the atmospheric condition so that the hydration reaction could grow continuously.

Figure 2. Appearance of the cracked specimen.

Figure 3 illustrates the compressive strength development of air-cured specimens at 7, 14, 28 and 56 days. The test results indicated a minimal difference between the four test values for each age, and the compressive strength of air-cured specimens increased with age, with compressive strengths of 4.71 MPa and 6.08 MPa at 28 days and 56 days, respectively. As waste gypsum, cement, and fly ash solidified in the specimen, a small number of components such as CaO and $Ca(OH)_2$ that might react with CO_2 were stabilized. C-S-H colloids continued to form more with age due to the solidification reaction with cement and fly ash, resulting in higher compressive strengths in CLSM specimens. The strength results of specimens exposed to the atmosphere (by air curing) indicated that CO_2 in the atmosphere would not react continuously with the waste gypsum in CLSM specimens, causing harmful phenomena such as surface layer cracking.

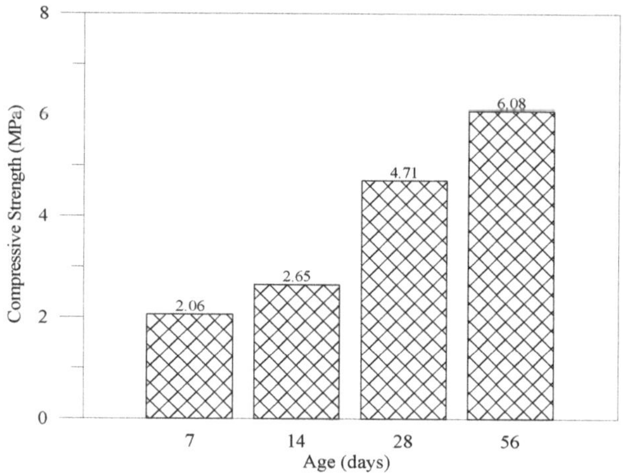

Figure 3. Compressive strength histograms.

3.2. Volume Stability

Among the tests measured at 28, 35, 42, 49 and 56 days, the length change curve along with age is shown in Figure 4. The length measurement value of the specimen at 28 days was used as the initial value for calculating the volume change. It was found that when exposed to an atmospheric CO_2 environment (atmospherically cured specimens), the length change of the specimens tended to be flat and did not change significantly. The average measured length change of the specimens was 0.0579% for 28 days of observation (age from 28 to 56 days). Based on the test results, it was assessed that CO_2 in the atmosphere would not react continuously with the waste gypsum in the specimens to compromise the volumetric stability of the CLSM specimens.

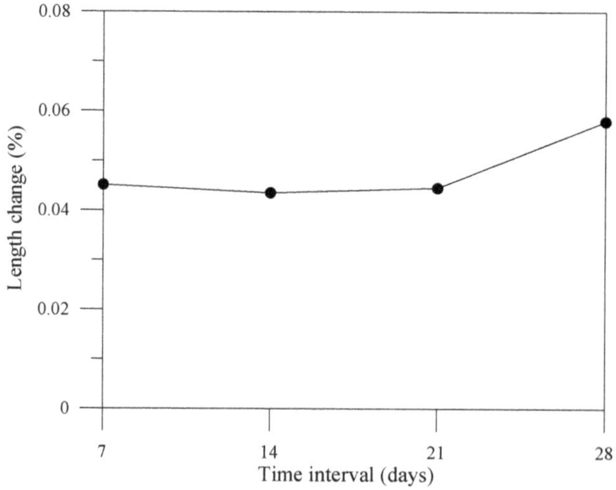

Figure 4. Length change curves.

After the specimen had been air-cured for 56 days, it was immersed in natural seawater, tap water and placed in the atmosphere. The length and weight variations at different ages of specimens immersed in tap water and seawater are shown in Figures 5 and 6, respectively. The results indicated that the average length variation of the specimens in seawater was 0.27%. In comparison, the length variation of the specimens before and after immersion in tap water was 1.00%. It can be deduced that there was no significant difference between the lengths of the specimens before and after immersion in seawater and tap water. After 28 days of immersion in seawater and tap water, the weight variations of the specimens were 3.20 and 2.93, respectively. Based on these results, it can be concluded that the weight fluctuations were insignificant. In addition, there was almost no change in the weight and volume of the specimens exposed to the atmosphere. Comparative appearances of the three immersed and exposed specimens are shown in Figure 7.

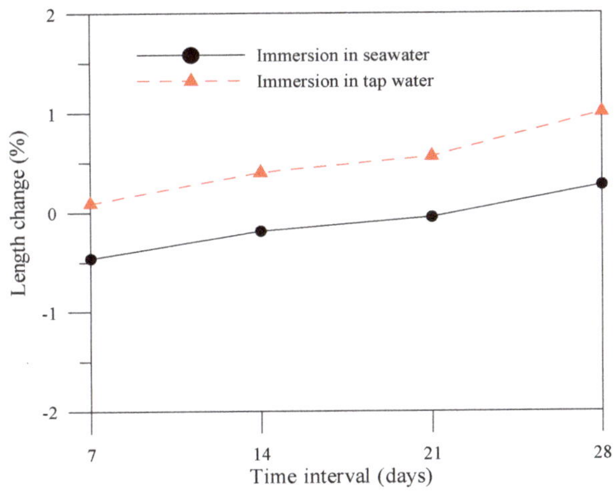

Figure 5. Trend of length change between seawater and tap water (black line: seawater).

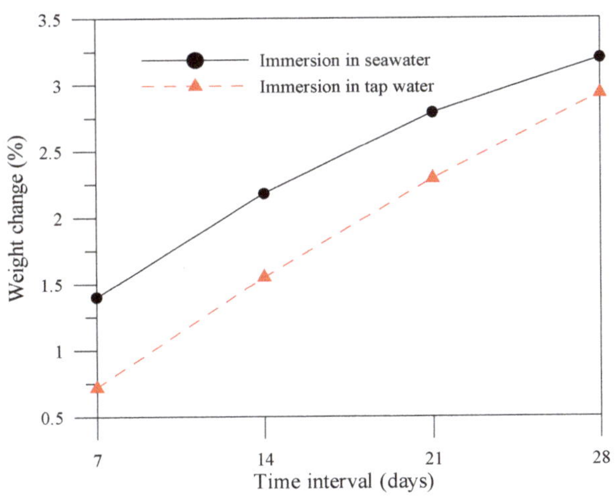

Figure 6. Trend of weight change between seawater and tap water (black line: seawater).

(**a**) atmosphere (**b**) seawater (**c**) tap water

Figure 7. Appearance of the three immersed and exposed specimens.

3.3. SEM Observations

Scanning electron microscopy photographs and corresponding energy dispersive X-ray (EDX) analyses are shown in Figure 8. Figure 8a,b show the surface microstructure of the waste gypsum specimens. It can be found in Figure 8a that hexagonal calcium hydroxide accounted for the majority of hydrates. In contrast, the larger particles of calcium hydroxide produced finer hydrates on the surface, and some areas of hydrates showed needle-like or column-like hydrates. Unhydrated fly ash particles were also observed on the surface of the microstructure. By magnifying the SEM photograph up to 10,000× (Figure 8b), it can be clearly observed that the interstices were filled with needle-shaped hydrides, which were presumed to be C-S-H and ettringite. The long needle-like crystals indicated the formation of large amounts of calcium–alumina hydrates (ettringite) in the waste gypsum specimens. In addition, since gypsum was the main component of the specimens, dispersed fibrous structures of the C-S-H gels were also observed, which is consistent with previous literature results [10,11]. The results of EDX in Figure 8c confirmed that the hydrates were primarily composed of Ca, Si and Al and that the hydrates were C-S-H or ettringite.

Figure 8d,e illustrate SEM images of the specimens after 56 days of curing, followed by 100 days of immersion in tap water and seawater. The specimen in Figure 8d revealed a large area of sodium hydroxide and needle-shaped hydrates (ettringite), and the water immersion facilitated the continued hydration of the specimens. In contrast, the unreacted fly ash particles continued to react to form C-S-H colloids, which filled the interface between the pores and the waste gypsum. Sodium hydroxide was observed in the specimen in Figure 8f; the specimen was immersed in seawater and the sulfide in the seawater was favorable for the growth of needle-like hydrates (ettringite). Dense ettringite could be seen filling the pores and interfaces on the surface of the specimens. From the EDX analysis results (Figure 8f), it can be verified that most of the hydrated reactants were ettringite (Ca-Si-Al crystals).

(**a**) Air-cured specimen at 56 days (×1000)

(**b**) Air-cured specimen at 56 days (×10,000)

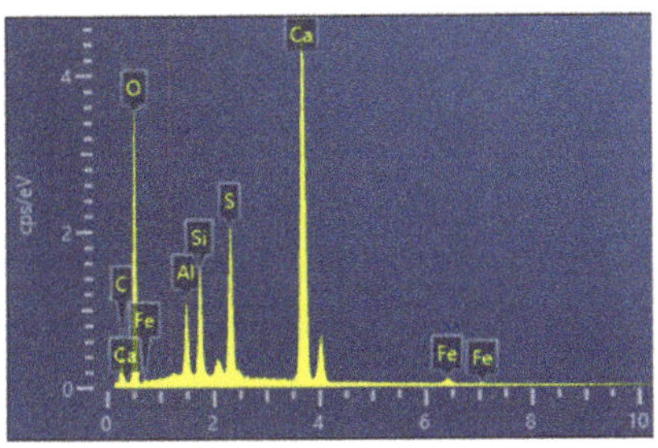

(**c**) EDX for ai-cured specimen at 56 days

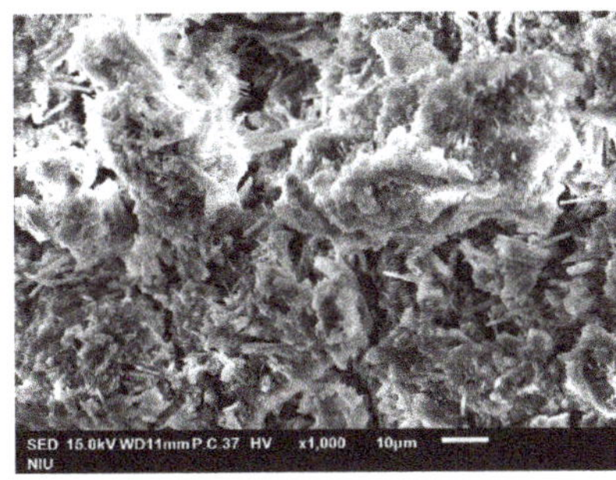

(**d**) Immersion in tap water for 100 days

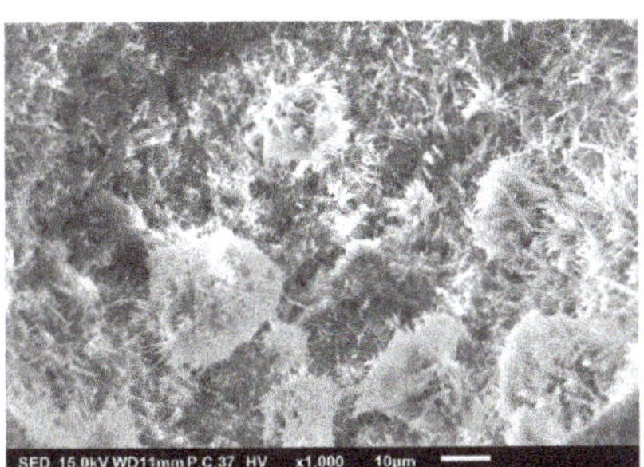

(**e**) Immersion in seawater for 100 days

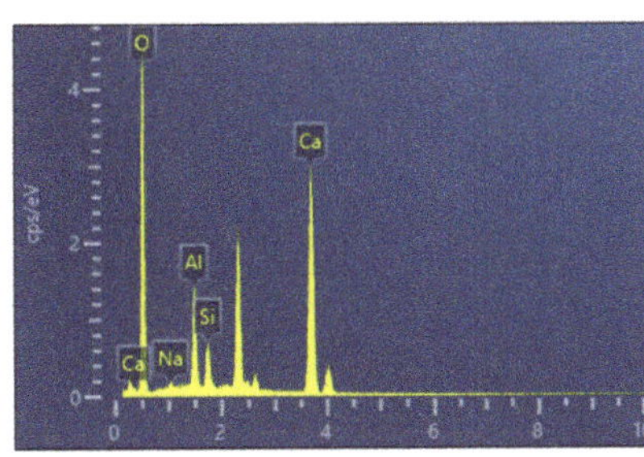

(**f**) EDX for specimen in seawater

Figure 8. SEM photos (×3000).

4. Conclusions

In this study, cement and fly ash were used as cementitious materials and waste gypsum was used as the fine aggregates, which were suitable for CLSM. The compressive strengths at 28 and 56 days were 4.71 and 6.08 MPa, respectively. The air-curing method should be applied to avoid abnormalities such as volume expansion after curing the specimen in water. With the hydrated blending reaction, the waste gypsum in the CLSM specimen continued to react with cement and fly ash. The reaction produced more ettringite, C-S-H colloids and other hydrates with age, increasing the compressive strength. In waste gypsum, some components react with CO_2 in the air, such as CaO and $Ca(OH)_2$, but they have become stable, allowing for the specimen to maintain a reasonable volume. In addition, the volume and weight of the specimen did not change significantly after being immersed in seawater and tap water, and the appearance of the specimen was intact.

Additionally, it was determined that waste gypsum can entirely replace fine aggregates in construction and can significantly increase the amount of reuse, thus achieving the effect of the green circular economy.

Author Contributions: Conceptualization, W.-T.L., A.C. and K.K.; methodology, K.K.; validation, W.-T.L., D.M. and M.Ł.; investigation, A.C.; resources, K.-L.L.; data curation, W.-T.L. and D.M.; writing—original draft preparation, W.-T.L. and K.K.; writing—review and editing, W.-T.L. and M.Ł.; visualization, A.C. and K.-L.L.; supervision, W.-T.L. and K.-L.L.; project administration, W.-T.L. and D.M. All authors have read and agreed to the published version of the manuscript.

Funding: The National Science and Technology Council (NSTC) in Taiwan supported this research with grant NSTC 111-2923-E-197-003-MY3 and the Polish National Centre for Research and Development in Poland, under the M-ERA.NET 3 program, grant number M-ERA.NET3/2021/144/3D-FOAM/2022.

Institutional Review Board Statement: Not applicable.

Informed Consent Statement: Not applicable.

Data Availability Statement: Not applicable.

Acknowledgments: Publication cost of this paper was covered with founds of the Polish National Agency for Academic Exchange (NAWA): "MATBUD'2023—Developing international scientific cooperation in the field of building materials engineering" BPI/WTP/2021/1/00002, MATBUD'2023.

Conflicts of Interest: The authors declare no conflict of interest.

References

1. Wei, Y.; Liu, G.; Fu, B.; Wang, B.; Liu, Y.; Xue, X.; Sun, M. Partitioning behavior of Pb in particulate matter emitted from circulating fluidized bed coal-fired power plant. *J. Clean. Prod.* **2021**, *292*, 125997. [CrossRef]
2. Zhang, W.; Gu, J.; Zhou, X.; Li, Y.; Wang, Y.; Xue, Y.; Liu, X. Circulating fluidized bed fly ash based multi-solid wastes road base materials: Hydration characteristics and utilization of SO_3 and f -CaO. *J. Clean. Prod.* **2021**, *316*, 128355. [CrossRef]
3. He, P.; Zhang, X.; Chen, H.; Zhang, Y. Waste-to-resource strategies for the use of circulating fluidized bed fly ash in construction materials: A mini review. *Powder Technol.* **2021**, *393*, 773–785. [CrossRef]
4. Xu, G.; Shi, X. Characteristics and applications of fly ash as a sustainable construction material: A state-of-the-art review. *Resour. Conserv. Recycl.* **2018**, *136*, 95–109. [CrossRef]
5. Gollakota, A.R.; Volli, V.; Shu, C.M. Progressive utilisation prospects of coal fly ash: A review. *Sci. Total Environ.* **2019**, *672*, 951–989. [CrossRef] [PubMed]
6. Ohenoja, K.; Pesonen, J.; Yliniemi, J.; Illikainen, M. Utilization of fly ashes from fluidized bed combustion: A review. *Sustainability* **2020**, *12*, 2988. [CrossRef]
7. Lin, W.T.; Lin, K.L.; Chen, K.; Korniejenko, K.; Hebda, M.; Łach, M. Circulation fluidized bed combustion fly ash as partial replacement of fine aggregates in roller compacted concrete. *Materials* **2019**, *12*, 4204. [CrossRef] [PubMed]
8. Lin, W.T.; Weng, T.L.; Cheng, A.; Chao, S.J.; Hsu, H.M. Properties of controlled low strength material with circulating fluidized bed combustion ash and recycled aggregates. *Materials* **2018**, *11*, 715. [CrossRef] [PubMed]
9. Chen, G.Y.; Huang, W.H. Activation of blast furnace slag with CFB fly ash as a supplementary binder material: Hydration products and effects of sulfate attack. *Crystals* **2022**, *12*, 41. [CrossRef]
10. Mota, B.; Matschei, T.; Scrivener, K. The influence of sodium salts and gypsum on alite hydration. *Cem. Concr. Res.* **2015**, *75*, 53–65. [CrossRef]
11. Zhang, Y.; Wan, Z.; Miranda de Lima Junior, L.; Çopuroğlu, O. Early age hydration of model slag cement: Interaction among C_3S, gypsum and slag with different Al_2O_3 contents. *Cem. Concr. Res.* **2022**, *161*, 106954. [CrossRef]

materials proceedings

MDPI

Proceeding Paper

The Influence of the Ageing Process on the Mechanical Properties of Cement Mortars with Nano-SiO_2 Admixture Initially Subjected to Thermal Treatment †

Elżbieta Horszczaruk [1,*] and Paweł Łukowski [2]

1 Department of Civil and Environmental Engineering, West Pomeranian University of Technology in Szczecin, al. Piastów 50a, 70-311 Szczecin, Poland
2 Department of Building Materials Engineering, Faculty of Civil Engineering, Warsaw University of Technology, 00-637 Warsaw, Poland
* Correspondence: elzbieta.horszczaruk@zut.edu.pl
† Presented at the 10th MATBUD'2023 Scientific-Technical Conference "Building Materials Engineering and Innovative Sustainable Materials", Cracow, Poland, 19–21 April 2023.

Abstract: The research aimed to evaluate the effect of nano-silica (NS) on the strength recovery of the previously heated cement mortars. One hundred twenty cement mortar specimens were prepared with 1% to 5% of NS. The specimens were heated at temperatures ranging from 200 °C to 800 °C. Half of the samples, after heating and cooling, were mechanically tested. The other half of the specimens were stored for 24 months in 90% relative humidity. The presence of NS fostered partial recovery of the compressive strength of the heated mortars. The cement mortars regained more than 40% of their compressive strength after heating at 400 °C and about 25% after heating at 600 °C. The strength recovery phenomenon faded away at 800 °C.

Keywords: cement mortars; compressive strength; high temperature; nano-silica

check for updates

Citation: Horszczaruk, E.; Łukowski, P. The Influence of the Ageing Process on the Mechanical Properties of Cement Mortars with Nano-SiO_2 Admixture Initially Subjected to Thermal Treatment. *Mater. Proc.* **2023**, *13*, 20. https://doi.org/10.3390/materproc2023013020

Academic Editors: Katarzyna Mróz, Tomasz Tracz, Tomasz Zdeb and Izabela Hager

Published: 14 February 2023

1. Introduction

Despite good thermal resistance, cement composites are subject to physical and chemical changes at high temperatures. These changes lead to the worsening of mechanical performance and structural defects. The research described in [1] and [2] has confirmed that a temperature lower than 300 °C does not affect the cement composites' strength. A significant strength drop is observed when the temperature reaches 400 °C [3], and at 800 °C, the strength decreases to be below 30% of the initial strength [4]. The limit temperature is 1000 °C, above which the cement composites lose most of their total strength [5]. The worsening of the mechanical performance of the cement composites, including concrete, at a high temperature is connected to the decomposition of hydrates, and mainly, the C-S-H phase and portlandite [6], a non-uniform thermal gradient, which causes shrinkage of the cement matrix and expansion of the coarse aggregate [7], as well as the loss of bound water, which creates voids [8].

Recent studies showed [9–12] that cement concrete damaged by fire can regain its original strength if exposed to moisture for a long time. The process of mechanical property recovery by concrete is called the after-fire hardening technique. The main reason for regaining the strength of the cement composites is the rehydration of the cement matrix dehydrated during the fire [11–13].

The tests on the cement mortars containing 1% to 5% of colloidal nano-silica (NS) have shown that NS can support the partial strength recovery of the cement composite following its heating at a high temperature.

2. Materials and Methods

2.1. Materials

The mortars to be tested were prepared with the use of Portland cement CEM I 42.5N, a magnetite aggregate with a grain size up to 2 mm and apparent density of 4.77 g/cm^3, and the water dispersion of nano-silica. The water dispersion of NS with the density of 1.4 g/cm^3 contained 50% of the colloidal NS by volume. The diameters of the NS particles were from 40 to 140 nm. The TEM images of NS particles are presented in Figure 1.

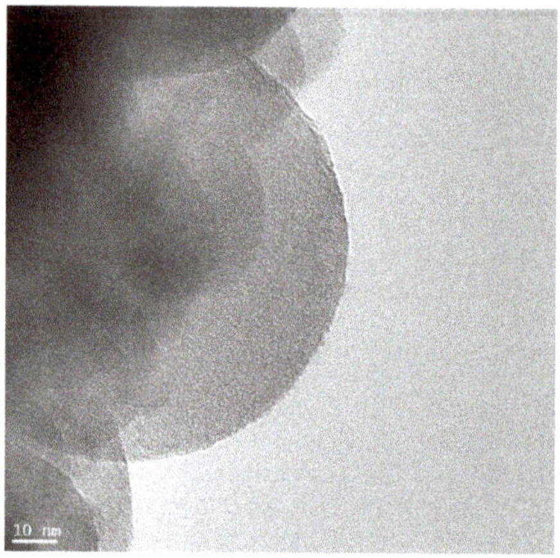

Figure 1. TEM images of nano-silica (NS) [14].

2.2. Samples and Their Storage

The tests were carried out on the cement mortars whose compositions are presented in Table 1. The water content was reduced in mortars containing nano-silica in the form of dispersion. Specimens of the size 40 mm × 40 mm × 160 mm were prepared according to the European Standard PN-EN 196-1 [15].

Table 1. Composition of the tested mortars.

Mortar Designation	Cement (kg/m^3)	Water (kg/m^3)	Magnetite (kg/m^3)	NS (% of Cement Mass)	NS Dispersion (kg/m^3)
M0	450	225	2439	0	0
M1	450	220.5	2439	1	9
M2	450	216	2439	2	18
M3	450	221.5	2439	3	27
M4	450	207	2439	4	36
M5	450	202	2439	5	45

The demoulded specimens were stored in a climate chamber on the grate above water at 20 ± 2 °C. After 28 days of curing, the specimens were heated. Then, part of the samples were tested, and the others were stored in the chamber for 24 months until further testing. In total, sixty specimens of the tested mortars were conditioned for two years, from which two specimens of each mortar were initially heated at the given temperature.

2.3. Test Methods

After 28 days of curing, the specimens were heated in the medium-temperature oven. In the first stage, the temperature in the oven rose at the rate of 1 °C/min until reaching the specified value. The specimens were heated at 200, 400, 600, and 800 °C. The final heating temperature was maintained for 60 min so that the entire volume of the specimens could achieve the proper uniform temperature. Then, the specimens were cooled down at the rate of 1 °C/min. The cooled samples were removed from the oven and weighed. The compressive strength of the remaining specimens was determined after 24 months of storage in the climate chamber. The compressive strength was determined according to the European Standard PN-EN 196-1 [15].

3. Results and Discussion

3.1. Compressive Strength after Heating

The results of the compressive strength testing after heating are presented in Figure 2. The positive effect of the NS admixture becomes visible with heating at an increasing temperature. The highest compressive strength within the tested temperature range was determined for M3 mortar specimens.

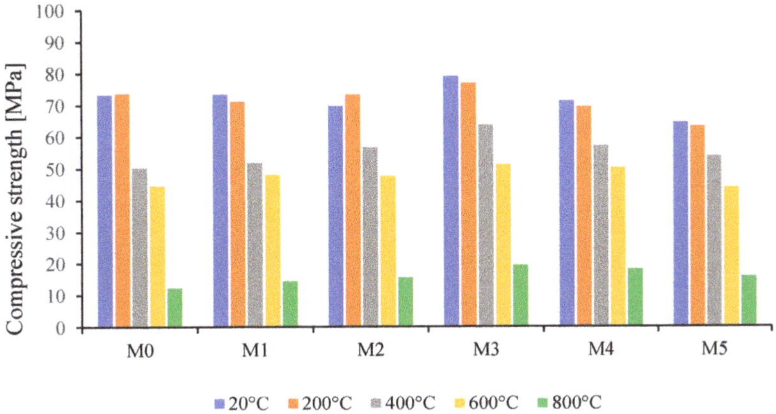

Figure 2. Compressive strength of the heated mortars.

3.2. Mechanical Property Recovery by Mortars Damaged at High Temperature

Some of the heated mortar specimens were stored for 24 months in the moisture conditions (see Point 2.2) and then tested. The results of compressive strength determination for these specimens are presented in Figure 3. The highest growth of the compressive strength was observed for M3 mortar specimens within the entire range of temperatures. However, the significant recovery of the strength after 24 months of storage was noted for all tested mortars.

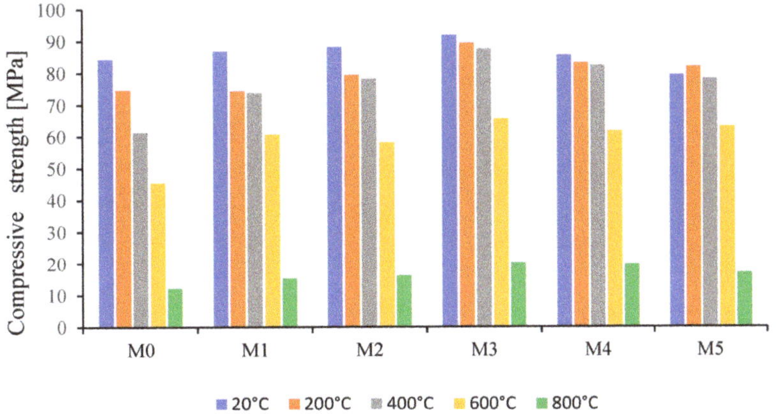

Figure 3. Compressive strength of the mortars after heating and curing for the next 24 months.

The recovery of the mechanical properties of the heated cement mortars after 24 months of storage in the moisture conditions is presented in Figure 4. The compressive strength determined after heating was accepted as the reference strength, and the strength after 24 months of storage was compared to it.

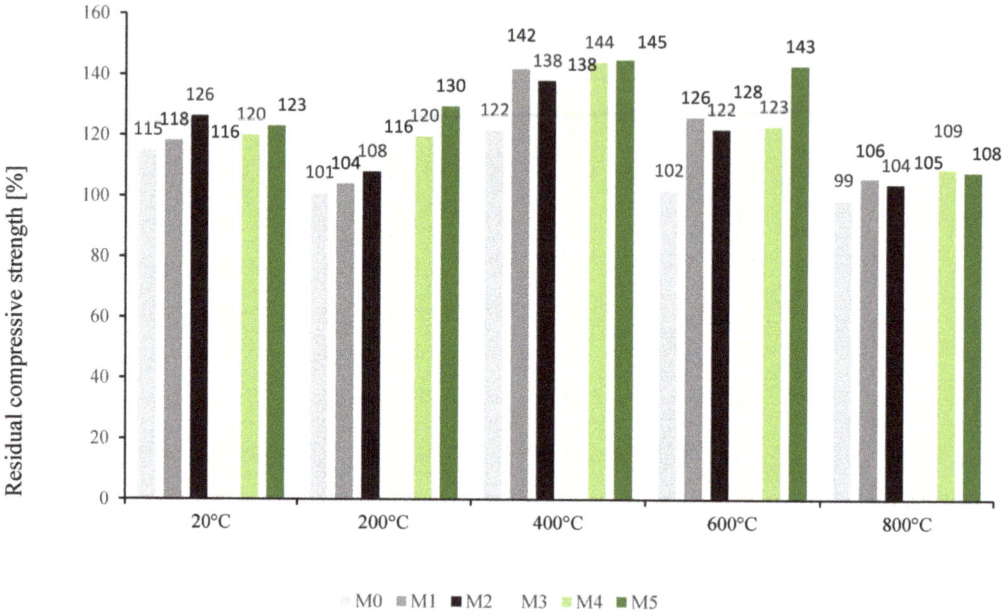

Figure 4. Relative compressive strength of the heated mortars after 24 months of further curing.

After 24 months of storage, all mortar specimens achieved a compressive strength higher than that obtained immediately after heating, except for the M0 mortar, which demonstrated a slight decrease in the average strength (by 1%) after heating at 800 °C. The positive effect of the NS's presence on the mortars' structure recovery is visible. The increase in the mortars' strength after heating at 200 °C was proportional to the NS content. The moisture delivered during storage in the climate chamber after heating made it possible for the NS to hydrate fully, which led to the growth of strength. The highest increase in the strength was observed for the specimens heated at 400 °C. The compressive strength of the mortars containing NS increased by 38% to 45%, while the compressive strength of the mortar M0 without NS increased by only 22%. The strength of the mortars with NS, heated at 600 °C, also rose after 24 months of storage. However, the increase was smaller (23% to 28%), except for in mortar M5, where a strength increase of 43% was noted. The phenomenon of a strength increase fades away after heating at 800 °C. Some compressive strength growth was observed only for the mortars with the higher NS content; it was 9% for the mortar M4 and 8% for the mortar M5. For the other mortars, the strength increased by about 5%, which is on the border of the measurement error.

The compressive strength increase in the heated cement mortars is caused by the rehydration of the C-S-H phase and hydration of other, previously not hydrated, components, such as cement grains, NS, and CaO [16,17]. The partial regaining of the compressive strength by the mortars containing the NS is possible when the volume of delivered water is sufficient for the hydration of yet unhydrated nano-silica. As shown in the study described in [18], the NS agglomerates forming in the cement matrix at the NS content above 3% are the reason for decreasing strength after heating the mortars compared to the mortars without nano-silica. The downfall of the strength is caused, among other reasons, by the air voids formed in the area of NS agglomerates. The tests presented in this paper have shown that sufficient water delivery leads to the hydration of yet unhydrated NS and, consequently, the filling of the created air pores and voids. The filling of the empty spaces causes, in turn, the higher compressive strength increase for the mortars M4 and M5. At the temperature of 800 °C, the damages in the material's internal structure are too big for significant strength recovery. The authors plan to continue the investigation using more sophisticated testing methods, including scanning electron microscopy, X-ray diffractometry, and X-ray computer tomography.

4. Conclusions

The presented research aimed to evaluate the nano-silica (NS) influence on the strength recovery of the previously heated cement mortars. The tests have shown that NS fosters compressive strength recovery by the mortars heated up to 600 °C. The composites regain more than 40% of the compressive strength after heating at 400 °C and about 25% after heating at 600 °C. The main reason for the strength recovery is the rehydration of the C-S-H phase and the hydration of other, previously not hydrated, components. Additionally, the hydration of yet unhydrated nano-silica contributes to regaining the compressive strength when the volume of delivered water is sufficient. The strength recovery phenomenon fades away at 800 °C when the damages in the material's internal structure are too big for significant strength recovery.

Author Contributions: Conceptualization, E.H. and P.Ł.; methodology, E.H. and P.Ł.; formal analysis, E.H.; investigation, E.H.; writing—original draft preparation, E.H.; writing—review and editing, P.Ł. All authors have read and agreed to the published version of the manuscript.

Funding: The publication cost of this paper was covered with funds from the Polish National Agency for Academic Exchange (NAWA): "MATBUD'2023—Developing international scientific cooperation in the field of building materials engineering" BPI/WTP/2021/1/00002, MATBUD'2023.

Institutional Review Board Statement: Not applicable.

Informed Consent Statement: Not applicable.

Data Availability Statement: Not applicable.

Conflicts of Interest: The authors declare no conflict of interest.

References

1. Korsun, V.I.; Vatin, N.; Korsun, A.; Nemova, D. Physical-Mechanical Properties of the Modified Fine-Grained Concrete Subjected to Thermal Effects up to 200 °C. *Appl. Mech. Mater.* **2014**, *633–634*, 1013–1017. [CrossRef]
2. Korsun, V.; Baranov, A. Mechanical Properties of High-Strength Concrete after Heating at Temperatures up to 400 °C. In Proceedings of the International Scientific Conference on Energy, Environmental and Construction Engineering, St. Petersburg, Russia, 19–20 November 2020; Springer International Publishing: Berlin/Heidelberg, Germany; pp. 454–463.
3. Handoo, S.K.; Agarwal, S.; Agarwal, S.K. Physicochemical, Mineralogical, and Morphological Characteristics of Concrete Exposed to Elevated Temperatures. *Cem. Concr. Res.* **2002**, *32*, 1009–1018. [CrossRef]
4. Xu, Y.; Wong, Y.L.; Poon, C.S.; Anson, M. Impact of High Temperature on PFA. *Cem. Concr. Res.* **2001**, *31*, 1065–1073. [CrossRef]
5. Husem, M. The Effects of High Temperature on Compressive and Flexural Strengths of Ordinary and High-Performance Concrete. *Fire Saf. J.* **2006**, *41*, 155–163. [CrossRef]
6. Alonso, C.; Fernandez, L. Dehydration and Rehydration Processes of Cement Paste Exposed to High Temperature Environments. *J. Mater. Sci.* **2004**, *39*, 3015–3024. [CrossRef]
7. Wang, H.Y. The Effects of Elevated Temperature on Cement Paste Containing GGBFS. *Cem. Concr. Compos.* **2008**, *30*, 992–999. [CrossRef]
8. Peng, G.-F.; Huang, Z.-S. Change in Microstructure of Hardened Cement Paste Subjected to Elevated Temperatures. *Constr. Build. Mater.* **2008**, *22*, 593–599. [CrossRef]
9. Henry, M.; Darma, I.S.; Sugiyama, T. Analysis of the Effect of Heating and Re-curing on the Microstructure of High-Strength Concrete Using X-ray CT. *Constr. Build. Mater.* **2014**, *67*, 37–46. [CrossRef]
10. Li, Q.; Yuan, G.; Shu, Q. Effects of Heating/Cooling on Recovery of Strength and Carbonation Resistance of Fire-Damaged Concrete. *Mag. Concr. Res.* **2014**, *66*, 925–936. [CrossRef]
11. Noman, M.; Yaqub, M. Restoration of Dynamic Characteristics of RC T-Beams Exposed to Fire Using Post Fire Curing Technique. *Eng. Struct.* **2021**, *249*, 113339. [CrossRef]

12. Usman, M.; Yaqub, M.; Auzair, M.; Khaliq, W.; Noman, M.; Afaq, A. Restorability of Strength and Stiffness of Fire Damaged Concrete Using Various Composite Confinement Techniques. *Constr. Build. Mater.* **2021**, *272*, 121984. [CrossRef]
13. Xuan, D.X.; Shui, Z.H. Rehydration Activity of Hydrated Cement Paste Exposed to High Temperature. *Fire Mater.* **2011**, *35*, 481–490. [CrossRef]
14. Brzozowski, P.; Strzałkowski, J.; Rychtowski, P.; Wróbel, R.; Tryba, B.; Horszczaruk, E. Effect of Nano-SiO_2 on the Microstructure and Mechanical Properties of Concrete under High Temperature Conditions. *Materials* **2022**, *15*, 166. [CrossRef] [PubMed]
15. *PN-EN 196-1:2016-07*; Methods of testing cement—Part 1: Determination of strength. Polish Committee of Standardization: Warsaw, Poland, 2016.
16. Poon, C.-S.; Azhar, S.; Anson, M.; Wong, Y.-L. Strength and durability recovery of fire-damaged concrete after post-fire-curing. *Cem. Concr. Res.* **2001**, *31*, 1307–1318. [CrossRef]
17. Henry, M.; Suzuki, M.; Kato, Y. Behavior of fire-damaged mortar under variable re-curing conditions. *ACI Mater. J.* **2011**, *108*, 281–289. [CrossRef]
18. Horszczaruk, E.; Sikora, P.; Cendrowski, K.; Mijowska, E. The effect of elevated temperature on the properties of cement mortars containing nanosilica and heavyweight aggregates. *Constr. Build. Mater.* **2017**, *137*, 420–431. [CrossRef]

materials proceedings

MDPI

Proceeding Paper

Characterization of a Coal Shale from Marcel Mining as Raw Material for Geopolymer Manufacturing †

Kinga Korniejenko [1,*], Beata Figiela [1], Kinga Pławecka [1], Alperen Bulut [2], Baran Şahin [2], Göksu Azizağaoğlu [2] and Michał Łach [1]

1 Faculty of Material Engineering and Physics, Cracow University of Technology, Jana Pawła II 37, 31-864 Cracow, Poland

2 Faculty of Engineering, İzmir Institute of Technology, Gülbahçe Kampüsü, 35430 Urla, İzmir, Türkiye

* Correspondence: kinga.korniejenko@pk.edu.pl; Tel.: +48-60997-4988

† Presented at the 10th MATBUD'2023 Scientific-Technical Conference "Building Materials Engineering and Innovative Sustainable Materials", Cracow, Poland, 19–21 April 2023.

Abstract: Implementing the idea of a circular economy is one way to reduce carbon emissions and, at the same time, the consumption of natural resources. The use of mining waste as a raw material helps meet the growing demand for construction materials with a smaller carbon footprint. The article shows the possibility of using a coal shale from Marcel mining to create new eco-friendly materials, geopolymers. The main aim of the presented research includes characteristics of raw material and synthesis of geopolymers based on mining waste (coal shale from Marcel mining) and next, investigations of the obtained materials. Geopolymer was prepared using a sodium activator plus milled and calcinated precursor materials. In this study, the following research methods were used: particle size analysis, XRD analysis, mechanical properties tests (compressive and bonding strength), and microstructure analysis—scanning electron microscopy. The results show potential for the extraction of waste from the Marcel company to obtain material for advanced applications in the geopolymerization process. The material had a compressive strength of 12.7 MPa and a bending strength of 3.4 MPa, which makes it possible for use in construction applications such as various types of foundations, walls, columns, lintels, terraces, stairs, ceilings, small building elements, and small architecture. The proposed process could be a promising alternative to current methods of managing waste rock, in particular hard coal mining.

Keywords: geopolymer; coal shale; mining waste; industrial by-products; circular economy

check for updates

Citation: Korniejenko, K.; Figiela, B.; Pławecka, K.; Bulut, A.; Şahin, B.; Azizağaoğlu, G.; Łach, M. Characterization of a Coal Shale from Marcel Mining as Raw Material for Geopolymer Manufacturing. *Mater. Proc.* **2023**, *13*, 21. https://doi.org/10.3390/materproc2023013021

Academic Editors: Katarzyna Mróz, Tomasz Tracz, Tomasz Zdeb and Izabela Hager

Published: 14 February 2023

1. Introduction

The concept of circular economy is expected to bring many benefits to both economic development and the environment. The potential environmental factors are those such as decreasing the use of natural resources, limitation of emissions, recycling, fewer material losses through recycling, etc. [1,2]. The circular economy can also be seen as the next possible step in the evolution of waste management systems, including the development of sustainable materials that will allow closed loops. It could be possible through the introduction of manufacturing processes by being able to reuse materials at the end of their useful life [2,3]. This issue is important in the context of a realistic introduction of the circular economy into existing systems [4,5]. Materials corresponding to the needs of this model are, among others, geopolymers.

Geopolymers are a class of synthetic inorganic aluminosilicate materials, usually formed as a result of the reaction of aluminosilicates (e.g., fly ash or metakaolin) with a solution of silicates under usually strongly alkaline or sometimes acidic conditions [5,6]. Their synthesis has been estimated to emit about five–six times less carbon dioxide than the production of Portland cement and needs significantly less energy for the manufacturing process [7,8], but it is not only the advantages of geopolymer materials that take into account the circular economy approach. One of the most important benefits is the possibility of using waste for production [5,7]. One of the most promising opportunities is mine tailings [7,9]. These types of wastes are reachable in aluminosilicate, which is a main component of the geopolymerization process [9,10].

In recent years, there has been a visible increase in interest in the use of mine tailings for the production of geopolymer materials in the literature [9,11,12]. Previous investigations were related to different kinds of waste, including coal shales [10], gold mining [13], tungsten mining waste mud [14], copper tailing [15], zinc mining [16], garnet tailings from molybdenum mines [17], and others [7,18]. From the point of view of Poland and neighboring countries, the promising materials are coal shales. Coal mine tailings represent a significant waste stream that is used for geopolymer manufacturing. The current work shows the possibility of using a coal shale from Marcel mining for this purpose. This mine tailing has not been previously analyzed as a raw material for geopolymerization. The main aim of the presented research is to characterize the raw material, design geopolymers based on mining waste from Marcel mining, and determine potential applications for them.

2. Materials and Methods

2.1. Materials

Geopolymer samples were produced based on mine tailings for coal production from a PGG Sp. z o.o. Branch KWK 'ROW' Ruch 'MARCEL' coal mine in the Silesia region (Radlin, Poland) and quartz sand as a fine aggregate. This mine was founded in 1949 and is still operating today. Currently, the mine employs 3,168 people, and its net daily output is 11,000 tons.

The main tailings were delivered in the form of rocks with dimensions ca. 20 cm. Firstly, they were fragmented into smaller particles by using a crushing machine. In the second step, they were ground using an ultra-centrifugal mill. The size reduction takes place by impact and shearing effects between the rotor and the fixed ring sieve. As a result of this operation, the fine powder (fraction < 40 µm) was obtained. In the next step, the obtained material was calcinated using a laboratory furnace. The calcinations were performed for 24 h at a temperature of 700 °C. The materials were heated and cooled together in a furnace. The main aim of this process was to increase the material reactivity by obtaining the appropriate microstructure. Additionally, the process made it possible to remove volatile substances, oxidize a portion of the mass of the material, and decrease the carbon content.

2.2. Samples Preparation

The calcined waste was mixed, in a ratio of 1:1, with quartz sand. The dry components were mixed in a slow-speed mixer (Geolab, Warsaw, Poland) cement mortar mixer for about 5 min. The prepared mix was activated with an activator solution consisting of a 10 M NaOH solution (technical sodium hydroxide flakes) and an aqueous sodium silicate solution—density 1.45 g/cm^3 (water glass solution, R-145)—in which the ratio of the sodium base solution to water glass solution was 1: 2.5. Tap water was used to prepare the solution. After the addition of liquid components, the process of mixing was continued for about 15 min, up to obtaining a homogeneous paste. The prepared geopolymer masses were poured into a set of molds with dimensions 50 mm × 50 mm × 200 mm (prismatic forms). Then, in order to remove air bubbles, the molds with the materials were placed on a vibrating table for 15 min. Subsequently, to prevent too-quick water evaporation, each mold was wrapped in a polyethylene film. For the acceleration of the curing process, the samples were placed at 75 °C for 24 h. After this time, they were cooled to room temperature, unmolded, and stored in laboratory conditions for 28 days at ambient temperature (ca. 23 °C). After this time, the mechanical properties were investigated—Figure 1.

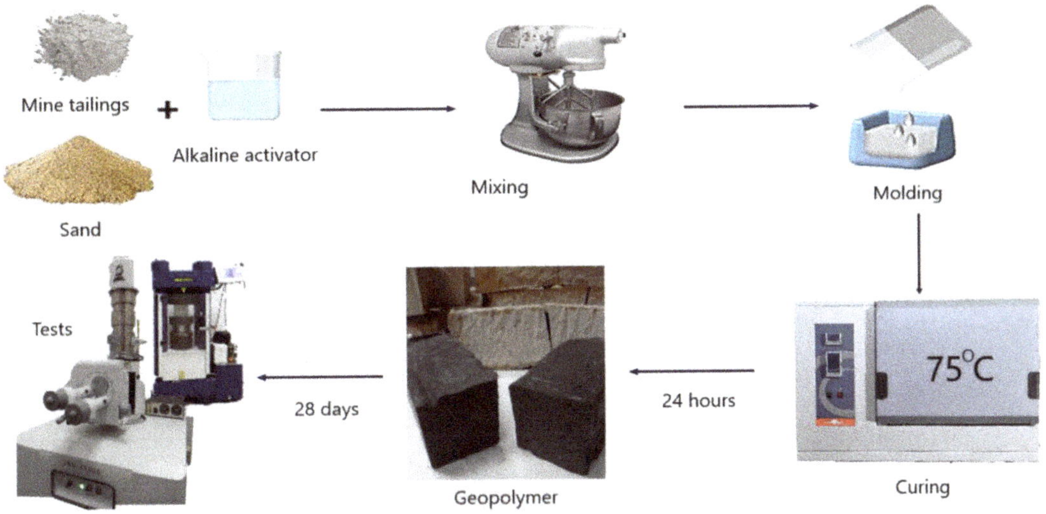

Figure 1. Scheme for sample preparation.

2.3. Methods

The particle size analysis was performed using a Particle Size Analyzer (AntonPaar GmbH, Graz, Austria) for the milled mine tailings after the calcination process to confirm the size of obtained grain and its distribution.

The mechanical properties of the samples were investigated using the MATEST 3000 kN test machine (Matest, Treviolo, Italy), according to the following regulations.

- PN-EN 12390-3:2019-07—Testing Hardened Concrete—Part 3: Compressive Strength of Test Specimens.
- PN-EN 12390-5:2019-08—Testing Hardened Concrete—Part 5: Flexural Strength of Test Specimens.

The applied speed in both cases was 0.05 MPa/s. The flexural strength was investigated in prismatic samples having dimensions: 50 mm × 50 mm × 200 mm and the distance between the supports was 150 mm (three-point bending test). The compressive strength was investigated in the samples after the flexural strength test with the use of a metal plate with dimensions: 50 mm × 50 mm.

The mineralogical composition was determined using X-ray diffraction XRD analysis. This investigation was carried out on powdered mine tailing before calcination using PANalytical AERIS (PANalytical, Almelo, The Netherlands). This device applies the X-ray method (Debye–Scherrer) for phase analysis. It was carried out using Cu-Kα radiation and components such as a nickel filter on the lamp, mask 13, and Szlot 1°. The following parameters were applied to this research—angle range: 9999–100°2Θ, measurement step: 0,0027166°2Θ, time of counts: 340,425 s, and total measurement time: 13 h 2 min 32 s. Next, the results were analyzed using HighScore Plus software, based on PDF4 + crystallographic database.

Microstructure investigation was performed on a JEOL JSM-IT200 scanning electron microscope (JEOL, Tokyo, Japan). This type of scanning electron microscope (SEM) has an energy-dispersive X-ray spectroscopy system (EDS), which allows for analyzing the chemical and oxygen composition of the investigated material. The research was provided on the material after mechanical properties tests. The samples were covered by gold plating for good conductivity. The cover layer was applied by a DII-29030SCTR Smart Coater (JEOL, Tokyo, Japan).

3. Results and Discussion

3.1. Physical Properties

The results of the particle size analysis, after grinding and milling, are presented in Figure 2 and Table 1.

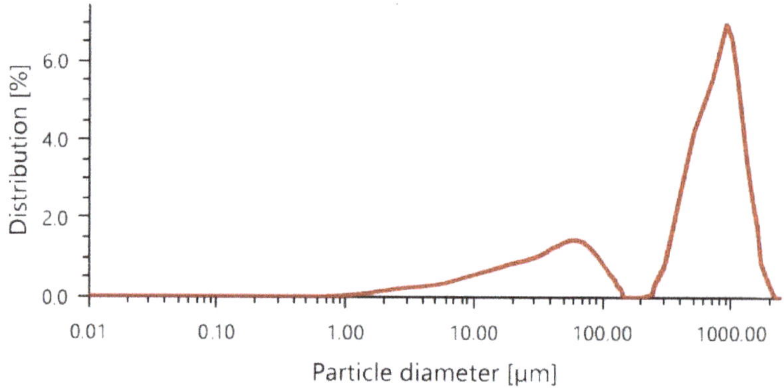

Figure 2. Particle size distribution.

Table 1. Basic parameters for the particle size distribution.

D_{10} (µm)	D_{50} (µm)	D_{90} (µm)	Mean Size (µm)	Mean Size (µm)
19.697	548.445	1150.324	600.729	2.062

The mean size of the particle is about 600 µm. In comparison to the other raw materials used in the geopolymerization process, such as metakaolin or fly ash, this particle size is quite large [2]. It should be noted that research shows that decreasing the particle size can positively influence the material reactivity [2], but at the same time increase the energy consumption for the whole process.

3.2. Mechanical Properties

The results for compressive and flexural strength are presented in Table 2.

Table 2. Mechanical properties.

Sample	Compressive Strength (MPa)	Flexural Strength (MPa)
50% mine tailings + 50% sand	12.7 (±3.3)	3.4 (±0.2)

The obtained values for mechanical properties are quite reasonable. The obtained results are not as good as for fly-ash-based geopolymers [5]. However, the values allow for planning some possible applications for this material, especially as a material for construction purposes, such as pavements onsite at the mining place [19].

3.3. Microstructure Investigation

The obtained diffractogram is presented in Figure 3. Based on this investigation a quantitative analysis was performed and the mineralogical composition was determined—Table 3.

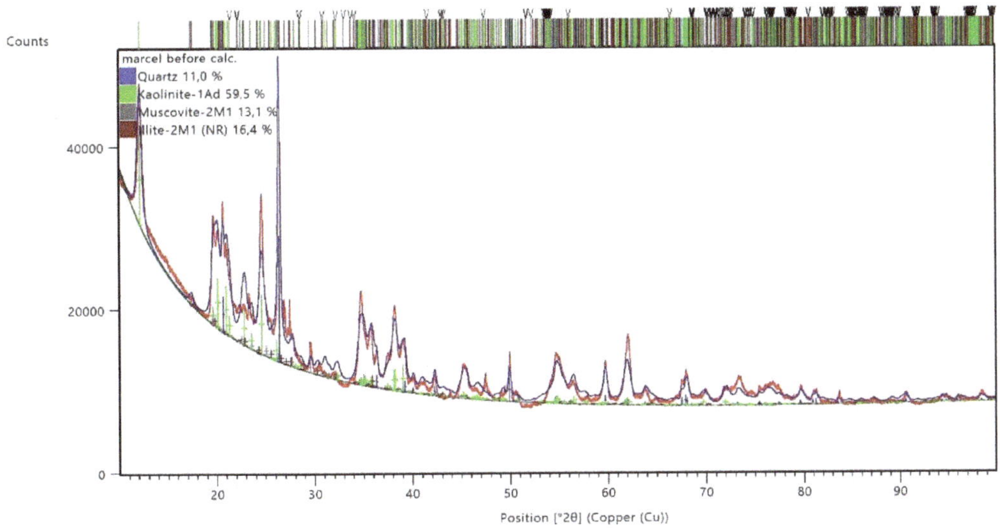

Figure 3. XRD analysis for coal shale from Marcel mining (before calcination).

Table 3. Phases identified along with their percentage share in the sample.

Identified Phases		Percentage (%)	Chart No.
Name	**Chemical Formula**		
Quartz	SiO_2	11.0	01-070-3755
Muscovite-2M1	$KAl_2(Si_3Al)O_{10}(OH,F)_2$	13.1	00-006-0263
Kaolinite-1Ad	$Al_2Si_2O_5(OH)_4$	59.5	00-058-2006
Illite-2M1	$(K,H_3O)Al_2Si_3AlO_{10}(OH)_2$	16.4	00-026-0911

The results show a relatively small amount of quartz, a high amount of kaolinite, and a slightly increased amount of illite and typical for coal shales amount of muscovite [20]. The relatively low mechanical properties of the designed geopolymers can be associated with a small amount of quartz that affects the mechanical properties [21]. At the same time, illite could also decrease the mechanical properties of geopolymer [21]. The common influence of a small amount of quartz and an elevated amount of illite probably decides the relatively low mechanical properties of the investigated material. The high amount of kaolinite, that under temperature is transformed into metakaolinite, should be a factor that makes the geopolymerization process possible. It is an advantage taking into consideration the geopolymerization process [22].

Moreover, SEM observation was provided for the investigated samples—Figure 4.

The visible microstructure of the material is typical for geopolymers [2,5]. In Figure 4a,c there is visible cracking which is an effect of previous mechanical strength tests. Additionally, an EDS analysis was performed for the selected points. Exemplary results are presented in Figure 5.

The oxide composition obtained as a result of EDS measurements is typical for a geopolymer material. It is characterized by a large amount of SiO_2, about 60%, and Al_2O_3 of more than 21% [2,5]. Another important compound, Na_2O at 17%, probably comes from the alkali activation process. There are also small amounts of MgO and K_2O (less than 1%), in the structure.

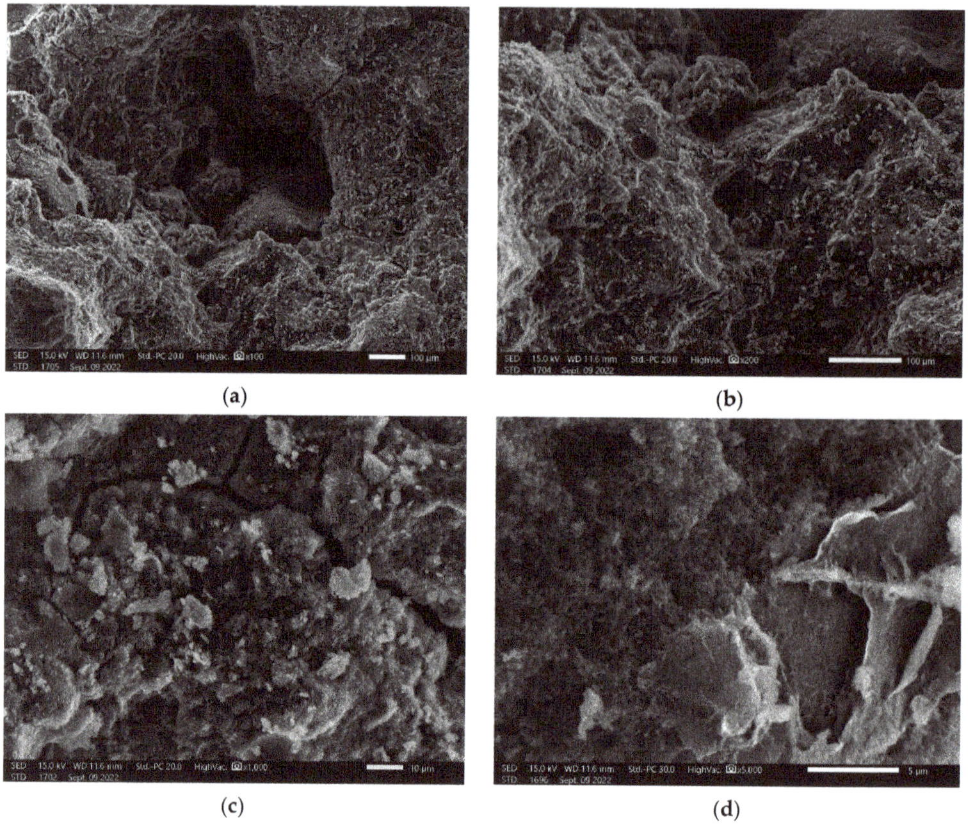

Figure 4. SEM images of geopolymer: (**a**) at 100× magnification; (**b**) at 200× magnification; (**c**) at 1000× magnification; and (**d**) at 2000× magnification.

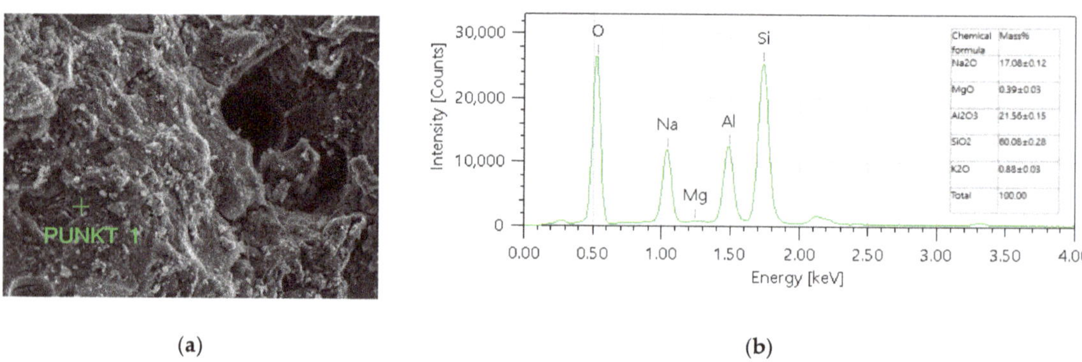

Figure 5. Geopolymer sample: (**a**) SEM image with the point of measurements; and (**b**) results of the EDS analysis.

4. Conclusions

The obtained results show the possibility of using a coal shale from Marcel mining to create new eco-friendly materials, geopolymers. The geopolymer was prepared using a sodium activator. The obtained material has a structure, chemical and mineralogical composition for geopolymer. The material had a compressive strength of 12.7 MPa and a bending strength of 3.4 MPa. These kinds of mechanical properties are relevant for construction applications such as various types of foundations, walls, columns, lintels, terraces, stairs, ceilings, small building elements, and small architecture. However, further applications required additional research, including durability studies. According to an environmental point of view, the most promising seems to be an application onsite for the mining industry to minimalize the cost of transportation.

Author Contributions: Conceptualization, K.K. and M.Ł.; methodology, K.K. and B.F.; validation, M.Ł. and K.P.; formal analysis, A.B., B.Ş. and G.A.; investigation, A.B., B.Ş., G.A. and K.P.; resources, B.F. and M.Ł.; data curation, A.B., B.Ş. and G.A.; writing—original draft preparation, K.K. and B.F.; writing—review and editing, M.Ł.; visualization, B.F. and K.P.; supervision, K.K.; funding acquisition, K.K. All authors have read and agreed to the published version of the manuscript.

Funding: This research was funded under the project: "Development of geopolymer composites as a material for protection of hazardous wrecks and other critical underwater structures against corrosion" under the M-ERA.NET 3 program by the Polish National Centre for Research and Development, grant number M-ERA.NET3/2021/71/MAR-WRECK/2022.

Institutional Review Board Statement: Not applicable.

Informed Consent Statement: Not applicable.

Data Availability Statement: Not applicable.

Acknowledgments: The publication cost of this paper was covered with funds of the Polish National Agency for Academic Exchange (NAWA): "MATBUD'2023—Developing international scientific cooperation in the field of building materials engineering" BPI/WTP/2021/1/00002, MATBUD'2023.

Conflicts of Interest: The authors declare no conflict of interest. The funders had no role in the design of the study; in the collection, analyses, or interpretation of data; in the writing of the manuscript; or in the decision to publish the results.

References

1. Vargas-Merino, J.A.; Rios-Lama, C.A.; Panez-Bendezú, M.H. Circular Economy: Approaches and Perspectives of a Variable with a Growing Trend in the Scientific World—A Systematic Review of the Last 5 Years. *Sustainability* **2022**, *14*, 14682. [CrossRef]
2. Figiela, B.; Brudny, K.; Lin, W.-T.; Korniejenko, K. Investigation of Mechanical Properties and Microstructure of Construction- and Demolition-Waste-Based Geopolymers. *J. Compos. Sci.* **2022**, *6*, 191. [CrossRef]
3. Eisenreich, A.; Füller, J.; Stuchtey, M.; Gimenez-Jimenez, D. Toward a circular value chain: Impact of the circular economy on a company's value chain processes. *J. Cleaner Prod.* **2022**, *378*, 134375. [CrossRef]
4. Khan, S.A.R.; Shah, A.S.A.; Yu, Z.; Tanveer, M. A systematic literature review on circular economy practices: Challenges, opportunities and future trends. *J. Entrepreneurship Emerg. Econ.* **2022**, *14*, 754–795. [CrossRef]
5. Kozub, B.; Bazan, P.; Gailitis, R.; Korniejenko, K.; Mierzwiński, D. Foamed Geopolymer Composites with the Addition of Glass Wool Waste. *Materials* **2021**, *14*, 4978. [CrossRef] [PubMed]
6. Pu, S.; Zhu, Z.; Song, W.; Huo, W.; Zhang, J. Mechanical and microscopic properties of fly ash phosphoric acid-based geopolymer paste: A comprehensive study. *Constr. Build. Mater.* **2021**, *299*, 123947. [CrossRef]
7. Figiela, B.; Korniejenko, K. The possibility of using waste materials as raw materials for the production of geopolymers. *Acta Innov.* **2020**, *36*, 48–56. [CrossRef]
8. McLellan, B.C.; Williams, R.P.; Lay, J.; van Riessen, A.; Corder, G.D. Costs and carbon emissions for geopolymer pastes in comparison to ordinary portland cement. *J. Clean. Prod.* **2011**, *19*, 1080–1090. [CrossRef]
9. Krishna, R.S.; Shaikh, F.; Mishra, J.; Lazorenko, G.; Kasprzhitskii, A. Mine Tailings-Based Geopolymers: Properties, Applications and Industrial Prospects. *Ceram. Int.* **2021**, *47*, 17826–17843. [CrossRef]
10. Sitarz, M.; Figiela, B.; Łach, M.; Korniejenko, K.; Mróz, K.; Castro-Gomes, J.; Hager, I. Mechanical Response of Geopolymer Foams to Heating—Managing Coal Gangue in Fire-Resistant Materials Technology. *Energies* **2022**, *15*, 3363. [CrossRef]
11. Qaidi, S.M.A.; Tayeh, B.A.; Zeyad, A.M.; de Azevedo, A.R.G.; Ahmed, H.U.; Emad, W. Recycling of mine tailings for the geopolymers production: A systematic review. *Case Stud. Constr. Mater.* **2022**, *16*, e00933. [CrossRef]
12. He, X.; Yuhua, Z.; Qaidi, S.; Isleem, H.F.; Zaid, O.; Althoey, F.; Ahmad, J. Mine tailings-based geopolymers: A comprehensive review. *Ceram. Int.* **2022**, *48*, 24192–24212. [CrossRef]
13. Opiso, E.M.; Tabelin, C.B.; Maestre, C.V.; Aseniero, J.P.J.; Park, I.; Villacorte-Tabelin, M. Synthesis and characterization of coal fly ash and palm oil fuel ash modified artisanal and small-scale gold mine (ASGM) tailings based geopolymer using sugar mill lime sludge as Ca-based activator. *Heliyon* **2021**, *7*, e06654. [CrossRef] [PubMed]

14. Silva, I.; Castro-Gomes, J.P.; Albuquerque, A. Effect of immersion in water partially alkali-activated materials obtained of tungsten mine waste mud. *Constr. Build. Mater.* **2012**, *35*, 117–124. [CrossRef]
15. Tian, X.; Xu, W.; Song, S.; Rao, F.; Xia, L. Effects of curing temperature on the compressive strength and microstructure of copper tailing-based geopolymers. *Chemosphere* **2020**, *253*, 126754. [CrossRef] [PubMed]
16. Wan, Q.; Rao, F.; Song, S.; Zhang, Y. Immobilization forms of ZnO in the solidification/stabilization (S/S) of a zinc mine tailing through geopolymerization. *J. Mater. Res. Technol.* **2019**, *8*, 5728–5735. [CrossRef]
17. Wang, A.; Liu, H.; Hao, X.; Wang, Y.; Liu, X.; Li, Z. Geopolymer Synthesis Using Garnet Tailings from Molybdenum Mines. *Minerals* **2019**, *9*, 48. [CrossRef]
18. Krishna, R.S.; Mishra, J.; Meher, S.; Das, S.K.; Mustakim, S.M.; Singh, S.K. Industrial Solid Waste Management through Sustainable Green Technology: Case Study Insights from Steel and Mining Industry in Keonjhar, India. In Proceedings of the Materials Today: Proceedings; Elsevier Ltd.: Amsterdam, The Netherlands, 2020; Volume 33, pp. 5243–5249. [CrossRef]
19. Zhou, C.; Liu, G.; Wu, S.; Lam, P.K.S. The environmental characteristics of usage of coal gangue in bricking-making: A case study at Huainan, China. *Chemosphere* **2014**, *95*, 274–280. [CrossRef] [PubMed]
20. Korniejenko, K.; Bazan, P.; Figiela, B.; Kozub, B.; Łach, M. Characterization of mine tailings as raw materials for geopolymer synthesis—Mineralogical composition. In Proceedings of the M2D2022: 9th International Conference on Mechanics and Materials in Design, Funchal, Portugal, 26–30 June 2022; pp. 1081–1094. Available online: https://paginas.fe.up.pt/~{}m2d/proceedings_m2d2022/index.htm (accessed on 12 October 2022).
21. Marsh, A.; Heath, A.; Patureau, P.; Evernden, M.; Walker, P. Alkali activation behaviour of un-calcined montmorillonite and illite clay minerals. *Appl. Clay Sci.* **2018**, *166*, 250–261. [CrossRef]
22. Mendes, B.C.; Pedroti, L.G.; Vieira, C.M.F.; Carvalho, J.M.F.; Ribeiro, J.C.L.; Albuini-Oliveira, N.M.; Andrade, I.K.R. Evaluation of eco-efficient geopolymer using chamotte and waste glass-based alkaline solutions. *Case Stud. Constr. Mater.* **2022**, *16*, e00847. [CrossRef]

materials proceedings

MDPI

Proceeding Paper

Sulfate Resistance in Carbonated Low-Calcium Silicate Cement Pastes †

Raikhan Tokpatayeva [1,*], Jan Olek [1] and Sadananda Sahu [2]

[1] Lyles School of Civil Engineering, Purdue University, West Lafayette, IN 47907, USA
[2] Solidia Technologies Inc., 11 Colonial Dr, Piscataway, NJ 08854, USA
* Correspondence: tokpatayeva@gmail.com

† Presented at the 10th MATBUD'2023 Scientific-Technical Conference "Building Materials Engineering and Innovative Sustainable Materials", Cracow, Poland, 19–21 April 2023.

Abstract: This paper focuses on the evaluation of sulfate resistance in carbonated pastes prepared from low-lime calcium silicates (CCS). The chemical interaction between the sulfate solution and paste powders was assessed by monitoring the leaching of the Ca and Si species, reduction in the content of carbonates and formation of gypsum. The analytical techniques used in the study included TGA, ICP-OES and IC. The results of the study revealed that the level of the resistance to the chemical effect of the sulfates depends on the type of the calcium silicate, degree of crystallinity of calcium carbonate, and the type of cation present in the sulfate solution.

Keywords: carbonation; calcium silicates; sulfate resistance

check for updates

Citation: Tokpatayeva, R.; Olek, J.; Sahu, S. Sulfate Resistance in Carbonated Low-Calcium Silicate Cement Pastes. *Mater. Proc.* **2023**, *13*, 22. https://doi.org/10.3390/materproc2023013022

Academic Editors: Katarzyna Mróz, Tomasz Tracz, Tomasz Zdeb and Izabela Hager

Published: 15 February 2023

1. Introduction

One of the growing areas of research interest is the development of eco-friendly alternatives to the OPC system. These low-lime calcium silicate cements harden and gain strength by carbonation (carbon mineralization), and thus store CO_2 permanently. The objective of the study presented in this paper was to evaluate the sulfate resistance of low-lime carbonated calcium silicate (CCS) pastes. The experimental plan involved exposing paste powders made out of different carbonated CCSs (wollastonite, amorphous CS, and two types of commercially produced low-calcium binders (Solidia^{TM} cements)) to sodium sulfate and magnesium sulfate solutions for a period of up to 120 days. At various times during the exposure period (0.5, 1, 5, 10, 20, 60, 90 and 120 days), the paste + sulfate solution slurries were filtrated and separated into solids and leachates. The solids were subjected to thermogravimetric analysis (TGA) in order to quantify the amount of gypsum formed. The leachates were analyzed by the ion-chromatography (IC) and by the inductively coupled plasma optical emission spectroscopy (ICP-OES) techniques, to determine changes in the concentrations of the Ca, Mg, Na, Si and sulfate species.

2. Description of the Analytical Methods

2.1. Thermogravimetric (TGA) Analysis

The thermogravimetric analysis of the pastes was conducted using the Q50 (TA Instruments) apparatus. The test involved placing a 15–20 mg powdered paste (passed through sieve #200) in an open platinum pan and heating it at a rate of 10 °C/min up to 900 °C in a nitrogen gas atmosphere (flow rate 60 mL/min).

2.2. Chemical Analysis of the Soak Solution

The collected soak solutions were first filtered using a 0.2 μm cellulose acetate filter to separate solids from the fluids. The resulting fluids were then analyzed using inductively coupled plasma, optical emission spectroscopy (ICP-OES, Thermo-Scientific iCAP 7400 spectrometer) and ion chromatography (IC, DIONEX ICS-900) techniques.

3. Materials

This study was conducted on pastes prepared from four different low-lime calcium silicate cements, referred to as #1, #2, #3 and #4 cements, respectively. The cements were characterized by an X-ray Fluorescence (XRF) technique and the results are shown in Table 1.

Table 1. XRF oxide composition of the cements.

Cement	Oxides, wt.%									
	CaO	SiO_2	Fe_2O_3	Al_2O_3	MgO	SO_3	Na_2O	K_2O	LOI	Total Alkalis
#1	46.01	51.47	0.77	0.87	0.08	0.01	0.16	0.04	0.72	0.18
#2	46.08	45.66	0.22	2.50	1.11	0.41	0.21	0.37	1.61	0.45
#3	43.64	44.0	1.79	5.13	1.15	0.16	0.33	1.94	0.71	1.61
#4	42.58	42.67	2.50	6.19	2.17	0.99	0.21	1.07	0.79	0.91

Comparing the oxide compositions of the cements, it can be seen that while they contained similar quantities of CaO (in the range of 43–46%), their SiO_2 content varied from 51.5% for cement #1, to 43% for cement #4. Regarding other minor oxides (Al_2O_3, Fe_2O_3, MgO, K_2O), the lowest amount was found in cement #1, whereas the highest amount was found in cement #4. All cements have similar values of LOI, except cement #2, which had an LOI about twice as high as the LOI values of other cements. The lowest alkali content was found in cement #1, and the highest was present in cement #3.

Carbonated cement paste samples were prepared by hand, mixing small amounts of cement powders with de-ionized water at water-to-cement ratio of 0.35. Once mixed, the pastes were placed in a thin (1.0–1.5 mm) layer in the VWR sterile polystyrene Petri dishes with diameter of 60 mm and the height of 1.5 mm. Immediately after completion of casting, the paste-containing Petri dishes were placed in VWR Symphony 1.4A CO_2 incubator, where they were carbonated for 5 days. The environmental conditions within the incubator were as follows: temperature 23 °C 90–100% relative humidity and 20% concentration of carbon dioxide. After the completion of carbonation, the paste samples were pulverized with mortar and pestle, and stored sealed in glass vials before being used for testing. Prior to being subjected to sulfate solution exposure, the powdered carbonated paste samples were characterized using thermal (TGA), XRD and FT-IR techniques. The results of these analyses are presented elsewhere [1].

4. Sulfate Exposure Experiment

4.1. Experimental Setup

In the current study, three different types of sulfate solutions were used: 0.35 M of sodium sulfate, magnesium sulfate and aluminum sulfate solutions. The pH values of the original sulfate solutions (i.e., before they contacted the paste powders) were in the range of 6–7 for sodium and magnesium sulfate solutions. On the other hand, the aluminum sulfate solution was rather acidic, with a pH level between 3 and 4.

All four carbonated paste powder samples were pulverized using mortar and pestle, passed through #200 (75 μm) sieve and submerged in sulfate solutions using a solid-to-liquid ratio of 1:4. The slurry samples were kept sealed in plastic centrifuge tubes with 50 mL capacity, and stored at room temperature (23 ± 5 °C). They were agitated periodically twice a week by hand. Due to its acidic nature, the mixing of the aluminum sulfate solution with paste powders resulted in the release of CO_2 gas bubbles from the decomposition of calcium carbonates. For this reason, the tubes containing aluminum sulfate solution were sealed only after the bubbling of CO_2 has ceased. In addition to the sulfate solutions, paste powder samples were also submerged in de-ionized water. Ultimately, the exposure test was held for 120 days. At different time periods (0.5, 1, 5, 10, 20, 60, 90 and 120 days) of exposure, 16 samples in total (four different paste sample slurries per each solution type) were taken from the set for further analysis.

4.2. Post-Exposure Analysis

After reaching the pre-designated period of exposure, the slurry samples were centrifuged and filtered using a 0.2 μm cellulose acetate filter to separate solids from the soak solutions. The solids were then oven-dried at 35 ± 5 °C for three days, pulverized using mortar and pestle and stored in sealed glass vials with a plastic cap. Afterwards, the powdered solids were subjected to thermogravimetric analysis. The filtered solutions were kept sealed in 15 mL plastic centrifuge tubes and stored in the laboratory refrigerator at ~4–5 °C, until they were removed for ICP and IC analyses.

5. Results and Discussion

5.1. Soak Solution Chemistry

The chemical analysis of the sulfate soak solutions indicated that the sulfate-rich environment significantly influences the amount of calcium and silica ions leached out from the CCS matrices. As shown in the graphs presented in the Figures 1–4, all sulfate solutions triggered a substantial increase (compared to deionized water soak solution) in the amount of calcium ions leached out from the CCS matrices. Depending on the solution, the observed increases were in the range of 10–15 times. There is also a noticeable trend of decay in Ca ion concentration with time, for all pastes exposed to aluminum sulfate, as well as cement #2 pastes exposed to sodium sulfate, and cement #3 and #4 pastes exposed to the magnesium sulfate solution. This might be due to the consumption of calcium ions during precipitation of new compounds, such as gypsum (see Section 5.2). With respect to the leaching of the silica species, the effect of sulfate solutions, especially aluminum and magnesium sulfates, appears to be quite opposite to that observed for the calcium species. As an example, for all four types of pastes, the exposure to aluminum sulfate resulted in only minimal levels of silica in the soak solution, indicating that this sulfate solution stabilizes silica species present in the specimens. Similarly, it also appears that magnesium sulfate stabilized silica in all paste samples, except that of the carbonated wollastonite paste (cement #1). However, the effect of sodium sulfate seems to be rather selective: it stabilized the leaching of silica compared to immersion in deionized water only for the cement #2 paste samples.

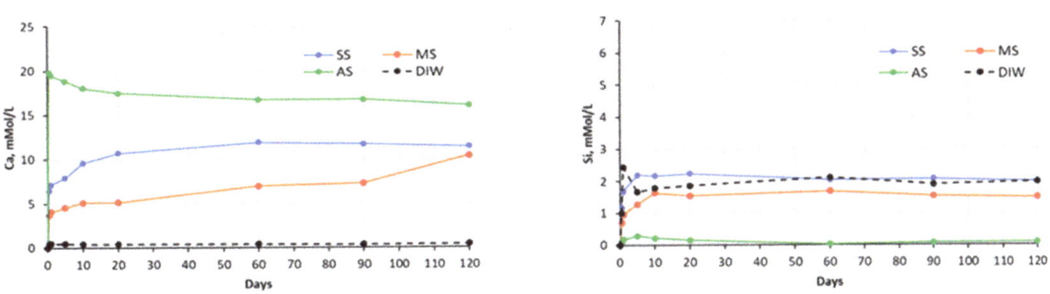

Figure 1. Changes in concentrations of: (**a**) Ca and (**b**) Si species present in sulfate solutions with time, in contact with cement #1 paste specimens (SS—sodium sulfate, MS—magnesium sulfate, AS—aluminum sulfate, DIW—de-ionized water).

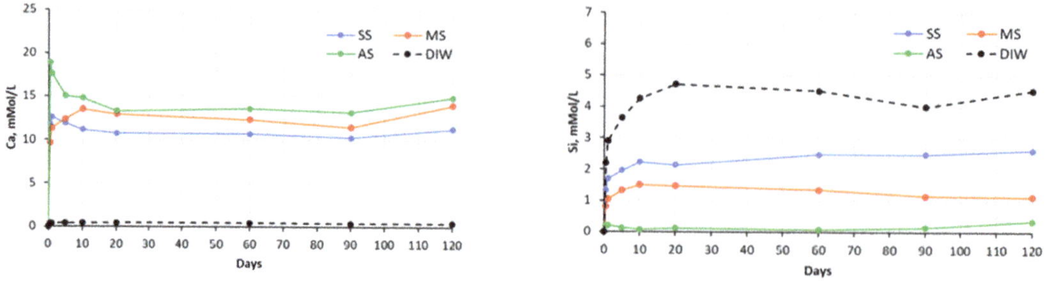

(**a**) (**b**)

Figure 2. Changes in concentrations of: (**a**) Ca and (**b**) Si species present in sulfate solutions with time, in contact with cement #2 paste specimens (SS—sodium sulfate, MS—magnesium sulfate, AS—aluminum sulfate, DIW—de-ionized water).

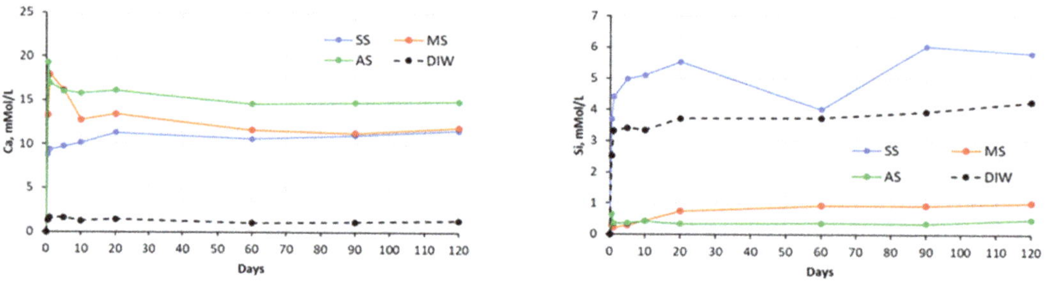

(**a**) (**b**)

Figure 3. Changes in concentrations of: (**a**) Ca and (**b**) Si species present in sulfate solutions with time, in contact with cement #3 paste specimens (SS—sodium sulfate, MS—magnesium sulfate, AS—aluminum sulfate, DIW—de-ionized water).

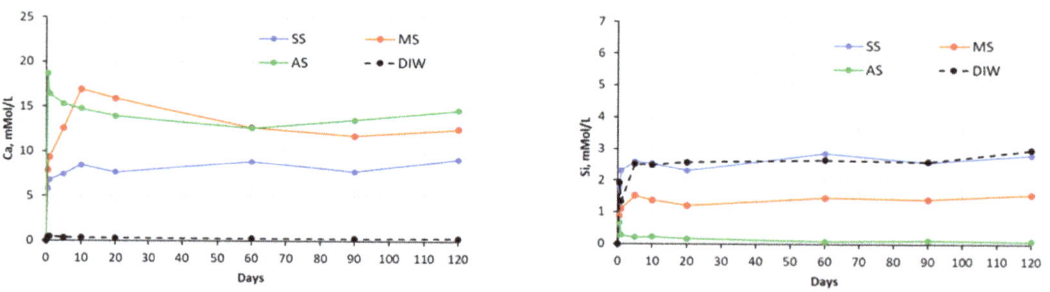

(**a**) (**b**)

Figure 4. Changes in concentrations of: (**a**) Ca and (**b**) Si species present in sulfate solutions with time, in contact with cement #4 paste specimens (SS—sodium sulfate, MS—magnesium sulfate, AS—aluminum sulfate, DIW—de-ionized water).

The change in concentrations of sodium and magnesium ions over the entire test period is presented in Figure 5. The concentrations of aluminum cations are not presented, as they are negligibly small. The concentrations of sodium ions remain essentially constant, except for a slight reduction in the case of cement #2 paste. In contrast, one can observe that over time, the concentration of magnesium ions reduced by as much as ~68% for the cement #3 paste sample. In the case of cement #2 and #4 paste samples, the reduction was about ~10%; whereas for cement #1 paste samples, that reduction was very small (~2%) and it was only detected after 120 days of exposure.

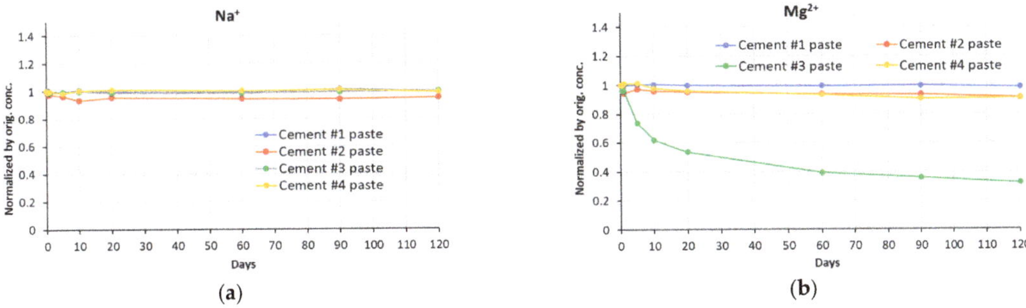

Figure 5. Normalized concentrations of Na^+ (**a**) and Mg^{2+} (**b**) ions in soak solutions of all cement paste samples, over a period of exposure.

The plots of normalized changes in concentration of sulfate ions (for sodium, magnesium and aluminum sulfate solutions) for soak solutions in contact with all cement paste samples used in this study are presented in Figure 6. The trends illustrated by the sulfate ion curves seem to be quite similar to those observed for the corresponding sodium and magnesium cations. In the case of the aluminum sulfate solution, the concentration of the sulfate ions in the soak solution after exposure of the paste samples (even at early days) was observed to be negligibly low as the concentration of Al^{3+} species.

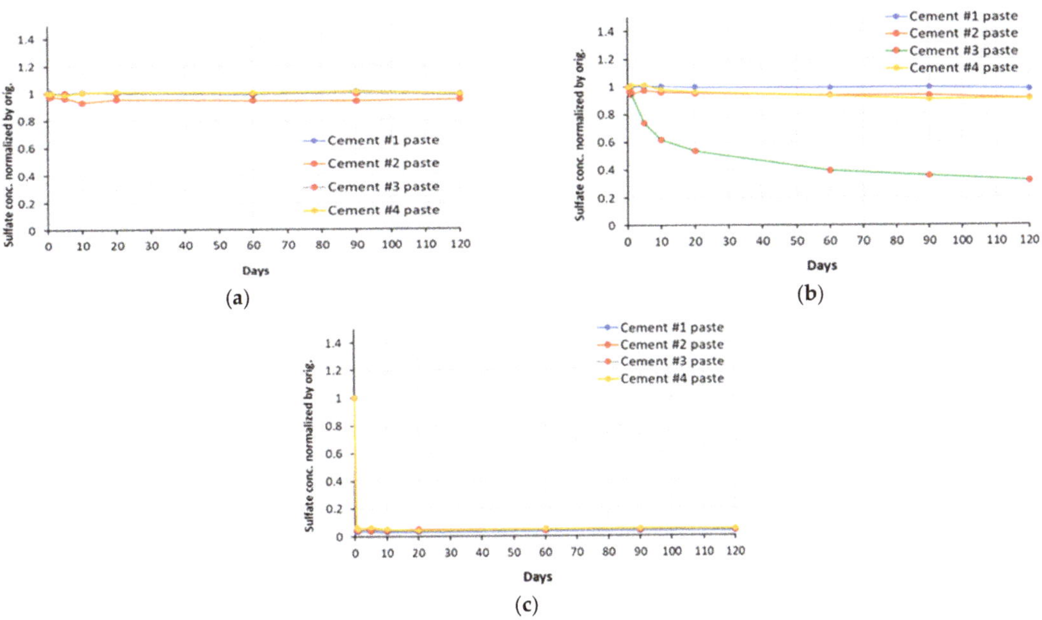

Figure 6. Normalized concentrations of sulfate ions in sulfate soak solutions ((**a**)—sodium sulfate, (**b**)—magnesium sulfate, (**c**)—aluminum sulfate) of all cement paste samples, over a period of exposure.

5.2. Thermal Analysis Results

The amounts of gypsum formed during the sulfate exposure period were quantified by TGA. To avoid an overestimation of the quantities of gypsum formed (due to the continuous dehydration of the amorphous silica gel in the temperature range of ~90–150 °C), the methodology, originally developed by Kim and Olek (2012) [2] for the estimation of the calcium hydroxide, was implemented. The results of the quantitative evaluation of the gypsum formed during the exposure test are presented in the graphs shown in Figure 7.

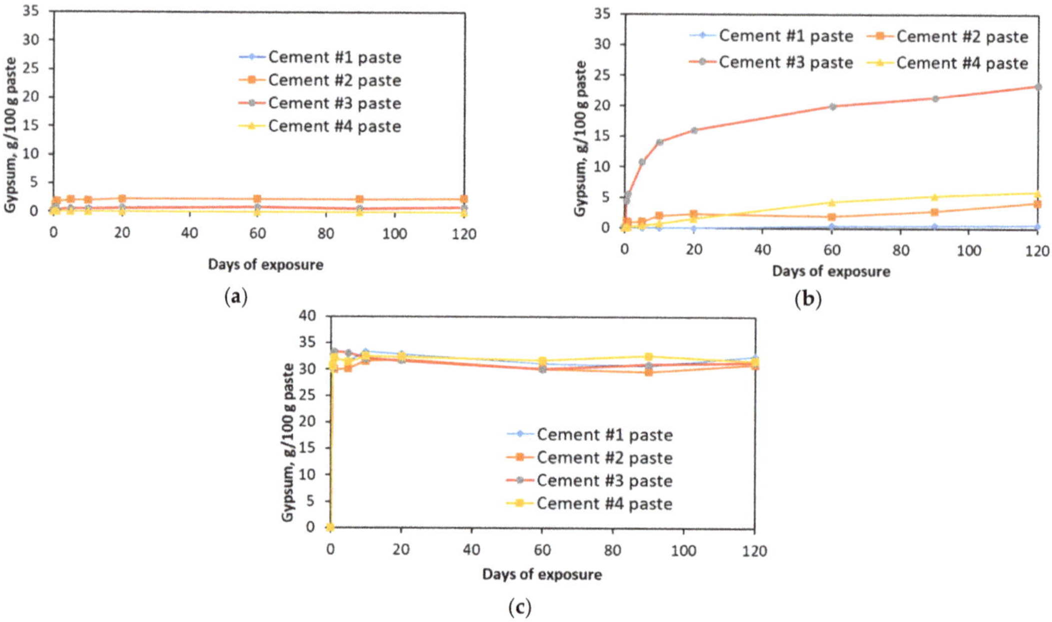

Figure 7. Amount of gypsum formed during exposure to (**a**) sodium, (**b**) magnesium and (**c**) aluminum sulfate solutions.

As it can be seen from these graphs, exposure to the sodium sulfate solution did not result in the formation of any significant amounts of gypsum compared to other sulfate solutions. Specifically, at the end of the exposure period, only a small amount (about 2.5 g of gypsum per 100 g of paste) was observed in cement #2 paste samples. In contrast, exposure to magnesium sulfate solution resulted in the formation of much more significant amounts of gypsum in all paste samples, except in the case of carbonated wollastonite, i.e., cement #1 samples (only about 0.5 g/100 g of paste was formed in this case). By the end of the 120 day exposure period, ~24 g of gypsum per 100 g of paste formed in cement #3 samples, and ~4–6 g of gypsum per 100 g of paste formed in cement #2 and #3 samples. In the case of the aluminum sulfate solution, about 30–32% (wt.) of gypsum formed in all paste samples within the first 12 h of exposure. This was due to the high acidity of the solution ($pH \approx 3$–4). Consequently, this resulted in the quick decomposition ("digestion") of the calcium carbonate phase, pH stabilizing at around 6–7, binding of released calcium ions by sulfate ions and precipitation of gypsum. That amount of gypsum remained more-or-less constant throughout the entire exposure period.

6. Conclusions

The main findings of the study can be summarized as follows:

- The resistance of CCS systems to sulfate attack was found to be influenced by both the type of cations present in the sulfate solution, and the composition of the low-lime calcium silicate cements itself. In particular, CCS showed a higher resistance to Na_2SO_4 solution than to $MgSO_4$, due to the absence of reactive phases, such as, e.g., $Ca(OH)_2$ and/or calcium aluminates. Nevertheless, interaction of some of the CCS samples with sodium sulfates resulted in the formation of gypsum (even though in negligible amounts). Regarding the aluminum sulfate solution, no resistance was observed in the case of all paste samples due to the acidic environment of the solution.

- Lower resistance to $MgSO_4$ was observed in CCS made of the cement containing more hydraulic calcium silicates (cement #3). In this case, the hydrated silica gel might be extensively modified with Ca. This can lead to a chemical reaction resulting in the formation of gypsum. Additionally, the consumption of magnesium ions and stabilization of silica indicate the interaction between hydrated silica gel and magnesium cations. This interaction, especially, might be due to cation exchange between calcium and magnesium ions. This, in turn, emphasizes the importance of the calcium ions and silica during the carbonation process and the level of pureness of the hydrated silica gel, with respect to sulfate resistance.
- The best resistance to sulfate attack was observed in CCS samples based on carbonated wollastonite, which contained higher amounts of crystalline calcium carbonate polymorph (calcite) and more highly polymerized silica gel.

Author Contributions: Conceptualization, R.T. and J.O.; methodology, R.T.; formal analysis, R.T. and J.O.; investigation, R.T.; writing—original draft preparation, R.T., writing—review and editing, J.O. and S.S., funding acquisition, J.O. All authors have read and agreed to the published version of the manuscript.

Funding: This research received no external funding.

Institutional Review Board Statement: Not applicable.

Informed Consent Statement: Not applicable.

Data Availability Statement: Not applicable.

Acknowledgments: SOLIDIA TECHNOLOGIES Inc. is acknowledged for providing the cement materials for the study. The publication cost of this paper was covered with funds from the Polish National Agency for Academic Exchange (NAWA): "MATBUD'2023—Developing international scientific cooperation in the field of building materials engineering" BPI/WTP/2021/1/00002, MATBUD'2023.

Conflicts of Interest: The authors declare no conflict of interest.

References

1. Tokpatayeva, R. Study of Sulfate Attack Resistance of Carbonated Low-Lime Calcium Silicate Systems. Ph.D. Thesis, Purdue University, West Lafayette, IN, USA, December 2020.
2. Kim, T.; Olek, J. Effects of Sample Preparation and Interpretation of Thermogravimetric Curves on Calcium Hydroxide in Hydrated Pastes and Mortars. *Transp. Res. Rec.* **2012**, *2290*, 10–18. [CrossRef]

materials proceedings

MDPI

Proceeding Paper

Surface Modification of Photocatalytic Cementitious Composites with Polyacrylic Superabsorbent Polymers (SAP) †

Maciej Kalinowski *, Piotr Woyciechowski and Wioletta Jackiewicz-Rek

Faculty of Civil Engineering, Warsaw University of Technology, 00-637 Warsaw, Poland

* Correspondence: maciej.kalinowski@pw.edu.pl

† Presented at the 10th MATBUD'2023 Scientific-Technical Conference "Building Materials Engineering and Innovative Sustainable Materials", Cracow, Poland, 19–21 April 2023.

Abstract: This study focuses on developing new surface modification methods for photocatalytic cementitious composites. The question that was investigated was if superabsorbent polymers can act as an intermediate environment, intensifying the incorporation of TiO_2 particles on the surface of the tested material. Four variants of surface modification were designed: water dispersions of TiO_2 with non-saturated SAP, with SAP in a hydrogel form, and two reference series. Due to the use of SAP, the efficiency in air purification from NO_x under different light conditions compared to the reference series increased significantly, exceeding a 75% increase under a UV-A light and a 600% increase under visible light.

Keywords: superabsorbent polymers; SAP; cement composites; titanium dioxide; gaseous pollutants; photocatalysis

Citation: Kalinowski, M.; Woyciechowski, P.; Jackiewicz-Rek, W. Surface Modification of Photocatalytic Cementitious Composites with Polyacrylic Superabsorbent Polymers (SAP). *Mater. Proc.* **2023**, *13*, 23. https://doi.org/10.3390/materproc2023013023

Academic Editors: Katarzyna Mróz, Tomasz Tracz, Tomasz Zdeb and Izabela Hager

Published: 15 February 2023

1. Introduction

Air pollution caused by human activity is one of the most worrying problems in urban areas nowadays [1,2]. Due to an increase in the concentration of harmful gaseous pollutants, mainly nitric oxides, ozone, and others, methods to reduce their presence need to be developed. As urban areas occupy significant areas of land, any active air purification system would require a vast amount of energy to operate and would contribute to a reduction in the quality of life in its vicinity (noise generated by high-efficiency air turbines). Moreover, most pollutants are produced due to the activity of transport systems within city boundaries, contributing to the differences in their concentrations over different areas (residential areas, roads, and other transport pavements, etc.) [3]. Therefore, different solutions to the problem have been developed over the years, with one of them being the use of photocatalytic materials embedded into cementitious materials—a passive system that requires only the presence of solar radiation to operate and purify the air from pollutants [4,5]. As a significant part of urban areas is covered with different types of pavements, mostly made of various cementitious materials and exposed to solar radiation, such a passive system would not require any changes to the existing infrastructure other than replacing non-photocatalytic surfaces with photocatalytic ones.

For a photocatalytic reaction to occur, a photocatalytic material (a semiconductor—for example, titanium dioxide) needs to be irradiated (for instance, with solar radiation). Due to energy introduction, electrons (e−) from the semiconductor shift from the valence band (vb) to the conduction band (cb), leading to the formation of electron holes h+ (Equation (1)). The resulting pairs of charges initiate a reduction–oxidation process (Figure 1). In the presence of adsorbed oxygen and water (strong oxidants) it is possible to decompose a wide range of air pollutants due to the formation of hydroxyl radicals ($OH^{\bullet}$) and superoxide radicals ($O_2^{\bullet -}$) (Equations (2) and (3)) [6].

$$TiO_2 \xrightarrow{h\nu} TiO_2 + h_{vb}^+ + e_{cb}^- \quad (1)$$

$$(H_2O)_{ads} + h^+ \rightarrow H^+ + OH^\bullet \quad (2)$$

$$(O_2)_{ads} + e^- \rightarrow O_2^{\bullet -} \quad (3)$$

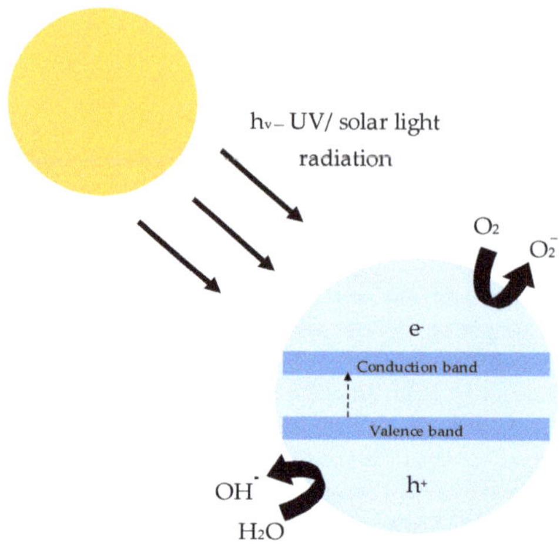

Figure 1. General mechanism of the photocatalytic reaction.

As shown, titanium dioxide grains, or other photocatalyst grains, must be exposed to UV/solar radiation for a photocatalytic reaction to occur. To accomplish this goal, cementitious materials incorporating that type of addition usually consist of two layers—one exposed to the external environment containing photocatalytic material and the other with no (or close to no) photocatalytic material. As photocatalysis occurs only in a near-surface material layer, dividing the composite into two parts reduces its overall cost without impacting its properties regarding air purification [7].

Due to its fine granulation (starting at 10 nm), titanium dioxide functions as nuclei for hydration products during the hydration of the binder. Due to this, the TiO_2 grains are usually covered with hydration products, limiting the effectiveness of the photocatalytic reaction of the hardened material (the TiO_2 grains must be exposed for it to occur) [8].

To improve the effectiveness of air purification from gaseous pollutants, the authors decided to investigate new methods of the surface modification of photocatalytic materials using polyacrylic superabsorbent polymers (SAP). Superabsorbent polymers can be described as cross-linked hydrogel networks which absorb water due to high osmotic pressure caused by the accumulation of ions within their structure [9]. Water absorption into the polymer network causes the SAP grains to increase their volume, pushing ions apart. With its network stretched out, the osmotic pressure decreases. Hydrogel obtained in that way can later desorb water stored in its volume under external pressure or due to changes in the properties of the desorption environment [10]. For that reason, SAP are usually used as an internal curing agent in cementitious materials, preventing the cement matrix's self-desiccation and promoting the hydration of the binder.

As the SAP hydrogel is introduced into the cement matrix, it desorbs water due to the high electrochemical activity of the environment. This process is prolonged (water desorbs from SAP during more than 30 days from its introduction into cementitious material) compared to water absorption time (usually 5–15 min in a water environment) as water desorption occurs during hydration and is limited by water pressure within the pore network and its microstructure [11].

The SAP also facilitates hydration in the direct vicinity of its grains. This effect was linked to the course of water desorption from the SAP grains. As water desorbs from SAP, it creates a thin film over the volume of a shrinking SAP hydrogel grain, which acts as an environment of intense hydration of the binder [9]. This phenomenon can be used to introduce various nanoparticles into cementitious material with the view of increasing their homogenous distribution within the cement matrix (Figure 2).

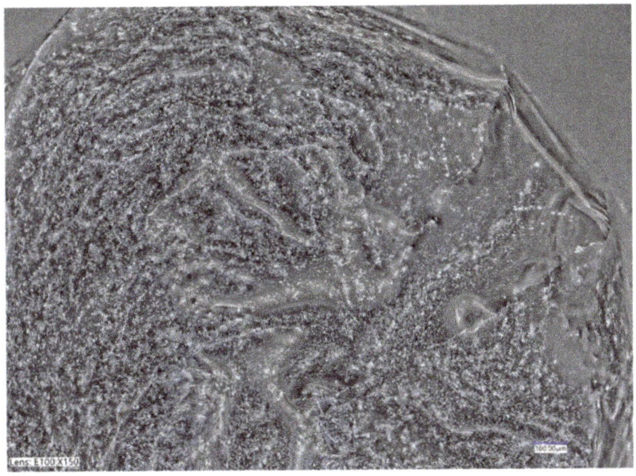

Figure 2. SAP hydrogel grain covered with nano-grains of titanium dioxide used in the study.

2. Materials and Methods

The main goal of the research was to increase the content of TiO_2 on the exposed surface of the material and obtain its homogenous distribution over the entire surface of the element. The phenomenon of adsorption of solid particles on the outer layer of the SAP grains was utilized to achieve it. In molds (in which samples for air purification tests were prepared), a thin layer of SAP (either in non-saturated or hydrogel form) was placed and sprayed over with 5–7 g of TiO_2 water dispersion. Next, the photocatalytic mortar of a proposed composition was poured into the mold (Figure 3).

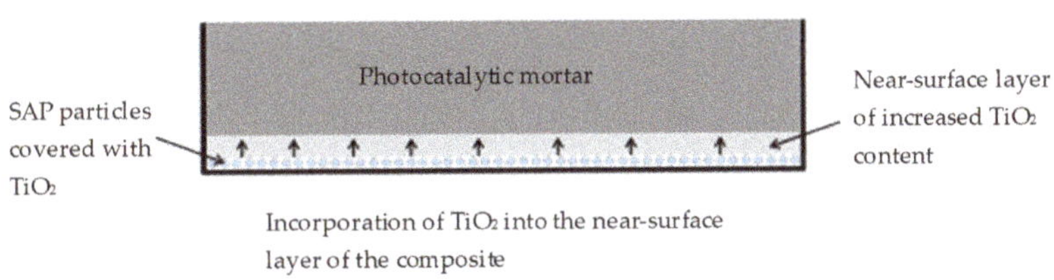

Figure 3. Method of surface modification with a reverse layer order—after demolding, the bottom surface was considered as an exposed surface and tested in air purification tests.

The ability to purify the air from harmful gaseous pollutants is an essential aspect of photocatalytic composites, so the viability of each solution was assessed through the air purification tests from NO and NO_x.

The cement used in this study—CEM II/A-S 52.5R (Ożarów, Poland)—met the requirements of EN 197-1 [12]. The study used two types of fire-dried quartz sand aggregates of different granulations—0.1/0.5 and 0.5./1.2 (Łajski, Poland)—which met the requirements of EN 13139 [13]. Two types of titanium dioxide were used: TiO_2 (A)—K7000 (Leverkusen, Germany) and TiO_2 (B)—P25 (Shanghai, PRC) of properties in the powder state presented in [6] (Table 1). TiO_2 (A) represented a VLA (visible light active) photocatalyst, while TiO_2 (B) represented a UV-A (UV light active) photocatalyst. VLA photocatalysts are active in both UV and visible light. However, their market price is several times higher than for

UV-A photocatalysts. Due to this fact, a mix of two photocatalysts was used so that the obtained photocatalytic material would have the capability to purify the air in both UV and visible light, with its overall price reduced.

Table 1. The content of crystalline phases and the size of crystallites (measured via XRD), and specific surface area (measured via BET).

Photocatalyst	Phase [%]		Size of Crystallites [nm]		Specific Surface Area [m^2/g]
	Rutile	Anatase	Rutile	Anatase	
TiO_2 (A)	-	100	-	10	246.8 ± 2.9
TiO_2 (B)	13	87	54	33	53.8 ± 0.2

Titanium dioxide was introduced into the mortar as a dispersion made of a portion of mixing water and a PCE superplasticizer (SP). As TiO_2 agglomerates in dry conditions, it was decided to use mechanical mixing and sonication to reduce its average grain size. Polyacrylic superabsorbent polymer (sodium acrylate and acrylic acid polymer) (Ludwigshafen, Germany) of properties described in [9] was used as a surface modification agent.

The water used in this study met the requirements of EN 1008 [14]. An additional mass of the PCE superplasticizer (SP) that met the EN 934-2 [15] requirements was added to mortars to modify their rheological properties (Myślenice, Poland). All described components were used to prepare a photocatalytic mortar. For each of the prepared series of samples, different surface modification techniques were applied (Table 2). The mass amount of dispersion was chosen so that the TiO_2 content in the composite would be 12.5 kg/m^3.

Table 2. Composition of the prepared mortar and information regarding the type of applied surface modification for each series of samples.

ID	Description	Cement [kg/m^3]	Water [kg/m^3]	0.1/0.05 [kg/m^3]	0.5/1.2 [kg/m^3]	TiO_2 Water Dispersion [kg/m^3]	SP [kg/m^3]
REF	Reference—no modification	780	255	574	574	72	3.3
REF—D	Reference—a layer of water TiO_2 dispersion						
SAP D	A layer of non-saturated ('dry') SAP covered in TiO_2 dispersion						
SAP H	A layer of water-saturated SAP hydrogel covered in TiO_2 dispersion						

The designed mortars were prepared according to the mixing procedure described in EN 196-1 [16]. The consistency of the mortar was tested based on EN 1015-3 [17]. Due to the low viscosity of the prepared mortar, only free flow was measured. After 24 h, samples were demolded and cured in the curing chamber (temperature 20 ± 2 °C, relative humidity RH ≥ 95%) until further tests. The tensile and compressive strength of the hardened mortar after two days were tested according to EN 196-1 [16] (Table 3) on 40 mm × 40 mm × 160 mm samples. As the considered surface modification techniques did not include the modification of the material's composition in its entire volume (only within an approx. 1–2 mm layer of the sample), it was assumed that the strength characteristics of the material would remain unchanged compared to the reference series.

The air purification from the nitrogen oxides test was performed according to the procedure developed during the 'Technology for the production of innovative self-cleaning prefabricated facade and surface elements that improve air quality' TECHMATSTRATEG-III/0013/2019 project. Tests were performed on 40 mm × 140 mm × 160 mm samples made from the designed photocatalytic mortar after 28 days of curing.

Table 3. Average compressive, tensile strength and consistency of prepared mortar.

Consistency—Free Flow [mm]	Compressive Strength [MPa]	Tensile Strength [MPa]
300	39.3 (CV = 4.4%)	5.9 (CV = 4.0%)

Before the test, the tested surface was cleaned of antiadhesive agent and other organic pollutants. It was sprinkled with distilled water, scrubbed, dried at 60 °C for two hours, and placed in the irradiation chamber for 16 h, with the test surface facing the light source, where the surface organic impurities were burned in UV-A radiation (irradiance 10 W/m^2). After that, the surface was cleaned with distilled water and dried at 60 °C for 2 h. The air purification tests were performed during the earliest two hours after the last drying cycle.

The sample was then placed in the glass reaction chamber with the tested surface facing the light source. The temperature in the glass reaction chamber during the experiment was kept at 25 ± 3 °C and the relative humidity at 40 ± 5%. The gas flow was maintained at a constant value of 2 L/min.

This study consisted of several stages, including filling the reaction chamber with nitrogen oxides to achieve a concentration of 100 ± 5 ppb (parts per billion, 10^{-9}), irradiating the sample with a UV-A light with an irradiance of 0.2 W/m^2, and measuring the concentration of nitrogen oxides while maintaining the gas flow. Then, the concentration of nitrogen oxides after switching off the light source while maintaining the gas flow was measured. In the next step, the reaction chamber was emptied of nitrogen oxides. Consecutively, the same procedure was performed for visible light with an irradiance of 115 W/m^2 and for a combined presence of UV-A and visible light with an irradiance of 0.2 W/m^2 and 115 W/m^2, respectively.

The nitrogen oxide concentration in the performed tests was approximately 100 ppb to model the actual NO_x concentration in Warsaw [6]. The UV-A and visible light irradiance were chosen as 0.2 and 115 W/m^2 to model the UV index of 8 (summer solar radiation conditions in the Polish territory) and the autumn/winter irradiance of visible light.

3. Results and Discussion

The introduction of an additional layer of SAP in either non-saturated or saturated (hydrogel) form significantly affected the distribution of TiO_2 on the surface of the mortar. An increase in the efficiency of air purification was observed for all light sources (Figure 4).

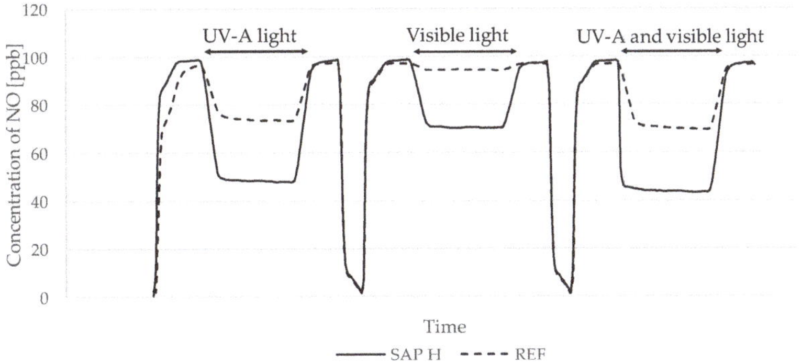

Figure 4. The efficiency of air purification from NO pollutants on a mortar sample formed in a mold with a layer of SAP hydrogel covered in TiO2 dispersion (SAP H) and reference series with no surface modification (REF). The test consisted of three parts: the study of the phenomenon of photocatalysis in UV-A light (first decrease in NO concentration), in visible light (second decrease in NO concentration), and in combined UV-A + visible light (third decrease in NO concentration). [ppb]—parts per billion, 10-9.

The photocatalytic mortar made with the SAP hydrogel layer showed the highest relative reduction of gaseous pollutants among the tested samples (Tables 4 and 5). The lowest coefficient of variation also characterized it. In the case of the photocatalysis intensity in visible light, compared to the reference series, the relative reduction of the NO concentration increased from 5.67% to 26.15%, and the relative decrease in NO_x concentration increased from 3.05% to 24.12%. The relative NO and NO_x concentration reductions increased to a slightly lower extent in the case of a layer made of non-saturated SAP. Spraying molds with dispersion alone also contributed to an increase in the relative reduction of the analyzed pollutants. However, this method was characterized by the highest coefficient of variation of all the tested series (from 18 to 37%). This observation suggests an uneven dispersion distribution on the surface of individual samples. The rheological properties of the dispersion itself are probably responsible for this phenomenon. When spraying the molds with a suspension with a viscosity similar to that of water, along with introducing the photocatalytic mortar into the mold, some of the dispersion previously placed flows out of the mold in an uncontrolled way, contributing to the uneven distribution of the photocatalyst on the surface.

Table 4. Relative reductions in NO concentrations in performed tests for all considered surface modification techniques.

ID	Description	The Relative Reduction in the Concentration of NO [%]					
		UV-A	CV [%]	Visible Light	CV [%]	UV-A + Visible Light	CV [%]
REF	No modification	26.85	10.58	3.50	16.85	30.87	9.36
REF—D	Water dispersion layer	33.64	18.02	5.67	24.95	37.77	18.75
SAP D	Non-saturated SAP with TiO_2 dispersion layer	48.05	7.11	24.41	16.89	52.93	6.01
SAP H	SAP hydrogel with TiO_2 dispersion layer	53.47	4.90	26.15	11.13	59.09	2.23

Table 5. Relative reductions in NO_x concentrations in performed tests for all considered surface modification techniques.

ID	Description	The Relative Reduction in the Concentration of NO_x [%]					
		UV-A	CV [%]	Visible Light	CV [%]	UV-A + Visible Light	CV [%]
REF	No modification	19.31	10.07	3.05	15.91	25.39	8.46
REF—D	Water dispersion layer	23.37	37.12	4.31	33.52	30.25	29.54
SAP D	Non-saturated SAP with TiO_2 dispersion layer	37.73	12.63	22.83	20.36	45.86	9.02
SAP H	SAP hydrogel with TiO_2 dispersion layer	44.23	12.01	24.12	13.31	51.88	4.12

The proposed surface modification method with SAP significantly affected the efficiency of purifying the air from NO and NO_x in visible light conditions. As one of the considered photocatalysts (K7000) is a VLA (visible light active) material, the SAP layer contributed to its better exposure to external radiation, increasing the overall efficiency of the material in purifying the air in conditions simulating common autumn/winter conditions.

Due to their properties, the SAP grains promote hydration in their direct vicinity [18]. This effect is linked with the kinetics of water desorption from the SAP structure—the formation of a thin water film on the outer layer of the SAP grain that acts as an environment of intense hydration/crystallization [9,19]. In the case of SAP covered with the TiO_2 grains, this phenomenon allowed for the immobilization of the TiO_2 grains in the near-surface layer and its crystallization on the surface, increasing its exposure to external radiation (Figure 5). However, due to water entrapment in the mold due to its continuous desorption from SAP during hydration, the porosity of the tested surfaces significantly increased. As the proposed surface modification method represents a first step in the development of surface modification methods for photocatalytic cementitious materials, in future iterations, the mold surface exposed to the SAP-TiO_2 layer is going to be prepared as a semi-permeable/sponge-like layer to limit the influence of the excess water released from SAP on the surface porosity.

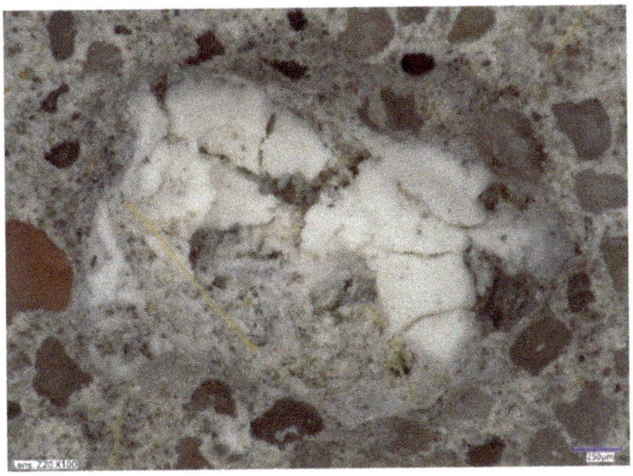

Figure 5. Micrograph of crystallized TiO_2 on the surface of the photocatalytic mortar.

4. Conclusions

Based on the performed research, several conclusions can be drawn:

- Modifying the surface of the composite solely with water dispersion of TiO_2 increased the efficiency of the photocatalytic reactions. However, due to the low viscosity of the modifying agent, this method of modification did not provide a homogenous distribution of TiO_2 on the surface of the cementitious composite (the coefficient of variation for that solution increased significantly);
- Surface modification with SAP, either in non-saturated or hydrogel form, increases efficiency in air purification from NO and NO_x pollutants under considered UV-A and visible light irradiance and allowed for the homogenous distribution of TiO_2 over the surface of cementitious material;
- Surface modification with the SAP hydrogel contributed to the highest reductions in concentrations of NO and NO_x under all considered light sources.

Author Contributions: Conceptualization, M.K., P.W. and W.J.-R.; methodology, M.K.; software, M.K.; validation, M.K., P.W. and W.J.-R.; formal analysis, P.W. and W.J.-R.; investigation, M.K.; resources, M.K., P.W. and W.J.-R.; data curation, M.K.; writing—original draft preparation, M.K.; writing—review and editing, M.K., P.W. and W.J.-R.; visualization, M.K.; supervision, P.W. and W.J.-R.; project administration, W.J.-R.; funding acquisition, W.J.-R. All authors have read and agreed to the published version of the manuscript.

Funding: This research was funded by the NCBiR (National Centre for Research and Development), Poland, under grant number TECHMATSTRATEG-III/0013/2019-01.

Institutional Review Board Statement: Not applicable.

Informed Consent Statement: Not applicable.

Data Availability Statement: Data is available on request.

Acknowledgments: The publication cost of this paper was covered with funds of the Polish National Agency for Academic Exchange (NAWA): "MATBUD'2023—Developing international scientific cooperation in the field of building materials engineering" BPI/WTP/2021/1/00002, MATBUD'2023.

Conflicts of Interest: The authors declare no conflict of interest.

References

1. Coelho, S.; Ferreira, J.; Rodrigues, V.; Lopes, M. Source apportionment of air pollution in European urban areas: Lessons from the Clair City project. *J. Environ. Manag.* **2022**, *320*, 115899. [CrossRef] [PubMed]
2. Seinfeld, J.H.; Pandis, S.N. *Atmospheric Chemistry and Physics: From Air Pollution to Climate Change*, 3rd ed.; John Wiley & Sons: Hoboken, NJ, USA, 2016; ISBN 978-1-118-94740-1.
3. Witkowski, H.; Jarosławski, J.; Tryfon-Bojarska, A. Application of Photocatalytic Concrete Paving Blocks in Poland—Verification of Effectiveness of Nitric Oxides Reduction and Novel Test Method. *Materials* **2020**, *13*, 5183. [CrossRef] [PubMed]
4. Bianchi, C.L.; Pirola, C.; Galli, F.; Cerrato, G.; Morandi, S.; Capucci, V. Pigmentary TiO_2: A challenge for its use as photocatalyst in NOx air purification. *Chem. Eng. J.* **2015**, *261*, 76–82. [CrossRef]
5. Rhee, I.; Lee, J.-S.; Kim, J.B.; Kim, J.-H. Nitrogen Oxides Mitigation Efficiency of Cementitious Materials Incorporated with TiO_2. *Materials* **2018**, *11*, 877. [CrossRef] [PubMed]
6. Witkowski, H.; Jackiewicz-Rek, W.; Jarosławski, J.; Chilmon, K.; Szkop, A. Ozone Formation during Photocatalytic Oxidation of Nitric Oxides under UV Irradiation with the Use of Commercial TiO_2 Photocatalytic Powders. *Materials* **2022**, *15*, 5905. [CrossRef] [PubMed]
7. Witkowski, H.; Jackiewicz-Rek, W.; Chilmon, K.; Jarosławski, J.; Tryfon-Bojarska, A.; Gąsiński, A. Air Purification Performance of Photocatalytic Concrete Paving Blocks after Seven Years of Service. *Appl. Sci.* **2019**, *9*, 1735. [CrossRef]
8. Shafaei, D.; Yang, S.; Berlouis, L.; Minto, J. Multiscale pore structure analysis of nano titanium dioxide cement mortar composite. *Mater. Today Commun.* **2020**, *22*, 100779. [CrossRef]
9. Kalinowski, M.; Woyciechowski, P. Chloride Diffusion in Concrete Modified with Polyacrylic Superabsorbent Polymer (SAP) Hydrogel—The Influence of the Water-to-Cement Ratio and SAP-Entrained Water. *Materials* **2021**, *14*, 4064. [CrossRef] [PubMed]
10. Kang, S.H.; Hong, S.G.; Moon, J. Absorption kinetics of superabsorbent polymers (SAP) in various cement-based solutions. *Cem. Concr. Res.* **2017**, *97*, 73–83. [CrossRef]
11. Kalinowski, M.; Woyciechowski, P. The course of water absorption and desorption from superabsorbent polymers (SAP) in cementitious environment. In Proceedings of the ICSBM 2019—The 2nd International Conference on Sustainable Building Materials, Eindhoven, The Netherlands, 12–15 August 2019; Volume 5, pp. 44–54.
12. *EN 197-1:2011*; Cement—Part 1: Composition, Specifications and Conformity Criteria for Common Cements. European Committee for Standardization: Brussels, Belgium, 2011.
13. *EN 13139:2002*; Aggregates for Mortar. European Committee for Standardization: Brussels, Belgium, 2002.
14. *EN 1008:2002*; Mixing Water for Concrete—Specification for Sampling, Testing and Assessing the Suitability of Water, Including Water Recovered from Processes in the Concrete Industry, as Mixing Water for Concrete. European Committee for Standardization: Brussels, Belgium, 2002.
15. *EN 934-2:2009+A1:2012*; Admixtures for Concrete, Mortar and Grout—Part 2: Concrete Admixtures—Definitions, Requirements, Conformity, Marking and Labelling. European Committee for Standardization: Brussels, Belgium, 2012.
16. *EN 196-1:2016*; Methods of Testing Cement—Part 1: Determination of Strength. European Committee for Standardization: Brussels, Belgium, 2016.
17. *EN 1015-3:1999*; Methods of Test for Mortar for Masonry—Part 3: Determination of Consistence of Fresh Mortar (by Flow Table). European Committee for Standardization: Brussels, Belgium, 1999.
18. Liu, J.; Farzadnia, N.; Shi, C. Microstructural and micromechanical characteristics of ultra-high performance concrete with superabsorbent polymer (SAP). *Cem. Concr. Res.* **2021**, *149*, 106560. [CrossRef]
19. Snoeck, D.; Velasco, L.F.; Mignon, A.; Van Vlierberghe, S.; Dubruel, P.; Lodewyckx, P.; De Belie, N. The effects of superabsorbent polymers on the microstructure of cementitious materials studied by means of sorption experiments. *Cem. Concr. Res.* **2015**, *77*, 26–35. [CrossRef]

materials proceedings

Proceeding Paper

Analysis of Thermal and Mechanical Parameters of the BFRP Bars †

Małgorzata Wydra [1,*], Piotr Dolny [1], Grzegorz Sadowski [1], Natalia Grochowska [2], Piotr Turkowski [3] and Jadwiga Fangrat [3]

1 Faculty of Civil Engineering, Mechanics and Petrochemistry, Warsaw University of Technology, Łukasiewicza 17, 09-407 Płock, Poland
2 Faculty of Physics, Warsaw University of Technology, Koszykowa 75, 00-662 Warsaw, Poland
3 Building Research Institute, Filtrowa 1, 00-611 Warsaw, Poland
* Correspondence: malgorzata.wydra@pw.edu.pl; Tel.: +48-505-225-741

† Presented at the 10th MATBUD'2023 Scientific-Technical Conference "Building Materials Engineering and Innovative Sustainable Materials", Cracow, Poland, 19–21 April 2023.

Abstract: Fibre-Reinforced Polymer (FRP) reinforcement bars are gaining interest in terms of using them as an internal reinforcement in concrete construction parts due to their high tensile strength, corrosion resistance, low weight, and electrical indifference. Nevertheless, low elasticity and difficulties related to a high reduction in mechanical properties at even slightly elevated temperatures seem to limit this potential, due to existing fire safety requirements for buildings. Basalt FRP, which is the subject of this experimental study, is a relatively new type of non-metallic bars, and their environmental friendliness has been underlined in previous studies. The aim of this study is to determine the mechanical properties of BFRP bars, such as tensile and compressive strength and elasticity modulus, at normal and elevated temperatures up to 200 °C. The medium values of compressive strength at room temperature were in the range of 441.2–466.8 MPa, and it was significantly lower than the tensile strength (930.5–1121.3 MPa). Additionally, low values of elasticity modulus, especially when comparing to steel bars (typically about 210 GPa), were found in both compression (mean values: 31.0–38.4 GPa) and tension (mean values: 43.3–44.6 GPa). Low elasticity modulus may lead to difficulties with excessive deflections and crack widths, when designing bent elements with such reinforcement. Moreover, reduced mechanical properties at compression should not be neglected when designing compressed parts. Additionally, the phase change parameters, e.g., glass transition temperature, have been determined by means of DMA method, and the glass transition temperature was found to be equal to 107.5 °C.

Keywords: Basalt Fibre-Reinforced Polymer; reinforcement; temperature

Citation: Wydra, M.; Dolny, P.; Sadowski, G.; Grochowska, N.; Turkowski, P.; Fangrat, J. Analysis of Thermal and Mechanical Parameters of the BFRP Bars. *Mater. Proc.* **2023**, *13*, 24. https://doi.org/10.3390/materproc2023013024

Academic Editors: Katarzyna Mróz, Tomasz Tracz, Tomasz Zdeb and Izabela Hager

Published: 15 February 2023

1. Introduction

Fibre-Reinforced Polymer (FRP) is a light, high-strength, and durable material. Its electric indifference, high corrosion resistance, high tensile strength, good damage tolerance, good fatigue performance and low energy consumption during the fabrication of raw materials should also be highlighted [1–8]. These advantages make them potentially attractive as an alternative to traditional reinforcement. Nevertheless, there are also important disadvantages, when comparing FRP to steel, which may significantly influence the performance of such a reinforcement in concrete building structures. One of these disadvantages is the fact that FRPs have much lower compressive strength than the tensile strength, another that they have a low elasticity modulus and, finally, poor mechanical performance at even slightly elevated temperatures.

This study aims to analyse thermal and mechanical properties of the basalt type of FRP, which is relatively new and has not as yet been sufficiently examined [9]. The environmental impact of BFRP composites (especially in terms of costs and amount of energy during production) should be emphasised, as it is lower when compared to CFRP [10].

2. Materials and Specimens

Three diameters of the same type of BFRP bars were tested: 8, 10, and 12 mm. The fibre content was equal to 77%, and epoxy type of matrix was used. Precise diameters of the bars were measured in five random locations along the bars and were equal to 8.1 ± 0.2, 9.2 ± 0.1 and 11.6 ± 0.3 (mean value ± standard deviation), respectively.

A cuboid specimen was cut out of the inner part of the Ø12 BFRP bar, with the cross section of 9.64 mm and 3.28 mm, on which Dynamic Mechanical Analysis was performed in order to determine the glass transition temperature of the analysed BFRP material.

Ø10 and Ø12 BFRP bars with the length of 1 m were tested in tension. In that case, 330 mm-long steel pipes were mounted at the end specimens in order to prevent crushing of the FRP in the grip of the hydraulic press (see Figure 1). Either epoxy resin (for Ø10 BFRP specimens) or expansive mortar (Ø10 and Ø12) was used to attach the FRP bars into the steel pipes.

Figure 1. Specimens for tension tests (Ø12 BFRP, steel caps mounted with the use of expansive mortar).

The specimens in compression at both room and elevated temperatures were tested with the use of steel caps (see Figure 2a), similarly to experiments performed by Khorramian and Sadeghian [11].

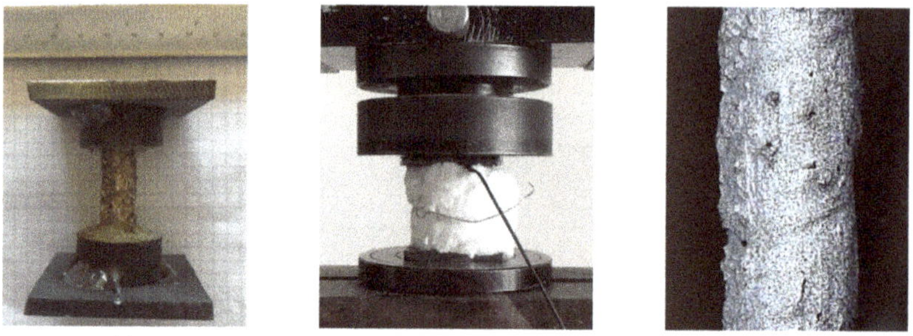

(a) (b) (c)

Figure 2. Specimen for compressive strength test (Ø12 BFRP) (**a**); specimen for compression test wrapped with two layers of ceramic wool (**b**); black–white pattern at the specimen's surface (**c**).

The steel plates (30 × 30 × 2 mm for Ø10 and 50 × 50 × 5 mm for Ø8 and Ø12) were welded with the round pipe pieces (Ø20.0 × 10 × 2 mm for Ø10 and Ø26.9 × 12 × 2 mm for Ø8 and Ø12). After preparation of the steel caps, they were attached at the ends of FRP bars with the use of epoxy resin and positioned with the use of a spirit level. The length of the bars was 4 cm. In the case of Ø12 bars, the method of specimens' preparation was improved in order to enable examination at higher temperatures. Therefore, cementitious expansive mortar was used instead of epoxy resin.

In the case of specimens tested at elevated temperatures, two layers of ceramic wool (see Figure 2b) were used to sustain the temperature after removing the specimens from the thermal chamber and placing them at the test stand.

In the case of specimens tested at room temperature in both compression and tension, a black–white pattern was added at the surface of the specimens (Figure 2c), so that Digital Image Correlation could be used to determine the strains during the tests, and as a result, moduli of elasticity (at compression and tension) could be calculated.

3. Methods

3.1. Glass Transition Temperature

Glass transition temperature was measured with the use of Discovery DMA 850 (TA Instruments) appliance in Oscillation Temperature Ramp. The specimen was submitted to simultaneous cyclic flexure with the amplitude of deflection equal to 0.8 μm and frequency of 1.0 Hz, and an increase in heating temperature up to 157 °C. The heating rate was equal to 2 °C/min.

The storage and loss moduli can be determined using following equations, representing the behaviour of viscoelastic material:

$$\sigma = \sigma_0 \sin(2\pi f t) \tag{1}$$

$$\varepsilon = \varepsilon_0 \sin(2\pi f t - \delta) \tag{2}$$

$$\tan\ \delta = E''/E' \tag{3}$$

where:

ε—strain;
ε_0—strain's amplitude;
σ—stress;
σ_0—stress's amplitude;
f—frequency;
t—time;
δ—phase lag between stress and strain;
E′—storage modulus;
E″—loss modulus.

3.2. Tensile Strength and Elasticity at Room Temperature

Average pace of tensile loading was equal to 6 MPa/s. Digital cameras had simultaneously been taking photos from one or two perpendicular directions in set up periods of time.

After the test, photos were analysed in DIC software and, for each specimen, three virtual tensiometers with the length of approximately 100 mm were set on each specimen to calculate strains. Elasticity moduli were calculated as the directional coefficients in linear approximation of stress–strain relations.

3.3. Compressive Strength and Elasticity at Room Temperature

Average pace of compressive loading was equal to 4 Mpa/s. The photos of the specimens were taken during the tests from one or two perpendicular directions in order to determine the strains with the use of DIC software after the tests. The virtual tensiometers with the length of approximately 10 mm were used.

The stress–strain relations for each specimen in compression were calibrated with the use of linear function, where modulus elasticity at compression was assumed as a directional coefficient of these functions. If two digital cameras were used, moduli of elasticity values were calculated on each direction separately.

Compressive strength was calculated as the force at failure divided by the cross-section area.

3.4. Compressive Strength at Elevated Temperatures

Specimens described in Section 2 were heated up till the predetermined value of temperature (up to 100 or 200 °C) was achieved at the surface of the bar. Temperature values were registered with the use of thermocouples mounted under the ceramic wool (see Figure 2b). After removing the specimens from heating chamber, compression tests were performed with the pace of 1.5 mm/min. Maximum forces and temperatures at the surfaces at failure time were registered for each specimen.

4. Results

Results from DMA tests were shown as storage and loss moduli changes (Figure 3). The glass transition temperature determined from loss modulus changes along the temperature growth was equal to 107.5 °C (Figure 3).

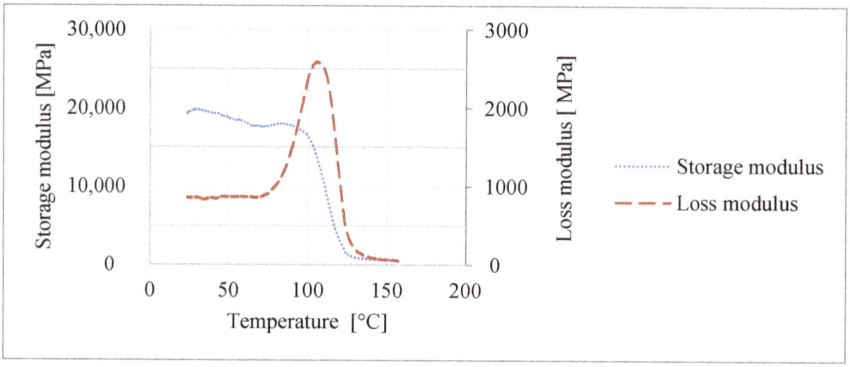

Figure 3. Results from DMA test.

The results from tensile strength tests were summarized (Table 1). The medium values of elasticity modulus were in the range from 43.3 to 44.6 MPa, while tensile strength medium values were 930.5 to 1073.1 MPa.

Table 1. Summary of the results—BFRP bars in tension.

Diameter	Parameter	Specimen No. 1	2	3	Mean Value
Ø10 (mounted with epoxy resin)	Tensile strength [MPa]	1143.3	977.8	1098.1	1073.1
	Modulus of elasticity [GPa]	45.6	41.3	47.0	44.6
Ø12 (mounted with epoxy resin)	Tensile strength [MPa]	1121.3	-	-	1121.3 [1]
	Modulus of elasticity [GPa]	44.1	44.1	46.1	44.8
Ø12 (mounted with expansive mortar)	Tensile strength [MPa]	908.4	946.2	936.8	930.5
	Modulus of elasticity [GPa]	45.2, 44.4	43.4, 43.7	42.0, 40.9	43.3 [2]

[1] determined only for one specimen. [2] mean value calculated on specimens tested from two directions.

Two out of the three Ø12 specimens with caps mounted with epoxy resin did not fail during the test as a result as achieving stresses equal to tensile strength, but the FRP bars with hardened epoxy resin started to slide out of the steel caps. Therefore, maximum strength values should not be considered as tensile strength in that case, and were excluded from the analysis. Nevertheless, elasticity moduli were calculated for these specimens. Figure 4 shows the typical mode of failure for the analysed specimens.

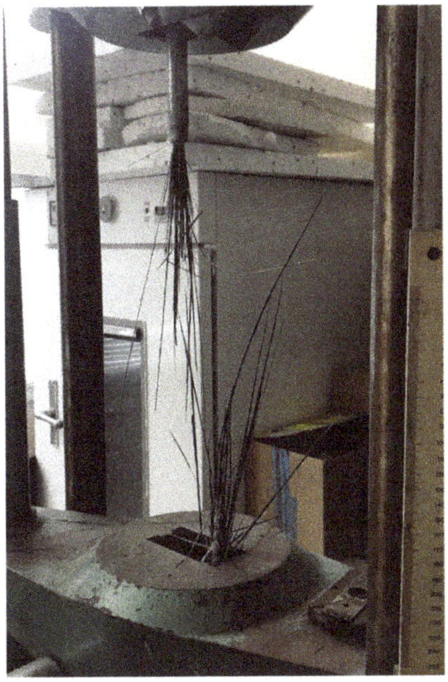

Figure 4. Specimen after failure in tension.

The results of the compressive tests at room temperature are summarized in Table 2, while Figure 5 shows the typical form of failure in compression. The medium values of compressive strength were in the range of 441.2 to 456.0 MPa, and medium values of elasticity modulus were in the range of 31.0 to 38.4 MPa.

Table 2. Summary of the results—BFRP bars in compression.

Diameter	Parameter	Specimen No. 1	2	3	Mean Value
Ø8	Compressive strength [MPa]	416.3	495.8	-	456.0
	Modulus of elasticity [GPa]	27.5, 35.4	35.8, 54.9	-	38.4 [1]
Ø10	Compressive strength [MPa]	434.7	517.5	448.3	466.8
	Modulus of elasticity [GPa]	34.6	28.2	30.2	31.0
Ø12	Compressive strength [MPa]	396.9	521.6	405.0	441.2
	Modulus of elasticity [GPa]	46.9, 50.0	24.5, 24.6	32.8, 37.0	35.1 [1]

[1] mean value calculated on specimens tested from two directions.

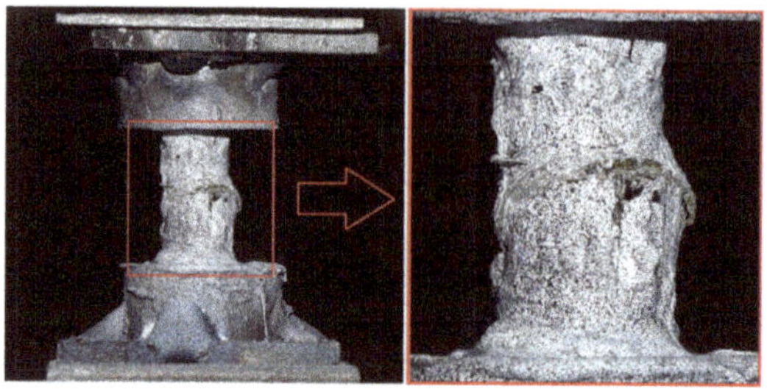

Figure 5. Specimen after failure in compression.

Compressive strength along with temperature at the surface of the specimen at failure time, at compressive strength tests at elevated temperatures (100 °C and 200 °C), are summarized in Table 3. Additionally, the results for four reference specimens tested without heating on the same day are included in this table. The strength retention ratio calculated for the medium temperature at failure equal to 97.3 °C was 24%, and for 191.0 °C, it was 8%.

Table 3. Compressive strength tests during heating results.

Temperature	Parameter	Specimen No. 1	2	3	4	Mean Value
20 °C (reference specimens)	Compressive strength [MPa]	497.7	519.5	528.0	472.2	504.3
100 °C	Compressive strength [MPa]	87.1	153.3	120.2	-	120.2
	Temperature at failure [°C]	97	98	97	-	97.3
200 °C	Compressive strength	43.5	49.2	34.1	-	42.3
	Temperature at failure [c]	183	198	192	-	191.0

5. Discussion

Tensile and compressive strength for BFRP bars may strongly vary depending on the type of used matrix, fibres, and volumetric proportions between matrix and fibres.

Basing on a comparison of the results from available experimental studies on mechanical properties of the BFRP bars ([2,3,12–16]—Table 4) the differences between tensile strength may vary from under 600 to even over 1500 MPa, which is a very wide range. In most cases, the tensile strength of the BFRP bars was higher than the typical value of tensile strength for steel reinforcement (about 500–600 MPa). However, no yielding occurs for non-metallic bars. As a result, rupture failure modes were noted in most cases in tension, which may result in a low safe reserve for design purposes.

In terms of tensile strength, the results from this study are similar to the works of Protchenko et al. [13], Urbanski et al. [14] and Włodarczyk and Trofimczuk [16].

Regarding compressive strength of the BFRP bars, there are few data available in the literature concerning this parameter. The reason for that may be the fact that reinforcement bars are typically submitted to tension during their lifecycle in most concrete structures. However, it is worth considering during designing that the compressive strength of the bars can be significantly lower than their tensile strength, and also lower than compressive strength of most of steel reinforcement bars (typically around 500–600 MPa). Moreover, similar results for compressive strength were noted within this study and by Thiyagarajan et al. [3] (about 450–500 MPa).

Elasticity modulus in tension measured within this study was equal to about 45 GPa, which is similar to other studies (38.34–52.0), apart from the Elgabbas et al. studies [2,15], in which the bars had a significantly higher stiffness (59.5–90.4 GPa). Even the highest value of elasticity modulus for BFRP bars amongst the available analysed literature (90.4 GPa) is much lower than typical values for reinforcement steel (about 210 GPa). This may lead to excessive deflections and crack propagation in bent concrete elements with non-metallic reinforcement.

Additionally, this study aimed to experimentally determine elasticity moduli at compression, which were 15–30% lower than the values measured in tension.

Further research will concern the examination of tensile strength at elevated and high temperatures in comparison to the available literature data [17–20]. The results from the current study will also be used for numerical modelling purposes regarding axially compressed concrete columns with BFRP reinforcement bars at room and high temperatures.

Table 4. Mechanical properties of the BFRP bars—comparison.

Reference	Diameters	Tensile Strength [MPa]	Compressive Strength [MPa]	Elasticity Modulus [GPa]
Thiyagarajan et al. [3]	8, 10 and 12 mm	1362.3–1585.6	470.2–495.3	48–52 (tension)
Fan and Zhang [12]	12 mm	569–681	-	-
Protchenko et al. [13]	8 mm	1103.3	-	43.9 (tension)
Urbanski et al. [14]	8 mm	1009.1–1089.4	-	38.34–40.72
Elgabbas et al. [2,15]	7–8 mm	1567–1680	-	59.5–69.0 (tension) 74.0–90.4 (flexion)
Włodarczyk and Trofimczuk [16]	8 and 10 mm	1103–1153	-	43.9–48.2 (tension)
This study	8, 10 and 12 mm	930.5–1121.3	441.2–466.8	43.3–44.6 (tension) 31.0–38.4 (compression)

6. Conclusions

The following conclusions can be drawn from this study:

1. The mechanical properties of the BFRP may strongly vary depending on many parameters, such as the type of matrix and fibres, and their volumetric proportions. However, there can be noted some trends that are similar to other studies, such as a significant reduction in compressive strength in reference to tensile strength (by over 60% in experiments performed by Thiyagarajan et al. [3] and about 40–50% in this study).
2. Elasticity modulus values determined with the use of Digital Image Correlation for BFRP bars were significantly lower than the values for traditional steel reinforcement (about 5 times lower in tension and 6 times lower in compression).
3. Low glass transition temperature (equal to 107.5 °C), at which structural changes in the material occurred, may lead to significant reduction in possible applicational areas. The same was confirmed in tests at elevated temperatures, as the retention ratio of compressive strength at about 100 °C was equal to 25%, and at 200 °C, it was 8%.

Author Contributions: Conceptualization, M.W. and J.F.; methodology and investigation, M.W., G.S., P.D., N.G. and P.T., writing—original draft preparation, M.W.; visualization, M.W.; supervision, J.F. All authors have read and agreed to the published version of the manuscript.

Funding: This research received no external funding.

Institutional Review Board Statement: Not applicable.

Informed Consent Statement: Not applicable.

Data Availability Statement: Data supporting findings of this study is contained within this article.

Acknowledgments: The publication cost of this paper was covered with funds provided by the Polish National Agency for Academic Exchange (NAWA): "MATBUD'2023—Developing international scientific cooperation in the field of building materials engineering" BPI/WTP/2021/1/00002, MATBUD'2023.

Conflicts of Interest: The authors declare no conflict of interest. The funders had no role in the design of the study; in the collection, analyses, or interpretation of data; in the writing of the manuscript; or in the decision to publish the results.

References

1. Burgoyne, C.; Byars, E.; Guadagnini, M.; Manfredi, G.; Neocleous, K.; Pilakoutas, K.; Taerwe, L.; Taranu, N.; Tepfers, R.; Weber, A. *Technical Report: FRP (Fibre Reinforced Polymer) Reinforcement in RC Structures*; Fédération Internationale du Béton (fib): Lausanne, Switzerland, 2007.
2. Elgabbas, B.; Ahmed, F.; Benmokrane, E. Basalt FRP reinforcing bars for concrete structures. In Proceedings of the 4th Asia-Pacific Conference on FRP in Structures, APFIS, Melbourne, Australia, 11–13 December 2013; Volume 440, pp. 11–13.
3. Thiyagarajan, P.; Pavalan, V.; Sivagamasundari, R. Mechanical characterization of basalt fibre reinforced polymer bars for reinforced concrete structures. *Int. J. Appl. Eng. Res.* **2018**, *13*, 5858–5862.
4. Bank, L.C. *Composites for Construction: Structural Design with FRP Materials*; John Wiley & Sons, Inc.: Hoboken, NJ, USA, 2006.
5. Pareek, K.; Saha, P. Basalt fiber and its composites: An overview. In Proceedings of the National Conference on Advances in Structural Technologies (CoAST-2019), Silchar, India, 1–3 February 2019.
6. Naser, M.Z.; Hawileh, R.A.; Abdalla, J.A. Fiber-reinforced polymer composites in strengthening reinforced concrete structures: A critical review. *Eng. Struct.* **2019**, *198*, 109542. [CrossRef]
7. Lau, D.; Qiu, Q.; Zhou, A.; Chow, C.L. Long term performance and fire safety aspect of FRP composites used in building structures. *Constr. Build. Mater.* **2016**, *126*, 573–585. [CrossRef]
8. Siddika, A.; Al Mamun, M.A.; Ferdous, W.; Alyousef, R. Performances, challenges and opportunities in strengthening reinforced concrete structures by using FRPs—A state-of-the-art review. *Eng. Fail. Anal.* **2020**, *111*, 104480. [CrossRef]
9. Fiore, V.; Scalici, T.; Di Bella, G.; Valenza, A. A review on basalt fibre and its composites. *Compos. Part B Eng.* **2015**, *74*, 74–94. [CrossRef]
10. Inman, M.; Thorhallsson, E.R.; Azrague, K. A mechanical and environmental assessment and comparison of Basalt Fibre Reinforced Polymer (BFRP) rebar and steel rebar in concrete beams. *Energy Procedia* **2017**, *111*, 31–40. [CrossRef]
11. Khorramian, K.; Sadeghian, P. Material Characterization of GFRP Bars in Compression Using a New Test Method. *J. Test. Eval.* **2021**, *49*, 20180873. [CrossRef]
12. Fan, X.; Zhang, M. Experimental study on flexural behaviour of inorganic polymer concrete beams reinforced with basalt rebar. *Compos. Part B Eng.* **2016**, *93*, 174–183. [CrossRef]
13. Protchenko, K.; Salha, M.; Urbański, M.; Narloch, P. Compressive Properties of BFRP and HFRP Bars. *IOP Conf. Ser. Mater. Sci. Eng.* **2021**, *1015*, 012091. [CrossRef]
14. Urbanski, M.; Lapko, A.; Garbacz, A. Investigation on concrete beams reinforced with basalt rebars as an effective alternative of conventional R/C structures. *Procedia Eng.* **2013**, *57*, 1183–1191. [CrossRef]
15. Elgabbas, F.; Ahmed, E.A.; Benmokrane, B. Physical and mechanical characteristics of new basalt-FRP bars for reinforcing concrete structures. *Constr. Build. Mater.* **2015**, *95*, 623–635. [CrossRef]
16. Włodarczyk, M.; Trofimczuk, D. *Prediction of Ultimate Capacity of FRP Reinforced Concrete Compression Members*; Innovations in Materials, Design and Structures Book of Abstracts for the 2019 fib International Symposium May 27–29; The Institute of Civil Engineering: Delhi, India, 2019; pp. 1353–1361.
17. Hajiloo, H.; Green, M.F.; Gales, J. Mechanical properties of GFRP reinforcing bars at high temperatures. *Constr. Build. Mater.* **2018**, *162*, 142–154. [CrossRef]
18. Rosa, I.C.; Firmo, J.P.; Correia, J.R.; Barros, J.A.O. Bond behaviour of sand coated GFRP bars to concrete at elevated temperature—Definition of bond vs. slip relations. *Compos. Part B Eng.* **2019**, *160*, 329–340. [CrossRef]

19. Ashrafi, H.; Bazli, M.; Najafabadi, E.P.; Vatani Oskouei, A. The effect of mechanical and thermal properties of FRP bars on their tensile performance under elevated temperatures. *Constr. Build. Mater.* **2017**, *157*, 1001–1010. [CrossRef]
20. Najafabadi, E.P.; Oskouei, A.V.; Khaneghahi, M.H.; Shoaei, P.; Ozbakkaloglu, T. The tensile performance of FRP bars embedded in concrete under elevated temperatures. *Constr. Build. Mater.* **2019**, *211*, 1138–1152. [CrossRef]

materials proceedings

MDPI

Proceeding Paper

Manufacturing of Low-Cost Bricks Using Waste Materials †

Kalaimani Ramakrishnan [1,*], Vigneshkumar Chellappa [2] and Subha Chandrasekarabarathi [1]

1 Ramco Institute of Technology, Rajapalayam 626117, India
2 Amity School of Design, Noida 201303, India
* Correspondence: kalaimani@ritrjpm.ac.in

† Presented at the 10th MATBUD'2023 Scientific-Technical Conference "Building Materials Engineering and Innovative Sustainable Materials", Cracow, Poland, 19–21 April 2023.

Abstract: Bricks consume a massive quantity of clay. Using clay bricks causes erosion, lowers the water table, and harms the environment. This research examined various waste materials, including fly ash, quarry dust, marble dust, eggshell powder (ESP), and rice husk ash (RHA), in varying percentages to avoid using clay in manufacturing bricks. Compressive strength and water absorption tests were conducted, and the results were compared with the specifications for traditional clay bricks. It was observed that the compressive strength and water absorption values met the relevant standards needed for standard construction bricks. Furthermore, the cost of manufacturing bricks from waste materials was estimated, and the findings show that manufacturing bricks from waste materials cost less than conventional bricks. Finally, it was concluded that the brick industry could become more sustainable and economically feasible by using specific waste materials in manufacturing bricks.

Keywords: clay bricks; waste materials; cost estimation; production

check for updates

Citation: Ramakrishnan, K.; Chellappa, V.; Chandrasekarabarathi, S. Manufacturing of Low-Cost Bricks Using Waste Materials. *Mater. Proc.* **2023**, *13*, 25. https://doi.org/10.3390/materproc2023013025

Academic Editors: Katarzyna Mróz, Tomasz Tracz, Tomasz Zdeb and Izabela Hager

Published: 15 February 2023

1. Introduction

Bricks are widely used in construction projects as they possess beneficial characteristics such as low cost, high durability, and ease of handling [1]. Clay is the primary raw material used in brick production. For brick manufacturing, almost 340 billion tons of clay are used annually [2]. The increasing clay use as a resource for manufacturing bricks has caused an alarmingly high degree of variance in this natural material [3]. This has therefore led several researchers to find alternative resources or ways to recycle the wastes produced by various industrial processes. As a result, many waste materials have been used in brick manufacturing, including marble dust [4], sludge from water treatment plants [5], fly ash [6], sugarcane bagasse ash [7], rice husk ash (RHA) [8], waste glass powder [9], sawdust [10], quarry dust [11], and eggshell powder (ESP) [12]. The properties of clay must be altered by the additives physically and chemically within a specific range, and they must not adversely affect the clay's strength and durability [13].

Fly ash is a typical clay component generated when coal is burned in coal-fired power plants. It is a varied substance with a glossy appearance made of mullite (alumina and silica) and iron oxides (hematite and magnetite) [14]. Its elemental composition is similar to that of brick soils. Bricks' strength can be increased and water absorption reduced by adding fly ash to clay [15,16]. RHA is a potential source of Amphorous reactive silica released by the combustion of rice hulls [17]. It has excellent thermal insulation and keeps the temperature stable. According to a study by Ramasamy [18], adding a small amount of RHA will increase the material's compressive strength. Food processing firms produce tons of eggshells as waste, which has led to environmental issues because of their inappropriate disposal in our environment [19]. The calcium carbonate in ESP dissolves in different acids and has a chemical composition similar to limestone [20]. ESPs are potential substitute raw materials for clay brick production [12].

A study by Dhanapandian and Gnanavel [21] indicated that up to 20% of waste marble powder improved water absorption. Adding waste marble to clay reduces the fire temperature without affecting the brick's properties [22] and thus lowers the production cost and saves energy. The bricks' physical strength at high temperatures was increased by including waste marble [23]. The fluxing effect of these wastes at higher temperatures boosted several brick properties, most notably flexural strength, compressive strength, and bulk density [4]. Past studies, e.g., from the authors of [24], have proven that waste marble powder may be used inexpensively in clay bricks. Quarry dust is produced by cutting and grinding the stone. Depending on the quarry fine dust's physical and chemical properties, it may be used to produce bricks as a filler, a replacement for clay, a colorant, a fluxing agent, or even body fuels [25]. Quarry dust addition improves the bricks' durability [26]. The potential use of naturally occurring wastes, readily accessible and inexpensive, is becoming increasingly more important in reducing building material costs without sacrificing quality. Given the above, this study investigated the behavior of bricks made by replacing clay with waste materials such as ESP, RHA, fly ash, quarry dust, and marble dust and estimated the cost of brick production.

2. Materials and Methods

2.1. Materials

The waste materials such as fly ash, quarry dust, marble dust, ESP, and RHA were collected from local dealers in Rajapalayam, Tamilnadu, India. Figure 1 shows the mixtures of all waste materials used in this study. The chemical composition of raw materials is shown in Table 1.

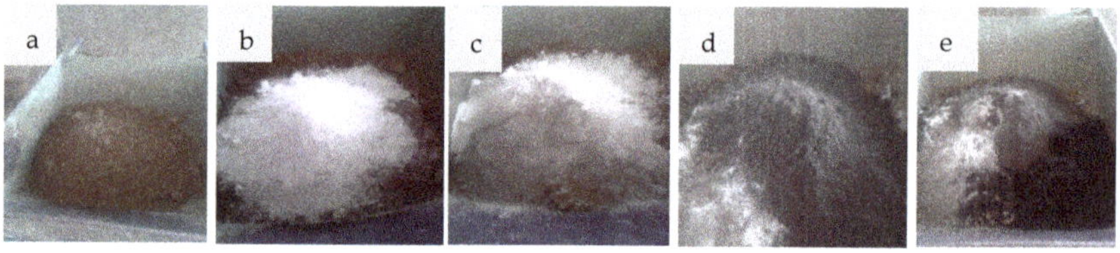

Figure 1. Waste materials: (**a**) class-C fly ash, (**b**) addition of ESP, (**c**) addition of marble dust, (**d**) quarry dust addition, and (**e**) RHA addition.

Table 1. Raw materials' chemical composition (% by weight).

Component	Fly Ash	Marble Dust	RHA	ESP	Quarry Dust	Cement
SiO_2	35–59	28.35	90.20	0.01	69.94	21.54
Al_2O_3	23–33	0.42	0.85	0.01	14.60	5.32
CaO	10–16	40.45	1.18	52.75	2.23	63.60
Loss on ignition	1–2	-	3.95	46.62	-	-
S	0.5–1.5	-	-	0.5-1.5	-	-
Fe	0.5–2	-	-	-	-	-
Fe_2O_3	-	9.7	1.38	0.01	2.16	3.6
MgO	-	16.25	1.21	0.51	0.38	1
SO_3	-	-	-	0.62	-	-

2.2. Mixing and Specimen Preparation Procedure

Different proportions of raw materials were used to make bricks, as shown in Table 2. The collected waste materials, cement, and water were mixed using a pan mixer until a consistent mixture was obtained. It was noted that to maintain the same amount of mixture consistency, for each trial, the amount of added water was 3 to 4 liters for a weight of 21 kg, and the amount of water used in each trial varied depending on the materials used. A conveyor belt was used to transport the mixed materials to the compression system for compacting, and bricks were cast using a hydraulic pressing machine with a 2200 psi pressure. The amount of compression that one brick was subjected to varied depending on the size of the materials. A total of 120 bricks (24 in each mixing ratio) were produced, each brick weighing approximately 3.5 kg. The produced bricks were taken and stacked for curing for 21 days to ensure a high-quality product [16]. A plastic sheet was laid in an area close to the brick-making unit and was covered immediately with a jute sack to prevent the bricks from drying too quickly. They were watered once every day for seven days, and to ensure a well-cured brick, they were left under plastic for eleven days [12]. Figure 2 depicts the specimen preparation process.

Table 2. Materials composition of various trails.

Materials	Trial 1	Trial 2	Trial 3	Trial 4	Trial 5
Fly ash (%)	40	40	40	40	40
Marble dust (%)	10	10	10	10	10
Cement (%)	5	5	5	5	5
RHA (%)	2	2	2	3	3
Quarry dust (%)	40	38	33	27	22
ESP (%)	3	5	10	15	20

(a) (b) (c)

Figure 2. Specimen preparation process: (**a**) mixing of materials, (**b**) casting of bricks, and (**c**) curing of bricks.

3. Experimental Study

Testing bricks for strength, durability, efflorescence, and dimensional tolerance is mandatory [14]. In this study, a dimension test, water absorption test, compressive strength test, and efflorescence test were carried out to determine the quality of bricks. Details of these tests are discussed in the following sections.

3.1. Dimension Test

A total of 20 samples were chosen randomly from each mixed ratio, and the blisters were removed using a trowel before conducting the test. The dimensions of bricks were tested per India's industry standards (IS: 1077(1992), clause 6.2). First, all the selected bricks were arranged lengthwise in five rows on a flat surface, and the length of the overall bricks was measured using tape. The values were recorded for each row. Similarly, the bricks were arranged widthwise, and the values were recorded. Finally, all the bricks were arranged on the basis of height, and the overall height was determined. The observed values were recorded, and the standard deviations are shown in Table 3. It was found that the dimension tolerance values for bricks were within the IS limits [27].

Table 3. Dimensions of bricks.

Trail	Dimension (mm)			Tolerance (%)		
	L *	B #	D +	L	B	D
1	4517	2158	1378	−1.8	−1.9	−1.57
2	4533	2163	1386	−1.46	−1.68	−1.0
3	4528	2159	1394	−1.57	−1.86	−0.43
4	4522	2165	1375	−1.69	−1.59	−1.78
5	4548	2168	1382	−1.13	−1.45	−1.28

* Length; # breadth; + depth.

3.2. Water Absorption Test

Water absorption affects the durability of bricks. To conduct this experiment, three specimens from each mixed ratio were selected. Samples were first placed in an oven at 105 °C until they reached a consistent density. After being overdried, the bricks were immersed in water for 24 hours. Then, the specimens were removed from the water and cleaned with a damp cloth. The percentage of samples' water content was determined using:

$$\text{water absorption in }\% = \frac{W2 - W1}{W1} \times 100, \quad (1)$$

where W1 is the brick's dry weight, and W2 is the brick's wet weight.

Figure 3 shows the average water absorption results for brick in each mixed ratio. According to the protocol outlined in IS [27], the bricks must absorb no more than 20% of their weight in water after 24 h of immersion in cold water. Figure 3 shows that the water absorption values for bricks were less than 7% in each trial, meeting the IS requirements [27]. A lower rate of water absorption was observed in trail 1 with the increase in the percentage of quarry dust.

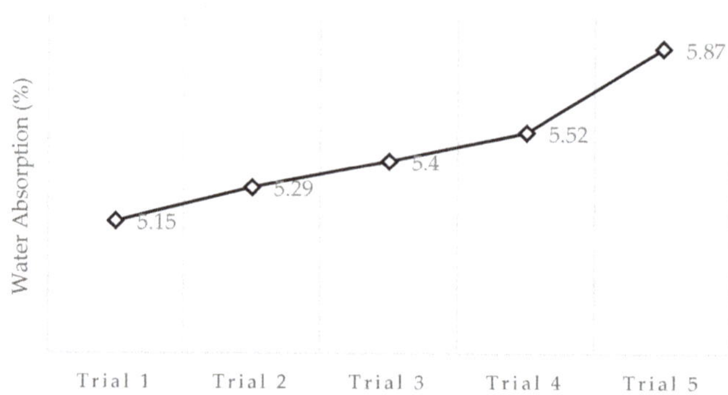

Figure 3. Water absorption results for brick samples.

3.3. Compressive Strength Test

The compression test machine measured the dry compressive strength of brick samples. Three random samples were taken from each mixed ratio, as in the water absorption test. The sample's face was subjected to a compression load with dimensions of 230 mm × 110 mm. The compressive strength was determined by dividing the overall load by the original cross-sectional area of the specimen. Figure 4 shows the compression strength test arrangement, and the average compressive strength results are displayed in Figure 5. From Figure 5, it was observed that the composition used in trial 3 achieved the highest compressive strength compared with other compositions used in other trials. However, the compressive brick strength results ranged from 7 to 14 N/mm^2, which could be classified as Grade A [28].

Figure 4. Compressive strength testing of samples.

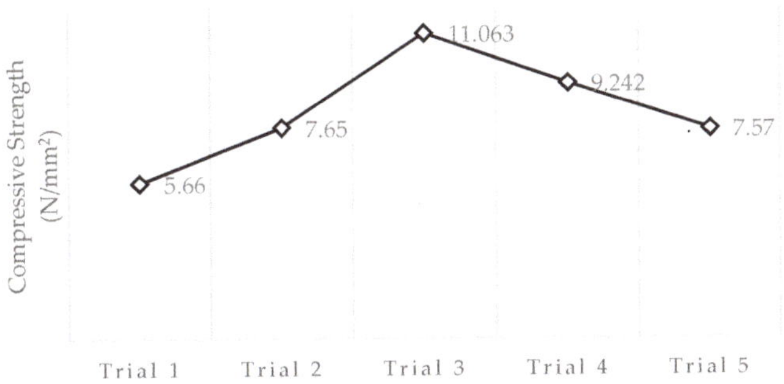

Figure 5. Compressive strength results for brick samples.

3.4. *Efflorescence Test*

According to Ukwatta et al. [29], salt deposits on brick surfaces lead to efflorescence, which detracts from the aesthetics of brick masonry structures. The samples were tested following the procedure given in IS 3495 (part 3) to determine the efflorescence. One random sample from each mixed ratio was selected. A shallow flat bottom tray was taken, and water was poured to a height of 2.5 cm. The samples were placed vertically in the tray and kept until they absorbed the water. When the bricks appeared to dry, the same quantity of water was poured into the tray and evaporated. Once the evaporation occurred, the samples' evaluation was tested on the basis of the classification provided in [28]. In this study, it was observed that there was no efflorescence in brick samples, indicating that none of the samples had been exposed to sulfate attack [28].

4. Cost Estimation

The cost of the waste materials per kg weight, including the shipping cost, was estimated in rupees (INR). The weight of the waste materials used in one brick is multiplied by the unit cost of the waste materials to calculate the total material cost. The cost of human resources and electricity used to manufacture the bricks were also considered. The total cost of brick production in the trials was estimated, as shown in Table 4. It was observed that during trial 1, the cost of production of one brick was INR 3.32, and the production cost gradually increased to INR 3.660 during trial 5. It is noteworthy that although the addition of quarry dust was reduced in each trial, the addition of ESP was simultaneously increased, affecting the cost of brick production. Table 4 shows that the optimum cost for brick manufacturing was the ratio used in trial 3, which indicates high compressive strength results (see Figure 5).

Table 4. Cost for production of bricks in each trial.

	Unit Cost (in INR)	Trial 1		Trial 2		Trial 3		Trial 4		Trial 5	
		Weight (kg)	Cost (INR)	Weight (kg)	Cost (INR)	Weight (kg)	Cost (INR)	Weight (kg)	Cost (INR)	Weight (kg)	Cost (INR)
					Materials						
Fly ash	0.7/kg	1.4	0.980	1.4	0.980	1.4	0.980	1.4	0.980	1.4	0.980
Quarry dust	0.42/kg	1.4	0.589	1.33	0.558	1.16	0.490	0.945	0.397	0.77	0.323
Marble powder	0.4/kg	0.35	0.140	0.35	0.140	0.35	0.140	0.35	0.140	0.35	0.140
Cement	6.50/kg	0.175	1.138	0.175	1.138	0.175	1.138	0.175	1.138	0.175	1.137
ESP	1/kg	0.105	0.105	0.175	0.175	0.35	0.350	0.525	0.525	0.7	0.700
RHA	0.10/kg	0.07	0.0007	0.07	0.0007	0.07	0.0007	0.105	0.010	0.105	0.010
					Others						
Labor/brick	0.3		0.300	0.3	0.300	0.3	0.300	0.3	0.300	0.3	0.300
Electric supply/brick	0.07		0.070	0.07	0.070	0.07	0.070	0.07	0.070	0.07	0.070
Total cost			3.322		3.361		3.475		3.560		3.660

5. Comparative Analysis

The manufactured bricks were compared with conventional clay bricks' specifications and manufacturing costs. First, the optimum proportion of manufactured bricks was selected among the different compositions of trials. The results yielded from water absorption (5.4%), and compressive strength (11.063 N/mm^2) tests indicate that the composition of trial three was the optimum proportion for producing high-quality bricks; hence, these results were compared with the conventional bricks.

Three random samples of clay bricks were collected from local manufacturers. The manufacturers reported that manufacturing one clay brick ranges from INR 5–6. The same procedures were followed to conduct the compressive strength and water absorption tests for testing the samples. The test results of conventional bricks are shown in Table 5. The average weight of clay bricks was 3.258 kgs. From Table 5, it was observed that the produced bricks were found to be economically feasible, resulting in lower water absorption, higher compressive strength, and lower manufacturing cost compared with conventional clay bricks.

Table 5. Test results for conventional clay bricks.

Experimental Test	Sample 1	Sample 2	Sample 3	Average
Water absorption (%)	12.715	10.170	9.906	10.930
Compressive strength (N/mm^2)	5.260	9.010	8.860	7.710

6. Conclusions

Clay is the primary raw material used in brick manufacturing. However, the use of clay causes erosion and lowers the water table. In this study, an attempt was made to incorporate industrial waste materials into brick manufacturing to avoid using clay. The readily available raw waste materials such as RHA, quarry dust, fly ash, marble powder, and ESP were used in different proportions for brick production. Compressive strength, water absorption, efflorescence, and dimension tests were conducted to determine the quality of the bricks. The results obtained from the tests were compared with the conventional clay bricks. Additionally, the manufacturing cost of bricks from waste materials was compared with clay bricks. The findings indicated that the bricks produced from waste materials were economically feasible and yielded high compressive strength. Future research will focus on performing the sustainability assessment of manufactured bricks from waste materials on economic, social, technical, and environmental aspects.

Author Contributions: Conceptualization, K.R. and S.C.; methodology, K.R. and V.C.; formal analysis, K.R. and S.C.; investigation, V.C.; writing—original draft preparation, K.R. and V.C.; writing—review and editing, V.C. All authors have read and agreed to the published version of the manuscript.

Funding: This research received no external funding.

Institutional Review Board Statement: Not applicable.

Informed Consent Statement: Not applicable.

Data Availability Statement: Not applicable.

Acknowledgments: The publication cost of this paper was covered with funds from the Polish National Agency for Academic Exchange (NAWA): "MATBUD'2023—Developing international scientific cooperation in the field of building materials engineering" BPI/WTP/2021/1/00002, MATBUD'2023.

Conflicts of Interest: The authors declare no conflict of interest.

References

1. Khitab, A.; Anwar, W. *Advanced Research on Nanotechnology for Civil Engineering Applications*; IGI Global: Hershey, PA, USA, 2016; pp. 1–27.
2. Pawar, A.S.; Garud, D.B. Engineering properties of clay bricks with use of fly ash. *Int. Res. J. Eng. Technol.* **2014**, *3*, 75–80.
3. Zhan, L. Production of bricks from waste materials. *Constr. Build. Mater.* **2014**, *47*, 643–655. [CrossRef]
4. Ricardo, R.J.; Brito, D.; Sardinha, M. Mechanical properties of structural concrete containing very fine aggregates from marble cutting sludge. *Constr. Build. Mater.* **2015**, *77*, 349–356.
5. Heniegal, A.M.; Ramadan, M.A.; Naguib, A.; Agwa, I.S. Study on properties of clay brick incorporating sludge of water treatment plant and agriculture waste. *Case. Stud. Constr. Mater.* **2020**, *13*, e00397. [CrossRef]
6. Murugesan, T.; Bahurudeen, A.; Sakthivel, M.; Vijay, R.; Sakthivel, S. Performance evaluation of burnt clay-fly ash unburnt bricks and precast paver blocks. *Mater. Today Proc.* **2017**, *4*, 9673–9679. [CrossRef]
7. James, J.; Pandian, P.K. A short review on the valorisation of sugarcane bagasse ash in the manufacture of stabilized/sintered earth blocks and tiles. *Adv. Mater. Sci. Eng.* **2017**, *2017*, 1706893. [CrossRef]
8. Sutas, J.; Mana, A.; Pitak, L.J.P.E. Effect of rice husk and rice husk ash to properties of bricks. *Procedia Eng.* **2012**, *32*, 1061–1067. [CrossRef]
9. Demir, I. Reuse of waste glass in building brick production. *Waste Manag. Res.* **2009**, *27*, 572–577. [CrossRef]
10. Okunade, E.A. The effect of wood ash and sawdust admixtures on the engineering properties of a burnt laterite-clay brick. *Res. J. Appl. Sci.* **2008**, *8*, 1042–1048. [CrossRef]
11. Gorai, S. Utilization of Fly ash for sustainable environment management. *J. Mater. Environ. Sci.* **2018**, *9*, 385–393.
12. Ngayakamo, B.H.; Bello, A.; Onwualu, A.P. Development of eco-friendly fired clay bricks incorporated with granite and eggshell wastes. *Environ. Chall.* **2020**, *1*, 100006. [CrossRef]
13. Christy, C.F.; Tensing, D. Greener building material with fly ash. *Asian J. Civ. Eng.* **2011**, *12*, 87–105.
14. Kayathri, K.; Vigneshkumar, C.; Gohila, M.R.; Karthik, K. Effect of copper slag, fly ash and granite power as a partial replacement in fine aggregate. *Int. J. Innov. Res. Technol. Sci. Eng.* **2014**, *2*, 439–443.
15. Shakir, A.A.; Naganathan, S.; Mustapha, K.N. Properties of bricks made using fly ash, quarry dust and billet scale. *Constr. Build. Mater.* **2013**, *41*, 131–138. [CrossRef]
16. Abbas, S.; Saleem, M.A.; Kazmi, M.S.; Munir, M.J. Production of sustainable clay bricks using waste fly ash: Mechanical and durability properties. *J. Build. Eng.* **2017**, *14*, 7–14. [CrossRef]
17. Jittin, V.; Bahurudeen, A.; Ajinkya, S.D. Utilisation of rice husk ash for cleaner production of different construction products. *J. Clean. Prod.* **2020**, *263*, 121578.
18. Ramasamy, V. Compressive strength and durability properties of rice husk ash concrete. *KSCE J. Civ. Eng.* **2012**, *6*, 93–102. [CrossRef]
19. Choudhary, R.; Koppala, S.; Swamiappan, S. Bioactivity studies of calcium magnesium silicate prepared from eggshell waste by sol–gel combustion synthesis. *KSCE J. Civ. Eng.* **2015**, *3*, 173–177. [CrossRef]
20. Hamada, H.M.; Tayeh, B.A.; Al-Attar, A.; Yahaya, F.M.; Muthusamy, K.; Humada, A.M. The present state of the use of eggshell powder in concrete: A review. *J. Build. Eng.* **2020**, *32*, 101583. [CrossRef]
21. Dhanapandian, S.; Gnanavel, B. An investigation on the effect of incorporation of granite and marble wastes in the production of bricks. *ARPN J. Eng. Appl. Sci.* **2009**, *4*, 46–50.
22. Segadães, A.M.; Carvalho, M.A.; Acchar, W. Using marble and granite rejects to enhance the processing of clay products. *Appl. Clay Sci.* **2005**, *30*, 42–52.
23. Rasool, M.A.; Hameed, A.; Qureshi, M.U.; Ibrahim, Y.E.; Qazi, A.U.; Sumair, A. Experimental study on strength and endurance performance of burnt clay bricks incorporating marble waste. *J. Asian Archit. Build. Eng.* **2022**, *22*, 240–255. [CrossRef]

24. Qureshi, M.U.; Al-Sawafi, B.; Al-Washahi, M.; Al-Saidi, M.; Al-Badi, S. The Sustainable Use of Fine Marble Waste Powder for the Stabilization of Desert Sand in Oman. In Proceedings of the 1st GeoMEast International Congress and Exhibition, Cairo, Egypt, 10–15 November 2017.
25. Katz, A.; Baum, H. Effect of high levels of fines content on concrete properties. *ACI Mater. J.* **2006**, *103*, 474.
26. Kumar, D.S.; Hudson, W.R. *Use of Quarry Fines for Engineering and Environmental Applications*; Special Report; University of Texas: Austin, TX, USA, 1992.
27. IS. Indian Standard 3495 (Parts 1 to 4), Methods of Tests of Burnt Clay Bricks. Available online: https://www.iitk.ac.in/ce/test/IS-codes/is.3495.1-4.1992.pdf (accessed on 12 December 2021).
28. IS. Indian Standard 1077, Common Burnt Clay Building Bricks—Specification. Available online: https://www.iitk.ac.in/ce/test/IS-codes/is.1077.1992.pdf (accessed on 12 December 2021).
29. Ukwatta, A.; Mohajerani, A.; Eshtiaghi, N.; Setunge, S. Variation in Physical and Mechanical Properties of Fired-clay Bricks Incorporating ETP Biosolids. *J. Clean. Prod.* **2015**, *119*, 76–85. [CrossRef]

materials proceedings

MDPI

Proceeding Paper

Physico-Chemical Properties of Sewage Sludge Ash and Its Influence on the Chemical Shrinkage of Cement Pastes †

Marcin Adamczyk *, Tomasz Zdeb and Tomasz Tracz

Faculty of Civil Engineering, Cracow University of Technology, ul. Warszawska 24, 31-155 Cracow, Poland

* Correspondence: marcin.adamczyk@pk.edu.pl

† Presented at the 10th MATBUD'2023 Scientific-Technical Conference "Building Materials Engineering and Innovative Sustainable Materials", Cracow, Poland, 19–21 April 2023.

Abstract: According to European Union regulations regarding the storage and disposal of industrial waste, waste generated during the treatment of municipal sewage should be managed in an environmentally friendly manner. In order to reduce environmental risks and minimise volume, sewage undergoes thermal treatment. The resulting ash contains toxic components, so it is essential to properly dispose of it. For this reason, the possibility of utilising this type of waste in cement composites has been subject to preliminary verification. To this end, so far, basic tests regarding the parameters such as water demand and setting time have been performed for cement pastes containing ash from the incineration of sludge produced from municipal sewage (Sewage Sludge Ash—SSA). The main purpose of this article is to describe the chemical shrinkage of cement pastes containing this additive. CEM I Portland cement and fly ash obtained at the Thermal Sludge Disposal Station at the Płaszów Sewage Treatment Plant were used in this study. Cement was replaced with ash in the rates of 10%, 20%, and 30%. The results confirm that with an increase in the SSA additive, a reduction of chemical shrinkage of 9.5%, 21.4%, and 31% is observed at 10%, 20%, and 30% content of the additive, respectively. The presented results confirm the possibility of using this material as a cement replacement in cement composites.

Keywords: sewage sludge ash (SSA); cement pastes; chemical shrinkage

check for updates

Citation: Adamczyk, M.; Zdeb, T.; Tracz, T. Physico-Chemical Properties of Sewage Sludge Ash and Its Influence on the Chemical Shrinkage of Cement Pastes. *Mater. Proc.* **2023**, *13*, 26. https://doi.org/10.3390/materproc2023013026

Academic Editors: Katarzyna Mróz and Izabela Hager

Published: 15 February 2023

1. Introduction

Progressive urbanisation is indispensable in any developing society. With it, sewage treatment is becoming an increasing problem. The result of treatment is a large quantity of sewage sludge (SS), the amount of which corresponds only 1–3% of the volume of flowing sewage. However, with increasing urbanisation and economic progress, the amount of generated waste is starting to become a serious problem. EU regulations require us to dispose of this waste properly, due to its specific properties. This type of waste contains heavy metals and pathogenic organisms, which can have a highly negative impact on the environment and human health. In order to protect the environment and individuals, and to minimise the volume of sludge generated, it is subjected to thermal treatment. However, the ash produced by the thermal disposal of sludge (SSA) still contains heavy metals. The use of this ash and other wastes in cement materials seems to make the most sense due to its high immobilisation properties [1–6]. The idea of this type of utilisation of sludge seems to be highly appropriate, since, as the research of the authors of the study [7] demonstrates, these composites have the ability to almost completely immobilise heavy metals. Heavy metals content in ash depends primarily on the urbanisation and industrialisation of the area. The amount of fly ash resulting from the thermal disposal of sewage sludge and contaminated with heavy metals is a major problem in many developing countries.

Literature reports on the physical and mechanical properties of hardened cement composites containing SSA are quite widely described [8–12], as well as in our previous studies [13], while the number of publications on their durability is limited [13,14]. So far, it is difficult to find publications devoted to the problems of chemical and autogenous shrinkage of cement binders containing ash formed after thermal disposal of sewage sludge.

The reaction of cement with water reduces the volume of the mixture. This reduction can be calculated from the density of substrates and products of cement hydration. Volume changes of this nature are called chemical shrinkage or contraction, and they involve a smaller volume of water in the hydrated phases than its volume in the liquid phase. Contraction depends on the mineral composition of the cement. The C_3A phase shows the greatest contraction, while the C_2S phase shows the least [15]. Yodsudjai et al. [16] demonstrated that varying levels of C_3A contained in different types of cement play a major role in the behaviour of cement paste during chemical shrinkage. For this reason, replacing cement with SSA, and thus reducing C_3A, may result in a reduction of chemical shrinkage. In addition, due to the high content of hematite in the ash in question, its hydration can also affect the volume of the reaction products of the blended cement paste.

2. Materials and Methodology

Ash produced during the incineration of sewage sludge generated at the Płaszów sewage treatment plant belonging to the Municipal Water and Sewage Company in Kraków was used as the material for this study. The treatment plant has its own sludge thermal disposal station with a capacity of about 64 tonnes of dry matter per day. The ash is produced by incineration of the pre-dried and centrifuged sludge in a fluidised bed furnace at a temperature of approx. 850 °C. In addition, pure CEM I 42.5R Portland cement was used in this study. The basic chemical composition of the materials used in this study is presented in Table 1. During testing of binders containing SSA, a series of cement pastes containing 0%, 10%, 20%, and 30% ash were made, replacing cement by weight.

Table 1. Chemical composition of SSA with the used CEM I 42.5R cement.

Component [%]	SiO_2	Al_2O_3	Fe_2O_3	CaO	Na_2O_e	SO_3	P_2O_5	MgO	Cl^-	LOI
SSA	29.9	9.6	22.5	13.5	1.0	0.1	18.0	-	0.15	0.3
Cement	19.8	4.9	2.7	64.6	0.48	2.9	-	1.0	0.024	2.9

The scope of the conducted measurements mainly focused on determining the value of chemical shrinkage during the early curing period of cement pastes containing SSA as an additive (7 days). Chemical shrinkage was determined based on an improved dilatometry method developed by the team of Zhang et al. [17]. The method was based on ASTM C 1608 Standard Test Method for Chemical Shrinkage of Hydraulic Cement Paste [18]. The measurement set-up consisted of a 500-mL conical flask that was set on a magnetic stirrer. The measurement set-up is presented in Figure 1.

The magnetic stirrer was introduced into a conical flask and 100 g of binder was introduced. Then, distilled water was poured in, after removing air bubbles. A vacuum chamber was used for this purpose, with internal pressure of 40 mb, and the time in which the water was subjected to the vacuum was 15 min. The entire set-up was placed on a magnetic stirrer constantly stirring the suspension. The flask was then plugged tightly with a rubber stopper with a 10-mL pipette with a measurement accuracy of 0.1 mL. The pipette was filled with oil to avoid water loss from the measuring system due to evaporation. The mixing process at 750 rpm continued for the entire period of conducting measurements, i.e., 7 days. The cement content of the binder was gradually replaced by the addition of SSA at 10%, 20%, and 30% by weight. Changes in the volume of liquid in the pipette were recorded every 30 min throughout the measurement period. In addition, the test stand was placed in an enclosed fume hood to maintain constant test conditions (Figure 1). A series of measurements were also performed for the properties of SSA. Chemical composition was determined by XRF X-ray fluorescence method. Phase composition was determined by XRD using a Philips PW1830 device. Grain size distribution was measured with the Malvern

Instruments Mastersizer 2000 laser diffraction particle size analyser. The specific surface area value was determined using the BET method with the Quantachrome NovaWin 2200 device. Other measurements of cement pastes, such as standard consistency and setting times were made in accordance with PN EN 196-3 [19]. Setting times were measured using a CONTROLS VICAMATIC 3 automatic Vicat apparatus.

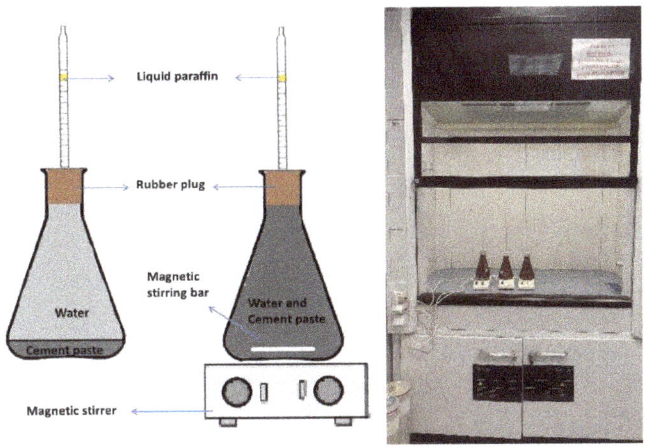

Figure 1. Chemical shrinkage measurement set-up.

3. Research Results and Discussion

The chemical composition of the ash mainly contains oxides of silicon (SiO_2), aluminium (Al_2O_3), and iron (Fe_2O_3), suggesting that the material has the potential to react with calcium hydroxide, and thus to exhibit pozzolanic properties. The sum of the masses of these oxides regardless of the sludge production season exceeded 60%.

A high content of P_2O_5 results in an increase in the setting time of the binders, as demonstrated later in this study. The presence of ions carrying corrosion hazards such as sulphates, chlorides, and alkalis appear to be at low and safe levels given the criteria included in the PN EN 450-1 [20] standard for fly ash. As a result of incineration, health-hazardous bacteria and organic matter are removed from SSA. Therefore, according to the 2012 Act on waste [21], it becomes stabilised.

According to the roentgenogram presented in Figure 2, three main crystalline phases can be distinguished due to the specific oxide composition. The dominant phase is inert silica, occurring mainly in the form of β-quartz. It is not an amorphized structure mainly resulting from the relatively low temperature during fluidised bed sludge treatment, i.e., at approx. 850 °C. Iron occurring mainly in the third oxidation number forms a hematite phase. In addition, due to the high content of phosphate oxides resulting from the high content of detergents in the sewage, the formation of a crystalline calcium-aluminium phosphate salt with the general formula of $Ca_9Al(PO_4)_7$ occurs.

The grain size distribution is unimodal with a distinct peak at diameters between 40 μm and 50 μm. The full range of grain sizes is from 0.3 μm to 100 μm. The grain sizes of d_{10}, d_{50}, and d_{90} are 1.8, 25.8, and 63.0 μm, respectively. The BET specific surface area, in turn, was 3900 m^2/kg. The specific density of SSA was at the level of 2.90 g/cm^3. Full granulometry is presented in Figure 3.

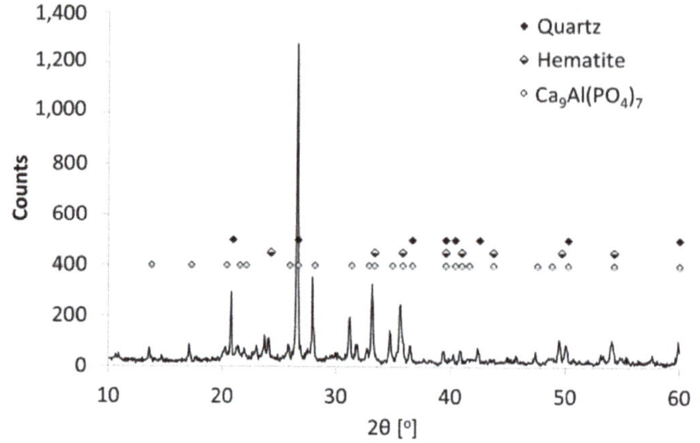

Figure 2. SSA phase composition.

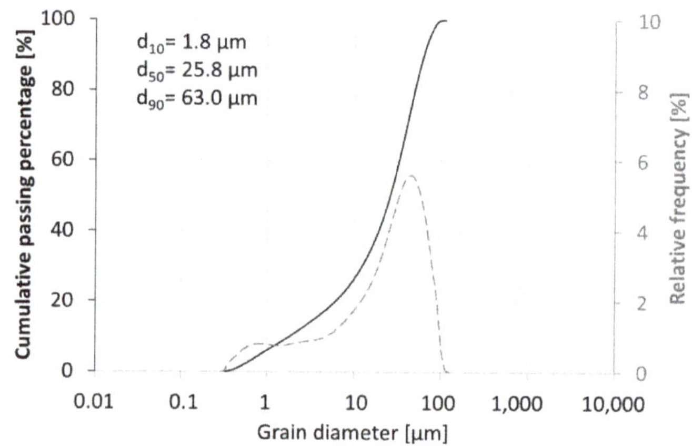

Figure 3. SSA grain size distribution.

Due to the high phosphate content in SSA, setting times were measured for binders containing SSA. Measurements of setting times were performed in accordance with PN EN 196-3 [19], which means that it had a different w/b ratio connected with standard requirements. The results are presented in Figure 4. The pastes prepared in this way were subjected to setting time tests. The addition of SSA at a level of 30% of the weight of the binder causes a sharp increase in the initial and the final setting time for the binder. The initial setting is doubled in length, while the final setting time is extended by approx. 60%. This phenomenon has been observed by a number of researchers [22–24], who attribute the setting delay properties of phosphate ions to the hindered hydration of cement grains. This is due to the fact that PO_4^{3-} ions react with Ca^{2+} ions in the liquid phase of the paste. The consequence of this process is the precipitation on the surface of the cement grains of calcium phosphate $Ca_3(PO_4)_2$ with low solubility. It takes the form of a finely crystalline and poorly permeable layer, the presence of which greatly impedes the hydration process. The crystallisation process can also occur in the pores of the paste, so that phosphate ions can not only affect the setting time but also the strength of the paste [22]. In addition, the grain size distribution and grain morphology of SSA, presented in Figure 3, results in a significant increase in the water demand of the analysed binders, as shown in Figure 5.

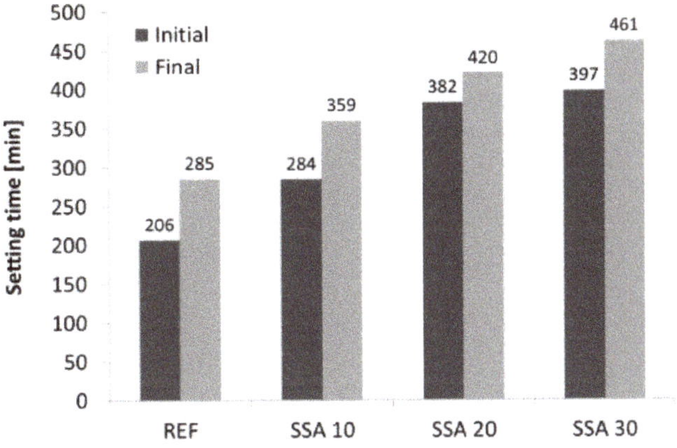

Figure 4. Setting time of cement binders with SSA content.

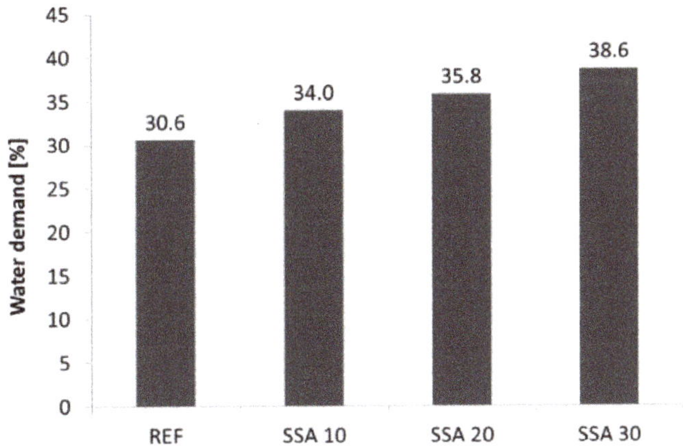

Figure 5. Water demand of the analysed binders.

The chemical shrinkage of cement binders containing various pozzolanic additives in their composition is well described in the literature [25–28]. However, it is difficult to find any reports on the effect of SSA addition on the chemical shrinkage of such binders. Cement pastes with the addition of highly reactive pozzolanic materials (silica fume), are characterised by high degree of chemical shrinkage. The main reason is the reaction, as a result of which calcium hydroxide reacting with pozzolana molecules forms an additional C-S-H phase. In the presence of reactive silica, calcium ions are steadily consumed, which maintains their concentration at a low level. The result is further dissolution of clinker grains and thus an increase in chemical shrinkage. In the absence of pozzolana, the concentration of Ca^{2+} ions stabilises at the level of saturated solution, thus inhibiting the dissolution of clinker grains and their hydration, and consequently reducing chemical shrinkage [26]. Materials with a lower specific surface area and thus lower pozzolanic reactivity, such as fly ash, used as cement additives, result in lowering the chemical shrinkage of such binders. These additives do not undergo a pozzolanic reaction in the initial hydration phase, but only act as fillers. The fine particles of the additives fill the free spaces between the cement grains and reduce the amount of water involved in the reaction. Therefore, the level of chemical shrinkage, due to the reduced amount of clinker grains replaced by pozzolanic material with lower reactivity, is reduced especially in the initial stages of hydration. The level of this shrinkage clearly increases as hydration progresses over the long term. This effect is associated with the formation of a C-S-H phase on the surface of the mineral additive grains. However, the content of hydrated calcium silicates is lower than in the reference binder (without mineral additive); hence, the final chemical shrinkage value remains lower [28].

The chemical shrinkage of binders with the addition of ash from sewage sludge incineration has been determined for pastes containing SSA in amounts ranging from 0 to 30%, in 10% increments. In addition, in order to obtain information regarding the behaviour of the SSA additive in the presence of water, the change in its volume without cement (SSA100) was measured. The results are presented in Figure 6.

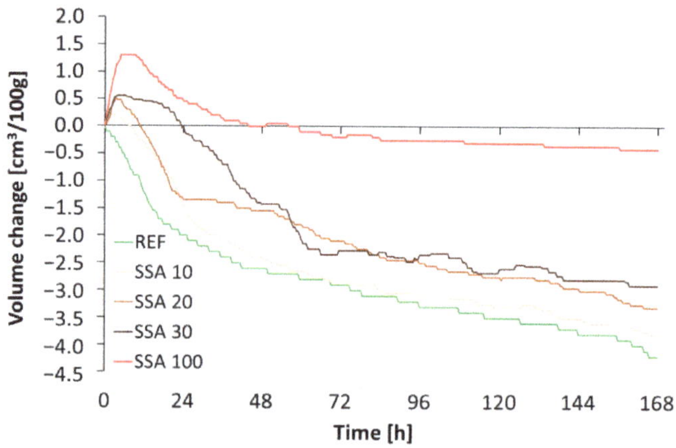

Figure 6. Chemical shrinkage of cement binders containing SSA additive.

The reference sample shows an increase in chemical shrinkage throughout the hydration period. The volume change was 4.2 $cm^3/100$ g of binder after 168 h of hydration. Samples containing SSA additive show a slight expansion during the initial hydration period. The level of this expansion increases with increasing amounts of additive used and reaches its highest value of 0.55 $cm^3/100$ g of binder at 30% SSA content. Pastes containing SSA additive undergo expansion in the initial stage of hydration, i.e., up to 6 h. It is likely that this phenomenon is influenced by two mechanisms occurring simultaneously. On the one hand, the presence of phosphate ions causes crystallisation of low solubility calcium phosphate salts on the surface of clinker grains, which entails a reduction of the rate of hydration and, at the same time, the phenomenon of contraction. The hydration of hematite (Fe_2O_3), a component of the SSA additive, occurs in parallel. Upon contact with water, hydration products appear on its surface [29], which, according to the authors of the publication [30], constitute the goethite (FeOOH) phase. The molar volume of goethite is almost 1.5 times that of its non-hydrated form as hematite. It is also worth mentioning that with the increasing degree of hydration of hematite, e.g., to $Fe(OH)_3$, the molar volume more than doubles [31,32]. In the next phase of the hydration of binders with SSA, the shrinkage of the tested binders becomes the dominant effect. Its course depends on the content of the SSA additive. Increasing the amount of the cement replacement additive reduces chemical shrinkage relative to the reference binder by 9.5%, 21.4%, and 31.0% at 10, 20, and 30% of SSA in the binder, respectively.

4. Conclusions

Despite the significant changes in the technological properties of SSA-containing binders, mainly concerning their setting time and water demand, this additive does not preclude its use in cement composites. SSA content is possible even at the level of 30% by the weight of cement causing an acceptable level of reduction of the performance of binders. Due to their high water demand, it becomes necessary to use superplasticizers during the preparation of mortars or concretes. The results of chemical shrinkage tests confirmed that the addition of SSA has a positive effect on this characteristic, reducing its value by up to approx. 30%. This is due to its rather low pozzolanic reactivity and relatively small specific surface area. Reducing the amount of cement in the paste further increases the reduction of chemical shrinkage. The effect of hydration of hematite grains in SSA, with simultaneous inhibition of the cement hydration process by the activity of phosphate ions in the initial phase of binder hydration, entails the occurrence of an expansion effect, facilitated by increasing amounts of added SSA.

Author Contributions: Conceptualization, M.A. and T.Z.; methodology, M.A.; validation, T.Z. and T.T.; investigation, M.A.; resources M.A and T.Z.; writing—original draft preparation, M.A.; writing—review and editing, T.Z. and T.T. All authors have read and agreed to the published version of the manuscript.

Funding: This research received no external funding.

Institutional Review Board Statement: Not applicable.

Informed Consent Statement: Not applicable.

Data Availability Statement: Not applicable.

Acknowledgments: Publication cost of this paper was covered with funds from the Polish National Agency for Academic Exchange (NAWA): "MATBUD'2023—Developing international scientific cooperation in the field of building materials engineering" BPI/WTP/2021/1/00002, MATBUD'2023.

Conflicts of Interest: The authors declare no conflict of interest.

References

1. Niu, M.; Li, G.; Wang, Y.; Li, Q.; Han, L.; Song, Z. Comparative study of immobilization and mechanical properties of sulfoaluminate cement and ordinary Portland cement with different heavy metals. *Constr. Build. Mater.* **2018**, *193*, 332–343. [CrossRef]
2. Vyšvařil, M.; Bayer, P. Immobilization of Heavy Metals in Natural Zeolite-blended Cement Pastes. *Procedia Eng.* **2016**, *151*, 162–169. [CrossRef]
3. Chen, Q.Y.; Tyrer, M.; Hills, C.D.; Yang, X.M.; Carey, P. Immobilisation of heavy metal in cement-based solidification/stabilisation: A review. *Waste Manag.* **2009**, *29*, 390–403. [CrossRef] [PubMed]
4. Belebchouche, C.; Moussaceb, K.; Tahakourt, A.; Aït-Mokhtar, A. Parameters controlling the release of hazardous waste (Ni2+, Pb2+ and Cr3+) solidified/stabilized by cement-CEM I. *Mater. Struct.* **2015**, *48*, 2323–2338. [CrossRef]
5. Pawelska-Mazur, M.; Kaszynska, M. Mechanical Performance and Environmental Assessment of Sustainable Concrete Reinforced with Recycled End-of-Life Tyre Fibres. *Materials* **2021**, *14*, 256. [CrossRef]
6. Skibicki, S.; Pułtorak, M.; Kaszyńska, M.; Hoffmann, M.; Ekiert, E.; Sibera, D. The effect of using recycled PET aggregates on mechanical and durability properties of 3D printed mortar. *Constr. Build. Mater.* **2022**, *335*, 127443. [CrossRef]
7. Giergiczny, Z.; Król, A. Immobilization of heavy metals (Pb, Cu, Cr, Zn, Cd, Mn) in the mineral additions containing concrete composites. *J. Hazard. Mater.* **2008**, *160*, 247–255. [CrossRef]
8. Lynn, C.J.; Dhir, R.K.; Ghataora, G.S.; West, R.P. Sewage sludge ash characteristics and potential for use in concrete. *Constr. Build. Mater.* **2015**, *98*, 767–779. [CrossRef]
9. Chang, Z.; Long, G.; Zhou, J.L.; Ma, C. Valorization of sewage sludge in the fabrication of construction and building materials: A review. *Resour. Conserv. Recycl.* **2020**, *154*, 104606. [CrossRef]
10. Garcés, P.; Carrión, M.P.; García-Alcocel, E.; Payá, J.; Monzó, J.; Borrachero, M.V. Mechanical and physical properties of cement blended with sewage sludge ash. *Waste Manag.* **2008**, *28*, 2495–2502. [CrossRef]
11. Payá, J.; Monzó, J.; Borrachero, M.V.; Morenilla, J.J.; Bonilla, M.; Calderón, P. Some Strategies for Reusing Residues From Waste Water Treatment Plants: Preparation of Binding Materials. In *International RILEM Conference on the Use of Recycled Materials in Building and Structures*; RILEM Publications SARL: Champs-sur-Marne, France, 2004; pp. 814–823, Print-ISBN: 2-912143-52-7.
12. Cyr, M.; Coutand, M.; Clastres, P. Technological and environmental behavior of sewage sludge ash (SSA) in cement-based materials. *Cem. Concr. Res.* **2007**, *37*, 1278–1289. [CrossRef]
13. Zdeb, T.; Tracz, T.; Adamczyk, M. Physical, mechanical properties and durability of cement mortars containing fly ash from the sewage sludge incineration process. *J. Clean. Prod.* **2022**, *345*, 131055. [CrossRef]
14. Rutkowska, G.; Wichowski, P.; Fronczyk, J.; Franus, M.; Chalecki, M. Use of fly ashes from municipal sewage sludge combustion in production of ash concretes. *Constr. Build. Mater.* **2018**, *188*, 874–883. [CrossRef]
15. Kurdowski, W. Deformation of the Paste. In *Cement and Concrete Chemistry*, 1st ed.; Springer: Dordrecht, The Netherlands; Heidelberg, Germany, 2014. [CrossRef]

16. Yodsudjai, W.; Wang, K. Chemical shrinkage behavior of pastes made with different types of cements. *Constr. Build. Mater.* **2013**, *40*, 854–862. [CrossRef]
17. Zhang, T.; Gao, P.; Luo, R.; Guo, Y.; Wei, J.; Yu, Q. Measurement of chemical shrinkage of cement paste: Comparison study of ASTM C 1608 and an improved method. *Constr. Build. Mater.* **2013**, *48*, 662–669. [CrossRef]
18. *ASTM C1608-07*; Standard Test Method for Chemical Shrinkage of Hydraulic Cement Paste. 2012. Available online: https://www.astm.org/c1608-07.html (accessed on 21 November 2022).
19. *PN EN 196-3*; Metody badania cementu, Część 3. Oznaczanie czasów wiązania stałości objętości. 2006. Available online: https://sklep.pkn.pl/pn-en-196-3-2016-12p.html (accessed on 15 November 2022).
20. *PN EN 450-1:2012*; Popiół lotny do betonu—Część 1: Definicje, specyfikacje i kryteria zgodności. 2012. Available online: https://sklep.pkn.pl/pn-en-450-1-2012p.html (accessed on 15 November 2022).
21. Internet System of Legal Acts. Available online: https://isap.sejm.gov.pl/isap.nsf/DocDetails.xsp?id=wdu20130000021 (accessed on 6 November 2022).
22. Tkaczewska, E. Wpływ jonów fosforanowych PO_4^{3-} na proces hydratacji cementu. *Cem. Wapno Bet.* **2012**, *17/79*, 401–408.
23. Cau Dit Coumes, C.; Courtois, S. Cementation of a low-level radioactive waste of complex chemistry: Investigation of the combined action of borate, chloride, sulfate and phosphate on cement hydration using response surface methodology. *Cem. Concr. Res.* **2003**, *33*, 305–316. [CrossRef]
24. Bénard, P.; Garrault, S.; Nonat, A.; Cau-Dit-Coumes, C. Hydration process and rheological properties of cement pastes modified by orthophosphate addition. *J. Eur. Ceram. Soc.* **2005**, *25*, 1877–1883. [CrossRef]
25. Bouasker, M.; Mounanga, P.; Turcry, P.; Loukili, A.; Khelidj, A. Chemical shrinkage of cement pastes and mortars at very early age: Effect of limestone filler and granular inclusions. *Cem. Concr. Compos.* **2008**, *30*, 13–22. [CrossRef]
26. Wang, J.; Cheng, Y.; Yuan, L.; Xu, D.; Du, P.; Hou, P.; Zhou, Z.; Cheng, X.; Liu, S.; Wang, Y. Effect of nano-silica on chemical and volume shrinkage of cement-based composites. *Constr. Build. Mater.* **2020**, *247*, 118529. [CrossRef]
27. Park, B.; Choi, Y.C. Effects of fineness and chemical activators on the hydration and physical properties of high-volume fly-ash cement pastes. *J. Build. Eng.* **2022**, *51*, 104274. [CrossRef]
28. Gao, Y.-L.; Zhou, S.-Q. Influence of ultra-fine fly ash on hydration shrinkage of cement paste. *J. Cent. South Univ. Technol.* **2005**, *12*, 596–600. [CrossRef]
29. Jang, J.-H.; Dempsey, B.A.; Burgos, W.D. Solubility of Hematite Revisited: Effects of Hydration. *Environ. Sci. Technol.* **2007**, *41*, 7303–7308. [CrossRef] [PubMed]
30. Langmuir, D. Particle size effect on the reaction goethite = hematite + water; correction. *Am. J. Sci.* **1972**, *272*, 972. [CrossRef]
31. Czarnecki, L.; Emmons, P.H. Mechanizmy destrukcji. In *Naprawa i ochrona konstrukcji betonowych*, 1st ed.; Polski Cement: Kraków, Poland, 2002; ISBN 83-91300-6-4.
32. Mancini, A.; Lothenbach, B.; Geng, G.; Grolimund, D.; Sanchez, D.; Fakra, S.; Dähn, R.; Wehrli, B.; Wieland, E. Iron speciation in blast furnace slag cements. *Cem. Concr. Res.* **2021**, *140*, 106287. [CrossRef]

materials proceedings

MDPI

Proceeding Paper

Increasing the Pozzolanic Reactivity of Recovered CDW Cement Stone by Mechanical Activation †

Roland Szabó *, Máté Szűcs, Mária Ambrus and Gábor Mucsi

Institute of Raw Material Preparation and Environmental Processing, University of Miskolc, Egyetemváros, 3515 Miskolc, Hungary
* Correspondence: roland.szabo@uni-miskolc.hu
† Presented at the 10th MATBUD'2023 Scientific-Technical Conference "Building Materials Engineering and Innovative Sustainable Materials", Cracow, Poland, 19–21 April 2023.

Abstract: The study focuses on enhancing the reactivity of the fine size fraction of construction and demolition waste (CDW) by mechanical activation in a stirred media mill. Systematic measurements were carried out to monitor the change in cement stone reactivity. The fine size fraction of CDW (<200 μm) was milled in a stirred media mill for 1, 3, 5, and 10 min. The dispersion characteristics (particle size distribution, specific surface area (SSA)) of the mechanically activated CDW powder were determined using a laser particle size analyzer. Changes in the structure of the mechanically activated CDW powder particles were determined by Fourier transform infrared spectroscopy (FTIR) measurements. The effect of the mechanical activation on the pozzolanic reactivity of CDW powder was measured by lime sorption test and compressive strength measurements. The results clearly show that Portland cement can be replaced with mechanically activated CDW powder; however, increasing its amount decreases the strength. Furthermore, the grinding fineness significantly influenced the pozzolanic reactivity of the mechanically activated CDW powder, and thus the strength of the specimens. The CDW powder milled for 10 min had 51% more lime uptake than the initial CDW sample, and the specimen strength at the age of 7 days was 23% higher using ground CDW powder than using initial CDW at a 20% cement replacement ratio.

Keywords: construction and demolition waste; cement substitution; mechanical activation; stirred media mill; reactivity

check for updates

Citation: Szabó, R.; Szűcs, M.; Ambrus, M.; Mucsi, G. Increasing the Pozzolanic Reactivity of Recovered CDW Cement Stone by Mechanical Activation. *Mater. Proc.* **2023**, *13*, 27. https://doi.org/10.3390/materproc2023013027

Academic Editors: Katarzyna Mróz, Tomasz Tracz, Tomasz Zdeb and Izabela Hager

Published: 15 February 2023

1. Introduction

The construction sector can be considered the largest resource user and waste producer in the world. About 3 billion tons of construction and demolition waste are generated annually in the world [1], while in the European Union (EU), 37.1% of total waste (approximately 800 million tons) was generated in the construction industry in 2020 [2], and this trend is constantly increasing. For this reason, the reuse and recycling of construction and demolition waste (CDW) is a key point not only for increasing resource efficiency, but also for reducing the use of large amounts of primary materials, energy consumption, and waste generation.

Despite the fact that many ways of utilizing CDW are known [3–8], there are many problems that prevent their large-scale use as a substitute for primary raw materials [9]. One of the largest components of CDW waste is concrete waste (other than excavated soil), approximately 70–80% of whose volume consists of fine and coarse aggregates, causing continuous depletion of natural resources [7]. Fine and coarse aggregates can be extracted from concrete waste using different processes (crushing, magnetic separation and sieving process), which can then be used as recycled concrete aggregates in different structures [7,10–14]. At the same time, the fine powder fraction (<150 μm) mainly from the cement stone part of concrete can also be used in recycled concrete or geopolymer [15–17].

However, in order to increase the reactivity of the CDW particles, milling may also be necessary, which can be performed in various mills (planetary mill [18], ball mill [17–19], vibratory mill [18], and stirred media mill [20]). The advantages and disadvantages of CDW powder fineness were summarized in a study by Tang et al. [21].

The aim of the research work is to increase the pozzolanic reactivity of the CDW powder fraction using a high energy-density mill (stirred media mill) and to study the relationships between its reactivity, structural characteristics, and the properties of the produced cement stone.

2. Materials and Methods

The base materials for the experiments were CDW with particle size <63 mm from Mento Ltd. (Bodrogkeresztúr, Hungary) and Portland cement (CEM II/A-S 42.5 R). As a first step, the CDW sample was crushed with a jaw crusher with a gap size of 20 mm. The fine fraction (<200 μm) was removed with sieving, and the 20-8 mm fraction was used in a Deval apparatus for autogenous milling for 60 min. This way, due to the abrasion of the CDW, a higher amount of cement stone could be obtained. The fine products of the previous steps were used as feed for the mechanical activation. The mineralogical composition of CDW was determined with a Bruker D8 Advance X-ray powder diffractometer (XRD) (Cu-Kα radiation, 40 kV, 40 mA) in parallel beam geometry (Göbel-mirror), as can be seen in Table 1. The high amount of quartz and carbonates come from the sand and hydrated cement. Mechanical activation was carried out in a stirred media mill with 530 cm^3 volume, using ZS type, Ø 1–1.2 mm ceramic beads and 5 m/s circumferential velocity. Both the material and grinding media filling ratio was 0.7. The applied milling times were 1, 3, 5 and 10 min. The 0 min milling refers to the pre-processed sample that was not mechanically activated in the stirred media mill.

Table 1. Mineralogical composition of CDW (wt. %).

Phase Name	Raw CDW
Quartz	54.4
Muscovite 2M1	4.0
Calcite	9.7
Vaterite	6.3
Albite	6.1
Orthoclase	3.0
Chlorite IIb	0.6
Ettringite	0.3
Portlandite	0.2
Actinolite	0.4
Amorphous	15.0

The particle size distributions and SSA of the mechanically activated samples were analyzed with a HORIBA LA950-V2 laser particle size analyzer. The pozzolanic reactivity was assessed based on the lime sorption test carried out according to the MSZ 4706-2:1998 standard. For the structural analysis, a Jasco FTIR-4200 analyzer equipped with a diamond ATR was used, with 4 cm^{-1} spectral resolution. The spectra were recorded between 400–4000 cm^{-1} and baseline-corrected.

To further test the reactivity of the mechanically activated CDW cement, samples were prepared to examine the possibility of Portland cement replacement using the mechanically activated CDW cement stones. In the specimens, the Portland cement was replaced with 0%, 10%, 20% and 30% mechanically activated cement stone, and a 0.33 water-cement ratio was applied to all mixtures. The uniaxial compressive strength of the 20 × 20 mm cubical specimens was measured with an SZF-1 type hydraulic press.

3. Results and Discussion

3.1. Mechanical Activation of CDW

3.1.1. Particle Size, SSA

The effect of mechanical activation on the particle size and geometric SSA of CDW is shown in Table 2. Milling resulted in a decrease in the particle size of CDW, while the specific surface area increased. After 10 min of milling, the specific surface area of CDW increased by 33%, but at the same time, no significant decrease in particle size was observed. This can presumably be attributed to the high quartz content (Figure 1), which hindered the efficiency of the milling.

Table 2. The characteristic particle size and SSA values of the ground cement stones.

Milling Time, min	x_{10}, μm	x_{50}, μm	x_{80}, μm	SSA, cm^2/cm^3
0	2.61	10.74	22.76	9500
1	2.31	8.27	23.65	10,961
3	2.25	8.23	24.97	11,096
5	2.24	7.38	20.1	11,715
10	2.02	6.99	19.62	12,601

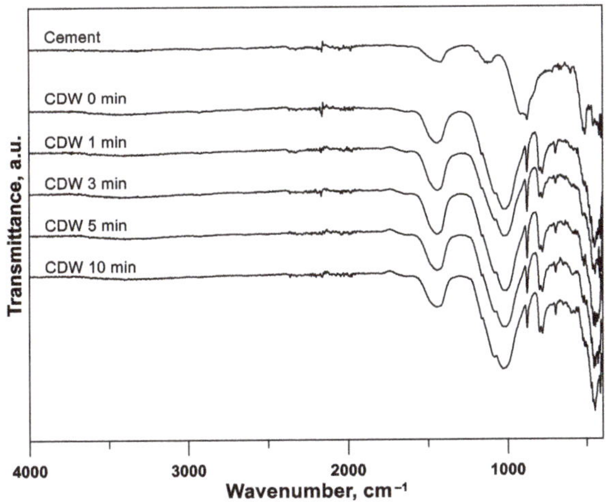

Figure 1. The FTIR spectra of the cement and ground CDW samples.

3.1.2. FTIR

The FTIR spectra of the cement and ground CDW can be seen in Figure 1. In the case of the raw cement spectrum, the transmittance bands at 1421, 874 and 713 cm^{-1} can be assigned to v3, v2, and v4 stretching modes of CO_3^{2-}, respectively. The weak band at 1124 cm^{-1} is generally the indication of Si–O–Si stretching vibrations and the band at 521 cm^{-1} corresponds to the Si–O deformation vibrations of the siliceous phases [22,23].

For the CDW samples, some new bands could also be observed compared to the cement sample. The broad band between ~3500–2800 cm^{-1} corresponds to the stretching O–H, and the weak band at 1614 cm^{-1} to the O–H bending mode, indicating the presence of a small amount of structural and weakly bound water in the samples. The bands centered between 1420–1450 cm^{-1} and at 875 cm^{-1} are associated with the calcite and other carbonate species. The broad band of 1010–1030 cm^{-1} originates from the asymmetric stretching vibrations of the C–S–H structure that was originally formed in the cementitious matrix. The bands at 795 and 695 cm^{-1} show the presence of quartz [20,24]. Comparing the spectra of the ground CDW samples, no significant changes occurred in the structure with the different milling times. After 3 min or longer mechanical activation, the intensity of the C–S–H band slightly increased and broadened, indicating the possible amorphization of the material.

3.2. CDW Reactivity

3.2.1. Pozzolanic Reactivity

Lime adsorption is an indicative measure of hydraulic/pozzolanic reactivity. The results of the lime sorption tests are shown in Table 3. Based on the Table 3, it can be stated that the SSA and CaO uptake values followed a similar trend after the mechanical activation: the increased SSA significantly increased the amount of adsorbed CaO. The lime uptake of CDW increased from the initial 126.9 mg CaO/g solid material to 192.1 mg CaO/g solid material. Thus, as a result of a 33% increase in SSA, the CaO uptake increased by over 50%.

Table 3. Lime adsorption of CDWs.

SSA of CDW (cm^2/cm^3)	∑Adsorbed CaO (mg/g)
9500	126.9
10,961	131.4
11,096	126.7
11,715	169.0
12,601	192.1

Figure 2 illustrates the volume change (swelling) during the reaction between CDW and lime. It can be clearly seen that the ground CDWs had a larger volume than the original at the end of the lime sorption test (after 30 days), indicating a greater degree of reactivity.

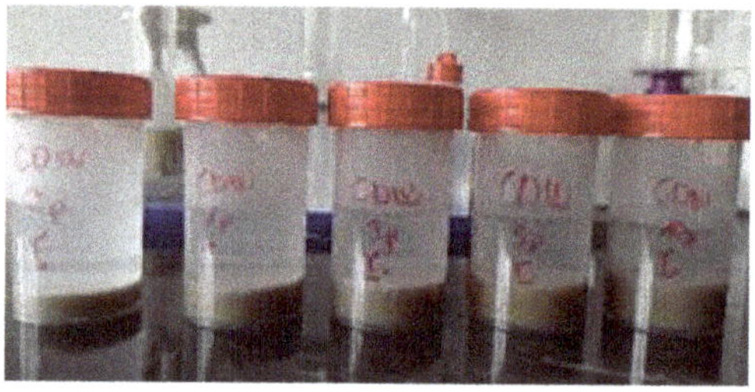

Figure 2. CDW samples after the lime sorption test (from left to right: 0 min, 1 min, 3 min, 5 min and 10 min).

3.2.2. Compressive Strength

Figure 3 shows the effect of CDW dosage and powder fineness on the compressive strength of the cement-based specimens. Based on the results, it can be concluded that the replacement of cement with CDW reduced the strength of the specimens. This correlates well with the results of other studies [16,20,25,26]. However, it is also seen that the ground CDWs showed better results compared to the unground (0 min) sample, at 20 and 30% cement replacement. Furthermore, specimens prepared using CDW ground for 5 and 10 min gave similar results at 20 and 30% cement replacement, which can be explained by their similar particle size distribution (Table 2). Generally, no clear correlation can be observed between the grinding time and replacement ratio. This can be explained by the heterogeneity and the high quartz content of the recycled cement stone [16,27].

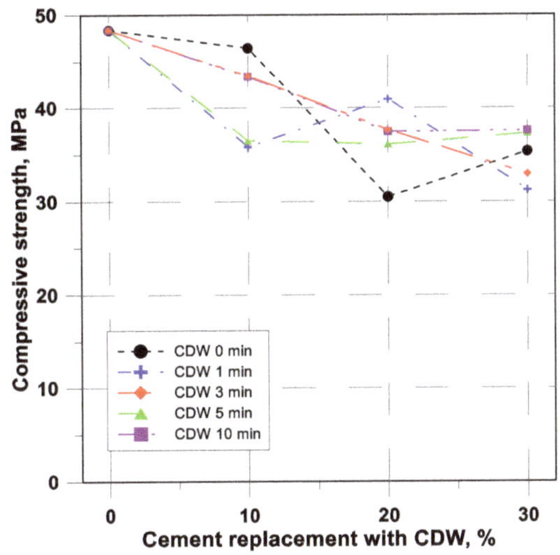

Figure 3. Results of compressive strength measurements.

3.2.3. FTIR

Figure 4 shows the FTIR spectra of the samples in the case of 0 min CDW dosing at different ratios. Based on Figure 4, it can be stated that the intensity of some bands decreased as a result of CDW dosing. The intensity of the bands around 1420 cm^{-1} and 874 cm^{-1}, which can be attributed to calcite and other carbonates, decreased to the greatest extent. In addition, the intensity of the band at 3640 cm^{-1} (which indicates the presence of portlandite ($Ca(OH)_2$) [28]) and the S–O stretching vibration of $[SO_4]^{2-}$ at 1150–1100 cm^{-1} also decreased with the addition of CDW.

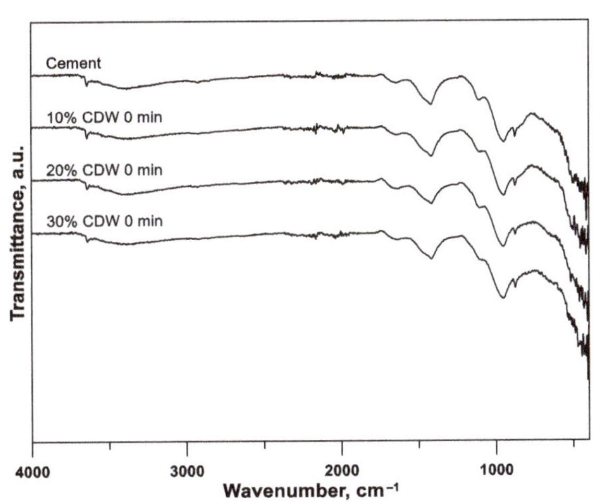

Figure 4. FTIR spectra of samples with different CDW content.

Figure 5 shows the FTIR spectra of the samples containing CDW with different powder fineness, produced with 20% cement replacement. Figure 5 shows that the same bands appeared for each sample. The intensity of the bands at 1420 cm^{-1} and 874 cm^{-1} belonging to calcite and other carbonates increased with the fineness of the CDW. The broad band at around 3400 cm^{-1} is attributed to the symmetric stretching vibration of the H_2O molecule, while the sharp, narrow band at 3640 cm^{-1} can be assigned to O–H stretching vibration (portlandite) [26,28], which showed a slight decrease by 10 min milling. The band at around 960 cm^{-1} implies the Si–O stretching vibrations, indicating a wide range of C–S–H. The band at 960 cm^{-1} is assigned to Si–O stretching vibrations, which is the result of the C–S–H phase with Ca/Si $\approx$ 2. When it reaches 1080 cm^{-1}, it indicates the formation of silica gel [29]. Higher intensity C–S–H related bands were observed with longer CDW cement stone milling times, indicating the effectiveness of mechanical activation.

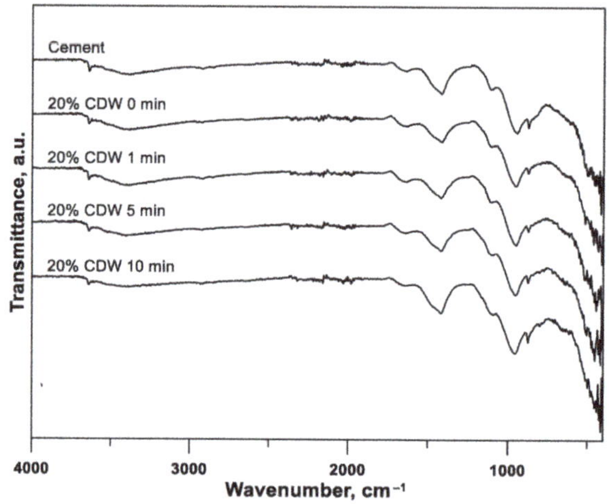

Figure 5. The FT-IR spectra of specimens (cement replacement with 20% CDW).

4. Conclusions

Based on the results, the following conclusions can be drawn:

- The increased specific surface area due to mechanical activation improved the pozzolanic reactivity of CDW dust (50% better CaO uptake after 10 min of milling).
- The use of CDW dust as a cement substitute reduced the strength of cement-based materials.
- The use of mechanically activated CDW led to a lower strength reduction than the unactivated sample.
- As a result, cement can be replaced with mechanically activated CDW in remarkable quantities (20 or 30%); thus it can be used primarily in areas where high structural performance is not required.
- No significant changes occurred in the structure due to MA. However, the increase in intensity of the C–S–H band indicates the possible surface amorphization of the particles.

Author Contributions: R.S.: conceptualization, methodology, investigation, data curation, writing—original draft preparation, and supervision. M.S.: investigation, writing—review and editing, data curation, formal analysis, and visualization. M.A.: writing—original draft preparation, formal analysis, data curation, and visualization. G.M.: conceptualization, methodology, formal analysis, writing—review and editing, and supervision. All authors have read and agreed to the published version of the manuscript.

Funding: Publication cost of this paper was covered with funds of the Polish National Agency for Academic Exchange (NAWA): "MATBUD'2023–Developing international scientific cooperation in the field of building materials engineering" BPI/WTP/2021/1/00002, MATBUD'2023.

Institutional Review Board Statement: Not applicable.

Informed Consent Statement: Not applicable.

Data Availability Statement: Not applicable.

Acknowledgments: The research work was performed in the Centre of Excellence in Sustainable Natural Resource Management at the Faculty of Earth Science and Engineering, University of Miskolc. Project no. 2019-2.1.7-ERA-NET-2022-00051 has been implemented with the support provided by the Ministry of Innovation and Technology of Hungary from the National Research, Development and Innovation Fund, financed under the 2019-2.1.7-ERA-NET funding scheme. The authors thank the Institute of Mining and Geotechnical Engineering of the University of Miskolc for their help during the mechanical strength tests and the Institute of Mineralogy and Geology for the XRD analysis of CDW.

Conflicts of Interest: The authors declare no conflict of interest.

References

1. Akhtar, A.; Sarmah, A.K. Construction and demolition waste generation and properties of recycled aggregate concrete: A global perspective. *J. Cleaner Product.* **2018**, *186*, 262–281. [CrossRef]
2. Eurostat, Waste Statistic–European Commission. 2022. Available online: https://ec.europa.eu/eurostat/statistics-explained/index.php?title=Waste_statistics#Total_waste_generation (accessed on 12 October 2022).
3. Gedik, A. A review on the evaluation of the potential utilization of construction and demolition waste in hot mix asphalt pavements. *Res. Con. Rec.* **2020**, *161*, 104956. [CrossRef]
4. Li, L.; Liu, Q.; Huang, T.; Peng, W. Mineralization and utilization of CO_2 in construction and demolition wastes recycling for building materials: A systematic review of recycled concrete aggregate and recycled hardened cement powder. *Sep. Pur. Technol.* **2022**, *298*, 121512. [CrossRef]
5. Bordoloi, S.; Afolayan, O.D.; Ng, C.W.W. Feasibility of construction demolition waste for unexplored geotechnical and geo-environmental applications—A review. *Constr. Build. Mater.* **2022**, *356*, 129230. [CrossRef]
6. Tavakoli Mehrjardi, G.; Azizi, A.; Haji-Aziz, A.; Asdollafardi, G. Evaluating and improving the construction and demolition waste technical properties to use in road construction. *Trans. Geotech.* **2020**, *23*, 100349. [CrossRef]
7. Mohammed, M.S.; ElKady, H.; Abdel-Gawwad, H.A. Utilization of construction and demolition waste and synthetic aggregates. *J. Build. Eng.* **2021**, *43*, 103207. [CrossRef]
8. Khodaei, H.; Olson, C.; Patino, D.; Rico, J.; Jin, Q.; Boateng, A. Multi-objective utilization of wood waste recycled from construction and demolition (C&D): Products and characterization. *Waste Man.* **2022**, *149*, 228–238. [CrossRef]
9. Luciano, A.; Cutaia, L.; Altamura, P.; Penalvo, E. Critical issues hindering a widespread construction and demolition waste (CDW) recycling practice in EU countries and actions to undertake: The stakeholder's perspective. *Sust. Chem. Pharm.* **2022**, *29*, 100745. [CrossRef]
10. Kumar, R. Influence of recycled coarse aggregate derived from construction and demolition waste (CDW) on abrasion resistance of pavement concrete. *Constr. Build. Mater.* **2017**, *142*, 248–255. [CrossRef]
11. Thomas, C.; Setién, J.; Polanco, J.A. Structural recycled aggregate concrete made with precast wastes. *Construct. Build. Mater.* **2016**, *114*, 536–546. [CrossRef]
12. López Gayarre, F.; Suárez González, J.; Blanco Viñuela, R.; López-Colina Pérez, C.; Serrano López, M.A. Use of recycled mixed aggregates in floor blocks manufacturing. *J. Clean. Prod.* **2017**, *167*, 713–722. [CrossRef]
13. Plaza, P.; Sáez del Bosque, I.F.; Frías, M.; Sánchez de Rojas, M.I.; Medina, C. Use of recycled coarse and fine aggregates in structural eco-concretes. Physical and mechanical properties and CO2 emissions. *Constr. Build. Mater.* **2021**, *285*, 122926. [CrossRef]
14. Tam, V.W.Y.; Soomro, M.; Evangelista, A.C.J. A review of recycled aggregate in concrete applications (2000–2017). *Constr. Build. Mater.* **2018**, *172*, 272–292. [CrossRef]
15. Braga, M.; de Brito, J.; Veiga, R. Incorporation of fine concrete aggregates in mortars. *Constr. Build. Mater.* **2012**, *36*, 960–968. [CrossRef]
16. Moreno-Juez, J.; Vegas, I.J.; Frías Rojas, M.; Vigil de la Villa, R.; Guede-Vázquez, E. Laboratory-scale study and semi-industrial validation of viability of inorganic CDW fine fractions as SCMs in blended cements. *Constr. Build. Mater.* **2021**, *271*, 121823. [CrossRef]
17. Tan, J.; Cizer, Ö.; De Vlieger, J.; Dan, H.; Li, J. Impacts of milling duration on construction and demolition waste (CDW) based precursor and resulting geopolymer: Reactivity, geopolymerization and sustainability. *Res. Con. Rec.* **2022**, *184*, 106433. [CrossRef]
18. Fediuk, R.S.; Ibragimov, R.A.; Lesovik, V.S.; Pak, A.A.; Krylov, V.V.; Poleschuk, M.M.; Stoyushko, N.Y.; Gladkova, N.A. Processing equipment for grinding of building powders. *IOP Conf. Ser. Mater. Sci. Eng.* **2018**, *327*, 042029. [CrossRef]
19. Li, S.; Li, Q.; Zhao, X.; Luo, J.; Gao, S.; Yue, G.; Su, D. Experimental Study on the Preparation of Recycled Admixtures by Using Construction and Demolition Waste. *Materials* **2019**, *12*, 1678. [CrossRef]
20. Mucsi, G.; Halyag Papné, N.; Ulsen, C.; Figueiredo, P.O.; Kristály, F. Mechanical Activation of Construction and Demolition Waste in Order to Improve Its Pozzolanic Reactivity. *ACS Sustain. Chem. Eng.* **2021**, *9*, 3416–3427. [CrossRef]
21. Tang, Q.; Ma, Z.; Wu, H.; Wang, W. The utilization of eco-friendly recycled powder from concrete and brick waste in new concrete: A critical review. *Cement Concr. Compos.* **2020**, *114*, 103807. [CrossRef]
22. Amor, F.; Diouri, A.; Ellouzi, I.; Ouanji, F. Development of Zn-Al-Ti mixed oxides-modified cement phases for surface photocatalytic performance. *Case Stud. Constr. Mater.* **2018**, *9*, e00209. [CrossRef]

23. Cho, J.; Waetzig, G.R.; Udayakantha, M.; Hong, C.Y.; Banerjee, S. Incorporation of Hydroxyethylcellulose-Functionalized Halloysite as a Means of Decreasing the Thermal Conductivity of Oilwell Cement. *Sci. Rep.* **2018**, *8*, 16149. [CrossRef] [PubMed]
24. Cristelo, N.; Fernández-Jiménez, A.; Vieira, C.; Miranda, T.; Palomo, Á. Stabilisation of construction and demolition waste with a high fines content using alkali activated fly ash. *Constr. Build Mater.* **2018**, *170*, 26–39. [CrossRef]
25. Pešta, J.; Ženíšek, M.; Kočí, V.; Pavlů, T. Environmental perspectives of recycled concrete powder as cement replacement. In *Special Concrete and Composites 2020: Proceedings of the 17th International Conference, Lísek, Czech Republic, 14–15 October 2020*; AIP Publishing LLC: Melville, NY, USA, 2021; Volume 2322, p. 020028. [CrossRef]
26. Lu, B.; Shi, C.; Zhang, J.; Wang, J. Effects of carbonated hardened cement paste powder on hydration and microstructure of Portland cement. *Constr. Build. Mater.* **2018**, *186*, 699–708. [CrossRef]
27. Monasterio, M.; Caneda-Martínez, L.; Vegas, I.; Frías, M. Progress in the influence of recycled construction and demolition mineral-based blends on the physical–mechanical behaviour of ternary cementitious matrices. *Constr. Build. Mater.* **2022**, *344*, 128169. [CrossRef]
28. Horgnies, M.; Chen, J.J.; Bouillon, C. Overview about the use of Fourier transform infrared spectroscopy to study cementitious materials. *WIT Trans. Eng. Sci.* **2013**, *77*, 251–262. [CrossRef]
29. Pan, X.; Shi, C.; Hu, X.; Ou, Z. Effects of CO_2 surface treatment on strength and permeability of one-day-aged cement mortar. *Constr. Build. Mater.* **2017**, *154*, 1087–1095. [CrossRef]

materials proceedings

Proceeding Paper

The Use of Calcined Diatomite as an Additive to Geopolymeric Materials †

Kinga Pławecka *, Agnieszka Bąk, Maria Hebdowska-Krupa and Michał Łach

Faculty of Material Engineering and Physics, Cracow University of Technology, Jana Pawła II 37, 31-864 Cracow, Poland
* Correspondence: kinga.plawecka@pk.edu.pl; Tel.: +48-576547854

† Presented at the 10th MATBUD'2023 Scientific-Technical Conference "Building Materials Engineering and Innovative Sustainable Materials", Cracow, Poland, 19–21 April 2023.

Abstract: Diatomites are mineral resources formed from diatoms. They are widely used in sorption processes, medicine, cosmetology, and in protecting animals from parasites. Attempts are being made to incorporate them into concretes and construction binders to improve various performance properties. This paper presents the results of analyses (particle size analysis, XRD, and SEM) of a fine fraction of non-calcined and calcined diatomite as an additive to geopolymers made from fly ash from lignite combustion. The fly ash was also analyzed in the same way. Diatomite was introduced in its calcined and non-calcined form at 10%, 15%, and 30% by weight, replacing parts of the filler sand. The geopolymer mixtures were activated with 10 and 14 M aqueous sodium hydroxide solution with sodium water glass. As a result, it was found that it was possible to obtain geopolymers with diatomite additives with a compressive strength of about 34 MPa. In addition, after the strength tests, the microstructure of the obtained geopolymers was analyzed by scanning electron microscopy.

Keywords: diatomite; geopolymer materials; ash from lignite combustion

Citation: Pławecka, K.; Bąk, A.; Hebdowska-Krupa, M.; Łach, M. The Use of Calcined Diatomite as an Additive to Geopolymeric Materials. *Mater. Proc.* **2023**, *13*, 28. https://doi.org/10.3390/materproc2023013028

Academic Editors: Katarzyna Mróz, Tomasz Tracz, Tomasz Zdeb and Izabela Hager

Published: 15 February 2023

1. Introduction

The production of Portland cement contributes about 7% of the world's CO_2 emissions, instigating the search for alternatives to this type of material [1]. Alkali-activated binders or geopolymers are examples of such alternatives. These materials are formed by the polycondensation of aluminosilicates (containing silicates with alkali) and are characterized by an amorphous or semi-crystalline structure [2,3]. In the technology of cement and concrete, in addition to the already used and standardized mineral additives, attempts are being made to use undeveloped industrial waste [4,5]. Lach et al. investigated the possibility of immobilizing waste from municipal waste incinerators in geopolymers. The results of the study showed a high level of immobilization of compounds and elements such as sulfates, chlorides, fluorides, zinc, and barium [6]. Fly ash and slag have an advantage over other materials because, as finished waste from energy processes, they do not require additional reactivation during geopolymerization [7]. Moreover, their properties depend on their composition, the technology of production, or the raw material used in the combustion process. In addition, their use as a by-product of the energy industry is in line with the policy of a closed-loop economy [8]. Another group of potential mineral additives is power industry co-products, commonly known as calcium fly ash [9]. Calcium fly ash has a more complex composition than silica fly ash. The vitreous phase of calcium fly ash is characterized by a rich content of silica and aluminum. In their structure, there may also be glasses of the C-F-S structure (CaO-Fe_2O_3-SiO_2) [10]. Among the main phase components of calcareous fly ash are gehlenite, anorthite, anhydrite, tricalcium aluminate, quartz, calcium sulfate-aluminate, and unbound calcium oxide (CaO_x) [11].

Over recent times, diatomite material has become increasingly popular due to its unique properties and various applications. For example, the diatomite material is used in filtration processes and the production of insulating or sound-absorbing materials [12,13]. However, the most well-known example of the material's applications is its ability to sorb various types of petroleum substances. Diatomite can also support the synthesis of zeolite structures [12]. Thus, it is possible to obtain zeolites from waste raw materials that have significant SiO_2 content in their composition [14,15]. Researchers have also focused on the use of diatomite as an additive to cement binders or concrete [16]. Investigations have shown that the addition of diatomite lowers the consistency of fresh concrete mixes. The initial mechanical properties of mortars modified with diatomite additives were low, but after 28 days of seasoning, the strength increased [17].

This paper presents a comparison of the properties of geopolymers based on lime fly ash from an incinerator in Belchatow and sand, modified with different contents of diatomite dust—10%,15%, and 30% by weight. Alkali activation was carried out using two different activators: aqueous sodium glass solution and 10 and 14 mol sodium hydroxide. Diatomite dust was introduced into geopolymer mixtures in calcined and non-calcined forms. Investigations into the produced geopolymer materials showed great potential in using diatomite as one of the precursors in geopolymers. These investigations are important for providing new knowledge not only on the possibility of using local resources to produce construction materials by geopolymerization but also on the possibility of using industrial waste. Activities such as these are part of the environmentally friendly policy of a closed-loop economy.

2. Materials and Methods

2.1. Materials and Samples Preparation

Fly ash was obtained from the combustion of lignite coal from the Belchatow Power Plant (Belchatow, Poland). These ashes are characterized by relatively high calcium content. Other precursors used in the production of geopolymers were construction sand (Swietochlowice, Poland) and diatomite dust, obtained from an open-pit diatomite mine in Jawornik Ruski (Zohatin, Poland). The studies were conducted on non-calcined and calcined diatomite dust at 650 °C.

Table 1 shows the results of the particle size distribution for raw diatomite and fly ash dust. Investigations were carried out using a particle size analyzer (AntonPaar GmbH, Graz, Austria).

Table 1. Particle size distribution.

Material	D_{10} [µm]	D_{50} [µm]	D_{90} [µm]	Mean Size [µm]
Diatomite dust	2.997	11.232	20.970	12.218
Fly ash from Belchatow	3.295	20.411	37.125	21.413

The average particle size for diatomite dust is smaller than for fly ash from Belchatow and is approximately 12 µm. For the fly ash from Belchatow, it is approximately 21 µm. Figure 1 shows the particles of fly ash from Belchatow from the lignite combustion process and the particles of diatomite dust.

The alkaline activator was a 10 and 14 mol sodium hydroxide solution (PCC Rokita SA, Brzeg Dolny, Poland), and the water glass was sodium R-145 (STANLAB, Gliwice, Poland) with a molar modulus of 2.5 and a density of about 1.45 g/cm^3; the Na/Al ratio was 1:2. To prepare the mass, the precursors were mixed with the activator for about 10 min and poured into molds. The molds were placed on a vibrating table to eliminate air bubbles. After preparing the masses, the samples were tightly covered with foil and then placed in a laboratory dryer (SLW 750 STD, Pol-Eko-Aparatura, Wodzislaw Slaski, Poland) for 24 h at 75 °C. After 24 h, the samples were unmolded and cured under laboratory conditions (temperature about 20 °C, relative humidity about 50%) for 28 days. Strength tests were carried out in the next stage. Table 2 shows the names of the samples to better systematize the mixing ratios.

(a) (b)

Figure 1. Particle morphology: (**a**) Belchatow fly ash (1000× magnification), (**b**) diatomite dust (1000× magnification).

Table 2. Composition of geopolymer samples based on fly ash from Belchatow, sand, and diatomite dust (non-calcined (DN) and calcined (DK)).

Index	Base Materials (S) [Weight Ratio]			Alkaline Activator (L)	Liquid/Solid Ratio [Weight Ratio]
	Fly Ash	Sand	Diatomite Dust		
R10	1	1	-		1:0.30
10MDN10%	1	0.9	0.10		1:0.35
10MDK10%	1	0.9	0.10		1:0.35
10MDN15%	1	0.85	0.15	10 M NaOH + sodium water glass (weight ratio: 1:2.5)	1:0.40
10MDK15%	1	0.85	0.15		1:0.40
10MDN30%	1	0.70	0.30		1:0.45
10MDK30%	1	0.70	0.30		1:0.45
R14	1	1	-		1:0.3
14MDN10%	1	0.9	0.10		1:0.35
14MDK10%	1	0.9	0.10		1:0.35
14MDN15%	1	0.85	0.15	14 M NaOH + sodium water glass (weight ratio: 1:2.5)	1:0.40
14MDK15%	1	0.85	0.15		1:0.40
14MDN30%	1	0.70	0.30		1:0.45
14MDK30%	1	0.70	0.30		1:0.45

The sand used in the produced geopolymers functioned as a filler. Furthermore, the amount of alkaline solution added increased with the percentage of diatomite dust in the mixture. Diatomite is commonly used as a sorbent [13]. A higher amount of alkaline solution was necessary to achieve the proper workability and consistency of the geopolymer mixture.

2.2. Research Methods

2.2.1. Phase Composition of Precursors

The PANalytical Aeris instrument (Malvern PANalytical, Lelyweg 1, Almelo, The Netherlands) was used to investigate the mineralogical composition of the geopolymer samples prepared. Quantitative analysis was carried out using the Rietveld method, which was implemented in HighScore Plus software (version: 4.8, Malvern PANalytical B.V.,

Almelo, The Netherlands). The International Centre for Diffraction Data (ICDD) PDF-4+ database was used during the analysis. Measurements were recorded in the range of 10–100°, with a step size of 0.003° (2θ) and a time per step of 340 s, using Cu Kα radiation.

2.2.2. Strength Tests

Compressive strength tests were carried out by EN 12390-3 ("Testing of hardened concrete. Compressive strength of specimens") on cubic specimens (50 × 50 × 50 mm) using a Matest 3000 kN universal testing machine (Matest, Treviolo, Italy). The test speed was set at 0.05 MPa/s. The test was carried out using a load cell designed for lower loads—300 kN. Strength tests were conducted after 28 days of conditioning of the specimens. For the compressive strength tests, 10 samples of each type of geopolymer were used to calculate the average of the results.

2.2.3. Microstructure

The geopolymer materials were microscopically observed to characterize the shaped structure. The study was carried out using a JEOL JSN5510LV scanning electron microscope (JEOL Ltd., Tokyo, Japan). Samples were used for the study after mechanical property tests. Before the examination, the surface of the sample was coated with a conductive gold layer on a JOEL JEE-4X vacuum evaporator (JEOL Ltd., Tokyo, Japan).

3. Results and Discussion

3.1. Phase Composition of Precursors Results

Table 3 shows the identified phases for calcium fly ash from Belchatow.

Table 3. The phase composition of fly ash from Belchatow.

Identified Phase	Gehlenite	Anhydrite	Anorthite	Mullite	Hematite	Ye'elimite	Chlormayenite	Lime	Quartz
Chemical formula	$Ca_2Al_2SiO_7$	$CaSO_4$	$CaAl_2Si_2O_8$	$Al_6Si_2O_{13}$	Fe_2O_3	$Ca_4Al_6(SO_4)$	$C_{12}Al_{14}O_{33}$	CaO	SiO_2
Percentage proportion [%]	30.9	16.2	15.5	14.1	10.2	5.7	3.2	2.9	1.3

Calcium fly ash from Belchatow Power Plant is a product of lignite coal combustion in pulverized coal furnaces. Due to the coal's mineral part composition, the ash has an aluminous–silicon–calcium character [18]. The material has a high gehlenite content (around 30%). Anhydrite, anorthite, and mullite oscillate within 14–16%. Hematite occurs within 10%. About 3% of the phase composition consists of chlormayenite and lime. The smallest percentage of the composition is occupied by quartz—about 1%. Similar phases have been shown by other researchers [11].

Table 4 shows the phase composition of non-calcined and calcined diatomite dust.

The analyzed diatomite dust was characterized by the presence of phases such as silicon oxide, kaolinite, albite, and aluminum oxide. Similar phases were obtained in the investigations by Ediz et al. and Ren et al. [19,20]. As a result of the calcination of diatomite, the percentage of albite increased to about 30%, while the percentage of kaolinite decreased to about 31%. In addition, in the calcined diatomite dust, the percentage of the silicon oxide phase increased to almost 38%. The non-calcined diatomite dust had a higher percentage of phases, such as kaolinite (about 49%) and aluminum oxide (0.5%), in its phase composition, compared to calcined diatomite dust.

Table 4. Phase composition of non-calcined and calcined diatomite dust.

Non-Calcined Diatomite Dust				
Identified phase	Silicon Oxide	Kaolinite—1A	Aluminum Oxide	Albite
Chemical formula	SiO_2	$Al_2Si_2O_5(OH)_4$	Al_2O_3	$NaAlSi_3O_5$
Percentage proportion [%]	32.5	49.1	0.5	17.9
Calcined Diatomite Dust				
Identified phase	Silicon Oxide	Kaolinite—1A	Aluminum Oxide	Albite
Chemical formula	SiO_2	$Al_2Si_2O_5(OH)_4$	Al_2O_3	$NaAlSi_3O_5$
Percentage proportion [%]	37.8	31.6	0.2	30.4

3.2. Mechanical Properties

Compressive strength is one of the basic methods for evaluating the proper course of the geopolymerization process. The strength results depend on several factors, such as the structure, the presence of a crystalline phase, and the distribution and hardness of insoluble Al-Si particles. In addition, the type of alkali used and the %CaO and %K_2O also affect the mechanical properties of the geopolymer composite [21,22]. Figure 2 shows a summary of the average results of compressive strength investigations performed for each type of geopolymer material.

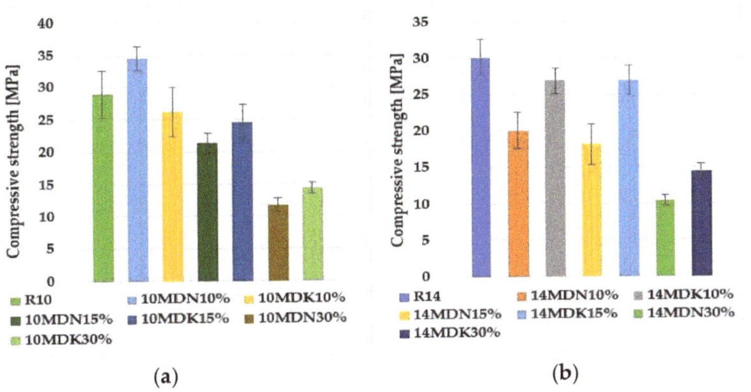

Figure 2. Compressive strength results for geopolymers based on calcium fly ash, sand, and calcined and non-calcined activated diatomite: (**a**) 10 mol alkali solution, (**b**) 14 mol alkali solution.

For geopolymers activated with a 10-mole alkali solution, the highest compressive strength value was obtained for the 10MDN10% sample—almost 35 MPa. The lowest value was obtained for the 10MDN30% material—about 12 MPa. The values of the average compressive strength for samples 10MDK10% and 10MDK15% oscillate at a similar level—about 25 MPa. The addition of 30% calcined diatomite (10MDK30%) caused a decrease in compressive strength by almost 50% compared to the reference sample (R10). The largest mean standard deviation was recorded for the reference material (R10) and 10MDK10%. However, the lowest values were for 10MDN30% and 10MDK30%.

For geopolymers activated with 14 molar alkali solution, the highest compressive strength value was obtained for sample R14 (reference material)—almost 30 MPa. The lowest value was obtained for the 14MDN30% material—about 11 MPa. The values of the average compressive strength for samples 14MDN10% and 14MDN15% oscillate at a similar level—about 20 MPa. The addition of 30% calcined diatomite (14MDK30%) caused a decrease in compressive strength by almost 50% compared to the reference sample (R14). The highest mean standard deviation, which remained similar, was recorded for the reference material (R14) and the 14MDN15% material. On the other hand, the smallest was for 14MDN30% and 14MDK30%.

3.3. *Microscopic Observations*

Scanning electron microscopy (SEM) allows a visual examination of the material to obtain morphological information and allows the evaluation of structures that cannot be revealed by other examination methods [23].

Figure 3 shows the microstructure of geopolymers based on limestone fly ash and sand activated with a 10 mol alkali solution (Figure 3a) and a 14 mol alkali solution (Figure 3b).

(**a**) (**b**)

Figure 3. Microstructure of reference geopolymers in magnifications 1000×: (**a**) 10R, (**b**) 14R.

In Figure 3a, the inconsistent structure of the porous material can be observed.

These are grains of unreacted limestone fly ash from Belchatow. However, as opposed to silica ash, limestone fly ash grains are characterized by very large particles of unburned carbon, porous and poorly sintered [11]. Furthermore, in addition to the geopolymer matrix, we can observe C-S-H. A similar structure was analyzed by Zhang et al. in their work [24].

Figure 3b shows the microstructure of a geopolymer based on lime fly ash from Belchatow and sand, activated with a 14 mol alkali solution. In this case, a much smaller amount of unreacted fly ash can be observed. The structure is more compact than that of the 10R material. Alehyen et al., in their work, focused on studying the microstructures of fly-ash-based geopolymer mortars. They described them as porous heterogeneous mixtures in which some of the ash grains did not react or had reacted partially. In addition, they indicated the possible presence of residual alkaline deposits and geopolymer gel [25].

Figure 4 shows the morphology of all geopolymer materials based on fly ash, sand, and diatomite dust (calcined and non-calcined).

As a result of the SEM analysis, it can be observed that materials based on calcium fly ash, sand, and diatomite dust (calcined and non-calcined), activated with a 10 mol alkali solution (Figure 4a–f), have more unreacted lignite particles (clusters of porous structures) in their structure. Compounds of geopolymers activated with 14 molar alkali solution show significantly fewer particles of unreacted ash (Figure 4g–l). According to the presented microstructure images, the fly ash-based materials are characterized by an amorphous structure and contain undecomposed fly ash particles. It was also noted that there is a changeable pore content in the microstructure of the materials.

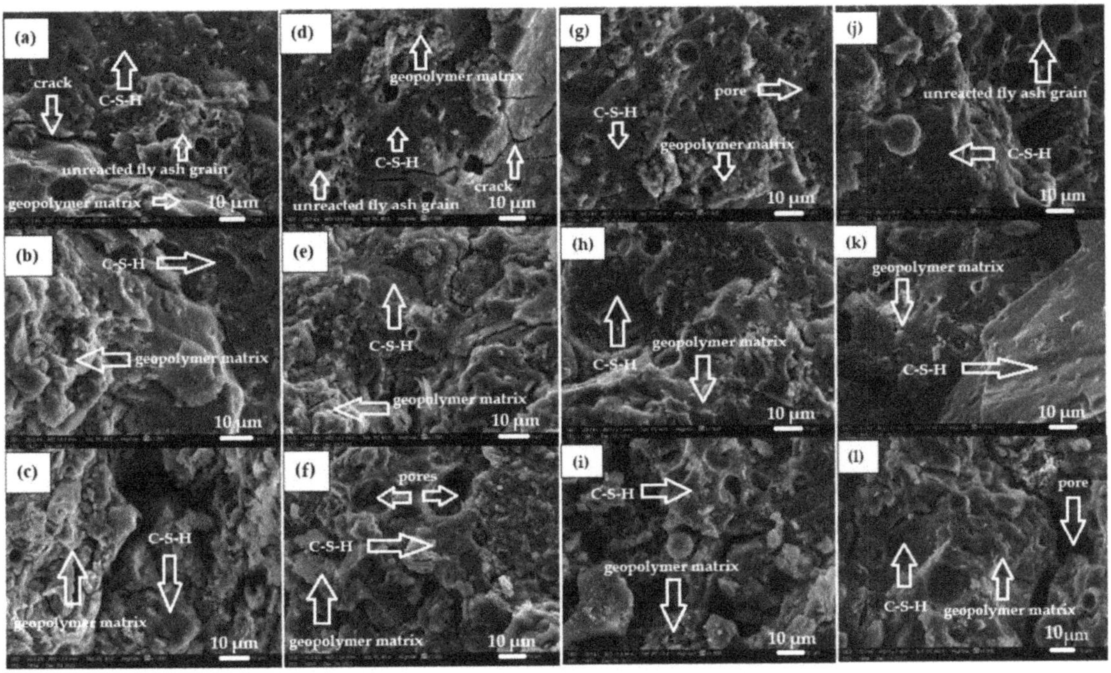

Figure 4. Microstructure of geopolymers in magnification 1000x: (**a**) 10MDN10%, (**b**) 10MDK10%, (**c**) 10MDN15%, (**d**) 10MDK15%, (**e**) 10MDN30%, (**f**) 10MDK30%, (**g**) 14MDN10%, (**h**) 14MDK10%, (**i**) 14MDN15%, (**j**) 14MDK15%, (**k**)14MDN30%, (**l**)14MDK30%.

4. Conclusions

The conducted strength and structural studies were aimed at evaluating the possibility of using potential precursors in the form of diatomite dust (calcined and non-calcined) as one of the components of geopolymer materials. The presented investigations allow us to conclude the following:

- The phase composition of calcined and non-calcined diatomite dust differs only in the percentage of phases—the same phases are present in both cases but in different percentage ratios.
- Mechanical investigations have shown that the addition of diatomite dust can positively affect the strength properties of the geopolymer.
- In addition to the percentage addition of diatomite dust, the mechanical properties of the tested geopolymer materials were influenced by the concentration of the alkali activator used.

Author Contributions: Conceptualization, K.P. and M.Ł.; methodology, K.P., A.B. and M.H.-K.; formal analysis, K.P., and M.Ł.; investigation, K.P. and M.H.-K.; resources, M.Ł. and M.H.-K.; writing—original draft preparation, K.P. and A.B.; writing—review and editing, M.Ł.; supervision, M.Ł. and M.H.-K. All authors have read and agreed to the published version of the manuscript.

Funding: This work has been financed by the National Centre for Research and Development in Poland under the grant: "Development and demonstration of technologies for the production of highly effective diatomite-based sorbents and diatomite fillers"; Project No.: POIR.04.01.04-00-0032/20.

Institutional Review Board Statement: Not applicable.

Informed Consent Statement: Not applicable.

Data Availability Statement: Not applicable.

Acknowledgments: The publication cost of this paper was covered with funds from the Polish National Agency for Academic Exchange (NAWA): "MATBUD'2023—Developing international scientific cooperation in the field of building materials engineering" BPI/WTP/2021/1/00002, MATBUD'2023.

Conflicts of Interest: The authors declare no conflict of interest.

References

1. Singh, N.B.; Middendorf, B. Geopolymers as an Alternative to Portland Cement: An Overview. *Constr. Build. Mater.* **2020**, *237*, 117455. [CrossRef]
2. Petrus, H.T.B.M.; Olvianas, M.; Shafiyurrahman, M.F.; Pratama, I.G.A.A.N.; Jenie, S.N.A.; Astuti, W.; Nurpratama, M.I.; Ekaputri, J.J.; Anggara, F. Circular Economy of Coal Fly Ash and Silica Geothermal for Green Geopolymer: Characteristic and Kinetic Study. *Gels* **2022**, *8*, 233. [CrossRef] [PubMed]
3. Nguyen, V.V.; Le, V.S.; Louda, P.; Szczypiński, M.M.; Ercoli, R.; Vojtěch, R.; Łoś, P.; Prałat, K.; Plaskota, P.; Pacyniak, T.; et al. Low-Density Geopolymer Composites for the Construction Industry. *Polymers* **2022**, *14*, 304. [CrossRef] [PubMed]
4. Miller, S.A.; Habert, G.; Myers, R.J.; Harvey, J.T. Achieving Net Zero Greenhouse Gas Emissions in the Cement Industry via Value Chain Mitigation Strategies. *One Earth* **2021**, *4*, 1398–1411. [CrossRef]
5. Plawecka, K.; Figiela, B.; Grela, A.; Buczkowska, K.E. Geopolymers Based on Plasma Incineration Waste as a Material for Circular Economy. In Proceedings of the IOP Conference Series: Earth and Environmental Science, Wrocław, Poland, 23–25 June 2021; Volume 942.
6. Łach, M.; Mierzwiński, D.; Korniejenko, K.; Mikuła, J.; Hebda, M. Geopolymers as a Material Suitable for Immobilization of Fly Ash from Municipal Waste Incineration Plants. *J. Air Waste Manag. Assoc.* **2018**, *68*, 1190–1197. [CrossRef]
7. Pławecka, K.; Bazan, P.; Lin, W.-T.; Korniejenko, K.; Sitarz, M.; Nykiel, M. Development of Geopolymers Based on Fly Ashes from Different Combustion Processes. *Polymers* **2022**, *14*, 1954. [CrossRef]
8. John, S.K.; Nadir, Y.; Girija, K. Effect of Source Materials, Additives on the Mechanical Properties and Durability of Fly Ash and Fly Ash-Slag Geopolymer Mortar: A Review. *Constr. Build. Mater.* **2021**, *280*, 122443. [CrossRef]
9. Nuaklong, P.; Jongvivatsakul, P.; Pothisiri, T.; Sata, V.; Chindaprasirt, P. Influence of Rice Husk Ash on Mechanical Properties and Fire Resistance of Recycled Aggregate High-Calcium Fly Ash Geopolymer Concrete. *J. Clean. Prod.* **2020**, *252*, 119797. [CrossRef]
10. Nuaklong, P.; Wongsa, A.; Sata, V.; Boonserm, K.; Sanjayan, J.; Chindaprasirt, P. Properties of High-Calcium and Low-Calcium Fly Ash Combination Geopolymer Mortar Containing Recycled Aggregate. *Heliyon* **2019**, *5*, e02513. [CrossRef]
11. Dąbrowski, M. *Effect of Addition of Lime Fly Ash on Microstructure and Frost Resistance of Composites with Cementitious Matrices*; Institute of Fundamental Technological Research, Polish Academy of Sciences, Laboratory of Strain Fields: Warsaw, Poland, 2016.
12. Ersoy, O.; Rençberoğlu, M.; Karapınar Güler, D.; Özkaya, Ö.F. A Novel Flux That Determines the Physico-Chemical Properties of Calcined Diatomite in Its Industrial Use as a Filler and Filter Aid: Thenardite (Na_2SO_4). *Crystals* **2022**, *12*, 503. [CrossRef]
13. Łach, M.; Pławecka, K.; Marczyk, J.; Ziejewska, C.; Hebdowska-Krupa, M.; Nykiel, M.; Hebda, M.; Miernik, K.; Mierzwiński, D.; Korniejenko, K.; et al. Use of Diatomite from Polish Fields in Sustainable Development as a Sorbent for Petroleum Substances. *J. Clean. Prod.* **2023**, *389*, 136100. [CrossRef]
14. Łach, M.; Grela, A.; Pławecka, K.; Guigou, M.D.; Mikuła, J.; Komar, N.; Bajda, T.; Korniejenko, K. Surface Modification of Synthetic Zeolites with Ca and HDTMA Compounds with Determination of Their Phytoavailability and Comparison of CEC and AEC Parameters. *Materials* **2022**, *15*, 4083. [CrossRef] [PubMed]
15. Łach, M.; Grela, A.; Bajda, T.; Mierzwiński, D.; Komar, N.; Mikuła, J. Production of Zeolite Sorbents from Burning and Co-Burning Biomass with Coal. *E3S Web Conf.* **2018**, *44*, 00097. [CrossRef]
16. Lv, Z.; Jiang, A.; Liang, B. Development of Eco-Efficiency Concrete Containing Diatomite and Iron Ore Tailings: Mechanical Properties and Strength Prediction Using Deep Learning. *Constr. Build. Mater.* **2022**, *327*, 126930. [CrossRef]
17. Kapeluszna, E.; Szudek, W.; Wolka, P.; Zieliński, A. Implementation of Alternative Mineral Additives in Low-Emission Sustainable Cement Composites. *Materials* **2021**, *14*, 6423. [CrossRef]
18. Formela, M.; Stryczek, S. Fly Ashes from the Combustion of Lignite as Additive to the Cement Slurry Used in the Work of Filling Voids in the Rock Mass. *Sci. J. Inst. Miner. Energy Econ. Pol. Acad. Sci.* **2017**, *97*, 117–134.
19. Ediz, N.; Bentli, I.; Tatar, I. Improvement in Filtration Characteristics of Diatomite by Calcination. *Int. J. Miner. Process.* **2010**, *94*, 129–134. [CrossRef]
20. Ren, Z.; Gao, H.; Zhang, H.; Liu, X. Effects of Fluxes on the Structure and Filtration Properties of Diatomite Filter Aids. *Int. J. Miner. Process.* **2014**, *130*, 28–33. [CrossRef]
21. Provis, J.L.; Lukey, G.C.; van Deventer, J.S.J. Do Geopolymers Actually Contain Nanocrystalline Zeolites? A Reexamination of Existing Results. *Chem. Mater.* **2005**, *17*, 3075–3085. [CrossRef]
22. Duxson, P.; Fernández-Jiménez, A.; Provis, J.L.; Lukey, G.C.; Palomo, A.; van Deventer, J.S.J. Geopolymer Technology: The Current State of the Art. *J. Mater. Sci.* **2007**, *42*, 2917–2933. [CrossRef]
23. Relucenti, M.; Familiari, G.; Donfrancesco, O.; Taurino, M.; Li, X.; Chen, R.; Artini, M.; Papa, R.; Selan, L. Microscopy Methods for Biofilm Imaging: Focus on SEM and VP-SEM Pros and Cons. *Biology* **2021**, *10*, 51. [CrossRef] [PubMed]

24. Zhang, J.; Li, S.; Li, Z.; Zhang, Q.; Li, H.; Du, J.; Qi, Y. Properties of Fresh and Hardened Geopolymer-Based Grouts. *Ceram. Silik.* **2019**, *63*, 164–173. [CrossRef]
25. Alehyen, S.; Achouri, M.E.; Taibi, M. Characterization, Microstructure and Properties of Fly Ash-Based Geopolymer. *J. Mater. Environ. Sci.* **2017**, *8*, 1783–1796.

materials proceedings

MDPI

Proceeding Paper

Flexural Performance of Chopped Basalt Fiber Reinforced Concrete Beams †

Birgir Leo Olafsson [1], Gudbjartur Jon Einarsson [2] and Eythor Rafn Thorhallsson [3,*]

1 The Icelandic Road and Coastal Administration (IRCA), IS-210 Gardabaer, Iceland
2 Landsvirkjun, IS-103 Reykjavik, Iceland
3 Department of Applied Engineering, Reykjavik University, IS-102 Reykjavik, Iceland
* Correspondence: eythor@ru.is; Tel.: +354-5996200

† Presented at the 10th MATBUD'2023 Scientific-Technical Conference "Building Materials Engineering and Innovative Sustainable Materials", Cracow, Poland, 19–21 April 2023.

Abstract: This article discusses the flexural strength of fiber reinforced concrete beams made by adding an optimal percentage of basalt fibers to the concrete mix. Two types of standard C30/37: XC1 concrete were used in this study, one with the aggregate size limit 19 mm and the other with the aggregate size of max. 4 mm. The basalt fibers used are two different types: Reforcetech minibar is a stiff basalt fiber, and its diameter is 0.72 mm and length 50 mm; Basaltex BCS17-25.4-KV1 is 30 mm in length with a flat cross-section of 0.017 × 4 mm. The test method measures the flexural efficiency of the strength parameters extracted from fiber reinforced concrete. The conclusion of this study is that the scale, quantity and type of basalt fibers have an impact on how the concrete acts under load. All types of fiber show promising results.

Keywords: concrete; basalt fibers; flexural behavior; tensile strength; cement

check for updates

Citation: Olafsson, B.L.; Einarsson, G.J.; Thorhallsson, E.R. Flexural Performance of Chopped Basalt Fiber Reinforced Concrete Beams. *Mater. Proc.* **2023**, *13*, 29. https://doi.org/10.3390/materproc2023013029

Academic Editors: Katarzyna Mróz, Tomasz Tracz, Tomasz Zdeb and Izabela Hager

Published: 16 February 2023

1. Introduction

The objective of this study was to investigate the effect of two types of basalt fibers in two different types of concrete mixes. The types of concrete that were used were of standard C30/37: XC1, containing aggregate size limits of maximum 19 and 4 mm. The basalt fibers used are Reforcetech minibar, which is a stiff basalt fiber of diameter 0.72 mm and length 50 mm, and Basaltex BCS17-25.4-KV1, which is 30 mm in length with a flat cross-section of 0.017 × 4 mm. The cement met the specifications of NS-EN 197-1:2011 for Portland cement CEM I 52.5 N.

1.1. Fibers in Concrete

While ordinary concrete is the world's most frequently used building material, its damaging structural qualities, such as brittle strain failure, must be improved. It has been established that adding fibers to the concrete mix improves the aforementioned properties by increasing the material's strength and decreasing brittleness.

The notion of reinforcing fibers is not a recent concept. Fibers have been used since ancient times as reinforcement; for example, horsehair has been used traditionally in mortar, and straw in mudbricks.

Asbestos fibers were used in buildings in the 1900s. Some of the topics of concern were the development of composite materials starting in the 1950s and fiber-reinforced concrete. When the health hazards related to asbestos were discovered, concrete and other products were required to remove the substance. In the 1960s, steel, glass (GFRC) and synthetic fibers (e.g., polypropylene) were used in concrete. Research on new concrete structures, made of fiber-reinforced concrete, continues today [1–3].

1.2. Basalt Fibers

Paul Dhé made the first attempts at basalt fiber production in the United States in 1923 and was later granted the US patent of 1.462.446. Scientists in the United States, Europe and the Soviet Union continued to develop these attempts in aerospace and military applications during the Second World War. Basalt fibers have been used since their declassification in 1995 in a broader range of civil applications.

Basalt fiber (BF) manufacturing technology is a single-stage process: melting, homogenization of basalt and extraction of fibers. Basalt is heated just once. BF is further refined into "cold processing" materials, with low energy costs [4].

BFs consist of one component, crushed basalt, from a carefully chosen source of a quarry. Basalt of high acidity (over 46% silica) and low iron element is considered suitable for fiber production. Basalt is an inert and naturally occurring material found around the world. Basalt products are environmentally friendly and not dangerous. Unlike other composites, such as glass fiber, no materials are substantially applied during manufacturing; the basalt is simply washed and then melted. The production of basalt fibers includes melting of the crushed and washed basalt rock at about 1.500 °C. The molten rock is extruded by small nozzles when the desired temperature is reached to create continuous basalt fiber filaments [5,6].

High strength and high modulus with excellent shock resistance are the main features of basalt fibers; they are a low-cost alternative to carbon fibers and can replace them in some applications. Basalt fiber reinforced concrete (BFRC) research has largely focused on fundamental mechanical properties: compressive, split tensile and flexural strength. In the case of BF, the research shows, in general, that the addition of fibers is beneficial up to approximately 0.3–0.5 percent by volume and detrimental thereafter [6]. A new basalt concrete reinforcement product, called minibars (MB), has recently been developed. The minibars are basically a scaled-down version of basalt fiber reinforced polymer rebar. In contrast to steel (SF) and fiberglass fibers, basalt is not affected by corrosion or acids. Basalt fibers even possess high corrosion and chemical durability to corrosive media such as salts, acid and alkali solutions. Basalt fibers have better mechanical properties than glass fibers in most cases, but in a few cases are somewhat worse than silica fibers. Due to high costs compared to other fibers, silica fibers are rarely used for concrete reinforcement. Basalt fibers are characterized by a high tensile strength from approximately 1700 MPa to 4800 MPa. The tensile strength of the basalt fibers (4800 MPa) is higher than that of the steel fibers [7].

2. Materials and Methods

2.1. The Study

The layout of the experiment consisted of the use of two types of basalt fibers manufactured by two different manufacturers, with the dose values for basalt fibers of 12 kg per cubic meter of concrete. The research method was based on "EN 14488-3 Testing sprayed concrete—Part 3: Flexural strengths (first peak, ultimate and residual) of fibre reinforced beam specimens".

The BFs used in this study were of two different kinds, one of stiff minibars and one of a soft fiber type. Table 1 shows the characteristics of each type.

Table 1. Technical characteristics of the basalt used in the study.

Material Properties	Reforcetech Minibars	Basalttex BCS17-25.4-KV1
Diameter (mm)	0.72	0.017 × 4
Length (mm)	50	25
Density (kg/m^3)	2100	2670
Melting point (°C)	1450	1450
Tensile strength (N/mm^2)	900	1350
Young's modulus (N/mm^2)	44,000	88,000

The Reforcetech bar is a stiff basalt fiber used in this study. Its diameter is 0.72 mm and length 50 mm, and the length factor 70. Basalt strands have been twisted together and mixed with Vinylester resin and, as a result, it has a coarse surface causing better bonding with concrete to prevent extraction. See Figure 1.

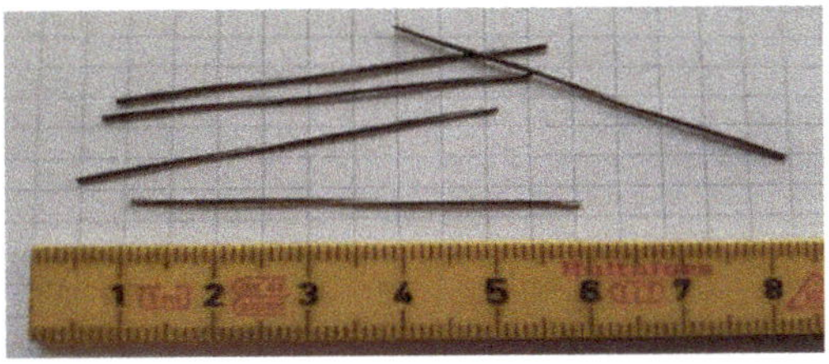

Figure 1. The Reforcetech bars.

Basaltex is the other type of basalt fiber used in the study. It is soft, composed of microfine strands of bundles that are slightly stiff but soften when treated and broken apart. It is about 30 mm in length with a flat cross-section already joined together in a bundle. See Figure 2.

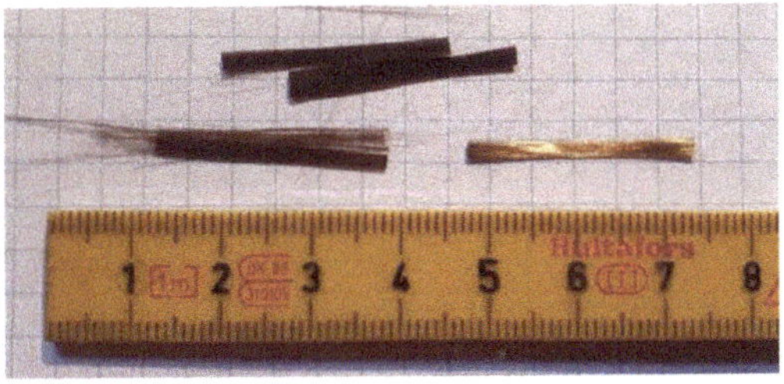

Figure 2. The Basaltex fibers.

Reference is made to ÍST EN 14488-3:2006 and the ACI study (American Concrete Institute), ACI 544.4R-18, which defines the ASTM C1609 M, for the results of the tensile ductility test. The key data and processes for the project were extracted from a four-point bend resistance test. The explanation for why a four-point test was used instead of a three-point test is that it gives a more accurate result when considering the effect of the fiber on the distribution of the load. The bending tensile strength shows a bending strain. In particular, the maximum post-peak load behavior shall be investigated.

2.2. The Test

The samples of max. grain 4mm concrete were 14 days old when tested, but the max. grain 19 mm concrete was 28 days old when the test took place. Even though compression strength tests indicated a full strength of the 14-day samples, the difference in age of the samples should be taken into account in direct comparison. The numbers of the samples are shown in Table 2.

Table 2. Samples.

Type	b × h × l [mm]	kg/m^3	Max Aggregate Size 19 mm	Max Aggregate Size 4 mm
Reforcetech Minibars	150 × 150 × 600	12	3	3
Basalttex BCS17-25.4-KV1	150 × 150 × 600	12	3	3

2.3. Equipment and Tools

The Tinius Olsen press that was used is computer-controlled with software from Horizon. The measurement frame includes two movement sensors on each side of the beam fastened on the frame, in accordance with ÍST EN 14488-3: 2006. See Figures 3 and 4.

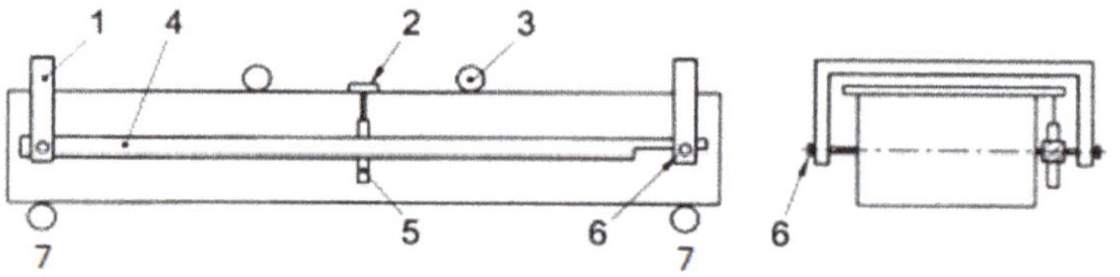

Figure 3. The measuring frame. 1. The steel frame; 2.. Movement sensors on each side. Independent of the load; 3. Load points; 4. The steel frame; 5. Movement sensors for the beam; 6. Free swing bolts; 7. Supports.

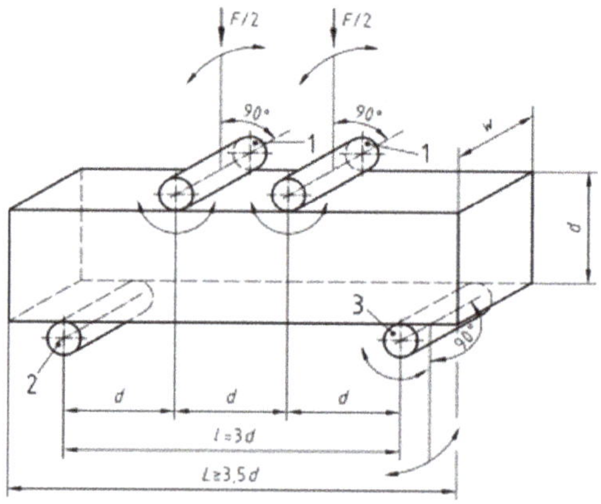

Figure 4. The set-up according to the IST EN 14488-3: 2006.

2.4. Installation and Implementation

Installation and implementation were in accordance with the standard IST EN 14488-3: 2006 and Guidance from ACI no. 544.4R-18. Beams were all constructed in the same form, 150 × 150 × 600 mm. Deviations were ±5 mm in length. Hardly any deviation was found in the height or width. In Figure 5 is a profile as shown in the IST EN 14488-3: 2006, d = 150 mm, w = 150 mm, L = 600 mm.

Figure 5. The actual setup.

2.5. The Procedure

The procedure of the experiment was as follows:

1. Beams measured and marked for easier installation in the press;
2. Beams placed in the press together with the measuring frame and sensors;
3. Transmission sensors and loads zeroed in the application;
4. Execution. The computer program created a log file that automatically ran until entry, and the load source was 5 mm. Variable speed is included in the entry standard 14488-3: 2006. Firstly 5 mm/min up to 250 N then 0.25 mm/min until the post came in 1mm. Thereafter, 1 mm/min to end when the load source shift reached 5mm;
5. When the execution was completed, the location of the first crack in the beams was measured and logged;
6. Removal of the beam, the crack located and the beam opened up to view the fiber distribution.

The cracking occurred in all cases from the bottom of the beam and upward, between the load points (rolls). From that, it is easy to recognize that this was a moment of failure in all cases. From that, the conclusion can be made that all tests are valid.

The Basalttex fibers were difficult to see in the fractured concrete but the Reforcetech fibers were more noticeable. Shown in Figure 6.

Figure 6. Reforcetech fibers are visible in the fracture but Basalttex fibers are less visible.

3. Results

After each execution, a cvs log file was created. It showed the load, the movement of the beam from the beginning of the test, movement speed, time and the two movement sensors. From each datafile, a graph was extracted for a visual comparison. The calculations were as follows Formula (1): The total average displacement was calculated using the two displacement sensors on the press, where $d1$ is the reading on displacement sensor 1 and $d2$ is the reading on displacement sensor 2. The $d1_0$ and $d2_0$ are the set displacements at the beginning of the test.

$$d_{tot} = \frac{((d1 - d2_0) + (d2 - d2_0))}{2} \tag{1}$$

Each flexural strength was calculated as an equivalent elastic tensile strength: flexural strength in MPa = P × l/(w × d^2) where:

- P is the load (P_{fp} or P_{ult}) defined in newtons, see Figure 7;
- l is the span (600 mm);
- w is the average beam width at the fracture plane (nominally 150 mm);
- d is the average beam depth at the fracture plane (nominally 150 mm).

As previously stated, the log file automatically ran until the entry of the load source was 5 mm. The entry standard 14488-3: 2006 includes variable speed. Firstly 5 mm/min up to 250 N, then 0.25 mm/min until the post was 1mm long. After that, 1 mm/min until the load sourceshift reached 5mm.

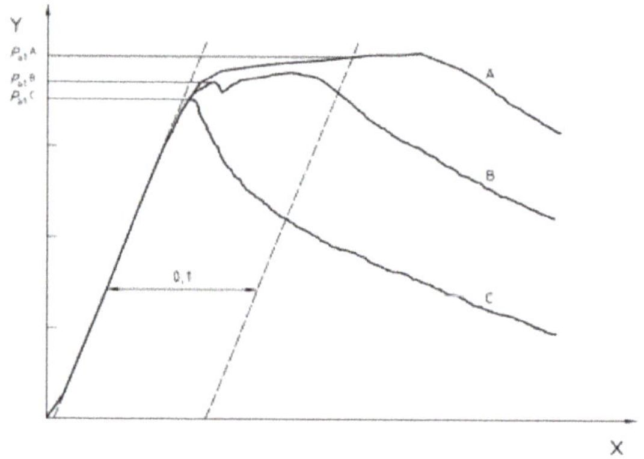

Key
X Central deflection in mm
Y Load in kN
$P_{a1}{}^{A}$ is the first peak load (P_{fp}) for curve A
$P_{a1}{}^{B}$ is the first peak load (P_{fp}) for curve B
$P_{a1}{}^{C}$ is the first peak load (P_{fp}) for curve C

Figure 7. Methodology of defining the load P_{fp} from IST EN 14488-3: 2006.

4. Conclusions

The BCS 17-25.4-KV1 generally has an interesting result. The negative part would be that it is porous and appears to soak up most of the water in the solution, so that the mix's workability is steadily decreasing. Additionally, in the bending tensile test, the impact of fiber varied significantly with the size of the aggregates in the concrete mix. The question raised here is whether the BCS distribution in the mixing process depends on when it is added to the mix, and furthermore, before applying to the blend, whether it would saturate the fibers in water, assist the process of even delivery and have less effect on the mix's water content. In all cases, the post-failure behavior is apparent. However, on the positive side of the BCS, it is important to see how the elongation persists in the 19 mm grain concrete with considerable force, whereas the 4 mm grain concrete is more brittle, fractures, and disintegrates directly after break. The peak load value of BCS in the 19 mm grain size combination is considerably higher than in any other sample type in this study.

The behavior of the RFT Reforcetech minibars after the first peak (failure point) is different from the BCS. The BCSs have a somewhat more definitive curve than the RFT fibers, and a quicker post failure force drop down. The RFT result shows that concrete containing RFT fibers has more strength than other fiber-mixed concrete, as shown in Figures 8 and 9. In RFT, the distribution is better than in BCS and the RFTs show similar results regardless of the mix's grain size. It can be concluded that RFT significantly increases the workability of concrete mixes.

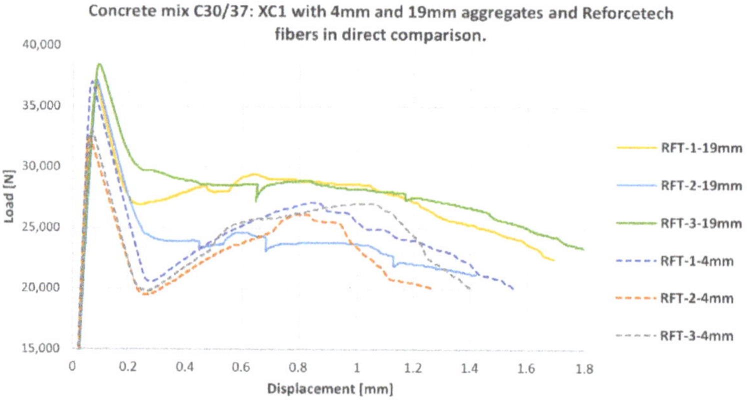

Figure 8. Reforcetech fibers load curves.

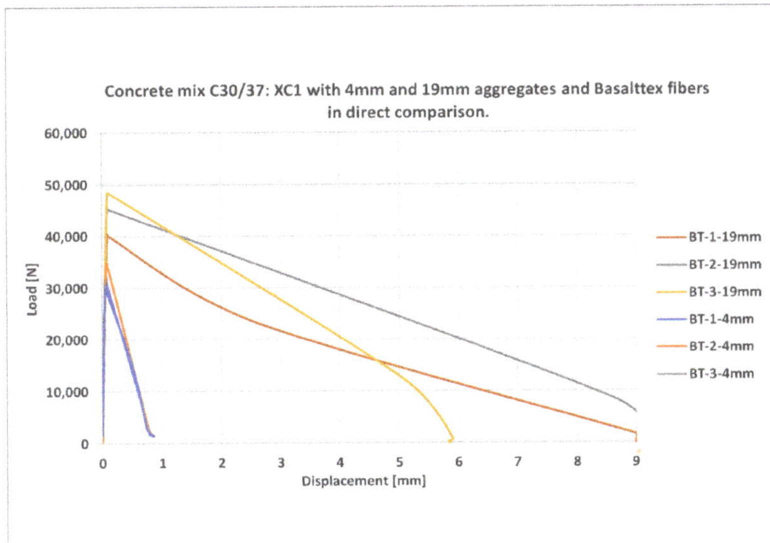

Figure 9. Basalttex fibers load curves.

The conclusion of this study is that the scale, quantity and type of basalt fibers have an impact on the outcome of how the concrete acts under load. The study also demonstrates how varying aggregate sizes affect how the beams behave under load. The study also demonstrates how varying aggregate sizes affect how the beams behave under load. All types of fiber show promising results, and a further analysis on variable quantities and fiber sizes would be worthwhile.

Author Contributions: Conceptualization and methodology, all authors; software, B.L.O.; validation, B.L.O. and G.J.E.; formal analysis, B.L.O.; investigation, B.L.O. and G.J.E.; resources, B.L.O. and G.J.E.; data curation, B.L.O.; writing—original draft preparation, B.L.O. and G.J.E.; writing—review and editing, all authors.; visualization, B.L.O.; supervision, G.J.E. and E.R.T. All authors have read and agreed to the published version of the manuscript.

Funding: This research received no external funding.

Informed Consent Statement: Not applicable.

Data Availability Statement: All data are stored by authors.

Acknowledgments: The publication cost of this paper was covered with funds from the Polish National Agency for Academic Exchange (NAWA): "MATBUD'2023—Developing international scientific cooperation in the field of building materials engineering" BPI/WTP/2021/1/00002, MATBUD'2023.

Conflicts of Interest: The authors declare no conflict of interest.

References

1. Johannesson, B.; Sigfusson, T.I.; Franzson, H.; Erlendsson, Ö.; Harðarson, B.S.; Thorhallsson, E.R.; Vehmas, T. *GREENBAS: Sustainable Fibres from Basalt Mining*; Nordisk Ministerråd: København K, Denmark, 2017; ISBN 978-92-893-4809-6. [CrossRef]
2. Thorhallsson, E.; Zhelyazov, T.; Gunnarsson, A.; Snaebjornsson, J.T. Concrete beams reinforced with prestressed basalt bars. In Proceedings of the Fib Symposium, Copenhagen, Denmark, 18–20 May 2015; Concrete Innovation and Design: Copenhagen, Denmark, 2015; pp. 277–278.
3. Thorhallsson, E.R.; Gudmundsson, S.H. Test of prestressed basalt FRP concrete beams with and without external stirrups. In Proceedings of the Fib Symposium Engineering a Concrete Future: Technology, Modelling & Construction, Tel Aviv, Israel, 22–24 April 2013; Dancygier, A.V., Ed.; Israeli Association of Construction and Infrastructure Engineers (IACIE) and Faculty of Civil and Environmental Engineering, Technion-Israel Institute of Technology: Tel Aviv, Israel, 2013; pp. 393–396.

4. Thorhallsson, E.R.; Snaebjornsson, J.T. Basalt Fibers as New Material for Reinforcement and Confinement of Concrete. *Solid State Phenom.* **2016**, *249*, 79–84. [CrossRef]
5. Thohallsson, E.R.; Konradsson, A. Renovation of Concrete Columns by Wrapping Basalt Fiber Sheets. *IABSE Symp. Rep.* **2013**, *99*, 903–910. [CrossRef]
6. Nováková, I.; Thorhallsson, E.R.; Wallevik, O.l. Influence of environmentally friendly basalt fibers on early-age strength of sprayed concrete. In Proceedings of the Fib Symposium 2019: Concrete Innovations in Materials, Design and Structures, Krakow, Poland, 27–29 May 2019; ISSN 2617-4820.
7. Thorhallsson, E.R.; Erlendsson, J.Ó. Continuous Basalt Fiber as Reinforcement Material in Polyester Resin. In Proceedings of the 2nd International Conference on Mechanics of Composites, Porto, Portugal, 11–14 July 2016.

materials proceedings

MDPI

Proceeding Paper

Lightweight Geopolymer Composites: The Impact of the Aggregate †

Zélia Alves *, João A. Labrincha and Rui M. Novais

Department of Materials and Ceramic Engineering, CICECO—Aveiro Instituto of Materials, University of Aveiro, Campus Universitário de Aveiro, 3810-193 Aveiro, Portugal

* Correspondence: zeliaralves@ua.pt

† Presented at the 10th MATBUD'2023 Scientific-Technical Conference "Building Materials Engineering and Innovative Sustainable Materials", Cracow, Poland, 19–21 April 2023.

Abstract: In this work, lightweight geopolymer composites were produced by varying the content of expanded perlite and geopolymer spheres, used as aggregates. Firstly, highly porous geopolymer spheres containing mostly red mud as the solid precursor were produced, which leads to the valorization of significant amounts of this metallurgical waste. Then, the incorporation of each aggregate in the slag matrix demonstrated that the geometric density, compressive strength, and thermal conductivity of the geopolymer composites decrease when their content increases. Composites with low density (0.8 g/cm^3) and low thermal conductivity (0.130 W/m K) were produced by using the expanded perlite. However, the sphere-containing composites showed higher specific strength and good thermal insulation properties. The results show an interesting potential of using a waste-derived aggregate, synthesized at much lower temperatures compared to expanded perlite, addressing the demand for developing novel energy efficient materials.

Keywords: slag; expanded perlite; geopolymer spheres; lightweight material; thermal conductivity

check for updates

Citation: Alves, Z.; Labrincha, J.A.; Novais, R.M. Lightweight Geopolymer Composites: The Impact of the Aggregate. *Mater. Proc.* **2023**, *13*, 30. https://doi.org/10.3390/materproc2023013030

Academic Editors: Katarzyna Mróz, Tomasz Tracz, Tomasz Zdeb and Izabela Hager

Published: 21 February 2023

1. Introduction

The production of ordinary Portland cement (OPC)-based concrete is responsible for a large amount of energy consumption, in addition to a very high share of the anthropogenic CO_2 emission (~8%), which is one of the major causes of climate change. Besides this, the energy inefficiency stemming from the heating and cooling needs of the building also raises concerns regarding sustainability in the building sector [1]. In this context, there is an urgent need for alternative and highly efficient new materials with a lower carbon footprint. Geopolymers emerge as one of the most promising alternatives due to their lower embodied CO_2, but also due to the possibility of using industrial wastes as precursors instead of non-renewable raw materials [1]. The production of lightweight and thermal insulating geopolymer concrete is a particularly interesting approach to increase the energy efficiency of buildings, as these materials can mitigate the energy losses [2,3]. One of the common routes involves incorporation of natural (e.g., cork [4]) or synthetic (e.g., expanded polystyrene, crumb rubber [5,6]) lightweight aggregates that ensure weight reduction. As the construction industry is striving to minimize consumption of virgin resources and precursors derived from fossil fuels, the production of aggregates from wastes or industrial by-products is given priority. In line with the above, highly porous red mud-based geopolymer spheres were synthesized in this work and used as lightweight aggregates to produce low density composite materials. Red mud is a by-product produced in large amounts during alumina production (0.8–1.5 tons of red mud per ton of alumina), which is considered an environmental hazard [7]. Then, the mechanical, thermal, and physical properties were determined to infer the influence of geopolymer spheres' incorporation content on the produced composites using slag as a binder. A comparison with a commercial and non-renewable lightweight aggregate was also performed, in this case by using expanded perlite. Geopolymer mortars containing expanded perlite have been proven to produce materials with better fire resistance and higher thermal insulation than siliceous sand [8].

2. Materials and Methods

2.1. Materials

The porous geopolymer spheres, to be used as synthetic lightweight aggregate, were produced using two solid precursors: red mud (RM), supplied by a bauxite mining company from Greece, and metakaolin (MK) (Argical™ M1200S from Univar®, Seattle, WA, USA). Prior to use, RM was milled and sieved to attain a particle size below 75 μm, while MK was sieved to reach a particle size below 63 μm. The chemical composition of RM (Table 1) determined via X-ray fluorescence (XRF) shows that this waste is an iron-rich material (41 wt.%) with a low silica and alumina content. For this reason, MK was employed to adjust the SiO_2/Al_2O_3 ratio. To perform the chemical activation of these solid precursors, a mixture of 100 g of industrial grade sodium silicate solution (Chem-Lab, Belgium) and 13.22 g of sodium hydroxide (ACS reagent, 97%; Sigma Aldrich, Saint Louis, MO, USA) were used. Sodium dodecyl sulfate (SDS) and polyethylene glycol (PEG 600) were supplied by Sigma Aldrich to be used as a pore foaming agent and consolidation medium of the geopolymer spheres, respectively. For the geopolymer composites synthesis, the main solid precursor in the binder's production was a commercial slag (Koranel®, Metallo, Beerce, Belgium) activated via an alkaline solution consisting of a mixture of 100 g of potassium silicate (K40, Quimialmel, Albergaria-a-Velha, Portugal) and 15.25 g of potassium hydroxide (85%; PanReac AppliChem, Castellar del Vallès, Spain). Expanded perlite (Figure 1a), with a geometric density of 0.11 g/cm^3 and particle diameter (D50) of 0.537 mm (Figure 1b), was used as a natural lightweight material and supplied by a Greek company.

Table 1. Chemical composition from XRF of metakaolin (MK), red mud (RM), and slag.

Oxides (wt.%)	MK	RM	Slag
SiO_2	54.40	9.20	26.92
Al_2O_3	39.40	18.19	8.36
CaO	0.10	9.12	3.04
K_2O	1.03	0.12	0.16
Fe_2O_3	1.75	40.66	51.27
MgO	0.14	0.27	0.93
Na_2O	-	4.72	1.97
TiO_2	1.55	5.68	0.32
SO_3	-	0.41	0.73
P_2O_5	0.06	0.11	0.71
MnO	0.01	0.05	0.77
LOI	2.66	10.71	0.08

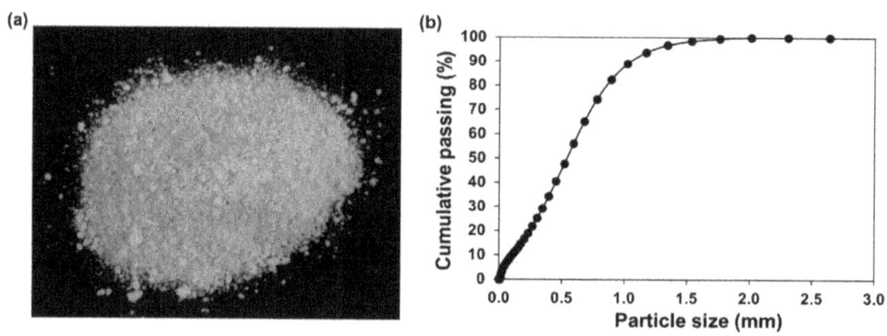

Figure 1. (**a**) Expanded perlite aggregate and its (**b**) particle size distribution.

2.2. Synthesis of the Geopolymer Spheres

The geopolymer spheres were prepared following the experience from previous investigations [9], in which 31 g of solid precursors (55 wt% of RM and 45 wt% of MK) were chemically activated with an alkaline solution (30 g) and distilled water (8.3 g). A planetary mixer was used to mix the precursors and the activator for 5 min. Then, the foamed slurry was obtained by adding the SDS (0.9 g) to the slurry and mixing it for 2 min. Finally, the geopolymer spheres were synthetized through the suspension solidification method by injecting the foamed slurry in a polyethylene glycol medium, which was maintained under the bath temperature of 80 ± 5 °C. After collecting the spheres, they were washed with water and cured in an oven at 40 °C for 24 h. Afterwards, the geopolymer spheres were cured for 28 days at room temperature.

2.3. Synthesis of the Geopolymer Composites

For the lightweight geopolymer composites preparation, firstly, the slag was activated with the potassium hydroxide solution with a liquid-to-solid ratio of 0.684. After mixing for 1 min, the lightweight aggregate was incorporated into the activated slag for 2 min. The volume ratios for each lightweight aggregate and the binder were firstly evaluated using preliminary tests. These tests showed that 80 vol.% and 85 vol.% were the maximum amount of expanded perlite and geopolymer spheres that could be added to the matrix, respectively, without disintegration of the composites. Three different compositions varying the amount of expanded perlite (70, 75, 80 vol.%) and other three varying the amount of geopolymer spheres (75, 80, 85 vol.%) were produced. Reference specimen was also produced without the addition of aggregates. The pastes were transferred to steel molds ($4 \times 4 \times 4$ cm^3), sealed to maintain the humidity, and cured for 24 h at room temperature. Then, the cubic specimens were removed from the molds and cured at room temperature until the 28th day.

2.4. Materials Characterization

The chemical composition of the solid precursors was determined via X-ray fluorescence (XRF- Philips X'Pert PRO MPD spectrometer, Malvern Panalytical, Malvern, UK). The morphology and porosity of the spheres were examined using an optical microscope (Nikon, H550S, Tokyo, Japan). The particle size distribution of expanded perlite was conducted using laser diffraction in the dry mode (HORIBA scientific LA-960V2, Kyoto, Japan). Scanning electron microscopy (SEM—Hitachi SU 70; energy dispersion spectroscopy—EDS Bruker) was also used to characterize the microstructure of the spheres. The geometric density was calculated via the ratio of dry mass to volume using three cubic specimens ($4 \times 4 \times 4$ cm^3) of each formulation cured for 28 days. The compressive strength of composite specimens ($4 \times 4 \times 4$ cm^3), measured 28 days after their synthesis, was performed using a Universal Testing Machine (Shimadzu AG-25 TA, Kyoto, Japan) at a 0.5 mm/min load rate in triplicate per batch, following the standard EN 1015-11:1999 [10]. The thermal conductivity was measured on three cubic specimens ($4 \times 4 \times 4$ cm^3) of each composition by using a heat flow meter apparatus following the ASTM C518-04 standard [11], where the sample is placed between two parallel plates, with a constant temperature gradient of 15 °C.

3. Results and Discussion

3.1. Geopolymer Spheres Characterization

Figure 2a shows the optical micrographs of RM-containing geopolymer spheres. The spheres have a spheroidal shape and narrow size distribution, with an average diameter of 2.9 ± 0.8 mm. The SEM micrographs of the geopolymer spheres' surface and interior microstructure are shown in Figure 2b. The spheres reveal a homogeneous smooth surface with small-sized pores, while their interior shows the presence of a large-sized and closed pore surrounded by mostly small-sized closed pores and a few open pores. The geometric density of geopolymer spheres was 0.57 g/cm^3, with this value endowing their use as a lightweight aggregate.

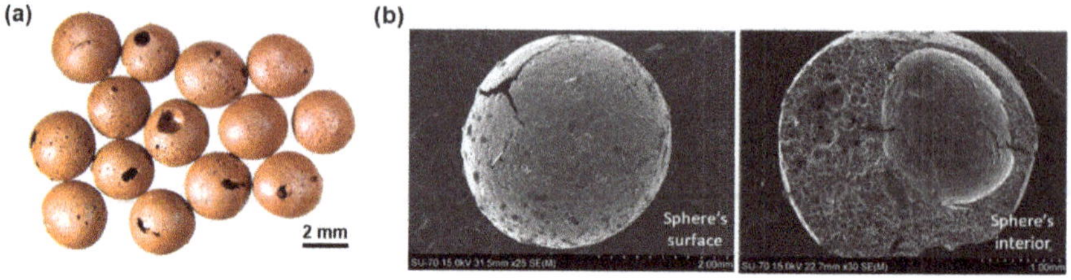

Figure 2. (**a**) Optical microscopy and (**b**) SEM micrographs of the red mud-based geopolymer spheres.

3.2. Geopolymer Composites Characterization

The geometric density of the geopolymer composites is reduced with the addition of both lightweight aggregates, reaching values below 1.40 g/cm^3 in all studied composites, which represent a reduction of at least 39% in comparison with the reference specimen prepared without the lightweight aggregate (Figure 3). A small density decrease is observed between the first and the twenty-eight day of cure as a result of the dehydration process during the polycondensation reaction. Noteworthy, a major reduction in the geometric density of composites is seen when increasing the amount of the aggregates, the lowest geometric density (0.75 g/cm^3) reached by the higher volume (80 vol.%) of expanded perlite, followed by the composite with 85 vol.% of spheres (0.84 g/cm^3). At 80 vol.% of aggregate, the geometric density of composite containing expanded perlite is roughly 1.3 times lower than the density with the same volume of spheres (1.00 g/cm^3). However, no significant difference is observed with 75 vol.% of each aggregate, even though the expanded perlite's geometric density is roughly five times lower than the density of the geopolymer spheres. This feature is attributed to the fragile nature of expanded perlite, which can be partially destroyed during the mixing step, leading to higher density values than those expected when considering the apparent density of this aggregate. In fact, this phenomenon led to specimens having heterogeneous mechanical properties, as can be observed in Figure 4.

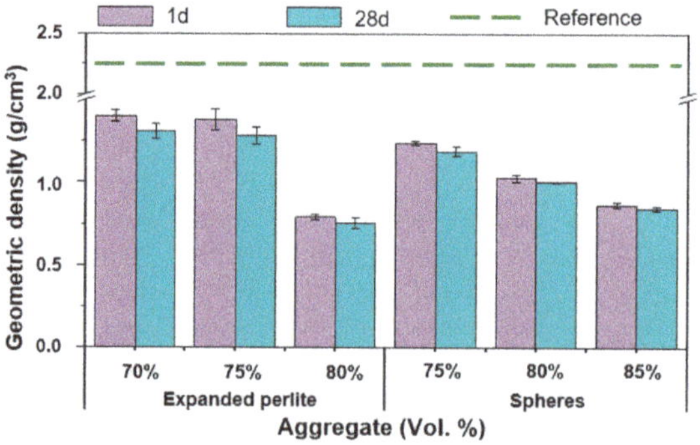

Figure 3. Geometric density of geopolymer composites varying the amount of expanded perlite and geopolymer spheres (measured on the 1st and 28th day). Reference means the control specimen without addition of lightweight aggregate.

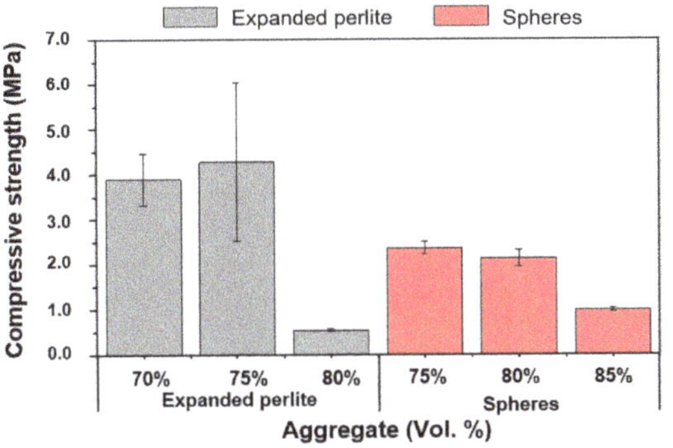

Figure 4. Compressive strength (MPa) of geopolymer composites varying the amount of expanded perlite and geopolymer spheres.

The mechanical performance under compression of all composites is shown in Figure 4. The compressive strength decline is more pronounced at the highest incorporation volume of the aggregates, with the lowest value for specimens composed with 80 vol.% of expanded perlite (0.5 MPa) followed by the one with 85 vol.% of spheres (1.0 MPa). Despite the small difference in their geometric densities, the composite with 85 vol.% of spheres has a compressive strength two times higher than composite containing 80 vol.% of expanded perlite. With 75 vol.% of expanded perlite, the compressive strength is approximately twofold higher than the composite using the same amount of spheres. However, the result seen for the perlite-containing composite is abnormal and should be considered with caution. These samples were heterogenous, possibly due to a partial destruction of perlite during the mixing step, and this explains the very high standard deviation of composites with 75 vol.% of expanded perlite. Nevertheless, it should be noted this feature was not observed when using higher volumes of this aggregate. Indeed, with the incorporation of 80 vol.% of aggregate, an opposite tendency occurs, where the composite with spheres has a compressive strength fourfold higher than the expanded perlite, which can be explained by its higher composite's geometric density. The specific strength of the specimens with lower densities was determined and showed the following: 80 vol.% of spheres (2.1 MPa cm^3/g) > 85 vol.% of spheres (1.8 MPa cm^3/g) > 80 vol.% of expanded perlite (0.7 MPa cm^3/g). The high specific strength of composites containing the spheres demonstrates the interesting properties of this material, enabling their use as a lightweight aggregate in the production of low density geopolymers.

The thermal conductivity of composites drops when the content of each lightweight aggregate rises, as shown in Table 2. The lowest value of thermal conductivity (0.130 W/m K) was observed in the composite with the highest volume of expanded perlite. For the same concentration, the addition of expanded perlite generates materials with lower thermal conductivity than the use of geopolymer spheres, but it is only significantly different at 80 vol.%. These results show that thermal conductivity can be controlled by the type of lightweight aggregate and by its content. The thermal conductivity value (0.175 W/m K) of composite containing the utmost volume of geopolymer spheres validates the strategy of using them as a lightweight aggregate. Moreover, their use has sustainable advantages over expanded perlite, since spheres are synthesized at 80 °C, this being much lower than the common temperatures involved in the expansion of perlite (850–1100 °C) [12,13]. Additionally, they are mostly produced using hazardous and abundant (150 Mt/year) industrial waste, contributing not only to reduce its stockpile but also to avoid the use of virgin raw materials. The lowest value of thermal conductivity obtained here (0.130 W/m K with expanded perlite) was compared with the literature and demonstrated to be smaller than geopolymer mortars incorporating expanded perlite (0.370 W/m K) [8] and crumb rubber (0.279 W/m K) [6], similar with the addition of waste-expanded polystyrene (0.121 W/m K) [5], but higher than those seen when using cork (0.072 W/m K) [4].

Table 2. Thermal conductivity (W/m K) of geopolymer composites produced with distinct expanded perlite and geopolymer spheres content.

Lightweight Aggregate	vol.%	Thermal Conductivity (W/m K)
Expanded perlite	70	0.240 ± 0.017
	75	0.215 ± 0.017
	80	0.130 ± 0.009
Spheres	75	0.234 ± 0.012
	80	0.215 ± 0.007
	85	0.175 ± 0.008

4. Conclusions

Geopolymer composites were produced here using expanded perlite or geopolymer red mud-based spheres as a lightweight aggregate. The effect of their content in the composite was investigated via the mechanical, thermal, and physical properties. Results show that by increasing the aggregate amount, the compressive strength, geometric density, and thermal conductivity tend to decrease, demonstrating that these properties can be tuned considering the application envisioned. The composition containing the highest expanded perlite volume achieved the lowest value of geometric density (0.8 g/cm^3) and of thermal conductivity (0.130 W/m K). Nevertheless, the use of red mud-based spheres resulted in a higher specific strength. These results suggest that this waste-derived aggregate can be a promising alternative in the development of sustainable and energy efficient geopolymer materials, also contributing to reduce the environmental impact associated with waste landfilling.

Author Contributions: Conceptualization, R.M.N. and J.A.L.; methodology, Z.A., R.M.N. and J.A.L.; software, Z.A.; validation, Z.A., R.M.N. and J.A.L.; formal analysis, Z.A.; investigation, Z.A.; resources, R.M.N. and J.A.L.; writing—original draft preparation, Z.A.; writing—review and editing, R.M.N. and J.A.L.; visualization, Z.A.; supervision, R.M.N. and J.A.L.; funding acquisition, R.M.N. and J.A.L. All authors have read and agreed to the published version of the manuscript.

Funding: This work was developed within the scope of the project CICECO-Aveiro Institute of Materials, UIDB/50011/2020 & UIDP/50011/2020, financed by national funds through the FCT/MEC and when appropriate co-financed by FEDER under the PT2020 Partnership Agreement. The authors would like to thank the project SMART-G (ERA-MIN/0001/2019), sponsored by FCT.

Institutional Review Board Statement: Not applicable.

Informed Consent Statement: Not applicable.

Data Availability Statement: Not applicable.

Acknowledgments: Publication cost of this paper was covered with founds of the Polish National Agency for Academic Exchange (NAWA): "MATBUD'2023–Developing international scientific cooperation in the field of building materials engineering" BPI/WTP/2021/1/00002, MATBUD'2023.

Conflicts of Interest: The authors declare no conflict of interest.

References

1. Zhang, Z.; Provis, J.L.; Reid, A.; Wang, H. Geopolymer Foam Concrete: An Emerging Material for Sustainable Construction. *Constr. Build. Mater.* **2014**, *56*, 113–127. [CrossRef]
2. Novais, R.M.; Buruberri, L.H.; Ascensão, G.; Seabra, M.P.; Labrincha, J.A. Porous Biomass Fly Ash-Based Geopolymers with Tailored Thermal Conductivity. *J. Clean. Prod.* **2016**, *119*, 99–107. [CrossRef]
3. Novais, R.M.; Ascensão, G.; Buruberri, L.H.; Senff, L.; Labrincha, J.A. Influence of Blowing Agent on the Fresh- and Hardened-State Properties of Lightweight Geopolymers. *Mater. Des.* **2016**, *108*, 551–559. [CrossRef]

4. Novais, R.M.; Senff, L.; Carvalheiras, J.; Seabra, M.P.; Pullar, R.C.; Labrincha, J.A. Sustainable and Efficient Cork—Inorganic Polymer Composites: An Innovative and Eco-Friendly Approach to Produce Ultra-Lightweight and Low Thermal Conductivity Materials. *Cem. Concr. Compos.* **2019**, *97*, 107–117. [CrossRef]
5. Colangelo, F.; Roviello, G.; Ricciotti, L.; Ferrándiz-Mas, V.; Messina, F.; Ferone, C.; Tarallo, O.; Cioffi, R.; Cheeseman, C.R. Mechanical and Thermal Properties of Lightweight Geopolymer Composites. *Cem. Concr. Compos.* **2018**, *86*, 266–272. [CrossRef]
6. Wongsa, A.; Sata, V.; Nematollahi, B.; Sanjayan, J.; Chindaprasirt, P. Mechanical and Thermal Properties of Lightweight Geopolymer Mortar Incorporating Crumb Rubber. *J. Clean. Prod.* **2018**, *195*, 1069–1080. [CrossRef]
7. Ye, N.; Chen, Y.; Yang, J.; Liang, S.; Hu, Y.; Hu, J.; Zhu, S.; Fan, W.; Xiao, B. Transformations of Na, Al, Si and Fe Species in Red Mud during Synthesis of One-Part Geopolymers. *Cem. Concr. Res.* **2017**, *101*, 123–130. [CrossRef]
8. Rashad, A.M. Insulating and Fire-Resistant Behaviour of Metakaolin and Fly Ash Geopolymer Mortars. *Proc. Inst. Civ. Eng. Constr. Mater.* **2019**, *172*, 37–44. [CrossRef]
9. Novais, R.M.; Carvalheiras, J.; Seabra, M.P.; Pullar, R.C.; Labrincha, J.A. Innovative Application for Bauxite Residue: Red Mud-Based Inorganic Polymer Spheres as PH Regulators. *J. Hazard. Mater.* **2018**, *358*, 69–81. [CrossRef] [PubMed]
10. *EN 1015-11:1999*; Methods of Test for Mortar for Masonary. Determination of Flexural and Compressive Strength of Hardened Mortar. British Standards Institution: London, UK, 1999.
11. *ASTM C518-04*; Standard Test Method for Steady-State Thermal Transmission Properties by Means of the Heat Flow Meter Apparatus. ASTM International: West Conshohocken, PA, USA, 2004.
12. Karakaş, H.; İlkentapar, S.; Durak, U.; Örklemez, E.; Özuzun, S.; Karahan, O.; Atiş, C.D. Properties of Fly Ash-Based Lightweight-Geopolymer Mortars Containing Perlite Aggregates: Mechanical, Microstructure, and Thermal Conductivity Coefficient. *Constr Build. Mater.* **2023**, *362*, 129717. [CrossRef]
13. Koukouzas, N.K.; Dunham, A.C.; Scott, P.W. Suitability of Greek Perlite for Industrial Applications. *Appl. Earth Sci. Trans. Inst. Min. Metall.* **2000**, *109*, 105–111. [CrossRef]

materials proceedings

MDPI

Proceeding Paper

Placing on the Market Modern Construction Products and Systems Contributing to Next Generation of Healthy, Nearly Zero-Energy Buildings †

Aneta Nowak-Michta

Chair of Building Materials Engineering, Faculty of Civil Engineering, Cracow University of Technology, 24 Warszawska St., 31-155 Cracow, Poland; aneta.nowak@pk.edu.pl

† Presented at the 10th MATBUD'2023 Scientific-Technical Conference "Building Materials Engineering and Innovative Sustainable Materials", Cracow, Poland, 19–21 April 2023.

Abstract: In Europe, the construction sector currently generates more than a third of CO_2 emissions. Moreover, residential buildings consume about 40% of the energy that we are able to produce. Construction product manufacturers implement innovative solutions that give us a chance to achieve more widespread construction of zero-emission and zero-energy buildings. However, they face legal barriers related to the lack of standardization procedures enabling the rapid placing of innovative construction products on the market to be achieved. The MEZeroE project ("Measuring Envelope products and systems contributing to next generation of healthy nearly Zero Energy Buildings"), funded by European Union Horizon 2020, aims to create a platform combining science and business to facilitate the development of nearly zero-energy buildings. The aim of this paper is to present the European procedure for placing innovative construction products on the market under the MEZeroE project.

Keywords: innovative construction products; placing on the market; standardization procedures; a platform connecting science and business; MEZeroE; nearly zero-energy buildings

Citation: Nowak-Michta, A. Placing on the Market Modern Construction Products and Systems Contributing to Next Generation of Healthy, Nearly Zero-Energy Buildings. *Mater. Proc.* **2023**, *13*, 31. https://doi.org/10.3390/materproc2023013031

Academic Editors: Katarzyna Mróz, Tomasz Tracz, Tomasz Zdeb and Izabela Hager

Published: 16 February 2023

1. Introduction

In accordance with the requirements of Directive 2010/31/EU [1], as amended, all buildings designed from 2021 onwards must be nearly zero-energy buildings (nZEBs). Buildings according to this standard should be characterized by low or zero demand for non-renewable primary energy and a high degree of use of energy from renewable sources produced on site or nearby. Climate-neutral construction is possible, but it requires innovative materials, construction techniques and legislation based on sustainable development. It also requires rising carbon costs to be reflected in building solutions, with no barriers to their adoption. Low-carbon solutions must become the norm and thus be more competitive than conventional solutions.

Demographic and urban trends indicate that over the next three decades, the number of people living in urban areas will increase by another two billion people and that 60% of the infrastructure that humanity needs is still to be built. In Europe, two-thirds of the population already live in cities. However, many urban centers are struggling with a housing shortage and the need to replace outdated infrastructure. Buildings are energy intensive and carbon intensive, and more than 30% of waste in European landfills comes from demolished buildings [2].

The challenge for the construction industry is to meet the growing demand for high-quality housing and infrastructure while transitioning to low-carbon construction and a circular economy. Currently, in Europe, the construction sector generates more than one-third of CO_2 emissions, and residential buildings consume around 40% of the energy that we are able to produce. Manufacturers of building materials implement innovative solutions that have a chance to achieve the widespread construction of zero-emission and zero-energy buildings. However, they face legal barriers related to the lack of standardization procedures enabling the rapid placing of innovative construction products on the market to be achieved [3].

The aim of the paper is to present the European platform developed and tested under the "Measuring Envelope products and systems contributing to next generation of healthy nearly Zero Energy Buildings" (MEZeroE) project [3], which is financed by European Union Horizon 2020 and supports the placing of innovative construction products on the market.

2. The Procedure for Placing Construction Products on the Market in the EU

Regulation EU No. 305/2011 of the European Parliament and of the Council of 9 March 2011 (CPR) [4] setting harmonized conditions for the marketing of construction products and repealing Council Directive 89/106/EEC [5] set out the conditions for placing construction products on the market or making them available on the market. It establishes harmonized rules for expressing the performance of construction products in relation to their essential characteristics and harmonized rules for the use of CE marking on these products.

In Poland, the regulation was introduced by the Act of 16 April 2004 on construction products [6], which defines the rules for placing construction products on the market. It approximates the executive acts and laws of all EU Member States relating to construction products. Pursuant to Article 5 of this Act, a construction product covered by a harmonized technical specification, i.e., a harmonized standard or European technical assessment (ETA), may be placed on the market or made available on the internal market only and exclusively in accordance with the provisions of the CPR [4].

According to the diagram in Figure 1, the European system [4] enables a construction product to be placed on the market on the basis of a harmonized standard or ETA. The CE marking and placing on the European market of construction products covered by harmonized standards have an established procedure and do not represent a problem for manufacturers. For construction products not covered or not fully covered by harmonized standards, an alternative route to CE marking is the route provided by the European Organization for Technical Assessment (EOTA). This route gives manufacturers the opportunity to apply for ETA. The manufacturer may issue a Declaration of Performance and affix the CE marking on the basis of the ETA. ETAs are issued by technical assessment bodies (TABs) on the basis of European assessment documents (EADs), developed by the EOTA and cited by the Commission in the Official Journal of the European Union.

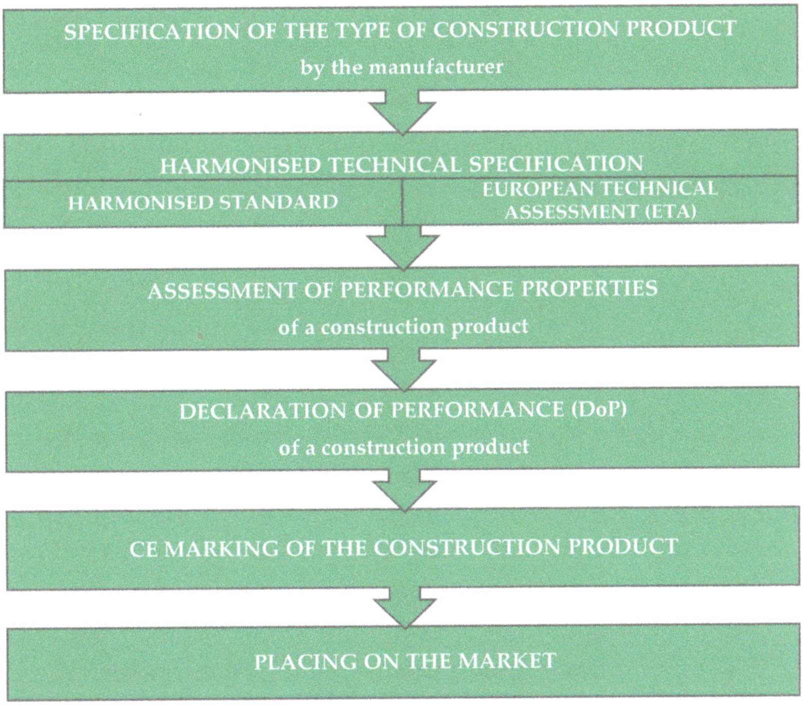

Figure 1. Diagram of placing construction products on the market [4].

3. Main Barriers to the Placing of Modern Construction Products on the Market

The Commission 2016 report on the implementation of the CPR identified some shortcomings in the process of implementation and significant challenges related to standardization, simplification for micro-enterprises, market surveillance and enforcement that merit further analysis and discussion. The evaluation of the CPR, the opinions of the REFIT Platform and the feedback from Member States and stakeholders have clearly identified barriers to the functioning of the single market for construction products and the ensuing failure to achieve the objectives of the CPR [2].

The "Clean energy for all Europeans" communication of November 2016 stresses the need to unlock the potential for growth and jobs by improving the functioning of the still fragmented single market for construction products. European Green Deal Communication, Circular Economy Action Plan and Renovation Wave Communication highlight the role of regulating the common provisions in the efforts towards energy- and resource-efficient buildings and renovations and in the sustainable development of construction products. The proposal for a revised directive on the energy performance of buildings emphasizes the importance of looking at the total greenhouse-gas emissions of buildings and construction materials throughout their life cycle in calculating the global warming potential of new buildings from 2030. EU Forest Strategy and Sustainable Carbon Cycle Communications were announced in the context of the revision of the regulation on construction products, developing a standard, robust and transparent methodology for quantifying the climate benefits of construction products and of carbon capture and utilization. In addition, both the European Parliament and the Council called for action to promote the circulation of construction products, to remove barriers to the single market for construction products and to contribute to the objectives of European Green Deal and the Circular Economy Action Plan [7].

Accordingly, two general objectives of the CPR review have been set [2]:

- Achieving a well-functioning single market for construction products.
- Contributing to the goals of the green and digital transitions, in particular, a modern, resource-efficient and competitive economy.

This review aims to address the following issues:

3.1. Lack of Harmonized Standards for Modern Products

So far, the standardization process underlying the CPR has not yielded satisfactory results. In recent years, draft harmonized standards developed by European standardization organizations have rarely been cited in the Official Journal of the EU, mainly due to legal issues such as contradiction with the requirements of the CPR or going beyond the scope of a mandate/request for standardization. The lack of citation of current harmonized standards for construction products is considered a key factor undermining the internal market, as outdated harmonized standards generate direct or indirect costs for businesses, especially SMEs. The standardization process is too slow to keep up with the development of the sector. The resulting standards may not always be appropriate for the market, nor may they meet the regulatory needs of Member States [2].

3.2. The EOTA Route

The EOTA route, which is the CPR alternative route for CE marking for construction products not covered, or not fully covered, by harmonized standards, also points to the poor performance of the standardization system. Poor performance results largely from long waiting time for the preparation of ETAs, as well as from problems with establishing EDOs for modern construction products [2].

3.3. Implementation Challenges at the National Level

The CPR Implementation Report identified some shortcomings in the functioning of Notified Bodies (NBs) and Notifying Authorities, indicating that the relevant provisions of the CPR would benefit from greater accuracy, among others, of requirements for NBs (Article 43), operational responsibilities of NBs (Article 52) and coordination of NBs (Article 55).

The simplification provisions in Annex 8 of the CPR, mainly aimed at small and medium companies, include Art. 5 (deviations from the preparation of the Declaration of Performance—DoP), Art. 36 (avoiding unnecessary repetition of tests), Art. 37 (simplified procedures for micro-enterprises) and Art. 38 (simplified procedures for products manufactured individually or to order in a non-series process). The use of these articles, with the exception of Art. 36, remains very limited. These provisions were intended to lead to a simplification effect, to reduce the administrative costs of placing construction products on the market without compromising the level of safety of construction works. However, various assessments and feedback from stakeholders indicate a low uptake of these simplification provisions, mainly resulting from the low awareness of their existence and from the lack of clarity, especially in specifying what "equivalent" documentation actually is. In conclusion, the attempt to level the playing field for smaller companies, especially through Art. 37, has failed. Additionally, the appropriateness of measures that allow certain manufacturers to implement such lighter procedures is questioned, given that this creates uncertainty for end-users who can reasonably expect that all CE-marked products conform to the same procedural requirements. The evaluation showed that while the CPR procedure was achieved, it did not meet expected levels.

While the CPR was expected to reduce costs and administrative burdens, it actually resulted in the opposite effect. The estimates from the study on the economic impact of the CPR show that the smallest businesses bear the greatest administrative burden.

Moreover, some provisions of the CPR are not sufficiently clear or overlap, either within the regulations themselves or between the CPR and other EU legislation. For example, Art. 9 sec. 2 of the CPR lists the information that must accompany the CE marking, most of which the manufacturer has already provided in the Declaration of

Performance (DoP). This has been the subject of strong criticism since the beginning of the implementation of the CPR. Overlap of the information required in the Declaration of Performance and CE marking creates unnecessary administrative and financial burdens and represents clear inefficiency. When it comes to clarity and coherence between the CPR and other EU legislation, there are a number of areas where they overlap or contradict each other, including Ecodesign Directive and several other product/technical directives. This results in a lack of clarity for the procedures established for construction products on the market, in particular with regard to parallel routes of CE marking [2].

3.4. CPR Misalignment on Green and Digital Transitions

In order to reduce emissions and achieve climate neutrality by 2050, the industry needs to be mobilized to make the transition to a climate-neutral and circular economy, especially in resource-intensive sectors such as construction. The building stock, which currently accounts for 40% of final energy and 36% of EU greenhouse-gas emissions, has a large cost-effective emission reduction potential. The European Climate Pact has recognized the need to use low-carbon materials to make buildings more climate friendly [2].

3.5. Fragmentation and Complex Structure of the Construction Market

The fragmentation and complex structure of the construction market are among the main barriers to technological innovation in this sector, due to the many disciplines involved. Design and production are typically distributed across a large number of enterprises, with a significant predominance of SMEs, and long and multi-faceted supply chains, making it difficult to bring together different specialists. The hallmark of an accepted technology is its inclusion in standards such as building codes and specifications, which provides designers, professionals, practitioners and contractors with certainty as to the profitability of the technology, the information necessary for its use and lower perception of the risk associated with its use [2].

4. Assumptions of the MEZeroE Platform [3,7]

The MEZeroE project aims to create an ecosystem combining infrastructural facilities and the knowledge of academic and research centers, and innovative solutions proposed by the industry in the form of a multilateral virtual market, open to the exchange and use of knowledge and experience between stakeholders from the construction industry. The ecosystem aims to provide ready-made services in the fields of modeling, testing and monitoring nZEB Enabler Envelope Technologies Solutions (nEESs), while creating a comprehensive knowledge management and training environment (Figure 2).

The task of the project is to take into account the trends of Industry 4.0 in the field of quick decision making and customer orientation open to innovation, with a focus on manufacturers of carbon-neutral construction products and a healthy indoor environment.

As part of the project, standard, modified and non-standard products are tested using state-of-the-art equipment. Support is also provided for manufacturers in the certification of the product and its placing on the market. Standard tests have been extended with tests in living laboratories located throughout Europe, where a new parameter of a variable expressing user feedback concerning the comfort of use has been introduced.

Access to MEZeroE is possible through the online platform (https://www.mezeroe.eu/, accessed on 1 December 2022), which is a multilateral virtual marketplace with a single entry point. This market includes nine pilot lines of measurement and verification (PM&VL) and three open innovation services (OISs) covering training, business model development, systematic intellectual property and knowledge management. The team of Cracow University of Technology constitutes the PM&VL7 line, which is composed of four sublines performing tests on durability, mechanical, vibroacoustic and thermal performance. MEZeroE will accelerate the placing of prototypes on the market as certified construction products.

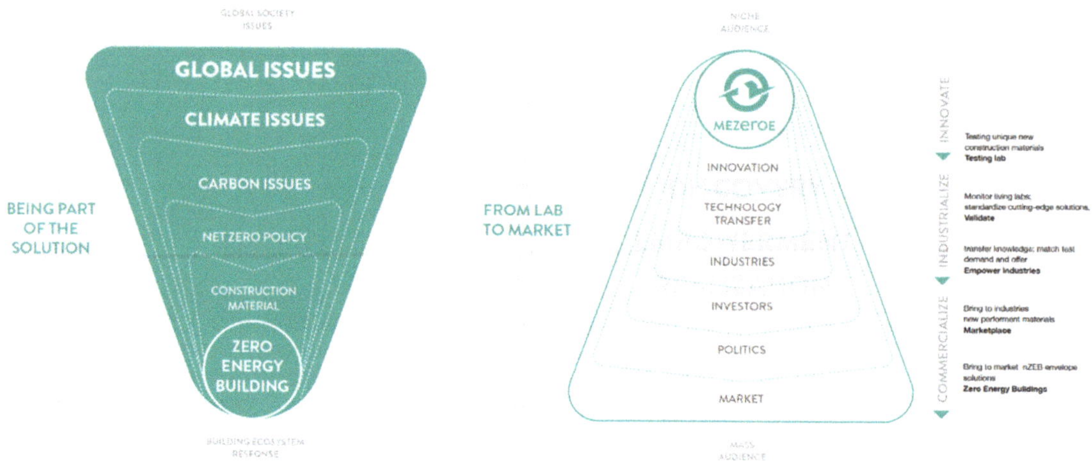

Figure 2. Diagram of the MEZeroE platform (schema by COMPAZ).

5. Open Innovation Services (OIS) in MEZeroE [3,7]

One of the key activities in the field of open innovation services of the OIS1 team is the creation of standard procedures for the certification and CE marking of construction products for modern construction products supplied by industrial partners. In addition, OIS1 is intended to create a roadmap for product certification and CE marking for groups of innovative products used in nZEBs and a general framework to support the creation of new roadmaps for specific products outside the consortium.

The innovation service is primarily addressed to manufacturers, but also in relation to ISO 17025, it aims to support laboratories through the standardization of performance evaluation. The service will enable the development of certification systems to be achieved in the field of technical guidelines, assessment and support in the certification of innovative construction products. The service is based on the following:

- Identifying the best methods for product characterization.
- Guiding product design by focusing on regulation.
- Defining lists of necessary tests for product certification.

These three pillars make a close link between OIS1 and construction product market possible.

The final goal of the project is to develop information paths in the field of the CE marking of various groups of construction products used in the building envelope. Tracks will include procedures in line with Regulation 305/2011 (CPR) [4] or its amendment, if approved before project completion [8], and ways to quickly standardize and place innovative products on the market using ETA. The tracks will include the following:

- Technical guidance for manufacturers and laboratories.
- Procedures for evaluating innovative products and (if provided for in CPR) possible product certification procedures.
- The service aims to provide support in the field of basic data to verify test results and assess the reliability of experimental methods, both when based on standards and other guidelines. Testing programs in the research and development phase will be verified in terms of the usability of the methods used for the subsequent certification process by including specific groups of stakeholders: certification bodies, technical assessment bodies and standardization bodies.
- Moving from sustainable structures to renewable buildings will be facilitated by providing ready-to-use approaches:
- Assessing the economic case for a circular economy.

- Environmental and social audits.
- Life cycle assessment (LCA).
- Life cycle cost analysis (LCC).

OIS1 is managed by specialists with experience in the European assessment of innovative products confirmed by participation in CEN and standardization in the national group, participation in numerous EOTA workshops and bodies, and development of EADs (European assessment documents) for individual innovative products for which there is a lack of coordinated technical references.

6. Summary

The article presents the European platform, developed and tested within the MEZeroE project, supporting the CE marking and placing of innovative construction products on the market.

The platform gives an opportunity to overcome legal barriers related to the lack of standardization procedures enabling the quick placing of innovative construction products on the market to be achieved. The platform is publicly available at https://www.mezeroe.eu/, accessed on 1 December 2022, where manufacturers of modern construction products can obtain additional information and join the MEZeroE community.

Funded by European Union Horizon 2020, the MEZeroE project aims to create a platform that combines science and business to achieve the development of nZEBs. The platform is building a comprehensive knowledge management environment as well as customized trainings. It will serve as an intermediary between stakeholder groups, from the design stage to the end user of nZEB envelope products.

Manufacturers of modern construction products implementing innovative solutions, through the MEZeroE platform, have a chance to speed up the process of marking their products with the CE mark and placing them on the market and thus facilitate the widespread construction of zero-emission and zero-energy buildings.

Funding: This research received no external funding.

Institutional Review Board Statement: Not applicable.

Informed Consent Statement: Not applicable.

Data Availability Statement: Not applicable.

Acknowledgments: The article was written based on the author's knowledge gained through her work on the MEZeroE Measuring Envelope project for products and systems contributing to healthy, nearly zero-energy buildings, which received funding from EU Horizon 2020 research and innovation program under subsidy No. 953157. The publication cost of this paper was covered with funding from Polish National Agency for Academic Exchange (NAWA): "MATBUD'2023—Developing international scientific cooperation in the field of building materials engineering" (BPI/WTP/2021/1/00002, MATBUD'2023).

Conflicts of Interest: The authors declare no conflict of interest.

References

1. Directive 2010/31/eu of the European Parliament and of the Council of 19 May 2010 on the Energy Performance of Buildings. Official Journal of the European Union. Available online: https://eur-lex.europa.eu/legal-content/EN/TXT/?uri=celex%3A32010L0031 (accessed on 1 December 2022).
2. Report from the Commission to the European Parliament and the Council on the Implementation of Regulation (eu) no 305/2011 of the European Parliament and of the Council of 9 March 2011 Laying Down Harmonised Conditions for the Marketing of Construction Products and Repealing Council Directive 89/106/eec. COM/2016/0445. 2016. Available online: https://eur-lex.europa.eu/legal-content/en/ALL/?uri=CELEX:52016DC0445 (accessed on 1 December 2022).
3. MEZEROE. Available online: https://www.mezeroe.eu/ (accessed on 1 December 2022).
4. Regulation (eu) no 305/2011 of the European Parliament and of the Councilof 9 March 2011 Laying Down Harmonised Conditions for the Marketing of Construction Products and repealing Council Directive 89/106/eec. L 88/5. Official Journal of the European Union. 2011. Available online: https://eur-lex.europa.eu/legal-content/EN/TXT/?uri=celex%3A32011R0305 (accessed on 1 December 2022).

5. Council Directive of 21 December 1988 on the Approximation of Laws, Regulations and Administrative Provisions of the Member States Relating to Construction Products. 89/106/EEC. Official Journal of the European Communities. 1988. Available online: https://eur-lex.europa.eu/legal-content/en/TXT/?uri=CELEX%3A31989L0106 (accessed on 1 December 2022).
6. Act of 16 april 2004 on construction products. Dz.U. 2004 nr 92 poz. 881 2004. Available online: https://isap.sejm.gov.pl/isap.nsf/DocDetails.xsp?id=wdu20040920881 (accessed on 1 December 2022).
7. COMPAZ. "Mezeroe Manifesto." 2021. Available online: https://www.mezeroe.eu/project-summary/manifesto (accessed on 1 December 2022).
8. Proposal for a Regulation of the European Parliament and of the Council laying Down Harmonised Conditions for the Marketing of Construction Products, Amending Regulation (eu) 2019/1020 and Repealing Regulation (eu) 305/2011. 2022/0094 (COD). 2022. Available online: https://eur-lex.europa.eu/legal-content/EN/TXT/?uri=celex%3A52022PC0144 (accessed on 1 December 2022).

materials proceedings

MDPI

Proceeding Paper

Influence of the Addition of Ground Granulated Blast Furnace Slag, Fly Silica Ash and Limestone on Selected Properties of Cement Mortars †

Wacław Brachaczek [1], Adam Chleboś [2,*], Magdalena Kupczak [1], Sebastian Spisak [1], Michał Stybak [1] and Katarzyna Żyrek [1]

1 Katedra Budownictwa, Wydział Inżynierii Materiałów, Budownictwa i Środowiska, Akademia Techniczno-Humanistyczna, 43-300 Bielsko-Biała, Poland
2 Sempre Farby; Interdyscyplinarna Szkoła Doktorska ATH, Akademia Techniczno-Humanistyczna, 43-300 Bielsko-Biała, Poland
* Correspondence: achlebos@sempref arby.pl
† Presented at the 10th MATBUD'2023 Scientific-Technical Conference "Building Materials Engineering and Innovative Sustainable Materials", Cracow, Poland, 19–21 April 2023.

Abstract: This publication presents the results of a study of the effect of partial replacement of cement with waste and natural materials: ground blast furnace slag, fly ash and lime meal on the strength properties of mortars compared to a reference mortar without these additives. In subsequent tests, these materials were gradually replaced with up to 10% CEM I cement. Their effects on consistency and compressive and flexural tensile strengths after 28 days of maturation were investigated. The tests were carried out on blocks measuring 40 mm × 40 mm × 160 mm according to the standard. In the range of tests analysed, the results obtained make it feasible to consider the possibility of partially replacing cement with silica fly ash and ground granulated blast furnace slag in cement mortars.

Keywords: ground blast furnace slag; silica fly ash; limestone powder; cement mortar; compressive strength

check for updates

Citation: Brachaczek, W.; Chleboś, A.; Kupczak, M.; Spisak, S.; Stybak, M.; Żyrek, K. Influence of the Addition of Ground Granulated Blast Furnace Slag, Fly Silica Ash and Limestone on Selected Properties of Cement Mortars. *Mater. Proc.* **2023**, *13*, 32. https://doi.org/10.3390/materproc2023013032

Academic Editors: Katarzyna Mróz, Tomasz Tracz, Tomasz Zdeb and Izabela Hager

Published: 16 February 2023

1. Introduction

The deteriorating air conditions and the significant environmental impact of the industry that causes further new restrictions on CO_2 emissions are introduced to achieve climate neutrality [1]. These restrictions strongly affect the cement industry, owing to the industry's significant carbon footprint [1]. To reduce the consumption of natural raw materials and the emissions needed to produce cement, clinker can be partially replaced by additives such as ground blast furnace slag, fly ash or limestone powder, which can reduce CO_2 emissions by up to 8% by 2050 [1]. The use of mineral materials of natural origin allows for higher strength properties, especially after a longer curing period.

One of the materials used to replace part of the cement was limestone powder. This material is obtained by grinding limestone. The main component of limestone used in the cement industry is calcite $CaCO_3$. Because of the high availability of this raw material, ground limestone as a type I additive is widely used in concrete technology departments, in addition to being used as a cement additive. Limestone powder is widely used because it has a beneficial effect on the properties of concrete, mainly owing to its high specific surface area—the additive reduces the distance between particles. The physical filling of the concrete structure makes the finished product stronger and improves its frost resistance and durability. The additive also contributes to the hydration process of the cement. Limestone particles can create additional foci of crystallisation, dynamising the setting time of the cement. The limestone powder in cement mortars disperses the grains of ground Portland clinker, facilitating the access of water to them, which increases their degree of reactivity [2].

In the cement industry, silica fly ash is also used, which is mainly obtained from the combustion of hard coal dust at 1300–1450 °C in pulverised-gravel furnaces [3]. Silica fly ash is characterised by a high content of glassy phase. It is widely used in the cement industry because of its high fineness and pozzolanic activity. The pozzolanic activity of silica fly ash means that on its own it does not exhibit binding properties, but after grinding on contact with moisture, it reacts with calcium hydroxide $Ca(OH)_2$ derived from the hydration of the silicate phases of Portland clinker to form products with binding and hydraulic properties [4,5]. The main components of fly ash are inorganic carbon constituents derived from the thermal decomposition of certain clay minerals, pyrite and calcite. Silica fly ash, according to the standard [6], is a type II concrete additive which means that it must not exceed 5% by weight of cement. It has the structure of fine dust with a spherical shape [4]. The undesirable components of silica fly ash used as a concrete additive include an excessively high sulphur content, free lime particles, unburned carbon and iron compounds: haematite and magnetite. These are formed on the surface of the fly ash grains and block the access of the liquid to its glassy phase, thus adversely affecting the pozzolanic reaction. The high roasting losses occurring especially in the production of raw material in less efficient older-type power plants and the associated increased content of unburned carbon cause an increase in the water content of the ash and lead to a decrease in the water and frost resistance of the finished cement product [5]. According to [5,7,8], the addition of silica fly ash V has a positive effect on reducing the overall porosity of cement mortars and contributes to a reduction in the dominant pore size. A characteristic feature of mortars with the addition of light silica fly ash is the low calcium hydroxide content [5]. The reduced porosity and $Ca(OH)_2$ content makes cement mortars with this additive show increased resistance to chemical corrosion, which is extremely important especially for materials intended for contact with water.

Ground granulated blast furnace slag according to the standard [6] is a type II concrete additive. Blast furnace slag is obtained during the smelting of pig iron in a blast furnace as a by-product. The component raw materials introduced into the furnace are iron ore, coke and fluxes, which lower the melting point of the ore and help to separate the metal from other admixtures contained in the ore. By burning the coke in hot air, the furnace is heated. This process takes place at a temperature of 1400–1600 °C. Molten blast furnace slag is obtained by melting the charge on the surface of the pig iron. The granulation process takes place after separation from the pig iron by rapid cooling with air or water. The formation of a microstructure that ensures an adequate level of activity is obtained by rapid cooling of the liquid slag. The high content of the glassy phase in the slag is responsible for the activity of granulated blast furnace slag [9]. Ground slag is a constituent of Portland multicomponent cements CEM II, metallurgical cements CEM III and multicomponent cements CEM V. It is classified as a material with latent hydraulic properties [9]. Ground granulated blast furnace slag consists mainly of calcium oxide (CaO), magnesium oxide (MgO), silicon dioxide (SiO_2) and aluminium oxide (Al_2O_3) [10]. In terms of radioactivity, according to the classification in the Ordinance of the Council of Ministers of 2 January 2007, Journal of Laws ne 4, item 29, blast furnace slags are classified in Group I, which means that these materials can be used in the production of building materials used in buildings for human and livestock residence [11]. Cements containing ground blast furnace slag are characterised by longer setting times, lower heat of hydration, better workability, significant strength gain over longer hardening periods and higher resistance to chemical aggression [12].

The aim of the study was to assess the feasibility of using ground blast furnace slag, silica fly ash and limestone powder in the production of cement mortars. A comparative analysis of the properties of fresh mortar and after hardening containing these materials was carried out. The design and testing procedures were based on the guidelines in the currently applicable national construction standards. Consistency tests were carried out in accordance with [13] and flexural and compressive strengths in accordance with [14].

2. Experimental Part

The tests were carried out at the Building Materials Laboratory at the University of Bielsko-Biała and at the R&D Laboratory of SEMPRE Farby Company in Bielsko-Biała. The design and testing procedures were based on the guidelines contained in the currently valid national construction standards. Consistency tests were carried out in accordance with the standard [2] and flexural tensile and compressive strength in accordance with [1]. The composition of the mortar on which the tests were carried out is shown in Tables 5–7. In all samples, the granulometric composition of the fine aggregate selected by sieve analysis was kept constant. The mortar binder was Portland cement CEM I 42.5 R Górażdże. The physical and mechanical properties of this cement are shown in Table 1.

Table 1. Physical and mechanical properties of the used cement.

Property	Unit	Value
Specific surface	cm^2/g	3800
Volume constancy	mm	0.2
Beginning of the setting time	min	202
Compressive strength after:		
2 days	MPa	28.8
28 days	MPa	58.2

Because of the study of the impact of the introduced changes mainly in terms of cement mortars, it was decided to use as the main aggregate dried quartz sand with a fraction of 0.0–0.5 mm typical for cement mortars, sourced from sand mine KOTLARNIA S.A. The grain size distribution of the sand is shown in Figure 1.

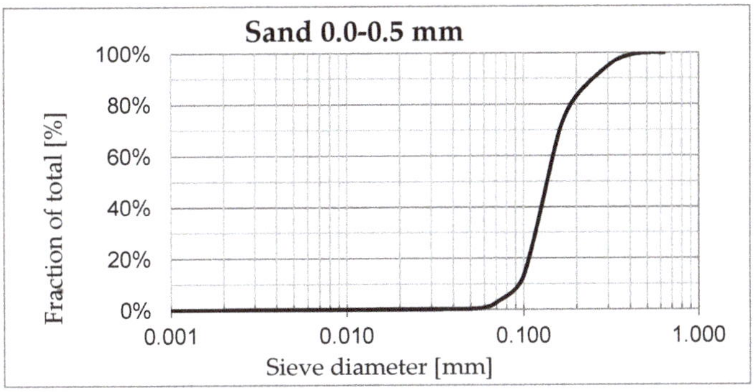

Figure 1. Particle size distribution of aggregates.

A constant quantity of batch water of 270 ml per kg of dry mix was maintained in all samples. Because of the influence of the additives used with different water content, the consistency was adjusted using the superplasticiser Melment F10—a polycondensate of sulphonated melamine and formaldehyde. The superplasticiser was dosed in such a way that the mortar maintained a constant flow measured on the flow table according to the standard [2]. With the use of recycled additives, the mass proportion of cement was reduced so that the sum of the cement and additive masses was constant. The maximum proportion of additives was 10% of the initial cement mass.

2.1. Materials and Methods

Waste and natural raw materials were used for the partial replacement of CEM I 42.5 R cement. The first of these was limestone powder from the production facilities in Bukowa. The specific surface area of the limestone powder according to Blaine method was approximately 5650 cm^2/g. Table 2. shows the chemical composition of this additive. The microstructure of the material is shown in Figure 2a.

Table 2. Chemical composition of the limestone powder used in the tests [% by weight].

SiO_2	Al_2O_3	Fe_2O_3	CaO	MgO	SO_3	Cl	Na_2O	K_2O
1.35	0.62	0.41	54.15	<0.01	0.70	<0.001	<0.01	<0.01

(a) (b) (c)

Figure 2. SEM photo of (**a**) limestone powder, (**b**) silica fly ash, (**c**) silica granulated ground blast furnace slag.

The silica fly ash used in the study was recovered from the combustion of hard coal in Opole. It conforms to the requirements of the standard [7]: owing to its roasting loss, it is classified in class A, whereas owing to its fineness, it is classified in category N. The specific surface area of the ash according to the Blaine method was approximately 4150 cm^2/g. The chemical composition of this material is shown in Table 3. The microstructure of the material is shown in Figure 2b.

Table 3. Chemical composition of the silica fly ash used in the tests [% by weight].

SiO_2	Al_2O_3	Fe_2O_3	CaO	MgO	SO_3	Cl	Na_2O	K_2O
51.2	25.6	6.01	3.58	2.64	0.39	<0.01	1.17	3.13

The granulated ground blast furnace slag used in the study came from the Katowice Steelworks. The specific surface area of the slag according to the Blaine method was approximately 3700 cm^2/g and was similar to the specific surface area of the cement used. Table 4 shows the results of the chemical composition analysis of this additive. The microstructure of the material is shown in Figure 2c.

Table 4. Chemical composition of the granulated ground blast furnace slag used in the study [% by weight].

SiO_2	Al_2O_3	Fe_2O_3	CaO	MgO	SO_3	Cl	Na_2O	K_2O
39.67	8.76	0.67	40.61	7.33	1.62	<0.001	<0.01	0.42

2.2. Formulations of the Test Samples

To compare the properties of cement mortars without additives and mortars containing waste products and lime meal in their composition, four types of mortar were prepared:

1. mortars without any additives,
2. mortars with limestone powder (Table 5),
3. mortars with silica fly ash (Table 6),
4. mortars with ground granulated blast furnace slag (Table 7).

Table 5. Composition of samples modified with limestone powder.

Sample No	Quartz Sand 0.0–0.5 mm [%Mass]	CEM I 42,5 R Portland Cement [%Mass]	Limestone Powder [%Mass]	Superplasticizer [%of Cement Weight]	Consistency with Superplasticizer [mm]
0	75	25.0	0	0	160
M2	75	24.5	0.5	0	156
M4	75	24.0	1.0	0.42	159
M6	75	23.5	1.5	0.43	162
M8	75	23.0	2.0	0.43	162
M10	75	22.5	2.5	0.66	166

Table 6. Composition of samples modified with silica fly ash.

Sample No	Quartz Sand 0.0–0.5 mm [%Mass]	CEM I 42,5 R Portland Cement [%Mass]	Silica Fly Ash [%Mass]	Superplasticizer [%of Cement Weight]	Consistency with Superplasticizer [mm]
0	75	25.0	0	0	160
P2	75	24.5	0.5	0.31	161
P4	75	24.0	1.0	0.52	166
P6	75	23.5	1.5	0.74	163
P8	75	23.0	2.0	0.98	165
P10	75	22.5	2.5	1.11	169

Table 7. Composition of samples modified with ground granulated blast furnace slag.

Sample No	Quartz Sand 0.0–0.5 mm [%Mass]	CEM I 42,5 R Portland Cement [%Mass]	Blast Furnace Grinding Slag [%Mass]	Superplasticizer [%of Cement Weight]	Consistency [mm]
0	75	25.0	0	0	160
Z2	75	24.5	0.5	0	162
Z4	75	24.0	1.0	0	160
Z6	75	23.5	1.5	0	159
Z8	75	23.0	2.0	0	159
Z10	75	22.5	2.5	0	158

In the samples, 2 per cent, 4 per cent, 6 per cent, 8 per cent and 10 per cent by weight of cement were replaced with additives of fly ash, limestone mortar and ground granulated blast furnace slag. In Tables 5–7, the proportion of additives is given in relation to the total weight of the dry ingredients, whereas the proportion of superplasticiser is given in relation to the weight of cement.

2.3. Samples Preparation

The samples were prepared by manually dispensing the dry ingredients and mixing the whole using an automatic mixer at a constant mixing time and speed. The whole dry mix was then poured into water and mixed with water according to the procedure in the standard [2]. The first mixing with water was done using an automatic mixer for 30 s. After this time, the mortar was mixed by hand for about one minute and again in the automatic mixer for one minute. After waiting for a mortar maturation time of 5 min for all samples, the mortar was mixed again in the mixer for 15 s. The consistency of the mortar was measured on a shaking table, 15 strokes per 15 s. Two perpendicular diameters of the cone spread were measured and the average of the two measurements expressed in mm was calculated.

The spread of the reference sample - without additives—was taken as the reference consistency. The amount of superplasticiser in subsequent samples depended on the consistency of the mortar—a spread of up to 10 mm more or less than that of the reference sample was considered acceptable. Once the amount of superplasticiser had been established, the whole mixture was mixed again in an identical manner. Once mixing with water was complete, samples were prepared for compressive and flexural strength measurements in the shape of a 40 mm × 40 mm × 160 mm cuboid according to [1]. After 28 days of seasoning under conditions according to [1], tests were performed to measure the bending tensile strength Rg and compressive strength Rc.

3. Test Results and Discussion

Flexural and compressive strength tests of the seasoned specimens were carried out using a laboratory testing machine according to the guidelines in the standard [1]. Three bending strength measurements and five compression strength measurements were made for each modification. The averaged results obtained are presented graphically in Figures 3–5.

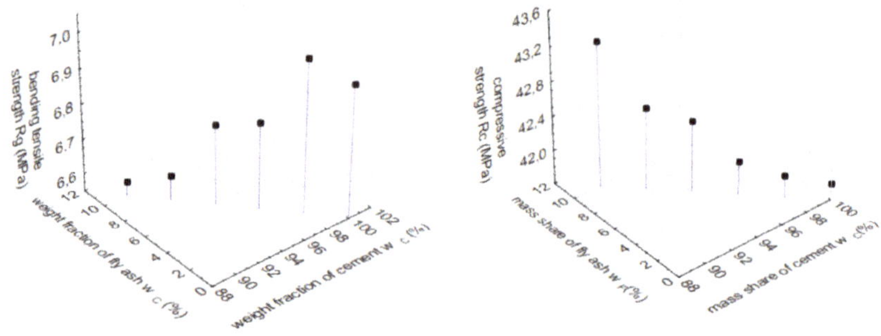

(**a**) (**b**)

Figure 3. (**a**) Dependence of average flexural tensile strength Rg, and (**b**) compressive strength Rc on the simultaneous change of cement and fly ash.

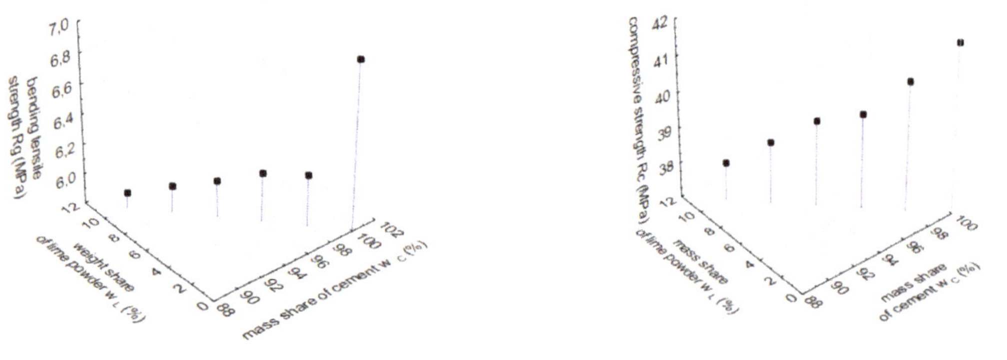

(**a**) (**b**)

Figure 4. (**a**) Dependence of mean flexural tensile strength, Rg, and (**b**) compressive strength, Rc, on the simultaneous change of cement and limestone powder.

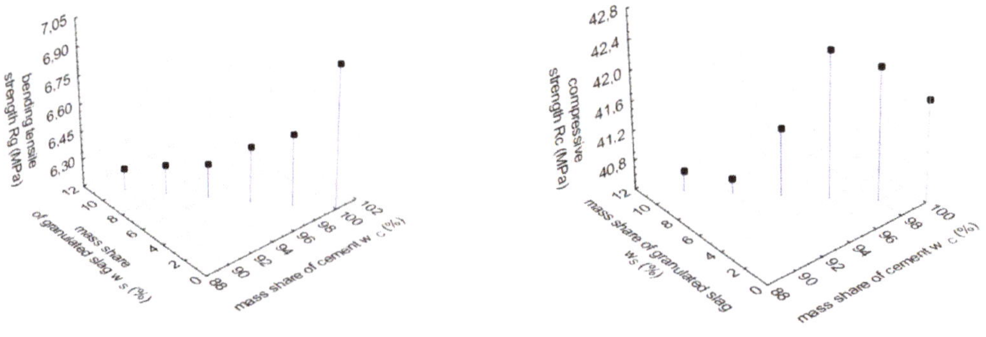

(a) (b)

Figure 5. (**a**) Dependence of mean flexural tensile strength, Rg, and (**b**) compressive strength, Rc on the simultaneous change of cement and granulated blast furnace slag.

Figure 3a shows the effect of the simultaneous change of cement and fly ash on the bending tensile strength, Rg. By analysing the result, in the analysed range of changes, the addition of the silica fly ash while reducing the proportion of cement reduces the strength, Rg. However, this change is small and did not exceed 5% when replacing cement with fly ash at 10%. When analysing the effect of these changes on the strength of Rc, it was found that the gradual substitution of cement with silica fly ash improves the strength of Rc. The observed trend of change can be explained by the properties of silica fly ash. Silica fly ash is classified as a type II additive. It consists mainly of fine spherical vitrified grains obtained from the combustion of coal dust, with pozzolanic properties, containing mainly SiO_2, Al_2O_3. The results obtained are consistent with those of Giergiczny in [5], as well as those of Rudkowska et al. [14].

The main substitutions that the use of fly ash entails are a change in water requirements and workability. In the case studied, the replacement of cement with silica fly ash was associated with the need to use a plasticizer to maintain a constant consistency (Figure 5a).

Figure 4a shows the effect of the simultaneous change of cement and limestone powder on the flexural tensile strength, Rg. By analysing the result, it can be seen that, in the range of changes studied, replacing the cement with limestone powder reduces the strength of Rg. However, this change is small and did not exceed 5% when replacing cement with limestone powder at 10%. A similar trend was observed when analysing the effect of these changes on the strength of Rc. Again, the gradual replacement of cement with limestone decreased the Rc strength. The decrease in Rc strength in this case was proportional to the amount of limestone powder and was approximately 10% when replacing the cement at 10% with limestone powder.

In the case studied, the replacement of cement with limestone powder was associated with the need to use a plasticizer in order to maintain a constant consistency (Figure 6b).

Figure 5a shows the effect of the simultaneous change of cement and granulated blast furnace slag on the bending tensile strength, Rg. By analysing the result, it can be concluded that, in the analysed range of changes, the replacement of cement with granulated blast furnace slag reduces the strength Rg. When replacing cement with 10% blast furnace slag, the change does not exceed 10%. When analysing the effect of replacing cement with blast furnace slag on Rc, it was found to be uneven. At slag amounts up to 6%, the strength increases slightly. However, a 6% substitution appears to be a limiting amount, as further increases in the proportion of slag in the binder composition are associated with a decrease in Rc strength. At a slag proportion of 10%, this decrease does not exceed 5%. It can therefore be assumed that an increase in the proportion of this additive in the binder slightly alters the strength properties of the mortars. Granulated blast furnace slag is a type II concrete additive alongside silica fly ash and silica dust. Granulated blast furnace slag is classified as a material with latent hydraulic properties and is therefore a basic component of multi-component cements. The similar compressive strength values of mortars containing up to 10% slag in the binder composition are also related to its specific surface area. In the case analysed, it was 3700 cm^2/g and was similar to the specific surface area of cement. The results obtained are consistent with those obtained in [12].

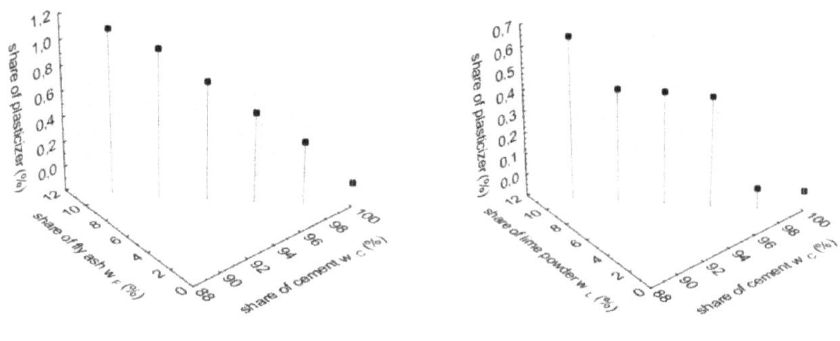

(a) (b)

Figure 6. The proportion of plasticizer as a percentage of cement to maintain consistency in mortars modified with (**a**) silica fly ash and (**b**) limestone powder.

In the analyzed case, in order to maintain a constant consistency, replacing cement with ground blast furnace slag did not require the use of a plasticizer in order to maintain a constant consistency. This was caused by the similar value of specific surfaces of cement and slag.

4. Summary

Silica fly ash from the Opole power plant is a good alternative to cement. It can be used as an additive in cement mortars. Its share of 10% in the binder had a negligible effect on the compressive strength. It should be noted that a slight increase was even observed in compressive strength. It should be noted, however, that in the case studied, little effect of this additive on the change in consistency was observed.

The blast furnace slag used in this study, which comes from the Katowice steelworks, can be used as an additive in cement mortar. The 6% proportion of this additive to the binder does not significantly alter the compressive strength. An additional advantage is that the use of this additive does not change the consistency of the mortar.

Of the additives analysed, the addition of limestone powder had the greatest effect on compressive strength. Each cement replacement with this additive was associated with a decrease in compressive strength. At a limestone powder proportion of ten per cent, the decrease in Rc was 10 per cent.

The results obtained make it possible to realistically consider the possibility of using silica fly ash and ground granulated blast furnace slag in cement mortar. Such action is in line with the sustainable development strategy. It makes it possible to recover valuable post-industrial by-products and reduce the use of pure, highly emissive CEM I cement. The implementation of similar solutions on a global scale would reduce dust and greenhouse gas emissions.

Author Contributions: Conceptualization, W.B. and A.C.; methodology W.B.; software W.B.; validation, S.S. and K.Ż.; formal analysis, M.K. and M.S.; investigation W.B. and A.C.; resources M.S. and S.S.; data curation K.Ż. and M.K.; writing—original draft preparation A.C.; writing—review and editing W.B. and A.C.; visualization W.B. and K.Ż.; supervision A.C.; project administration A.C.; funding acquisition. All authors have read and agreed to the published version of the manuscript.

Funding: This research received no external funding.

Institutional Review Board Statement: The study did not require ethical approval.

Informed Consent Statement: The study did not involve humans.

Data Availability Statement: Not applicable.

Acknowledgments: Publication cost of this paper was covered with funds from the Polish National Agency for Academic Exchange (NAWA): "MATBUD'2023—Developing international scientific cooperation in the field of building materials engineering" BPI/WTP/2021/1/00002, MATBUD'2023.

Conflicts of Interest: The authors declare no conflict of interest.

References

1. Środa, B. Przemysł cementowy na drodze do Zielonego Ładu. *Bud. Technol. Archit.* **2020**, *7*, s.78.
2. Wang, D.; Shi, C.; Farzadnia, N.; Shi, Z.; Jia, H. A review on effects of limestone powder on the properties of concrete. *Constr. Build. Mater.* **2018**, *192*, 153–166. [CrossRef]
3. Giergiczny, Z. Właściwości popiołu lotnego a trwałość betonu. *Bud. Technol. Archit.* **2007**, *7*, 44–48.
4. Giergiczny, Z. *Popiół Lotny w Składzie Cementu i Betonu*; Wydawnictwo Politechniki Śląskiej: Gliwice, Poland, 2013.
5. Giergiczny, Z. *Rola Popiołów Lotnych Wapniowych i Krzemionkowych w Kształtowaniu Właściwości Współczesnych Spoiw Budowlanych i Tworzyw Cementowych*; Wydawnictwo PK: Kraków, Poland, 2006.
6. *PN-EN 206+A2*; Beton: Wymagania, Właściwości Użytkowe, Produkcja i Zgodność. Polski Komitet Normalizacyjny: Warszawa, Poland, 2021.
7. *PN-EN 450-1*; 2012 Popiół Lotny do Betonu. Część 1: Definicje, Specyfikacje i Kryteria Zgodności. Polski Komitet Normalizacyjny: Warszawa, Poland, 2014.
8. Giergiczny, Z.; Ostrowski, M.; Baran, T. Wpływ popiołów lotnych krzemionkowych kategorii S na wybrane właściwości kompozytów cementowych. In Proceedings of the Międzynarodowa Konferencja Popioły z Energetyki, Zakopane, Poland, 21 October 2016; pp. 19–21.
9. Granulowany Żużel Wielkopiecowy Składnikiem Cementu i Spoiw Drogowych. Available online: https://www.gorazdze.pl/pl/granulowany-zuzel-wielkopiecowy-skladnikiem-cementu-i-spoiw-drogowych (accessed on 14 February 2023).
10. Ahmad, J.; Kontoleon, K.J.; Majdi, A.; Naqash, M.T.; Deifalla, A.F.; Ben Kahla, N.; Haytham, F.I.; Qaidi, S.M. A comprehensive review on the ground granulated blast furnace slag (GGBS) in concrete production. *Sustainability* **2022**, *14*, 8783. [CrossRef]
11. Giergiczny, Z.; Góralna, K. Mielony granulowany żużel wielkopiecowy-dodatek do betonu typu II. *Bud. Technol. Archit.* **2008**, *1*, 56–59.
12. Saafan, M.A.; Zeinab, A.E.; El Lakany, D. Microstructure and durability of ground granulated blast furnace slag cement Mortars. *Iran. J. Sci. Technol. Trans. Civ. Eng.* **2021**, *45*, 1457–1465. [CrossRef]
13. *PN EN 1015-3:2000*; Metody Badań Zapraw do Murów—Określenie Konsystencji Świeżej Zaprawy (za Pomocą Stolika Rozpływu). Polski Komitet Normalizacyjny: Warszawa, Poland, 2000.
14. *PN EN 196-1:2016-07*; Metody Badania Cementu. Część 1: Oznaczanie Wytrzymałości. Polski Komitet Normalizacyjny: Bielsko-Biała, Poland, 2018.

materials proceedings

MDPI

Proceeding Paper

Effect of Fungi Removal Method on the Mechanical Properties of Polymer Composites Reinforced with Oat and Millet Husks †

Maria Goron * and **Ewa Sudoł**

Instytut Techniki Budowlanej, 00-611 Warsaw, Poland

* Correspondence: m.goron@itb.pl

† Presented at the 10th MATBUD'2023 Scientific-Technical Conference "Building Materials Engineering and Innovative Sustainable Materials", Cracow, Poland, 19–21 April 2023.

Abstract: Polymer composites reinforced with crop husks show susceptibility to fungi of the *Agaricomycotina* subdivision, particularly *Coniophora puteana*. The material's resistance to the fungi is evaluated by exposing specimens to the fungus for 16 weeks, and then determining the mass loss and changes in mechanical properties. An important pre-testing step is cleaning the samples of mycelium. The study compares the effects of the manual cleaning method with a soft brush and with water under pressure. The aim of the study is to select a method that is non-destructive to the material. The results will be evaluated by mass loss, bending strength and modulus of elasticity.

Keywords: polymer composites; husks; oat; millet; fungi; mycelium removal

Citation: Goron, M.; Sudoł, E. Effect of Fungi Removal Method on the Mechanical Properties of Polymer Composites Reinforced with Oat and Millet Husks. *Mater. Proc.* **2023**, *13*, 33. https://doi.org/10.3390/materproc2023013033

Academic Editors: Katarzyna Mróz, Tomasz Tracz, Tomasz Zdeb and Izabela Hager

Published: 16 February 2023

1. Introduction

Recently, in the construction industry, great attention has been paid to various ecological aspects. As a result, materials that are easily biodegradable and durable are in demand [1,2]. In some applications such as terraces and façades, wood has been replaced by composites. Composite materials are made of polymers (PP, PVC or HDPE) acting as matrices and lignocellulosic particles as fillers. Natural fillers in the form of flour, fibers or shavings, as well as fibers from crop husks, for example, rice or oat [3–5], are used in composite materials. Natural fibers have a low density, are biodegradable, widely available and inexpensive. Therefore, they are viewed as more attractive materials than commonly used synthetic fibers. However, they also have various disadvantages, such as low thermal stability and susceptibility to moisture absorption or microorganisms [6–8]. It was thought that a thermoplastic material completely encapsulates the natural fiber component of the composite, protecting it from moisture and fungal decay. However, some publications show that fully encapsulating natural fibers with a polymer matrix is not possible [9,10].

Some publications prove that biodeterioration has a negative impact on the mechanical and physical properties of natural fiber composites [7,9,11–13]. This is determined by weight loss and changes in material strength [1,3,12,14]. Composite materials used in outdoor conditions are exposed to many factors, such as UV radiation, freezing, leaching, temperature changes and biological agents, so it is difficult to determine which factor has a decisive effect on degradation, since they all act simultaneously [3,15].

One of the most important steps in preparing composite specimens for testing after exposure to fungi is mycelium removal. Following exposure to fungi, specimens usually show significant mycelium overgrowth, which can interfere with the testing equipment and impair measurements. In addition, mycelium can cause damage to human health, so it should be removed before further tests. Often, this step is skipped in method descriptions in composite-related publications [6,13,16,17]. Some authors shortly describe the cleaning process. Composites can be cleaned with a soft sponge [8] or by sanding [18].

As composite materials are now widely used, this study aims to explore the resistance to fungal decay of the *Agaricomycotina* subdivision [19] and the effect of the method of mycelium removal after microbiological exposure on their mechanical properties. The subject of the study is composites reinforced with natural fibers, including oat and millet husks.

2. Materials

2.1. Composite Materials

The subject of the study is two types of composite construction profiles made of the following materials:

1. A composite with a PVC matrix and pulverized oat husk filler;
2. A composite with a PVC matrix and pulverized millet husk filler.

The profiles are designed for outdoor flooring, produced from the mixtures described in Table 1.

Table 1. Composition of composites used in the production of profiles.

Marking	Matrix	Mineral Filler (Type/Share [1])	Natural Filler (Type/Share [1])	Other [2]
Oat	PVC	$CaCO_3$/ 50 phr	Pulverized oat husks/ 30 phr	Physical modifiers, stabilizers; polyethylene wax
Millet			Pulverized millet husks/ 30 phr	

[1] The proportion of mixture components is given in parts per 100 phr (parts per hundred rubber). [2] Detailed data are not available due to company secrecy.

The profiles were manufactured by extrusion under industrial conditions. The usable surfaces were mechanically treated by brushing, which is standard for construction profiles.

2.2. Testing Specimens

Test specimens (80 × 10 × 5 mm) were cut from the central part of the profiles. The following series of specimens were used in the tests:

1. Reference specimens;
2. Test specimens (exposed to water and fungi);
3. Control specimens for the determination of dry mass.

Specimens were cut from each composite in the following quantities:

- Ten specimens for series 1 and 3;
- Forty specimens for series 2, including two methods of cleaning (by brushing and with water under pressure) and two methods of testing (loss in mass and flexural strength), which required 10 specimens for each version.

3. Methods

The resistance of the composites to a fungus of the *Agaricomycotina* subdivision was assessed using two methods. After exposure to the *Coniophora puteana* fungus and the application of two methods of mycelium removal, the mass loss and changes in flexural strength of the specimens were determined.

3.1. Mass Loss of Composites after Exposure to Coniophora Puteana

3.1.1. Pre-Conditioning by Soaking in Water

Series 2 specimens cut from both types of composite profiles were subjected to the aging cycle by soaking. The specimens were soaked in distilled water for 14 days. The water was changed 9 times during the cycle. The specimens were later dried in an environment-controlled room with a temperature of 20 °C and a humidity of 65%.

3.1.2. Determination of Initial Dry Mass

Series 3 specimens were also soaked in distilled water for 14 days, which was changed 9 times. Ten specimens were dried in an oven at 103 °C to constant weight. A moisture coefficient of each material was determined using these measurements. The initial dry mass for the remaining series was calculated by multiplying the moisture coefficient by their weight after soaking in water for 14 days.

3.1.3. Preparing Fungi for the Test

A medium of 40 g of malt extract, 35 g of agar and 1000 mL of distilled water was prepared in Kolle flasks. The medium was sterilized in an autoclave at 121 °C and 1 atm pressure. After solidifying, the medium was inoculated with hyphae of the fungus *Coniophora puteana*. The flasks were stored in an incubating chamber at 22 °C and 70 % humidity until the surface of the medium was completely covered by the mycelium.

3.1.4. Assessment of Fungal Activity

Using previously prepared test cultures of Coniophora puteana, the fungal activity of the specimens was verified. Specimens of Scots pine wood (15 × 25 × 50 mm) were placed in Kolle flasks on medium covered by the mycelium. The flasks were stored in an incubating chamber at 22 °C and 70% humidity for 16 weeks. After removal of the mycelium, the wood specimens were weighed and the mass loss was determined. It was confirmed that they achieved the required loss in mass over 20%.

3.1.5. Mass Loss Test

The composite specimens of series 2 were placed on the mycelium with the usable surface in direct contact with it. The flasks with specimens were transferred to the incubating chamber for 4 months. After the incubation period, the specimens were removed from the flasks. Half of the specimens underwent manual mycelium removal with a soft brush. The other half underwent mycelium removal using pressurized water and were dried for 7 days at 40 °C. Furthermore, the specimens were dried at 103 °C to determine the final dry mass after their exposure to fungus. Loss in mass was calculated according to (1). The result is expressed in %.

$$\Delta m_d = (m_i - m_c)/m_c \cdot 100, \tag{1}$$

where m_i—initial dry mass of the specimen, in grams; m_c—final dry mass of the specimen, in grams.

3.2. Bending Strength and Modulus of Elasticity after Exposure to Coniophora Puteana

The preparation and exposure to fungus of the specimens used to determine the flexural strength of the composites after fungal decay were carried out in the same way for the mass loss specimens (3.1.) After the incubation period of 4 months, the specimens were removed from the flasks, cleaned of mycelium by two methods and dried in an environment-controlled room at 20 °C and 65% humidity.

The specimens were tested for flexural strength and modulus elasticity. The specimens were placed with the usable side down to apply tension to the fungus-exposed surface. The load was applied at a rate of 2 mm/min until destruction.

The results obtained for the specimens exposed to fungi were related to the results obtained for the specimens in the initial state. The initial state results were obtained from previous tests for the same materials [20]. This made it possible to determine, according to (2), the effect of this exposure on the tested materials. The result is expressed in %.

$$\Delta\sigma_d = (\sigma_d - \sigma_i)/\sigma_i \cdot 100, \Delta E_d = (E_d - E_i)/E_i \cdot 100, \tag{2}$$

where σ_d, E_d—the flexural strength and modulus of elasticity after fungi exposure, in MPa; σ_i, E_i—the initial flexural strength and modulus of elasticity, in MPa.

4. Results and Discussion

Figures 1a and 2a illustrate the specimens of composites after exposure to *Coniophora puteana* over 16 weeks. The oat husk-reinforced composites demonstrated higher resistance to fungal overgrowth. The millet husk-composites showed significant susceptibility to *Coniophora puteana*. The mycelium was removed manually from both types of specimens using a soft brush (Figures 1b and 2b). The specimens, which were cleaned by water, looked the same, but had to be dried for 7 days at 40 °C to remove excess moisture.

(a)

(b)

Figure 1. Oat husk-reinforced composite specimens after (**a**) exposure to *Coniophora puteana*; (**b**) mycelium removal.

(a)

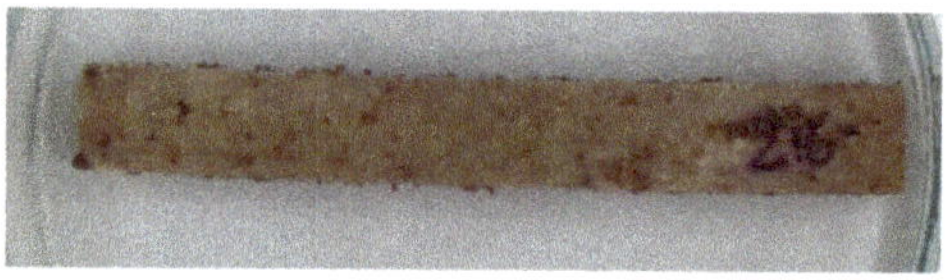

(b)

Figure 2. Millet husk-reinforced composite specimens after (**a**) exposure to *Coniophora puteana*; (**b**) mycelium removal.

The mass loss test results of the composites with oat filler obtained after 14 days of leaching in water, followed by 4 months of exposure to fungi at 22 °C and at 70% humidity, are identical for both methods of mycelium removal (Figure 3). The mass loss for this material is 0.3% compared to the initial state, demonstrating stable results regardless of the cleaning method. The composites reinforced with millet husks show a higher loss in mass and a larger difference between the two methods of mycelium removal (0.8% following cleaning with a brush and 2.3% following cleaning with water). The microstructure of the millet filler composite is irregular with large pores [20] that may cause fragments of filler to be pushed out by water under pressure. Soft brushing protects the specimens from this process, but also may affect the mass loss.

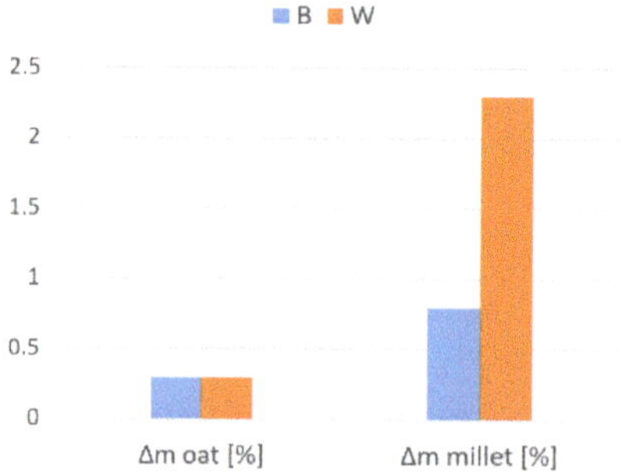

Figure 3. Mass loss test results of oat and millet husk-reinforced composites after exposure to Coniophora puteana and manual mycelium removal with a soft brush (B) and with water under pressure (W). The results are expressed in %.

Bending Strength and Modulus of Elasticity after Exposure to Coniophora Puteana

The test results of the specimens' mechanical properties show that in the initial state, the composites reinforced with oat and millet husk had a bending strength of 44 MPa and 31 MPa, respectively, while the specimens' modulus of elasticity was 3790 MPa and 2870 MPa, respectively (Figure 4). It can be noticed that the properties of the composite with oat husks are more useful than those of the composite with millet husks in the initial state. The reason for this may be the influence of the shape and wetting degree of the filler. Oat fillers have an elongated shape, which is advantageous for their mechanical properties. In general, millet fillers have irregular shapes [20]. In addition, fillers naturally have different moisture absorption capacities.

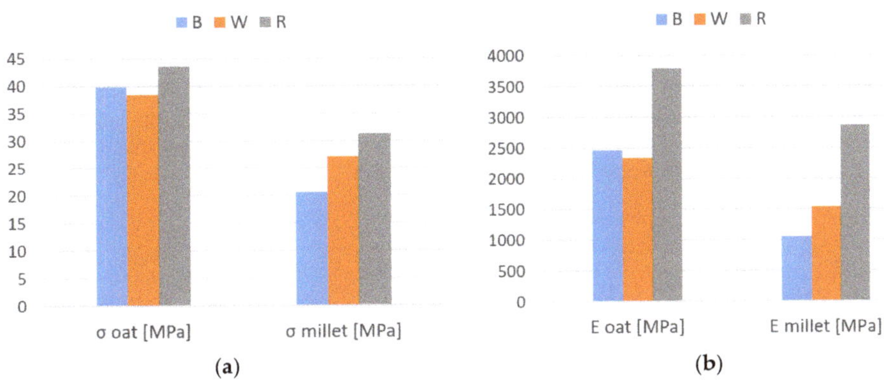

Figure 4. Test results of (**a**) bending strength and (**b**) modulus of elasticity of oat and millet husk-reinforced composites after exposure to *Coniophora puteana* and manual mycelium removal with a soft brush (B) and with water under pressure (W). Test results of reference specimens in initial state (R) have been also presented. The results are expressed in MPa.

The test results of the composites obtained after 14 days of soaking in water, followed by 4 months of exposure to fungi at 22 °C and at 70% humidity, show that these conditions reduce their flexural strength and modulus of elasticity, compared to the reference material. This was also reported in other papers [8]. The method of cleaning had a small effect on the oat husk composite. After exposure to fungus and cleaning with a brush, the oat husk composite had a bending strength of 40 MPa and a modulus of elasticity of 2460 MPa, while after water cleaning, the bending strength of this composite was 38 MPa and a modulus of elasticity was 2340 MPa (Figure 4).

On the other hand, the effect of the mycelium removal method on the composite with millet filler was significant in comparison to the oat filler composite. After exposure to *Coniophora puteana* and cleaning with a brush, the composite had a bending strength of 21 MPa and a modulus of elasticity of 1050 MPa, while the composite cleaned with water had a bending strength of 27 MPa and a modulus of elasticity of 1530 MPa (Figure 4). In relation to the mass loss test of the composite with millet husks, the reason for the difference may be the lack of fillers in the pores, as well as the higher efficiency of the mycelium removal method with water.

5. Conclusions

Based on the results of the oat husk-reinforced composite tests, the following conclusions can be drawn:

- The mass loss regardless of the mycelium removal method was 0.3% compared to the initial state.
- Brushed specimens showed slightly higher flexural strength and a higher modulus of elasticity than the specimens cleaned with water under pressure.

- The method of mycelium removal does not significantly affect the physical and mechanical properties of the oat filler composite.

The conclusions based on the results of the millet husk-reinforced composite are as follows:

- The mass loss of the specimens varies depending on the method of mycelium removal. Cleaning with water under pressure resulted in the highest mass loss at 2.3%, while the brush-cleaned specimens had a mass loss of 0.8%.
- The results of the flexural strength and modulus of elasticity tests also confirm the loss in mass of the specimens. The millet husk-composite after water cleaning showed higher results, including after brushing, compared to the oat husk composite.
- The differences in the results with different cleaning methods may be caused by the microstructure of millet husk fillers.

It was concluded that the effects of mycelium removal methods depend on the type of filler. It is still difficult to determine which was the most effective method of cleaning the specimens due to the constrained conditions and few composite fillers used. Further research including extended parameters that affect the mechanical and physical properties of the natural filler-reinforced composites is planned.

Author Contributions: Conceptualisation, M.G.; methodology, M.G.; formal analysis, E.S.; investigation, M.G.; writing—original draft preparation, M.G.; writing—review and editing, E.S.; visualization, M.G. All authors have read and agreed to the published version of the manuscript.

Funding: This research was funded by the Ministry of Education and Science as part of the project LN-002.

Institutional Review Board Statement: Not applicable.

Informed Consent Statement: Not applicable.

Data Availability Statement: The data presented in this study are available on request from the corresponding author.

Acknowledgments: The publication cost of this paper was covered with funds provided by the Polish National Agency for Academic Exchange (NAWA) ("MATBUD'2023—Developing international scientific cooperation in the field of building materials engineering" BPI/WTP/2021/1/00002, MATBUD'2023).

Conflicts of Interest: The authors declare no conflict of interest. The funders had no role in the study's design, in the collection, analyses, or interpretation of data, in the writing of the manuscript, or in the decision to publish the results.

References

1. Chmielnicki, B.; Jurczyk, S. Kompozyty WPC jako alternatywa dla wytworów z drewna. *Przetwórstwo Tworzyw* **2013**, *19*, 477–484.
2. Kaczmar, J.W.; Pach, J.; Kozłowski, R. Wykorzystanie włókien naturalnych jako napełniaczy kompozytów polimerowych. *Polimery* **2006**, *51*, 722–726. [CrossRef]
3. Faruk, O.; Błędzki, A.K.; Fink, H.P.; Sain, M. Biocomposites reinforced with natural fibers: 2000–2010. *Prog. Polym. Sci.* **2012**, *37*, 1552–1596. [CrossRef]
4. Penczek, S.; Pretula, J.; Lewiński, P. Polimery z odnawialnych surowców, polimery biodegradowalne. *Polimery* **2013**, *58*, 835–846. [CrossRef]
5. Nijjar, S.; Sudhakara, P.; Sharma, S.; Saini, S. Overview on the latest trend and development on mechanical, tribological & microstructural properties of natural fibre polymer composites. *Mater. Today Proc.* **2022**, *63*, 663–672.
6. Naumann, A.; Seefeldt, H.; Stephan, I.; Braun, U.; Noll, M. Material resistance of flame retarded wood-plastic composites against fire and fungal decay. *Polym. Degrad. Stab.* **2012**, *97*, 1189–1196. [CrossRef]
7. Tascioglu, C.; Goodell, B.; Lopez-Anido, R.; Peterson, M.; Halteman, W.; Jellison, J. Monitoring fungal degradation of E-glass/phenolic fiber reinforced polymer (FRP) composites used in wood reinforcement. *Int. Biodeterior. Biodegrad.* **2003**, *51*, 157–165. [CrossRef]
8. Ashori, A.; Behzad, H.M.; Tatmain, A. Effects of chemical preservative treatments on durability of wood flour/HDPE composites. *Composites Part B* **2013**, *47*, 308–313. [CrossRef]

9. Kamdem, D.P.; Jiang, H.; Cui, W.; Freed, J.; Matuana, L.M. Properties of wood plastic composites made of recycled HDPE and wood flour from CCA-treated wood removed from service. *Composites Part A* **2004**, *35*, 347–355. [CrossRef]
10. Yildiz, U.C.; Yildiz, S.; Gezer, E.D. Mechanical properties and decay resistance of wood-polymer composites prepared from fast growing species in Turkey. *Bioresour. Technol.* **2005**, *96*, 1003–1011. [CrossRef] [PubMed]
11. Schirp, A.; Ibach, R.; Pendleton, D.; Wolcott, M. Biological Degradation of Wood-Plastic Composites (WPC) and Strategies for Improving the Resistance of WPC against Biological Decay. In *Proceedings of the Development of Wood Preservative Systems*; American Chemical Society: San Diego, CA, USA, 2008.
12. Friedrich, D.; Luible, A. Investigations on ageing of wood-plastic composites for outdoor applications: A meta-analysis using empiric data derived from diverse weatherimg trials. *Constr. Build. Mater.* **2016**, *124*, 1142–1152. [CrossRef]
13. Barton-Pudlik, J.; Czaja, K.; Grzymek, M.; Lipok, J. Evaluation of wood-polyethylene composites biodegradability caused by filamentous fungi. *Int. Biodeterior. Biodegrad.* **2017**, *118*, 10–18. [CrossRef]
14. Coffta, G.; Borysiak, S.; Doczekalska, B.; Garbrczyk, J. Resistance of polypropylene-wood composites to fungi. *Polimery* **2006**, *51*, 276–279. [CrossRef]
15. Falk, R.; Lundin, T.; Felton, C. The Effects of Weathering on Wood-Thermoplastic Composites Intended for Outdoor Applications. In Proceedings of the 2nd Annual Conference on Durability and Disaster Mitigation in Wood-Frame Housing, Madison, WI, USA, 6–8 November 2000.
16. Altuntas, E.; Yilmaz, E.; Salan, T.; Alma, M.H. Biodegradation properties of wood-plastic composites containing high content of lignocellulosic filler and zinc borate exposed to two different brown-rot fungi. *BioResources* **2017**, *12*, 7161–7177.
17. Xu, K.; Feng, J.; Zhong, T.; Zheng, Z.; Chen, T. Effects of volatile chemical components of wood species on mould growth susceptibility and termite attack resistance of wood plastic composites. *Int. Biodeterior. Biodegrad.* **2015**, *100*, 106–115. [CrossRef]
18. Hosseinaei, O.; Wang, S.; Taylor, A.M.; Kim, J.W. Effect of hemicellulose extraction on water absorption and mold susceptibility of wood-plastic composites. *Int. Biodeterior. Biodegrad.* **2012**, *71*, 29–35. [CrossRef]
19. Kirk, P.M.; Cannon, P.F.; Minter, D.W.; Stalpers, J.A. *Dictionary of the Fungi*, 10th ed.; ResearchGate: Wallingford, UK, 2008.
20. Sudoł, E.; Kozikowska, E.; Goron, M. Odporność profili budowlanych z kompozytów polimerowych zbrojonych łuskami zbóż na grzyby domowe. *Mater. Budowlane* **2022**, *601*, 70–75. [CrossRef]

materials proceedings

MDPI

Proceeding Paper

Investigation of the Electrical Properties of Graphene-Reinforced Geopolymer Composites †

R. S. Krishna [1,2,*], Suman Saha [3], Kinga Korniejenko [4], Tanvir S. Qureshi [5] and Syed Mohammed Mustakim [6]

1 Department of Mechanical Engineering, Birla Institute of Technology and Science Pilani Hyderabad Campus, Secunderabad 500078, Telangana, India
2 Department of Mechanical Engineering, Veer Surendra Sai University of Technology, Burla 768018, Odisha, India
3 Department of Civil Engineering, National Institute of Technology Durgapur, Durgapur 713209, West Bengal, India
4 Faculty of Materials Engineering and Physics, Cracow University of Technology, Jana Pawła II 37, 31-864 Cracow, Poland
5 Department of Engineering Design and Mathematics, The University of the West of England, Bristol BS16 1QY, UK
6 Department of Environment & Sustainability, CSIR—Institute of Minerals and Materials Technology, Bhubaneswar 751013, Odisha, India
* Correspondence: 2021proj051@hyderabad.bits-pilani.ac.in or krishnaskull@gmail.com; Tel.: +91-7008601754
† Presented at the 10th MATBUD'2023 Scientific-Technical Conference "Building Materials Engineering and Innovative Sustainable Materials", Cracow, Poland, 19–21 April 2023.

Abstract: Geopolymer composites provide an environmentally friendly alternative to cement-based composites in the construction industry. Due to their distinctive material composition, geopolymers also exhibit electrically conductive properties, which permit their application as a functional material. The current work aims to study the distinctive electrical properties of fly-ash-based geopolymer composites. Varying dosages of graphene oxide (i.e., 0, 0.1, 0.2, 0.3, 0.4% (by wt. of binder)) were introduced into the geopolymer matrix to enhance electrical conductivity. While GO (graphene oxide) is typically less conductive, the interaction of GO sheets with the alkaline solution during geopolymerisation reduced the functional groups and produced cross-linked rGO (reduced graphene oxide) sheets with increased mechanical and electrical conductivity properties. Solid-state impedance spectroscopy was used to characterize the electrical properties of geopolymer composites in terms of several parameters, such as impedance, electrical conductivity and dielectric properties, within the frequency ranging from 10^1 to 10^5 Hz. The relationship between the electrical properties and graphene oxide reinforcement can effectively establish geopolymer composite development as smart materials with desirable functionality. The results suggest an effective enhancement in electrical conductivity of up to 7.72×10^{-13} $\Omega{\cdot}mm^{-1}$ and the dielectric response performance of graphene-reinforced fly-ash-based geopolymer composites.

Keywords: geopolymer composite; graphene; electrical conductivity; electrical property; microstructure

check for updates

Citation: Krishna, R.S.; Saha, S.; Korniejenko, K.; Qureshi, T.S.; Mustakim, S.M. Investigation of the Electrical Properties of Graphene-Reinforced Geopolymer Composites. *Mater. Proc.* **2023**, *13*, 34. https://doi.org/10.3390/materproc2023013034

Academic Editors: Katarzyna Mróz, Tomasz Tracz, Tomasz Zdeb and Izabela Hager

Published: 15 February 2023

1. Introduction

Geopolymers—at the micro-level—are primarily composed of amorphous materials of the long matrix, cross-linked polymer chains of tetrahedral AlO_4 and SiO_4 units [1,2]. Geopolymers are usually dielectric materials due to the silica content and alkali metal ions, which could work as ionic conductors via an applied electric field [3,4]. Although pure geopolymers are electrically conductive due to the availability of water molecules and hydroxide in their composition, the open pore networks in the matrix produce conductivity [5,6]. Therefore, the introduction of filler materials becomes predominately necessary to improve electrical conductivity [7,8]. The most common additives preferred for the fabrication of conductive geopolymer composites include carbon and metallic fillers, such as carbon fibres [5,9], nanotubes [10,11], graphite [8], graphene derivatives [7,8], steel fibres, etc.

Several research projects have been carried out to study and comprehend the influence of conductive fillers in the geopolymer matrix system. In one such examination, Payakaniti et al. optimized the inclusion of carbon fibre and stated that 0.5 wt.% provided superior electrical conductivity as well as mechanical properties [12,13]. Similar examinations were also carried out with carbon nanotubes [11,14]. The incorporation of 1 wt.% carbon nanotube in the geopolymer matrix enhanced the electrical conductivity performance by almost three times. Carbon nanotubes also reduce the electric resistance and impedance of associated composites [11,15]. Thus, it can be argued that the physical characteristics and the synthesis mechanisms of the conductive fillers play a vital role in the enhancement of geopolymer properties.

Research on the electrical properties of geopolymer composites is still a novel, diverse and challenging area, resulting in a lack of a universal approach and indicating the high complexity of the underlying problem [16]. Presently, one of the most promising multi-functional composites seems to be geopolymers incorporating graphene derivatives. The current study is therefore focused on the investigation of different effects of GO dosages (0, 0.1, 0.2, 0.3 and 0.4 wt.%) on the electrical properties and the dielectric response of fly-ash-based GRGC (graphene-reinforced geopolymer composite) specimens.

2. Experimental Procedure

2.1. Materials

Fly ash was procured from NTPC, Kaniha, in Odisha. The chemical composition of the FA (fly ash) mostly contained SiO_2 and Al_2O_3 with 60.34 and 30.83 wt.%, respectively. The particle size of FA was 19.18804 μm (median) and 36.73520 μm (mean). Low-cost GO synthesized via mechanical exfoliation with a layer thickness (>10 stacking layers) was given by CSIR-IMMT, Odisha. The particle size of the GO varied from 3 to 200 nm. NaOH flakes were of 99.6% purity, and the Na_2SiO_3 solution consisted of Na_2O (15.85 wt.%), SiO_2 (32.15 wt.%), and H_2O (52 wt.%).

2.2. GRGC Fabrication

A NaOH solution of 12M was prepared by mixing 480 g of NaOH flakes and 1000 mL of tap water, resulting in an exothermic reaction. GRGC specimens were prepared in 5 different batches with varying GO additions (0, 0.1, 0.2, 0.3 and 0.4) wt.% of FA, as shown in Table 1. The alkali activator solution was mixed via a magnetic stirrer at low rpm. GO was consequently introduced to the solution carefully to achieve maximum dispersion and to avoid the agglomeration phenomenon. Later, the solution was ultrasonicated for 30 min and mixed with FA for 10-15 min. The geopolymer slurry was poured into custom molds (10×3) mm^3 and cured at 25 °C (±3 °C) for 24 hr. The GRGC specimens were de-molded and cured in ambient conditions for 28 days. Figure 1a illustrates the cured GRGC specimens prior to SSIS (solid state impedance spectroscopy) characterization studies. The specimens were polished with emery paper at different grift sizes to obtain the necessary dimensions for the experimental setup, as shown in Figure 1a,b, for precise results.

Table 1. GRGC mixture composition.

Mixture	FA (g)	NaOH Soln. (g)	Na_2SiO_3 Soln. (g)	$NaOH/Na_2SiO_3$	Liquid/Binder	GO (wt.%)
GRGC0	100	16.66	33.33	0.5	0.5	0
GRGC1	100	16.66	33.33	0.5	0.5	0.1
GRGC2	100	16.66	33.33	0.5	0.5	0.2
GRGC3	100	16.66	33.33	0.5	0.5	0.3
GRGC4	100	16.66	33.33	0.5	0.5	0.4

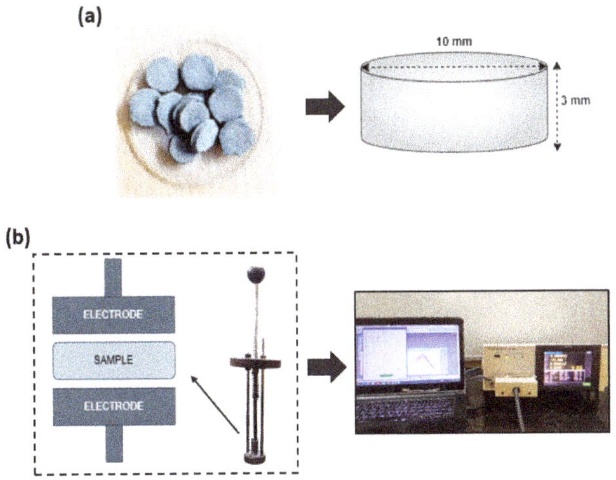

Figure 1. GRGC characterization process: (**a**) specimen specifications and (**b**) SSIS measurement methodology.

2.3. Testing Methods

Solid State Impedance Spectroscopy

The SSIS results were obtained from HIOKI IMPEDANCE ANALYZER IM3570 (AC) using LCR sample application software. The GRGC specimens were investigated using the two-probe method with electrodes, as depicted in Figure 1b, in a dry state under ambient laboratory conditions. The values for different parameters, such as impedance, capacitance and tan delta over a frequency range of 10^1 to 10^5 Hz, were obtained from the instrument. The values of the dielectric constant, dielectric loss and conductivity were calculated from Equations (1) to (4) [17–19]:

$$\varepsilon_r = \frac{C_p t}{\varepsilon_0 A} \quad (1)$$

$$D = \frac{\varepsilon_r''}{\varepsilon_r} \quad (2)$$

$$\omega = 2\pi f \quad (3)$$

$$\sigma = \varepsilon_0 \varepsilon_r'' \omega \quad (4)$$

where

C_p = capacitance of the specimen;
t = thickness of the specimen;
ε_0 = permittivity of free space constant (8.854×10^{-12} F/m);
ε_r = dielectric constant
ε_r'' = dielectric loss
A = area of the electrode (113.09 mm^2);
f = frequency;
ω = angular frequency;
σ = *conductivity*.

3. Results and Discussion

3.1. Solid-State Impedance Spectroscopy

3.1.1. Capacitance and Impedance/Resistance

The degree of (C_p) and (Z) of the GRGC specimens can be perceived in Figure 2a,b. At a lower frequency (10^1 Hz), higher capacitance values ranging from 9.26×10^{-10} to

4.5×10^{-9} were obtained, but the (C_p) values tend to drop and become constant (almost zero) at higher frequencies (i.e., specifically after 10^3 Hz). The incorporation of GO in the geopolymer composites results in higher (C_p) values, whereas in an opposite trend, lower (Z) was observed with a higher dosage of GO (4.36×10^5 Ω). The impedance of all geopolymer composites became almost constant at higher frequencies, but higher impedance values were observed at lower frequencies for GRGC0 specimens with no addition of GO (5.32×10^6 Ω). This could suggest that GRGC4 specimens offer minimal resistance and accelerate the movement of ions in the GRGC matrix in contrast with the GRGC0 specimen. This could also be signified by the acceleration effect of GO on the polycondensation reaction of the geopolymer composites [7]. The reduction in functional groups from GO may have enhanced the electrical conductivity properties of reduced GO (i.e., rGO), as similarly noted in the study by Saffi et al. [20].

3.1.2. Dielectric Constant, Dielectric Loss and Tangent Loss/Tan Delta

Figure 2c,d show the resulting values of dielectric constant (ε_r) and dielectric loss (ε_r'') for different GRGC specimens when measured at different frequencies. The incorporation of GO in the geopolymer mix leads to higher (ε_r) and (ε_r'') values of (10^{-5}–10^{-4}) and (10^{-4}–10^{-3}) at low frequencies, respectively. Both (ε_r) and (ε_r'') show a significant decrease with the increase in frequency initially. However, the decrease rate of (ε_r) and (ε_r'') was reduced at a higher frequency range and reached an approximately constant value. This phenomenon of the dielectric properties of GRGC specimens can be attributed to the polarization relaxation of molecules in the GRGC matrix. Initially, at lower frequencies, the molecules in the matrix have sufficient time and start orienting in the direction of the applied current. Consequently, at higher frequencies, the re-orientation is limited due to which the values of both (ε_r) and (ε_r'') are reduced drastically [21]. The polarization of the molecules can be firmly influenced by different aspects of the geopolymer mix, i.e., alumino-silicate gel, unreacted particles, impurities available in the composite mix, etc. A study by Hanjitsuwan et al. observed a similar phenomenon and described the rationale as electrode/specimen interfacial polarization and double-layer polarization [17,18]. Alternative polarization mechanisms include ionic, dipolar or molecular, electronic and atomic mechanisms [22].

The (D) curves for the GRGC specimens are exhibited in Figure 2e. The (D) curve peaks ranged between 3.06 and 12.7 and ought to be related to the trend of the dielectric properties. However, GRGC3 obtained the highest (D) value: 10^1 Hz. The (D) values decreased with an increase in frequency, and the curves became almost constant at higher frequencies for all GRGC specimens. This exception can be closely related to the compactness of the specimens and the function of GO in enhancing the strength of the composites primarily via their pore-filling characteristics. Earlier investigations also indicated the limit of GO (0.1–0.3 wt.%) in improving the strength of geopolymer composites, as higher dosages tend to decrease the compactness due to the consequence of agglomeration [16,23–26].

3.1.3. Conductivity

Figure 2f illustrates the conductivity (σ) results of the GRGC specimens. The conductivity values increase with the increase in GO dosage and range between 6.05×10^{-14} and 7.72×10^{-13} Ω·mm^{-1} for GRGC0 and GRGC4 at low frequencies of (10^1) Hz, respectively. The in situ reduction of GO improved electrical conductivity properties and contributed to enhancing GRGC conductive properties. At higher frequencies (i.e., 10^5 Hz), (σ) increased for all GRGC specimens, and the corresponding values include 1.18×10^{-12}–2.27×10^{-12} Ω·mm^{-1}. At lower frequencies, the (σ) of the GRGC specimens remained approximately constant. However, the increment in conductivity is significantly higher when the frequency is more than 10^4 Hz. This occurrence could be associated with the geopolymerisation reaction. The study by Hanjitsuwan et al. detected the same trend of increased conductivity for the geopolymer pastes with increased frequencies [17,18]. The molecular structure of geopolymeric gel is attributed to the increment of electrical conductivity at higher frequencies, mostly via the Na^+ ion hopping mechanism between the cation sites. A higher dosage of graphene in geopolymer composites leads to higher conductivity. Still, a significant increase in conductivity was observed at higher frequencies, which could be explained via the combination of Na^+ ion hopping and the electronic conductivity of in situ reduced GO, leading to the shortening of conduction distance. Thus, GO could be considered an effective agent for improving the (σ) of geopolymer composites for different applications [12]. The relationship between the (Z) and (σ) curves demonstrates the homogenous dispersion of GO in the geopolymer matrix since the agglomeration of GO could lead to improper conductivity in the GRGC specimens. The results also suggest that there might be a percolation threshold between 0.1 and 0.2 wt.% GO addition in the GRGC matrix. The manifestation is evident via the slight difference in the behaviour of curves in Figure 2b,f. Therefore, it can be considered that conductive fillers such as GO largely facilitate conductivity in geopolymer composites and can be tailored for appropriate piezoresistive responses according to the required applications.

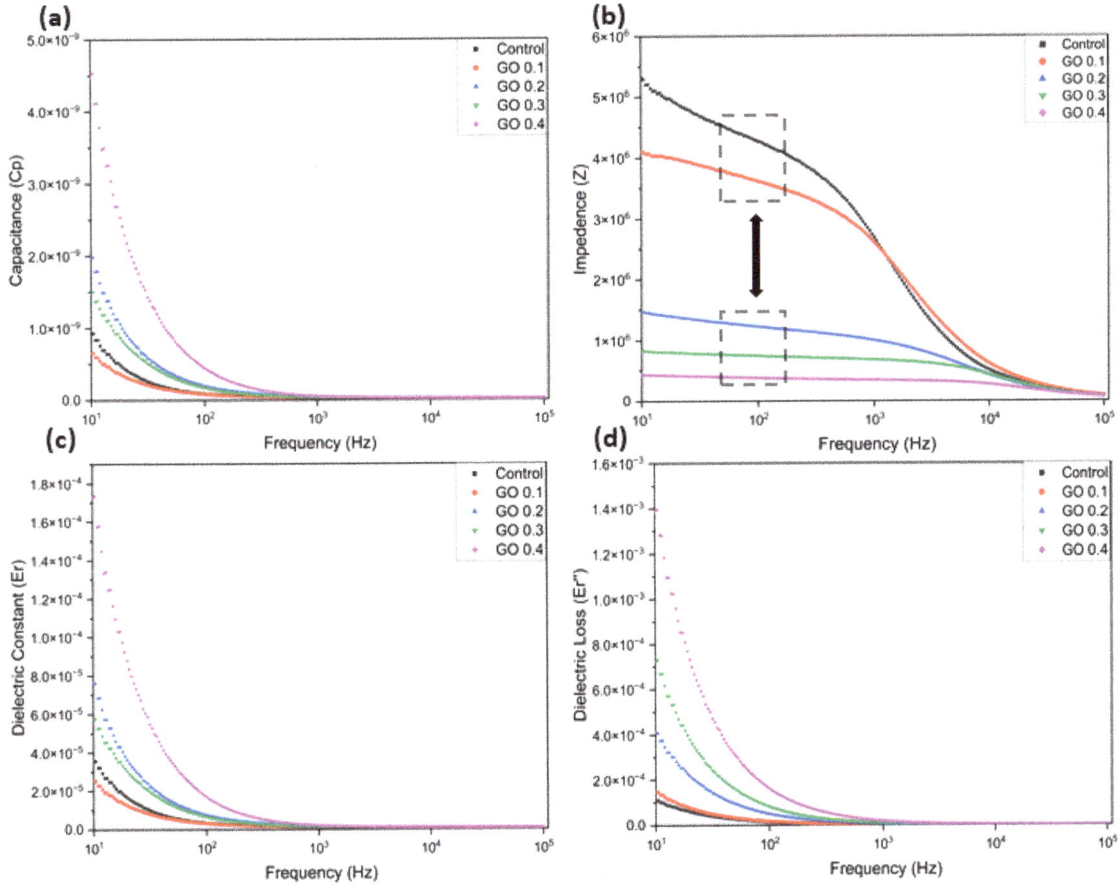

Figure 2. *Cont.*

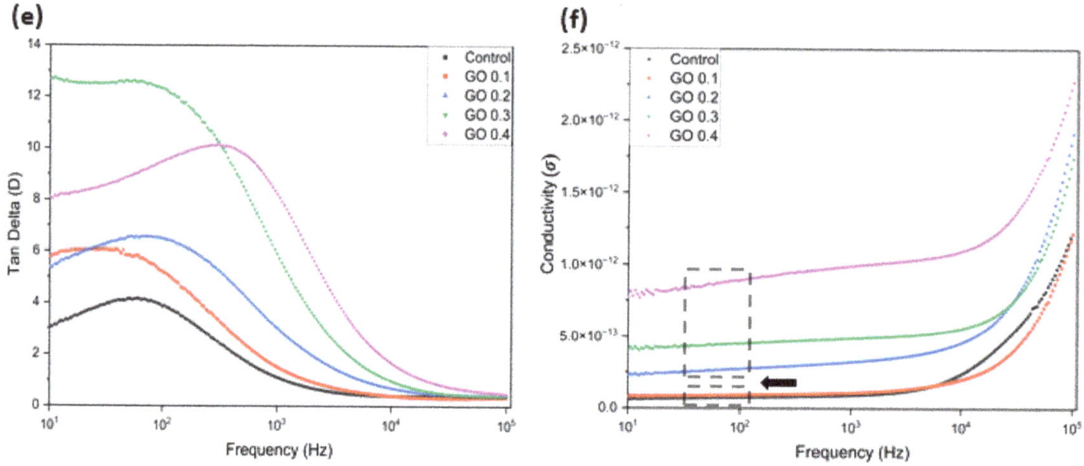

Figure 2. Influence of GO on different electrical parameters of geopolymer composites: (**a**) capacitance, (**b**) impedance, (**c**) dielectric constant, (**d**) dielectric loss, (**e**) tan delta and (**f**) conductivity.

4. Conclusions

In this study, different dosages of low-cost GO were incorporated in geopolymer composites to fabricate sustainable GRGC specimens for various smart applications. SSIS investigations were conducted at room temperature to characterize and assess the significant dielectric properties of GRGC specimens, which are concluded as follows.

The interaction of the GO sheets with the alkaline activator in geopolymeric reactions produced highly reduced and cross-linked GO sheets, enhancing the electrical conductivity properties of the composites.

At 10^1 Hz, GRGC specimens with 0.4 wt.% GO obtained a maximum ionic/electrical conductivity of 7.72×10^{-13} $\Omega \cdot mm^{-1}$ and a minimum impedance of 4.36×10^5 Ω, suggesting desirable low-frequency-based applications.

A percolation threshold was observed between 0.1 and 0.2 wt.% of GO introduction in the geopolymer matrix.

Increasing the GO dosage up to 0.4 wt.% aided in reducing the electrical impedance of GRGC specimens up to 91.81%.

Author Contributions: Conceptualization: R.S.K.; methodology: R.S.K.; formal analysis: R.S.K.; investigation: R.S.K.; resources: S.M.M.; writing—original draft preparation: R.S.K., S.S. and K.K.; writing—review and editing: T.S.Q. All authors have read and agreed to the published version of the manuscript.

Funding: The first author would like to acknowledge the funding of 500$ received through the ASTM International Student Project Grant Award 2021, USA.

Institutional Review Board Statement: Not applicable.

Informed Consent Statement: Not applicable.

Data Availability Statement: Not applicable.

Acknowledgments: The authors would like to thank Ashok Kumar Sahu and Muhammad Shahid Anwar of CSIR—Institute of Minerals and Materials Technology (CSIR-IMMT)—India, for the materials, facility and support necessary for the experimental work, along with the proofreading of the article. The authors are also grateful for Tanvir Qureshi's IFA New Starters Award at UWE Bristol, UK. The publication cost of this paper was covered with funds from the Polish National Agency for Academic Exchange (NAWA): "MATBUD'2023—Developing international scientific cooperation in the field of building materials engineering" BPI/WTP/2021/1/00002, MATBUD'2023.

Conflicts of Interest: The authors declare that they have no known competing financial interests or personal relationships that could have appeared to influence the work reported in this paper.

References

1. Krishna, R.S.; Mishra, J.; Meher, S.; Das, S.K.; Mustakim, S.M.; Singh, S.K. Industrial Solid Waste Management through Sustainable Green Technology: Case Study Insights from Steel and Mining Industry in Keonjhar, India. In Proceedings of the 2nd International Conference on Processing and Characterization of Materials, Rourkela, India, 12–14 December 2019; Volume 33, pp. 5243–5249.
2. Mishra, J.; Kumar Das, S.; Krishna, R.S.; Nanda, B.; Kumar Patro, S.; Mohammed Mustakim, S. Synthesis and Characterization of a New Class of Geopolymer Binder Utilizing Ferrochrome Ash (FCA) for Sustainable Industrial Waste Management. *Mater Today Proc.* **2020**, *33*, 5001–5006. [CrossRef]
3. Cai, J.; Pan, J.; Li, X.; Tan, J.; Li, J. Electrical Resistivity of Fly Ash and Metakaolin Based Geopolymers. *Constr. Build. Mater.* **2020**, *234*, 117868. [CrossRef]
4. Essaidi, N.; Nadir, H.; Martinod, E.; Feix, N.; Bertrand, V.; Tantot, O.; Lalande, M.; Rossignol, S. Comparative Study of Dielectric Properties of Geopolymer Matrices Using Different Dielectric Powders. *J. Eur. Ceram Soc.* **2017**, *37*, 3551–3557. [CrossRef]
5. Wanasinghe, D.; Aslani, F.; Ma, G. Effect of Carbon Fibres on Electromagnetic-Interference-Shielding Properties of Geopolymer Composites. *Polymers* **2022**, *14*, 3750. [CrossRef] [PubMed]
6. Nguyen, V.V.; Le, V.S.; Louda, P.; Szczypiński, M.M.; Ercoli, R.; Růžek, V.; Łoś, P.; Prałat, K.; Plaskota, P.; Pacyniak, T.; et al. Low-Density Geopolymer Composites for the Construction Industry. *Polymers* **2022**, *14*, 304. [CrossRef]
7. Long, W.-J.; Zhang, X.-H.; Dong, B.-Q.; Fang, Y.; Ye, T.-H.; Xie, J. Investigation of Graphene Derivatives on Electrical Properties of Alkali Activated Slag Composites. *Materials* **2021**, *14*, 4374. [CrossRef] [PubMed]
8. Mizerová, C.; Kusák, I.; Rovnaník, P. Electrical properties of fly ash geopolymer composites with graphite conductive admixtures. *Acta Polytech CTU Proc.* **2019**, *22*, 72–76. [CrossRef]
9. Farhan, K.Z.; Johari, M.A.M.; Demirboğa, R. Impact of Fiber Reinforcements on Properties of Geopolymer Composites: A Review. *J. Build. Eng.* **2021**, *44*, 102628. [CrossRef]
10. Ranjbar, N.; Zhang, M. Fiber-Reinforced Geopolymer Composites: A Review. *Cem. Concr. Compos.* **2020**, *107*, 103498. [CrossRef]
11. Su, Z.; Hou, W.; Sun, Z. Recent Advances in Carbon Nanotube-Geopolymer Composite. *Constr. Build. Mater.* **2020**, *252*, 118940. [CrossRef]
12. Payakaniti, P.; Pinitsoontorn, S.; Thongbai, P.; Amornkitbamrung, V.; Chindaprasirt, P. Electrical Conductivity and Compressive Strength of Carbon Fiber Reinforced Fly Ash Geopolymeric Composites. *Constr. Build. Mater.* **2017**, *135*, 164–176. [CrossRef]
13. Payakaniti, P.; Pinitsoonthorn, S.; Thongbai, P.; Amornkitbamrung, V.; Chindaprasirt, P. Effects of Carbon Fiber on Mechanical and Electrical Properties of Fly Ash Geopolymer Composite. *Mater. Today Proc.* **2018**, *5*, 14017–14025. [CrossRef]
14. Górski, M.; Czulkin, P.; Wielgus, N.; Boncel, S.; Kuziel, A.W.; Kolanowska, A.; Jędrysiak, R.G. Electrical Properties of the Carbon Nanotube-Reinforced Geopolymer Studied by Impedance Spectroscopy. *Materials* **2022**, *15*, 3543. [CrossRef]
15. Kusak, I.; Lunak, M.; Rovnanik, P. Electric Conductivity Changes in Geopolymer Samples with Added Carbon Nanotubes. *Procedia Eng.* **2016**, *151*, 157–161. [CrossRef]
16. Krishna, R.S.; Mishra, J.; Nanda, B.; Patro, S.K.; Adetayo, A.; Qureshi, T.S. The Role of Graphene and Its Derivatives in Modifying Different Phases of Geopolymer Composites: A Review. *Constr. Build. Mater.* **2021**, *306*, 124774. [CrossRef]
17. Hanjitsuwan, S.; Chindaprasirt, P.; Pimraksa, K. Electrical Conductivity and Dielectric Property of Fly Ash Geopolymer Pastes. *Int. J. Miner. Metall. Mater.* **2011**, *18*, 94–99. [CrossRef]
18. Hanjitsuwan, S.; Hunpratub, S.; Thongbai, P.; Maensiri, S.; Sata, V.; Chindaprasirt, P. Effects of NaOH Concentrations on Physical and Electrical Properties of High Calcium Fly Ash Geopolymer Paste. *Cem. Concr. Compos.* **2014**, *45*, 9–14. [CrossRef]
19. Sellami, M.; Barre, M.; Toumi, M. Synthesis, Thermal Properties and Electrical Conductivity of Phosphoric Acid-Based Geopolymer with Metakaolin. *Appl. Clay Sci.* **2019**, *180*, 105192. [CrossRef]
20. Saafi, M.; Tang, L.; Fung, J.; Rahman, M.; Liggat, J. Enhanced Properties of Graphene/Fly Ash Geopolymeric Composite Cement. *Cem. Concr. Res.* **2015**, *67*, 292–299. [CrossRef]
21. Chuewangkam, N.; Nachaithong, T.; Chanlek, N.; Thongbai, P.; Pinitsoontorn, S. Mechanical and Dielectric Properties of Fly Ash Geopolymer/Sugarcane Bagasse Ash Composites. *Polymers* **2022**, *14*, 1140. [CrossRef]
22. Malkawi, A.B.; Al-Mattarneh, H.; Achara, B.E.; Mohammed, B.S.; Nuruddin, M.F. Dielectric Properties for Characterization of Fly Ash-Based Geopolymer Binders. *Constr. Build. Mater* **2018**, *189*, 19–32. [CrossRef]
23. Krishna, R.S.; Mishra, J.; Das, S.K.; Mustakim, S.M. An Overview of Current Research Trends on Graphene and It's Applications. *Worls Sci. News* **2019**, *132*, 206–219.
24. Krishna, R.S.; Mishra, J.; Adetayo, A.; Das, S.K.; Mustakim, S.M. Green Synthesis of High-Performance Graphene Reinforced Geopolymer Composites: A Review on Environment-Friendly Extraction of Nanomaterials. *Iranian J. Mater. Sci. Eng.* **2020**, *17*, 10–24.

25. Krishna, R.S.; Mishra, J.; Das, S.K.; Nanda, B.; Patro, S.K.; Mustakim, S.M. A Review on Potential of Graphene Reinforced Geopolymer Composites. In *Tailored Functional Materials*; Mukherjee, K., Layek, R.K., De, D., Eds.; Springer: Singapore, 2022; Volume 15, pp. 43–60.
26. Das, S.K.; Krishna, R.S.; Mishra, S.; Mustakim, S.M.; Jena, M.K.; Tripathy, A.K.; Sahu, T. Future Trends Nanomaterials in Alkali-Activated Composites. In *Handbook of Sustainable Concrete and Industrial Waste Management*; Colangelo, F., Cioffi, R., Farina, I., Eds.; Woodhead Publishing: Cambridge, UK, 2021.

materials proceedings

MDPI

Proceeding Paper

Early-Age Creep and Shrinkage Properties of Printed and Cast Cement Composite †

Rihards Gailitis [1,*], Pauls Pudans [1], Krists Ziemelis [2], Girts Bumanis [2] and Andina Sprince [1]

1 Institute of Structural Engineering, Riga Technical University, Kipsalas 6A, LV-1048 Riga, Latvia
2 Institute of Materials and Structures, Faculty of Civil Engineering, Riga Technical University, Kipsalas 6A, LV-1048 Riga, Latvia
* Correspondence: rihards.gailitis@rtu.lv
† Presented at the 10th MATBUD'2023 Scientific-Technical Conference "Building Materials Engineering and Innovative Sustainable Materials", Cracow, Poland, 19–21 April 2023.

Abstract: In recent years, 3D printing has been more and more used in the development of buildings and building elements. Mostly-printed structures are subjected to compression that is oriented perpendicular to the layer laying direction. When applying load in this way, the printed structure exhibits characteristics similar to masonry structures. However, as the technology and application of 3D printing develop, the structures also become more complicated and subjected not only to direct compression but other stresses as well. In this paper, long-term properties together with compressive strength were determined for 3D-printed specimens with load applied in the same direction as the layers are laid. The long-term and mechanical properties were compared with cast same-composition specimens. Results show that for the printed specimens, the compressive strength was more than two times lower than cast specimens, while the creep properties were slightly lower for the printed specimens.

Keywords: cement composite long-term properties; 3D-printed cement composite; tensile stress

Citation: Gailitis, R.; Pudans, P.; Ziemelis, K.; Bumanis, G.; Sprince, A. Early-Age Creep and Shrinkage Properties of Printed and Cast Cement Composite. *Mater. Proc.* **2023**, *13*, 35. https://doi.org/10.3390/materproc2023013035

Academic Editors: Katarzyna Mróz, Tomasz Tracz, Tomasz Zdeb and Izabela Hager

Published: 17 February 2023

1. Introduction

Additive manufacturing, more frequently known as 3D printing, is the fabrication method of objects using a print head, nozzle, or other printer technologies to deposit material to the print surface [1]. The 3D printing process has been successfully used in many disciplines, such as aerospace, automotive, biomedical, and food industries. It is claimed to allow quicker and cheaper production of an object, especially intricate and delicate objects with complex geometry. Unlike other manufacturing means, 3D printing has a much higher automation level that further contributes to labor and cost reduction, reducing production time [2–5].

The civil engineering field also has developed to a stage where the design of the structure has become much more complex, while load-bearing necessities have not been reduced. Furthermore, to meet the demand for residential buildings as well as infrastructure objects, there is a high need to build faster and reduce building costs. There are high hopes that 3D printing, due to success in other fields, will also bear fruit in civil engineering applications. It has been estimated that 3D printing will reduce construction waste by 30 to 60%, decrease production time by 50 to 70%, and drop labor costs by 50 to 80% [6]. To show these improvements, cementitious material compositions are specifically designed to flow evenly and have proper layer adhesion to one another. Furthermore, 3D-printed concrete sections are more prone to acid attacks as well as shrinkage [7]. The weakest part of the 3D-printed section is claimed to be the layer connecting zone. They also show anisotropic behavior and insufficient insulating properties that lead to possible heat loss [8,9].

As 3D printing of structures are more frequently used to develop a structure where layers are put on top of one another and subjected to direct compression, it was predicted that the structure would work similarly to masonry structures. However, as not all the structures are subjected to direct compression, there is a need to gain data and knowledge on how printed structures act under other stresses. For instance, retaining walls have load applied on the side of them. Therefore, tensile stresses significantly affect the structure's load-bearing capacity.

This article investigates the early-age creep and drying shrinkage properties of 3D-printed cement composite that are loaded in the direction parallel to layers and cast cement composite and compares them to one another.

2. Materials and Methods

Printed beam shape specimens were prepared for tensile stress impact on printed specimens to relate creep and shrinkage property determination. Each printed specimen had four layers. They were modeled to be 40 × 40 × 1000 mm, but due to the cement composite mix flowability, the width of the specimens at the base was 83 mm, and 67 mm at the top. The used cement composite composition was similar to cement mortar. Due to the fact that the composition of the used cement composite is a trade secret, specific amounts and types of materials cannot be disclosed in this article. The mass partition of the used cement composite is shown in Table 1. The specimen printing process is shown in Figure 1.

Table 1. Used cement composite composition partitions.

Material	Partition, %
Portland cement	33
Quartz sand 0/0.4 mm	49
Water	18

Figure 1. Test specimen printing process.

When specimens were printed, they were left overnight to set. At the same time, cast specimens were prepared. They were poured into a steel prismatic cast of 40 × 40 × 160 mm. A day later, cast specimens were unmolded, and printed specimens were cut to the same shape (40 × 40 × 160 mm) as the cast specimens. They all were placed in an aqueous environment for 25 days. After 25 days, all specimens were prepared for the creep and drying shrinkage tests.

The compressive strength was determined for four specimens (prismatic specimens 40 × 40 × 160 mm) out of each specimen type. The load was applied to the specimens for creep tests according to the determined compressive strength values. Compressive strength and creep and drying shrinkage (further in the text referred to as shrinkage) tests were conducted using the identical shape specimens so that creep specimens would not have shape factor impact (if the compressive strength specimens were larger or smaller than creep specimens) in the load that they were subjected on the test stands. All of the specimens intended for creep testing were loaded with 20% of the ultimate compressive strength value. Specimen placement into the creep test stand and shrinkage stand is shown in Figure 2.

(a) (b)

Figure 2. The 3D-printed and cast cement composite shrinkage (**a**) and creep (**b**) test setup.

Creep and shrinkage tests were carried out for 28 days. Creep and shrinkage testing procedure, except testing time, was performed according to RILEM TC 107 recommendations [10]. The laboratory conditions for the creep and shrinkage tests were 24 ± 1 °C and 30 ± 3% relative humidity.

After creep and shrinkage tests to determine the reasons for inequal creep and shrinkage strains to the specimen sides, as well as to see the 3D-printed layer adhesion, quantitative image analysis was performed to the specimen's polished sections that were prepared and made according to the [11] used process.

3. Results and Discussion

The compressive strength was determined before the creep tests. Four specimens were used to determine compressive strength values for printed and cast specimens. The specimen's age at the time of testing was 28 days. The compressive strength values are shown in Figure 3.

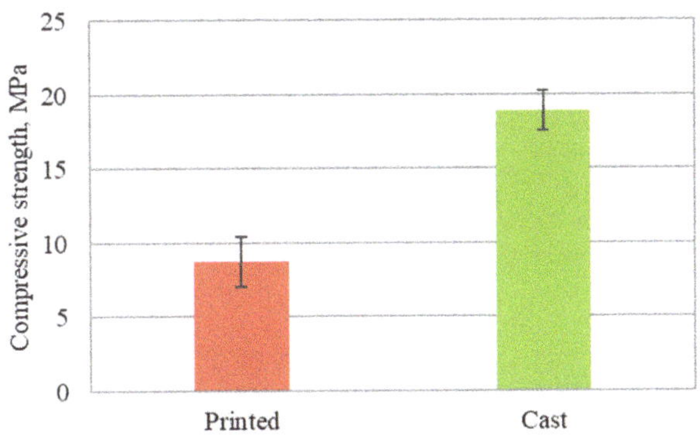

Figure 3. The 3D-printed and cast cement composite compressive strength values with measurement errors.

As visible from the compressive strength diagram, printed specimens loaded longitudinally to their layer placement exhibit more than two times lower compressive strength than cast specimens; furthermore, their standard measurement error is 17.9% larger than cast specimens.

Afterward, 28 day-long early creep and shrinkage tests were run, and the resulting readings are compiled in Figure 4.

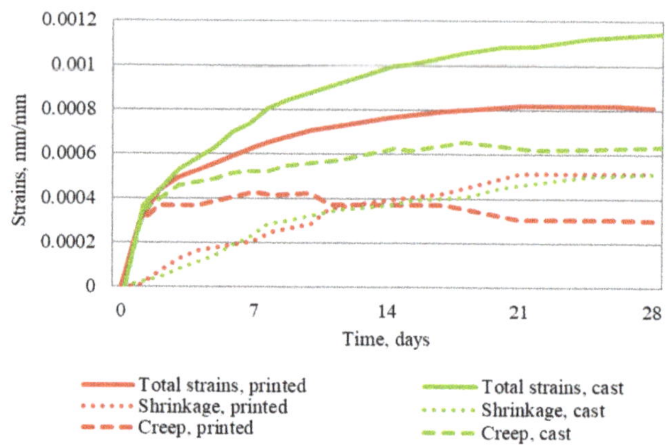

Figure 4. The 3D-printed and cast cement composite total recorded, creep, and shrinkage strains.

As it is clear from the curves in Figure 4, the shrinkage strains for printed and cast specimens are very close, or even identical. However, the creep strains have significant differences. Printed specimens show, at the peak values, 28.3% less creep strains. Additionally, the creep curve of printed specimens shows strain-decreasing relations starting from day 12 until day 28, while creep strains in cast specimens rise until day 18 and then exhibit a slight decrease. This implies that there must be some layer adhesion issue or that the load impact to the specimen resulted in the degradation of the structure. As the printed specimens have significantly lower compressive strength than cast specimens, it is necessary to calculate specific creep to see the creep strains without applied stress impact.

Specific creep values are calculated according to the equation, and the results are in Figure 5:

$$\chi_{cr}(t, t_0) = \frac{\varepsilon_{cr}(t, t_0)}{\sigma} = \frac{\varepsilon_{kop}(t) - \varepsilon_{sh}(t) - \varepsilon_{el}(t, t_0)}{\sigma} = \frac{1}{E_{cr}(t, t_0)} \quad (1)$$

where:

$\chi_{cr}(t, t_0)$ is the specific creep,
$\varepsilon_{cr}(t, t_0)$ is the creep strain,
$\varepsilon_{kop}(t)$ is the total strain,
$\varepsilon_{sh}(t)$ is the shrinkage strain,
$\varepsilon_{el}(t, t_0)$ is the elastic strain,
σ is the compressive stress,
and $E_{cr}(t, t_0)$ is the modulus of creep.

In Figure 5, it is clear that 3D-printed cement composites exhibit significantly higher specific creep; in other words, they are more willing to creep. On average, they have 32.8% higher specific creep than cast specimens. Furthermore, their specific creep appears within a couple of days, while it develops during the first 21 days of testing in cast specimens.

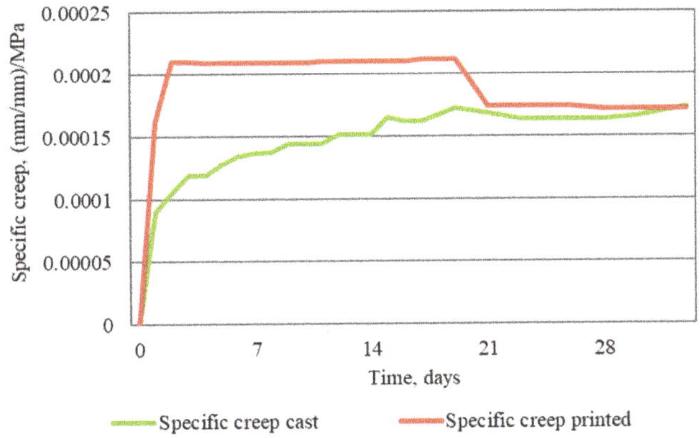

Figure 5. The 3D-printed and cast cement composite specific creep.

To further elaborate the assumption that printed specimens have some issues in the printed layers, the strain readings were divided into those that were measured to the top surface layer and those in which the surface consists of layer-side surfaces (placement and description shown in Figure 6).

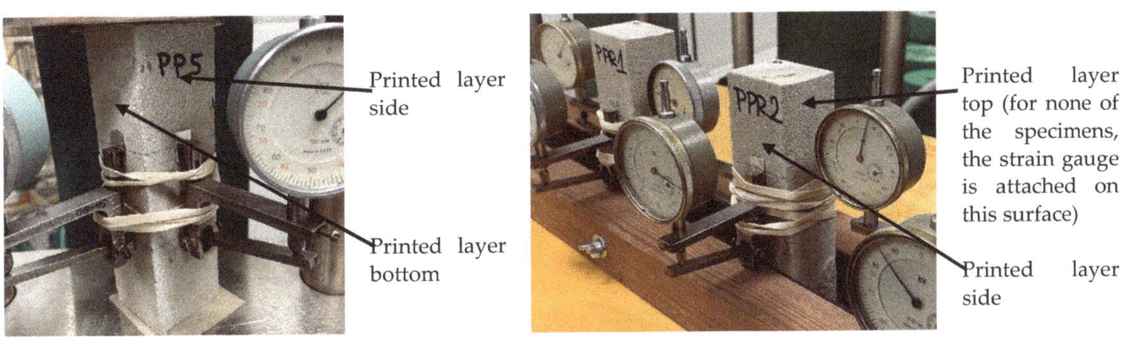

(**a**) (**b**)

Figure 6. The 3D-printed cement composite strain gauge placement on layer sides and surface for creep (**a**) and shrinkage (**b**) specimens.

The long-term shrinkage and creep strain curves according to strain gauge fitment are shown in Figure 7.

Here, it becomes clear that while long-term strain curves in the relation are similar, the creep strains and shrinkage strains are very different. While creep and shrinkage curves rise steadily to the layer top surfaces, the sides seem to have deterioration due to shrinkage. As the specimens were tested at the age of 28 days, the main shrinkage effect came from drying shrinkage. The shrinkage strain curves from the layer side surfaces lead to the conclusion that layers have been partially separated. To further elaborate, printed specimens after long-term tests were saturated in epoxy resin and used to make polished section specimens that then had their microstructure examined. It was determined that for all specimens, one side of the layer was more porous (see Figure 8) than the rest of the polished section surface.

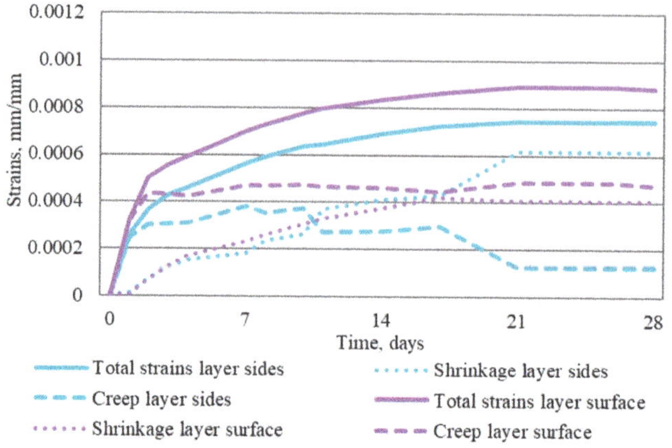

Figure 7. The 3D-printed cement composite total recorded, creep, and shrinkage strains according to strain gauge placement.

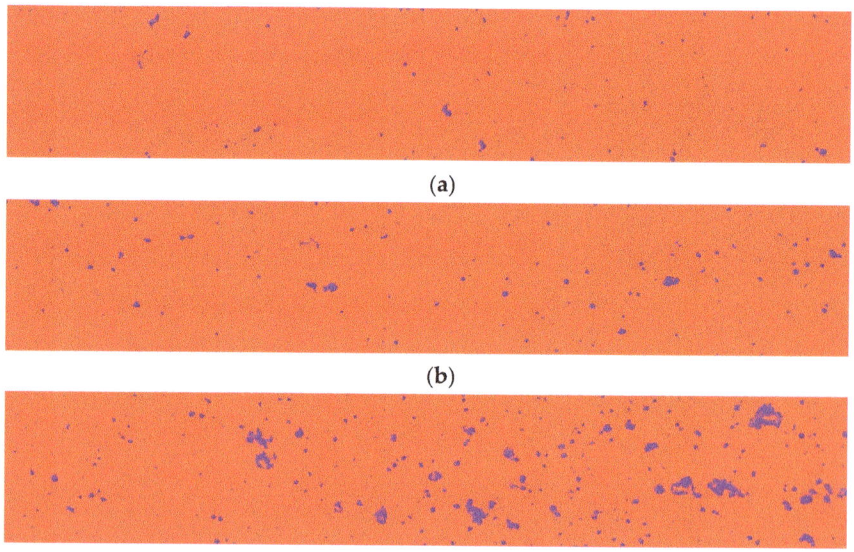

Figure 8. Quantitative image analysis images with matrix and filler part (red) and air void parts (blue) to the left side (**a**), middle part (**b**), and right side (**c**) of the specimen cross-section.

The specimens were printed using a plastic nozzle that had been printed on the plastic 3D printer. It had a stitched part that, as it turns out, frothed up part of the cement composite that interacted with this part of the nozzle. The model of the nozzle is shown in Figure 9. The white part in the model is the part where plastic layers are connected, and the stitch is developed.

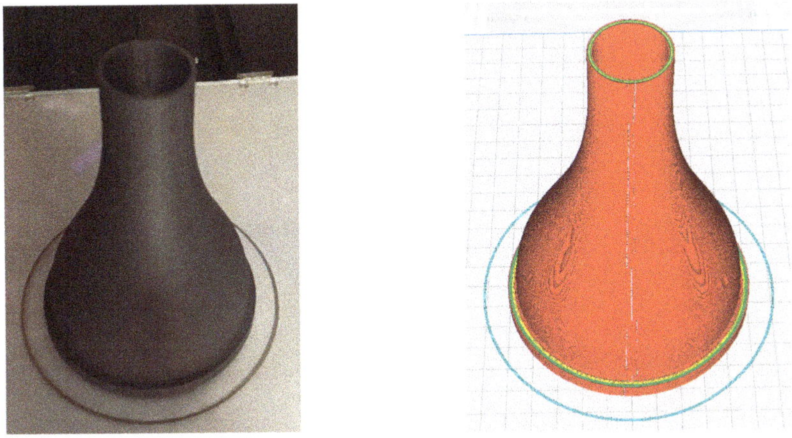

Figure 9. The 3D printer nozzle printed and modeled part.

4. Conclusions

Early-age creep and shrinkage property tests were performed on the prismatic specimens at the age of 28 days. The long-term property tests were carried out for 28 days. Prior to the long-term tests, compressive strength was determined for the cast and printed specimens. Following conclusions are:

1. According to the subjected stress state of the specimens, the printed cement composite specimens exhibit 53.7% lower compressive strength than cast cement composites. They also have a 17.9% higher standard error than cast specimens.
2. Printed specimens that have had load applied in the same way as the layers are laid show 28.3% fewer creep strains. Shrinkage strains are the same for printed and cast specimens.
3. Printed specimens are more prone to creep, as printed specimens' specific creep value is 32.8% higher on average than cast specimens.
4. Shrinkage strains in printed specimens have a significant role, and due to drying, shrinkage specimens show significant increases in shrinkage. It is very likely that this is due to the specifics of the used nozzle geometry. An increase in porosity in the specimen layer sides was observed. According to microstructure evaluation on specimen sides, there are 12% and 18% more pores to the shrinkage and creep specimens, correspondingly, than in the middle of the specimens.

Author Contributions: Conceptualization, R.G., P.P. and K.Z.; methodology, R.G. and K.Z.; software, R.G.; validation, P.P. formal analysis, R.G. and P.P.; investigation, R.G. and P.P.; resources, A.S. and G.B.; data curation, R.G. and P.P.; writing—original draft preparation, R.G.; writing—review and editing, R.G.; visualization, R.G.; supervision, A.S. and G.B.; project administration, R.G. and K.Z.; funding acquisition, R.G. All authors have read and agreed to the published version of the manuscript.

Funding: This publication was supported by Riga Technical University's Doctoral Grant program.

Institutional Review Board Statement: Not applicable.

Informed Consent Statement: Not applicable.

Data Availability Statement: Not applicable.

Acknowledgments: This work was supported by the European Social Fund within the Project No. 8.2.2.0/20/I/008 "Strengthening of PhD students and academic personnel of Riga Technical University and BA School of Business and Finance in the strategic fields of specialization" of the Specific Objective 8.2.2 "To Strengthen Academic Staff of Higher Education Institutions in Strategic Specialization Areas» of the Operational Programme «Growth and Employment". This work has been supported by the Latvian Council of Science within the scope of the project "Foamed Geopolymer Made by Additive Manufacturing for the Construction Technology Ap-plications (3D-FOAM)" No. project9608. Publication cost of this paper was covered with founds of the Polish National Agency for Academic Exchange (NAWA): "MATBUD'2023—Developing international scientific cooperation in the field of building materials engineering" BPI/WTP/2021/1/00002, MATBUD'2023.

Conflicts of Interest: The authors declare no conflict of interest.

References

1. Li, Z.; Hojati, M.; Wu, Z.; Piasente, J.; Ashrafi, N.; Duarte, J.P.; Nazarian, S.; Bilén, S.G.; Memari, A.M.; Radlińska, A. Fresh and hardened properties of extrusion-based 3D-printed cementitious materials: A review. *Sustainability* **2020**, *12*, 5628. [CrossRef]
2. Wong, K.V.; Hernandez, A. A Review of Additive Manufacturing. *ISRN Mech. Eng.* **2012**, *2012*, 1–10. [CrossRef]
3. Buswell, R.A.; Soar, R.C.; Gibb, A.G.F.; Thorpe, A. Freeform Construction: Mega-scale Rapid Manufacturing for construction. *Autom. Constr.* **2007**, *16*, 224–231. [CrossRef]
4. Wegrzyn, T.F.; Golding, M.; Archer, R.H. Food Layered Manufacture: A new process for constructing solid foods. *Trends Food Sci. Technol.* **2012**, *27*, 66–72. [CrossRef]
5. Nematollahi, B.; Xia, M.; Sanjayan, J. Current progress of 3D concrete printing technologies. In Proceedings of the 34th International Symposium on. Automation and Robotics in Construction and Mining, Taipei, Taiwan, 28 June–1 July 2017; pp. 260–267.
6. Zhang, J.; Wang, J.; Dong, S.; Yu, X.; Han, B. A review of the current progress and application of 3D printed concrete. *Compos. Part A Appl. Sci. Manuf.* **2019**, *125*, 105533. [CrossRef]
7. Baz, B.; Aouad, G.; Kleib, J.; Bulteel, D.; Remond, S. Durability assessment and microstructural analysis of 3D printed concrete exposed to sulfuric acid environments. *Constr. Build. Mater.* **2021**, *290*, 123220. [CrossRef]
8. Ding, T.; Xiao, J.; Zou, S.; Zhou, X. Anisotropic behavior in bending of 3D printed concrete reinforced with fibers. *Compos. Struct.* **2020**, *254*, 112808. [CrossRef]
9. Wang, L.; Tian, Z.; Ma, G.; Zhang, M. Interlayer bonding improvement of 3D printed concrete with polymer modified mortar: Experiments and molecular dynamics studies. *Cem. Concr. Compos.* **2020**, *110*, 103571. [CrossRef]
10. Acker, P.; Agullo, L.; Auperin, M.; Carol, I.; Carreira, D.J.; Catarino, J.; Chem, J.-C.; Chiorino, M.A.; Dougill, J.W.; Huet, C.; et al. Rilem Tc 107-Csp: Creep and Shrinkage Prediction Models: Principles of Their Formation Recommendation Measurement of time-dependent strains of concrete. *Mater. Struct.* **1998**, *31*, 507–512.
11. Gailitis, R.; Sprince, A.; Kozlovskis, T.; Radina, L.; Pakrastins, L.; Vatin, N. Long-Term Properties of Different Fiber Reinforcement Effect on Fly Ash-Based Geopolymer Composite. *Crystals* **2021**, *11*, 760. [CrossRef]

materials proceedings

MDPI

Proceeding Paper

Use of Bonded Joints for Fastening Sheet-Metal Components to Contemporary Facades Fitted with an External Thermal Insulation Composite System with Thin-Layer Acrylic Plaster †

Jiří Šlanhof, Aleš Průcha *, Barbora Nečasová and Adam Boháček

Faculty of Civil Engineering, Brno University of Technology, Veveří 331/95, 602 00 Brno, Czech Republic
* Correspondence: ales.prucha@vut.cz; Tel.: +420-541-147-989
† Presented at the 10th MATBUD'2023 Scientific-Technical Conference "Building Materials Engineering and Innovative Sustainable Materials", Cracow, Poland, 19–21 April 2023.

Abstract: This paper deals with the issue of fastening sheet-metal components on the facades of contemporary buildings that are massively insulated with external thermal insulation composite systems. This research focused on facades with thin-layer acrylic plaster and sheet-metal components made of aluminium, copper and hot-dipped galvanized sheet metal. Two different test methods and test sample geometries were used to determine the most critical aspects for the studied application sectors. Surprisingly high tensile properties as well as shear stresses in the bonded joints were recorded for all the selected combinations. The presented results confirmed the assumed benefits for the construction industry and the future practical use of this technology in construction, although the durability of a bonded joint will always depend mainly on the quality of the bonded substrate.

Keywords: adhesion; adhesive; joint; sheet-metal construction; facade; sheet-metal components; thin-layer plaster

Citation: Šlanhof, J.; Průcha, A.; Nečasová, B.; Boháček, A. Use of Bonded Joints for Fastening Sheet-Metal Components to Contemporary Facades Fitted with an External Thermal Insulation Composite System with Thin-Layer Acrylic Plaster. *Mater. Proc.* **2023**, *13*, 36. https://doi.org/10.3390/materproc2023013036

Academic Editors: Katarzyna Mróz, Tomasz Tracz, Tomasz Zdeb and Izabela Hager

Published: 16 February 2023

1. Introduction

Saving energy has been a major issue in recent years as part of the efforts to maintain the sustainable development of society. In 2022, the situation in Europe changed dynamically due to the military conflict in Ukraine, which resulted in the virtual collapse of energy markets and an uncontrollable rise in gas and electricity prices. Countries were forced to search for solutions in an unprecedented situation and the pressure for energy saving was intensified. The greatest potential of the construction industry lies in savings in heating buildings [1]. In new buildings, the desire to reduce heat loss is reflected both in the design of the building structure [2] and in the thermal parameters of the structures. However, the greatest potential, regarding the number of buildings, lies in existing structures, where the most common way of improving the energy balance has been additional insulation [3]. This can be implemented in two ways. The first is the use of suspended facade systems, which consist of a load-bearing (usually aluminium) structure, mineral wool thermal insulation, a windproof diffusion-open membrane, a ventilated air gap and a facing cladding board [4,5]. The second option is external thermal insulation using a composite system that is in full contact with the original surface of the facade and consists of thermal insulation (most often expanded polystyrene or mineral wool, or other materials that are used less frequently) and reinforcing layers of cement screed with fibreglass mesh, coloured primer and coloured thin-layer plaster [6,7]. Both options include a thick layer of thermal insulation that can be sized as required to provide the desired energy savings. Each of the two options has both advantages and disadvantages. In general, however, contact insulation is significantly more widespread. This is primarily due to its lower cost, which represents approximately one-third of a suspended ventilated facade system. The authors of this paper focused on the issue of sheet-metal facade components in contact-insulated facade systems, where secure fastening to the substrate becomes increasingly difficult due to the increasing thickness of the thermal insulation.

Currently, 200 mm thick layers of thermal insulation are common, and 300 mm thick insulation is not an exception. These surfaces are largely unsuitable for plumbing work, mainly due to the impossibility of firm fastening, as the outer plaster layer is too thin, and the insulation itself is not a load-bearing substrate. While there are system solutions for connecting sheet-metal components to multi-layer plaster, e.g., parapet connection profiles with integrated adhesive surfaces for bonding metal parapets or connection profiles for inserting vertical facade flashings, these are always components intended to be fitted during the installation of the insulation system. Using these solutions requires precise coordination during the construction process. Moreover, they are not always applicable as they do not allow the fitting of sheet-metal structures during additional work onto older, previously insulated facades. This article deals with the issue of fastening sheet-metal components by bonding them to the surface of thin-layer plaster. It presents an initial series of tests to verify the perspectives of this technology for further research. The technology of bonding sheet-metal components to thin-layer plaster, if properly validated, would be widely applicable, even outside the field of contact thermal insulation systems.

2. Materials and Methods

2.1. Materials

Materials commonly used for sheet-metal components such as aluminium alloy, copper alloy and hot-dipped galvanized steel were tested. These materials are used in the form of thin sheets on construction sites; however, this research required the production of solid circular plates (hereinafter referred to as targets) and strips to be bonded to the substrate. The metallic materials used were aluminium alloy EN AW-2011 according to the EN 573-3 standard [8] with a tensile strength of 295.0 MPa and a yield strength of 195.0 MPa, copper alloy CW004A-R250—SH (Cu-ETP) according to the EN 13,601 standard [9] with a tensile strength of 250.0 MPa and a yield strength of 200.0 MPa, and hot-dipped galvanized steel brand S235JR (1.0038) according to the EN 10025-2 standard [10] with a tensile strength of min. 360.0 MPa and a yield strength of 235.0 MPa. The thixotropic silane modified polymer adhesive ARDEX CA 20 P for external applications produced by Ardex Baustoff GmbH was chosen for bonding to the substrate. No mechanical properties were provided by the manufacturer. The following materials were used for the production of the solid plasterboard test specimens for bonding the metal tear-off elements:

- White pasty structural acrylic dispersion plaster with the brand name PROFI AKRYLATPUTZ produced by the manufacturer Profibaustoffe CZ, s.r.o., with a bond strength of min. 0.30 MPa;
- Universal primer coating used under the white pasty plaster with the brand name PROFI UNI Putzgrund produced by the manufacturer Profibaustoffe CZ, s.r.o., with a styrene–acrylate binder;
- An adhesive and screed compound for facade thermal insulation boards with the brand name PROFI UNI AM produced by the manufacturer Profibaustoffe CZ, s.r.o., with cement binder and a bonding strength to the solid substrate of min. 0.25 MPa;
- Fibreglass fabric for screed compound with the brand name R117 A101 produced by Saint-Gobain Vertex, s.r.o., with a tensile strength of 2100 N/50 mm;
- A smooth-surfaced cement-bonded particleboard with a thickness of 16 mm with the brand name CETRIS BASIC produced by the manufacturer CIDEM Hranice, a.s., with a transverse tensile strength of min. 0.63 MPa.

2.2. Methods

2.2.1. Preparation of Test Specimens

The test specimens consisted of a base plate, adhesive and a metal target of a circular shape with a bonding area of 2500 mm^2 for the tensile tests and a metal strip of 100 × 25 mm with a bonding area of 625 mm^2 for the shear tests. The base plate consisted of a layer that corresponded to the actual composition used in facades with contact insulation. The soft and easily deformable thermal insulator was replaced by a solid cement-bonded particleboard for laboratory purposes. The intention was to measure only the deformation of the tested adhesive without the undesired influence of any deformation of the thermal insulation. It was assumed that the adhesive would be torn off either in the plaster or in the screed. The cement-bonded particleboards were cut into 100 × 100 mm squares from a production size of 3350 × 1250 mm, and then cement screed with a thickness of 3 mm was applied to them in two layers. A fibreglass fabric was fitted between the first and second layers of the cement screed. After the screed had cured the surface was coated with a primer, and a final 1.5 mm thick layer of acrylic plaster, was applied. The composition of the test specimen for the tensile test is shown in Figure 1, and in Figure 2 the test specimen for the shear test can be seen. Close attention was paid to the production of the bonded joint, where a constant adhesive thickness of 3 mm had to be maintained. Inaccuracies of this basic parameter would adversely affect the measurement results [11].

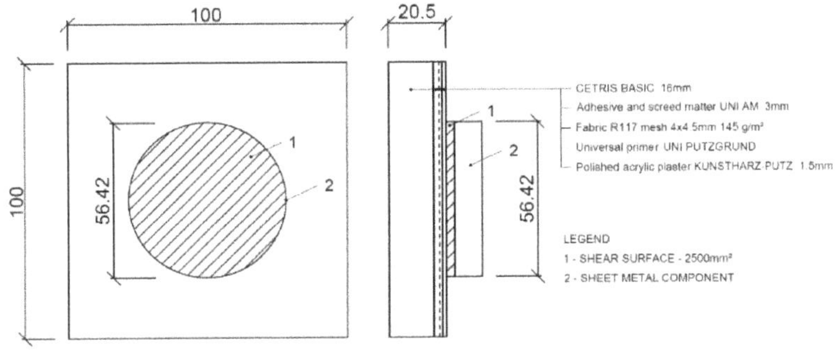

Figure 1. Test specimen for tensile test.

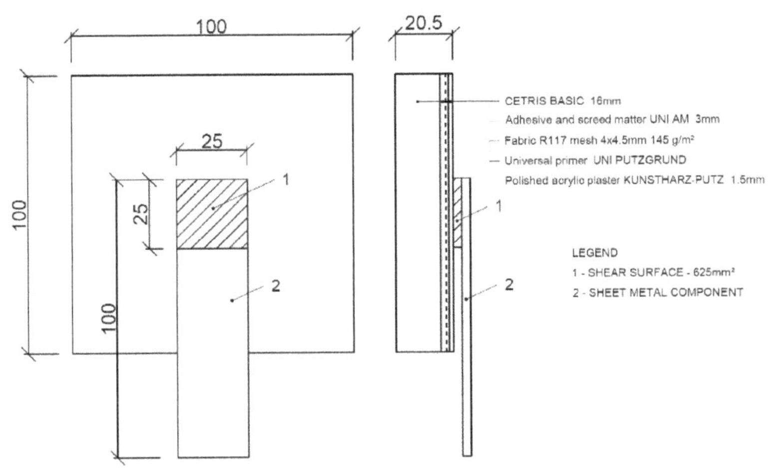

Figure 2. Test specimen for shear test.

2.2.2. Adhesion Tests

The tests were carried out on a LaborTech E.2 measuring device fitted with L06 vice jaws under normal laboratory conditions. The adhesion tests included both tensile and shear stresses. The basis of both the chosen test methods was the recording of the force course required to tear the adhered metal component of the substrate as a function of the elongation of the adhesive under testing. The loading was carried out at a rate of 5 mm per minute until failure, defined by a drop in force to 5% of the maximum achieved force for the tensile test and 30% for the shear test. The most important values for the evaluation were the maximum achieved force F (in N) and the corresponding elongation Δl (in mm). The tensile test was based on the procedure defined by the Czech technical standard ČSN 732,577 [12]. The principle of the test was to measure the force required to break away from the adhering body by tension applied perpendicular to the substrate. The tensile strength σ_{adh} (in MPa) was subsequently calculated according to the following relation:

$$\sigma_{adh} = \frac{F}{A} = \frac{F}{2500} \quad (1)$$

The shear test was based on the procedure defined by the European technical standard EN 1465 [13]. The principle of the test was to measure the tensile force acting parallel to the surface of the bonded joint. The shear adhesion (in MPa) was calculated according to the formula:

$$\tau = \frac{F}{A} = \frac{F}{625} \quad (2)$$

The elongation (in %) was subsequently calculated from the increment of the length Δl (in mm) to the original length l (in mm):

$$\varepsilon = \frac{\Delta l}{l} * 100 \quad (3)$$

3. Results

The results were used to generate a number of charts showing the dependence of the elongation Δl on the applied force F. An example of a chart for the tensile test is shown in Figure 3, and an example of a chart for the shear test is in Figure 4.

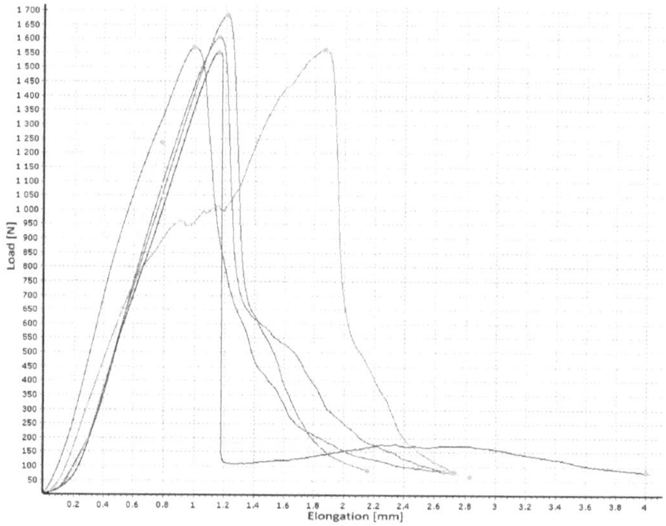

Figure 3. Example of tensile test results for measuring adhesive deformation on aluminium targets.

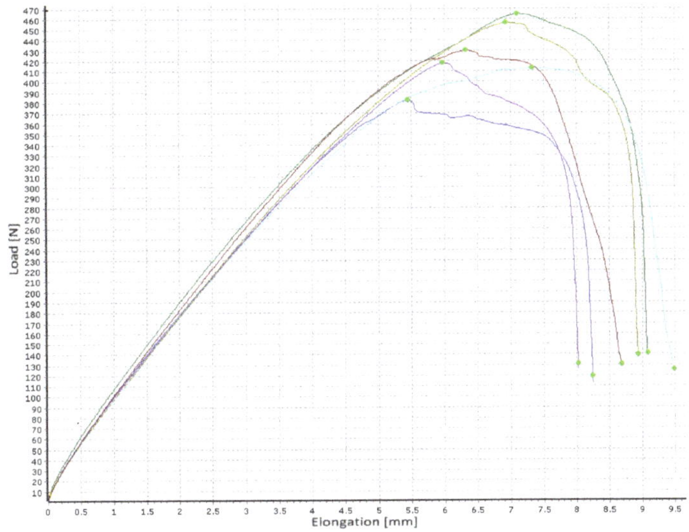

Figure 4. Example of shear test results for measuring adhesive deformation on galvanised strips.

The data from the deformation diagrams were tabulated in a clear way, with the decisive point for the evaluation being the maximum tensile or shear force and the corresponding elongation. Based on this data, the tensile adhesion σ_{adh} was calculated according to Formula (1), the shear adhesion τ was calculated according to Formula (2) and the elongation ε for both tensile and shear adhesion was calculated according to Formula (3). Statistical evaluation was performed for all the calculated parameters in the form of calculating the standard deviation s_x and the coefficient of variation v_x. The tensile test results can be seen in Table 1, and the shear test results are presented in Table 2.

Table 1. Results of tensile adhesion σ_{adh} (in MPa) and elongation ε (in %), standard deviation s_x (in MPa for σ_{adh}, in % for ε) and coefficient of variation v_x (in %) [1] for aluminium alloy (Al), galvanised steel (FeZn) and copper alloy (Cu) targets compared to results of adhesive alone (AD) on aluminium targets.

	Al	FeZn	Cu	AD
σ_{adh}	0.614	0.694	0.585	0.849
s_x	0.056	0.138	0.074	0.035
v_x	9.2	19.8	12.7	4.1
ε	39.8	35.5	33.3	90.1
s_x	11.1	8.3	10.4	10.2
v_x	27.8	23.4	31.3	11.3

[1] The average values are a sum of six measurements conducted for each tested combination.

A comparison of the adhesion and elongation in both tests can be seen in Figure 5. Possible types of destruction are shown in Figure 6. The most common type of destruction in the adhesion tests was the tearing of the metal target or strip, including the adhesive from the plaster surface. The second most common type of destruction was the tearing of the plaster, including the covering layer of the screed to the layer of the reinforcing fibreglass fabric. The least common destruction type presented only in copper strips in the shear test was the peeling of the adhesive from the copper material.

Table 2. Results of shear adhesion τ (in MPa) and elongation ε (in %), standard deviation s_x (in MPa for τ, in % for ε) and coefficient of variation v_x (in %) [1] for aluminium alloy (Al), galvanised steel (FeZn) and copper alloy (Cu) targets compared with the results of adhesive alone (AD) on aluminium strips.

	Al	FeZn	Cu	AD
τ	0.562	0.684	0.634	0.786
s_x	0.075	0.044	0.054	0.080
v_x	13.3	6.5	8.5	10.2
ε	223.3	217.0	200.0	403.3
s_x	18.6	22.1	24.2	69.2
v_x	8.3	10.2	12.1	17.2

[1] The average values are a sum of six measurements conducted for each tested combination.

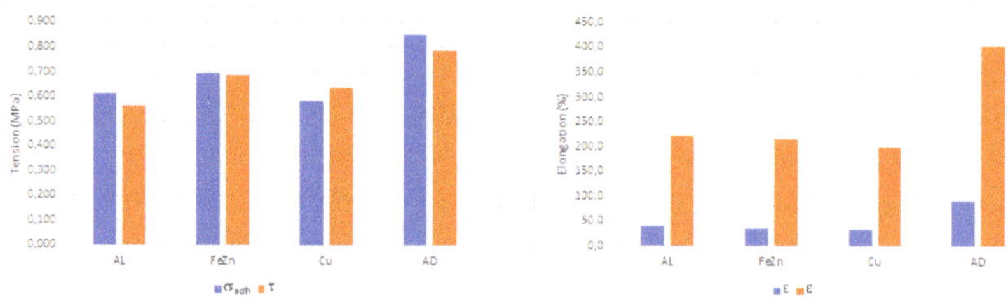

(a) (b)

Figure 5. Comparison of test results: (**a**) σ_{adh} (in MPa) for tensile test (blue) and τ (in MPa) for shear test (orange); (**b**) ε (in %) for tensile test (blue) and shear test (orange).

(a) (b) (c)

Figure 6. Possible types of destruction: (**a**) tearing of the metal strip including the adhesive from the plaster surface; (**b**) tearing of the plaster including the covering layer of the screed from the layer of reinforcing fibreglass fabric; (**c**) peeling off the adhesive from the copper strip.

4. Discussion

For the steel and aluminium components, no metal came off the adhesive in any of the measurements; the destruction always occurred in the underlying plaster. In the case of the copper alloy, a difference was determined between the tensile and shear tests. While in the tensile test the adhesive was torn off the plaster in all cases, in the shear test, on the other hand, all the measurements showed a complete peeling of the adhesive from the copper surface, and only in one case did a partial peeling of the adhesive from the copper surface combined with destruction in the plaster occur, as can be seen in Figure 5. The declared strength of the plaster by the manufacturer was 0.3 MPa. The real values of the observed tensile and shear adhesive strengths essentially exceed this value two-fold, which was clearly due to the higher strength of the plaster and its inherent adhesion to the underlying screed. In terms of the measured data, there was no significant difference in whether the destruction occurred in the surface thin-layer plaster or the screed, and the values of the measured forces and the calculated stresses were similar. However, lower values were obtained in the case of copper in the shear test, where the destruction occurred almost exclusively by peeling off the copper strip.

In addition to the tests of the bonded joint between the metal element and the plaster substrate, the tensile and shear properties of the adhesive itself were measured. The measurements were carried out using a standard tensile and shear test, where the aluminium targets and strips were bonded together, and subsequent destruction in the adhesive itself occurred during the measurement. Copper and steel zinc-plated elements were not used to measure the adhesive parameters. As expected, the adhesive itself performed better in both tests in terms of both adhesion and elongation (see Figure 5). In terms of tensile adhesion, the adhesive performed better by 34% on average, and in terms of shear adhesion, it performed better by 25%. In the case of elongation, the adhesive parameters were on average 150% better in the tensile test and 89% better in the shear test. However, elongation is not as important as adhesion in the case of bonding sheet-metal components to a facade. In this respect, the results demonstrated the fundamental fact that, under standard laboratory conditions, the tensile- and shear-load capacities of the adhesive itself were not significantly better than those of the bonded joint, which involved a metal element bonded to the surface of the used multi-layer facade plaster. It should be noted that the research did not include other important factors such as the effect of freezing and thawing, the effect of moisture, sudden temperature changes as described by Franco and Royer-Carfagni [14], adhesive thickness or ageing of the multi-layer plaster [15–19]. In practice, it would be glued to facades already subjected to external environmental influences, as described by Bochen [20].

5. Conclusions

The aim of the research was to verify in laboratory conditions whether it is reasonable to consider the technology of bonding sheet-metal components to the thin-layer plaster of thermal insulation systems. The start of the research was accompanied by a preliminary concern that the load capacity of the spot-bonded components would be low, which was estimated to the units of kilograms. The measured forces required to tear off a 2500 mm^2 bonded target were surprisingly high, ranging across all the materials tested from 99.3 kg to 199.5 kg of static load. For the 625 mm^2 adhesive strips, the range of forces required to tear it off ranged from 25.4 kg to 46.5 kg. The following conclusions were drawn based on the results obtained:

- The research was carried out exclusively under laboratory conditions and it provided interesting results at this initial stage;
- Even small bonded areas could carry significant tensile and shear loads sufficient for the anticipated loads, and the required load capacity of the bonded joint could be determined depending on the bonded area;
- The load capacity of the bonded joint will always depend on the quality of the substrate. In the case of a new plaster implemented in accordance with technological procedures, the use of the technology of additional bonding of any components to the surface of the multi-layer facade may be considered. However, the technology used must be properly tested;
- The obtained results do not yet allow the combinations of the materials used in this research to be declared a proven technology. The validation will continue with the inclusion of other influences acting on facades in outdoor environments. The results will be published separately.

The bonding technology of the sheet metal components and the issue of bonding to the surface of insulated facades, in general, appears to be promising for the future; however, further research is needed.

Author Contributions: Conceptualization, J.Š. and A.P.; methodology, J.Š. and B.N.; validation, J.Š., A.P. and A.B.; formal analysis, B.N.; investigation, J.Š. and A.P.; resources, A.P.; data curation, A.B.; writing—original draft preparation, J.Š.; writing—review and editing, A.P. and B.N.; visualization, J.Š.; supervision, B.N.; funding acquisition, J.Š. All authors have read and agreed to the published version of the manuscript.

Funding: This research was funded by an Internal Grant Programmes of the Brno University of Technology under No. FAST-S-22-7865 and No. FAST-J-22-7871.

Institutional Review Board Statement: Not applicable.

Informed Consent Statement: Not applicable.

Data Availability Statement: Not applicable.

Acknowledgments: The publication cost of this paper was covered with funds from the Polish National Agency for Academic Exchange (NAWA): "MATBUD'2023—Developing international scientific cooperation in the field of building materials engineering" BPI/WTP/2021/1/00002, MATBUD'2023

Conflicts of Interest: The authors declare no conflict of interest.

References

1. Nicklisch, F.; Weller, B.; Hommer, E.; Haberzettl, M. Evaluation of joining methods for novel timber–aluminum composite profiles for innovative louver windows and facade elements. *Wood Mater. Sci. Eng.* **2018**, *14*, 201–211. [CrossRef]
2. Chen, D. Heat Loss via Concrete Slab Floors in Australian Houses. In Proceedings of the 10th International Symposium on Heating, Ventilation and Air Conditioning (ISHVAC), Jinan, China, 19–22 October 2017. [CrossRef]
3. Sezer, F.S. Importance of Heat Insulation for Creating Energy Efficiency in Current Buildings: Bursa, Nilufer/Turkey Example. *Eur. J. Sustain. Dev.* **2017**, *6*, 57–68. [CrossRef]
4. Nečasová, B.; Liška, P. Research Summary on Characterizing Impact of Environment on Adhesion of Sealed Joints in Façade Applications. *Materials* **2020**, *13*, 4847. [CrossRef] [PubMed]
5. Nečasová, B.; Liška, P.; Novotný, M. Stress/Strain Behaviour of Mechanical and Adhesive Joints in Timber Façade Applications. *Key Eng. Mater.* **2020**, *868*, 142–149. [CrossRef]
6. Ślusarek, J.; Orlik-Kożdoń, B.; Bochen, J.; Muzyczuk, T. Impact of the imperfection of thermal insulation on structural changes of thin-layer façade claddings in ETICS. *J. Build. Eng.* **2020**, *32*, 101487. [CrossRef]
7. Olsson, L. Rain resistance of façades with façade details: A summary of three field and laboratory studies. *J. Build. Phys.* **2018**, *41*, 521–532. [CrossRef]
8. *ČSN EN 573-3*; Aluminium and Aluminium Alloys—Chemical Composition and Form of Wrought Products—Part 3: Chemical Composition and form of Products. Czech Standards Institute: Prague, Czech Republic, 2020.
9. *ČSN EN 13601*; Copper and Copper Alloys—Copper Rod, Bar and Wire for General Electrical Purposes. Czech Standards Institute: Prague, Czech Republic, 2021.
10. *ČSN EN 10025-2*; Hot Rolled Products of Structural Steels—Part 2: Technical Delivery Conditions for Non-Alloy Structural Steels. Czech Standards Institute: Prague, Czech Republic, 2020.
11. Davies, P.; Sohier, L.; Cognard, J.-Y.; Bourmaud, A.; Choqueuse, D.; Rinnert, E.; Créac'hcadec, R. Influence of adhesive bond line thickness on joint strength. *Int. J. Adhes. Adhes.* **2009**, *29*, 724–736. [CrossRef]
12. *ČSN 732577*; Test for Surface Finish Adhesion of Building Structures to the Base. Sorting No. 32691. Czech Office for Technical Approvals and Measurements: Prague, Czech Republic, 1981; p. 4.
13. *EN 1465*; Adhesives—Determination of Tensile Lap-Shear Strength of Bonded Assemblies. Sorting No. 66 8510. Czech Office for Standards, Metrology and Testing: Prague, Czech Republic, 2009; p. 12.
14. Franco, A.; Royer-Carfagni, G. Contact stresses in adhesive joints due to differential thermal expansion with the adherends. *Int. J. Solids Struct.* **2016**, *87*, 26–38. [CrossRef]
15. Machalická, K.V.; Vokáč, M.; Pokorný, P.; Pavlíková, M. Effect of various artificial ageing procedures on adhesive joints for civil engineering applications. *Int. J. Adhes. Adhes.* **2020**, *97*, 102476. [CrossRef]
16. Machalická, K.V.; Vokáč, M.; Kostelecká, M.; Eliášová, M. Durability of structural adhesive joints for facade applications exposed to the extended cataplasm test. *J. Adhes.* **2019**, *95*, 632–652. [CrossRef]
17. Machalická, K.; Vokáč, M.; Eliášová, M. Influence of artificial aging on structural adhesive connections for façade applications. *Int. J. Adhes. Adhes.* **2018**, *83*, 168–177. [CrossRef]
18. Chew, M.Y.L.; Zhou, X. Enhanced resistance of polyurethane sealants against cohesive failure under prolonged combination of water and heat. *Polym. Test.* **2002**, *21*, 187–193. [CrossRef]

19. Nečasová, B.; Liška, P.; Novotný, M. Ageing of Adhesive Joints for Façade Applications—Comparison of Artificial and Real Weathering Conditions. *MATEC Web Conf.* **2019**, *279*, 02013. [CrossRef]
20. Bochen, J. Study on the microstructure of thin-layer facade plasters of thermal insulating system during artificial weathering. *Constr. Build. Mater.* **2009**, *23*, 2559–2566. [CrossRef]

materials proceedings

MDPI

Proceeding Paper

Physico-Chemical Modelling of Chloride Migration in Cement-Based Materials Considering Electrode Processes †

Zine-Eddine Kribes, Rachid Cherif * and Abdelkarim Aït-Mokhtar

LaSIE, UMR, CNRS 7356, La Rochelle University, 17000 La Rochelle, France
* Correspondence: rachid.cherif@univ-lr.fr
† Presented at the 10th MATBUD'2023 Scientific-Technical Conference "Building Materials Engineering and Innovative Sustainable Materials", Cracow, Poland, 19–21 April 2023.

Abstract: A multispecies model for chloride migration in cementitious materials was proposed. The model is applied to simulate chloride migration in the concrete submitted to the chloride migration test, considering the electrode processes. These processes are very rarely integrated into the literature, despite their impact on the chloride migration and the electroneutrality in the pore solution of the material. The generation of OH^- in the cathode and H^+ in the anode allows for the monitoring of the electroneutrality. The modelling considers all the ions of the pore solution. Ion fluxes are calculated using Nernst–Planck equation. The Langmuir model is used to simulate the chloride isotherms. The thermodynamic equilibrium in the material is considered, which reflects the liquid–solid interactions during the migration. The ion profiles with and without considering the electrode processes are presented.

Keywords: durability; chloride migration; modelling; multispecies interactions; electrode processes

check for updates

Citation: Kribes, Z.-E.; Cherif, R.; Aït-Mokhtar, A. Physico-Chemical Modelling of Chloride Migration in Cement-Based Materials Considering Electrode Processes. *Mater. Proc.* **2023**, *13*, 37. https://doi.org/10.3390/materproc2023013037

Academic Editors: Katarzyna Mróz, Tomasz Tracz, Tomasz Zdeb and Izabela Hager

Published: 20 February 2023

1. Introduction

The corrosion of steel in reinforced concrete (RC) structures in coastal zones is mainly induced by chloride ingress, which represents a permanent risk of degradation. Chlorides penetrate into the cover concrete under a concentration gradient between seawater and the concrete pore solution and/or liquid pressure gradient in partially saturated concrete when heat and moisture transfers occur (tidal areas or marine fogs). When chlorides reach the rebars with a threshold concentration, they generate rebar depassivation and corrosion [1]. Given the high cost of maintenance/repair, nowadays the durability of RC structures in their environments is one of the main challenges mentioned in the specifications of construction.

In the last decades, several experimental and numerical studies were developed to propose methods and tools for predicting chloride transport in cementitious materials. The standard migration test is used to determine the chloride diffusion coefficient in a steady state or non-study state [2–8]. Moreover, single-species modelling was used to predict chloride transport in cementitious materials [2]. After that, many multispecies approaches were developed, considering several ions in the pore solution and more chemical and physical interactions during the transport [9–20]. Xia and Li [21] proposed numerical modelling of ion transport in saturated cementitious materials based on Poisson–Nernst–Planck (PNP) equations, considering the chemical interaction between the monovalent ions in the pore solution in order to monitor the impact of the interactions on the chloride ingress. Fenaux et al. [22] proposed a chloride transport modelling in saturated concrete taking into account monovalent and divalent ions of the pore solution: Cl^-, Na^+, K^+, OH^- and Ca^{2+}. The diffusion, migration and chemical activity were considered. The chemical activity was calculated using the Pitzer model. The numerical results highlighted the influence of the composition of the pore solution and the chemical activity on the chloride penetration.

Furthermore, recent research discussed the impact of the thermodynamic equilibria on the chloride reactive transport in cementitious materials [23–27]. Yu and Zhang [28]

proposed a model to predict the leaching of cement paste in an ammonium nitrate solution taking into account ion transport, chemical kinetics and thermodynamic equilibria. The ion transport was calculated by the PNP equation, while, the chemical activity was calculated using the Davies model. Tran et al. [29,30] developed a coupling between chloride transport and thermodynamic equilibrium, considering kinetic control to predict chloride fixation in concrete. Jensen et al. [31,32] developed a multispecies approach for reactive mass transport in a saturated mortar exposed to chlorides for 180 days, including the chemical equilibria. Cherif et al. [33] proposed a coupling between multispecies transport and thermodynamic equilibria in cementitious materials containing mineral additions. The modelling considered all the ions of the pore solution, the portlandite dissolution and Friedel's and Kugel salt precipitation during chloride transport. Ion fluxes were calculated using Nernst–Planck equation, while the thermodynamic coupling was based on the low mass action and the rate of dissolution/precipitation of the solid phases.

The models mentioned above are generally based on the PNP equation with a limited number of ions in the pore solution, whose concentrations are considered significant. The other models consider all the multispecies interactions in the material, but they are based on Fick's law, which misdescribes the chloride ion transport. Researchers dealing with chloride migration, considering electrode processes remain very little in the literature.

Motivated by this need, multispecies modelling of chloride migration is proposed in this study, considering the electrode processes. The modelling is applied on Portland cement submitted to standard chloride migration test. The electrode processes reflect the generation of OH^- and H^+ in the cathode and anode, respectively. The electrode processes ensure the electroneutrality in the migration cell (sample and compartments). The concentrations of OH^- and H^+ are calculated from the current density measured during the test, using Faraday's law. The charge passed is deduced from the current density measured. Ion fluxes are calculated by the Nernst–Planck equation, which describes the diffusion and migration of the species. The Langmuir model is used to simulate the chloride chemical fixation by the material (chloride isotherms). The chemical activity is neglected according to [3]. The considered ions are Cl^-, Na^+, K^+, OH^-, H^+, Ca^{2+} and $SO_4{}^{2-}$. The migration cell used is composed of two compartments: (1) upstream containing 25 mM NaOH and 83 KOH and 500 mM NaCl; (2) downstream containing only 25 mM NaOH and 83 KOH (boundary conditions). The composition of the pore solution of the material tested was considered the initial condition. An electrical field of 300 $V{\cdot}m^{-1}$ was applied at the sample boundaries and monitored using two calomel reference electrodes. The latter were placed at each side of the sample tested in order to maintain the electrical field constant modelling outputs are as follows:

- The ion profiles in the material that can be used for the calculation of the ion diffusion coefficients in the non-steady state (from the ion-penetration depth), with and without the electrode processes.
- The electroneutrality in the sample tested with and without integrating the electrode processes in order to highlight the need for the consideration of the electrode processes in the chloride migration modelling proposed.

2. Methodology

2.1. Modelling Principle

The time evolution of the ion concentration (C_i) during the migration test is calculated by using the mass balance equation (Equation (1)), which takes into account the porosity of the material tested (φ), the chloride concentration bonded to the cement matrix ($C_{i,b}$) calculated by the Langmuir's model and the ion flux (J_i) calculated by the NP equation (see Equation (2)). The internal electrical potential between ions is neglected in front of the applied electrical field of 300 $V{\cdot}m^{-1}$. The mass exchange term (q_i), added to the mass balance equation, describes the ion gain/loss in the pore solution due to the dissolution/precipitation of the solid phases considered (C-S-H, portlandite, monosulfoaluminates and trisulfoaluminates). Further details about the calculations of the term (q_i) and the thermodynamic equilibrium constants used are shown in [8]. The ions considered in this study are: Cl^-, Na^+, K^+, OH^-, H^+ Ca^{2+} and SO_4^{2-}. Note that the proposed modelling concerns ion transport in saturated materials that do not require coupling with convection and moisture transfer.

$$\frac{\partial C_i}{\partial t} + \frac{((1-\varphi)C_{i,b})}{\partial t} = -div(J_i) - \frac{\partial q_i}{\partial t} \quad (1)$$

$$J_i = -D_{E,i}\left(\nabla C_i + \frac{z_i C_i F E}{RT} + C_i \nabla ln\gamma_i\right) \quad (2)$$

where $D_{E,i}$ [$m^2.s^{-1}$] is the effective diffusion coefficient of the ion i, z_i is the valence of the ion i, F [$C\cdot mol^{-1}$] is the Faraday constant, E [$V\cdot m^{-1}$] is applied electric field, R [$J\cdot K^{-1}\cdot mol^{-1}$] is the ideal gas constant and T [K] is the temperature γ_i is the ion activity coefficient.

The electrode processes responsible for the generation of OH^- in the catholyte (upstream) and H^+ in the anolyte (downstream) are given in the following. Note that non-corrodible Platine electrodes were used.

$$\text{Cathode: } 2\,H_2O + 2\,e^- \rightarrow H_2 + 2\,OH^- \quad (3)$$

$$\text{Anode: } H_2O \rightarrow 0.5\,O_2 + 2\,H^+ + 2\,e^- \quad (4)$$

2.2. Case Study

The multispecies approach proposed is used to simulate all the ion movements in the migration cell during the standard migration test on ordinary concrete for 15 days. The composition of the concrete used is given in Table 1. The chemical composition of the pore solution of the concrete used is taken from Refs. [8,34] (see Table 2). The standard migration test simulated is composed of two compartments: the upstream and downstream compartments containing a basic solution of 25 mM NaOH and 83 KOH. In addition, 500 mM of NaCl is added to the upstream. An electrical field of 300 $V\cdot m^{-1}$ is applied at the sample boundaries. Concrete samples of 1 cm thickness are used (1D modelling). In this study, the numerical results show the ion profiles in the material tested that are useful for the calculation of the diffusion coefficient at the non-steady state [4,35].

Table 1. The composition of the concrete used.

	Cement [$kg\cdot m^{-3}$]	Sand [$kg\cdot m^{-3}$]	Coarse Aggregate [$kg\cdot m^{-3}$]	Water [$kg\cdot m^{-3}$]	W/C
Concrete	300	710	1242	180	0.6

Table 2. Initial and boundary conditions used.

	Boundary Conditions		Initial Conditions
	Upstream (x = 0, t) [$mol\cdot m^{-3}$]	Downstream (x = L, t) [$mol\cdot m^{-3}$]	Sample (t = 0, 0 < x < L) [$mol\cdot m^{-3}$] [8,34]
$C_{Cl}{}^-$	500	0	4
$C_{Na}{}^+$	525	25	51
$C_K{}^+$	83	83	117
$C_{OH}{}^-$	108	108	158
$C_{Ca}{}^{2+}$	0	0	2
$C_{SO4}{}^{2-}$	0	0	2

3. Results and Discussion

Figures 1 and 2 show the simulated profiles of Cl^-, Na^+, K^+, OH^-, H^+, Ca^{2+} and SO_4^{2-} in the sample at the end of the migration test (after 15 days) with and without considering the electrode processes, respectively. The free chloride concentration in the pore solution is maximum along the sample depth (~459 mol·m^{-3}) because of their migration from the upstream to the sample. The max concentration is relatively different compared to the literature's data without considering the thermodynamic equilibrium (the participation of chloride with the other ions in the pore solution to form salts). For these models in the literature, the max concentration is equal to the boundary condition on the side of the upstream (500 mol·m^{-3}) [2]. Bulleted lists look like this:

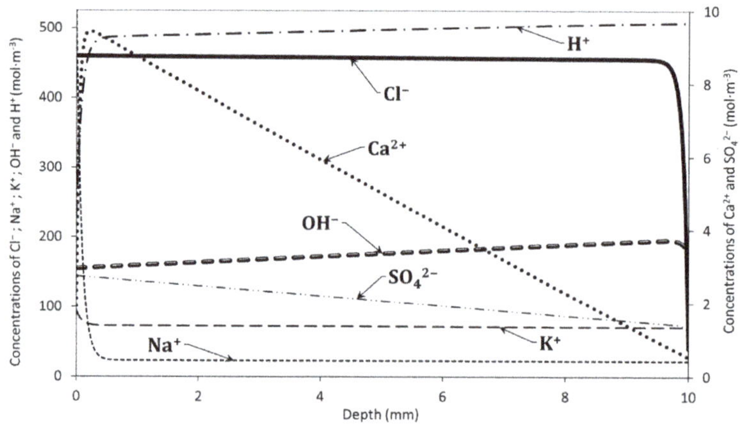

Figure 1. Profiles of Cl^-, Na^+, K^+, OH^-, H^+, Ca^{2+} and SO_4^{2-} in the sample at the end of the migration test (15 days) considering electrode processes.

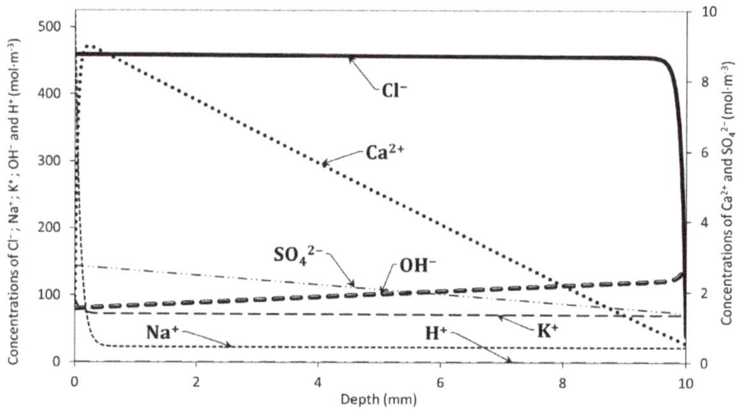

Figure 2. Profiles of Cl^-, Na^+, K^+, OH^-, H^+, Ca^{2+} and SO_4^{2-} in the sample at the end of the migration test (15 days) without considering electrode processes.

Moreover, we noted an increase in the concentrations of Na^+ and K^+ in the pore solution due to their migration from the downstream to the sample tested. The increase in Ca^{2+} and SO_4^{2-} concentrations are due to the dissolution of the portlandite, monosulfoaluminates, and trisulfoaluminates under the electrical field [36,37].

Finally, a difference between the concentration of H^+ and OH^- with and without considering the electrode processes is noticed. This is reflected by the electroneutrality ensured in the case of the migration modelling with electrode processes and not ensured in the case of the modelling without electrode processes (see Figure 3). This impacts also the concentration of the other ions in the pore solution. The electroneutrality was calculated using Equation (5). The results obtained confirm the need of considering the electrode processes in the chloride migration modelling.

$$\sum C_i z_i = 0 \quad (5)$$

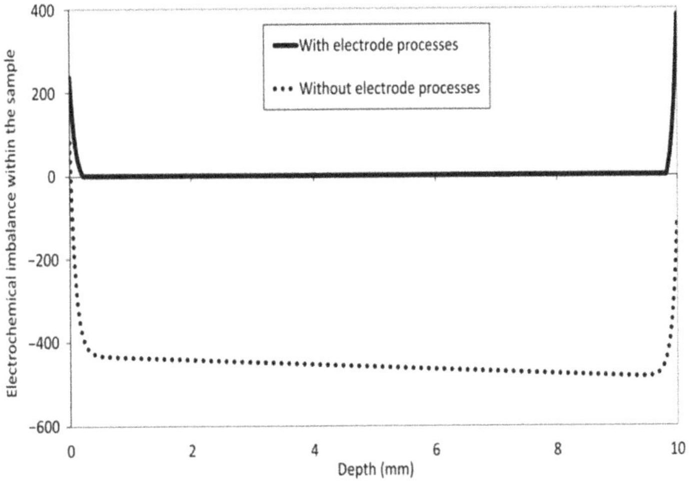

Figure 3. Electrochemical imbalance in the pore solution of the sample, deduced from ion concentrations in the steady state with and without considering electrode processes.

4. Conclusions

From the obtained results, the following main conclusions are made:

- The modelling proposed allowed us to simulate the standard migration test at the steady and non-steady states considering the real pore solution of the material tested and the dissolution/precipitation phenomena during the migration. The modelling was applied to OPC-based materials.
- Outputs of the modelling proposed are the ion profiles in the material tested during the migration test.
- The numerical results show the need of considering the electrode processes in the chloride migration modelling in order to better simulate the standard migration test. The model proposed could be improved by considering more solid phases of the material such as C-S-H, oxychloride, etc.

Author Contributions: Conceptualization, R.C. and A.A.-M.; methodology, Z.-E.K., R.C. and A.A.-M.; software, Z.-E.K. and R.C.; validation, Z.-E.K., R.C. and A.A.-M.; formal analysis, R.C. and A.A.-M.; resources, A.A.-M.; data curation, R.C. and A.A.-M.; writing—original draft preparation, Z.-E.K.; writing—review and editing, Z.-E.K., R.C. and A.A.-M.; visualization, R.C. and A.A.-M.; supervision, A.A.-M.; project administration, A.A.-M.; funding acquisition, A.A.-M. All authors have read and agreed to the published version of the manuscript.

Funding: This research was funded by the project ANR El-TORO (French agency of research).

Institutional Review Board Statement: Not applicable.

Informed Consent Statement: Not applicable.

Data Availability Statement: Data are available from the authors on request.

Acknowledgments: The authors would like to thank the project ANR El-TORO (French agency of research) for funding this work on durability and corrosion of cementitious materials. Publication cost of this paper was covered with founds of the Polish National Agency for Academic Exchange (NAWA): "MATBUD'2023—Developing international scientific cooperation in the field of building materials engineering" BPI/WTP/2021/1/00002, MATBUD'2023.

Conflicts of Interest: The authors declare no conflict of interest.

References

1. Poupard, O.; Aït-Mokhtar, A.; Dumargue, P. Impedance spectroscopy in reinforced concrete: Experimental procedure for monitoring steel corrosion part II: Polarization effect. *J. Mat. Sci.* **2003**, *38*, 3521–3526. [CrossRef]
2. Amiri, O.; Aït-Mokhtar, A.; Seigneurin, A. A complement to the discussion of Xu A. and Chandra S. about the paper "Calculation of chloride coefficient diffusion in concrete from ionic migration measurements" by Andrade C. *Cem. Concr. Res.* **1997**, *27*, 951–957. [CrossRef]
3. Truc, O.; Ollivier, J.P.; Carcassès, M. A new way for determining the chloride diffusion coefficient in concrete from steady state migration test. *Cem. Concr. Res.* **2000**, *30*, 217–226. [CrossRef]
4. Castellote, M.; Andrade, C.; Alonso, C. Measurement of the steady and non-steady-state chloride diffusion coefficients in a migration test by means of monitoring the conductivity in the anolyte chamber. Comparison with natural diffusion tests. *Cem. Concr. Res.* **2001**, *31*, 1411–1420. [CrossRef]
5. Yang, C.C.; Cho, S.W.; Chi, J.M.; Huang, R. An electrochemical method for accelerated chloride migration test in cement-based materials. *Mater. Chem. Phys.* **2003**, *77*, 461–469. [CrossRef]
6. Aït-Mokhtar, A.; Amiri, O.; Poupard, O.; Dumargue, P. A new method for determination of chloride flux in ce-ment-based materials from chronoamperometry. *Cem. Concr. Compos.* **2004**, *26*, 339–345. [CrossRef]
7. Friedmann, H.; Amiri, O.; Aït-Mokhtar, A.; Dumargue, P. A direct method for determining chloride diffusion coefficient by using migration test. *Cem. Concr. Res.* **2004**, *34*, 1967–1973. [CrossRef]
8. Cherif, R.; Hamami, A.E.A.; Aït-Mokhtar, A.; Siddique, R. Chemical properties of pore solution of hardened cement pastes with mineral additions. *Adv. Cem. Res.* **2021**, *33*, 331–341. [CrossRef]
9. Li, Y.L.; Page, C.L. Modelling of electrochemical chloride extraction from concrete: Influence of ionic activity coefficients. *Comput. Mater. Sci.* **1998**, *9*, 303–308. [CrossRef]
10. Samson, E.; Marchand, J. Numerical Solution of the Extended Nernst–Planck Model. *J. Colloid Interface Sci.* **1999**, *215*, 1–8. [CrossRef]
11. Amiri, O.; Aït-Mokhtar, A.; Dumargue, P.; Touchard, G. Electrochemical modelling of chloride migration in cement-based materials: Part I: Theoretical basis at microscopic scale. *Electrochim. Acta* **2001**, *46*, 1267–1275. [CrossRef]
12. Amiri, O.; Aït-Mokhtar, A.; Dumargue, P.; Touchard, G. Electrochemical modelling of chlorides migration in cement-based materials. Part II: Experimental study—Calculation of chlorides flux. *Electrochim. Acta* **2001**, *46*, 3589–3597. [CrossRef]
13. Amiri, O.; Friedmann, H.; Aït-Mokhtar, A. Modelling of chloride-binding isotherm by multi-species approach in cement mortars submitted to migration test. *Mag. Concr. Res.* **2006**, *58*, 93–99. [CrossRef]
14. Samson, E.; Marchand, J. Modeling the transport of ions in unsaturated cement-based materials. *Comput. Struct.* **2007**, *85*, 1740–1756. [CrossRef]
15. Friedmann, H.; Amiri, O.; Aït-Mokhtar, A. Modeling of the Electrical Double Layer Effects on Multispecies Ions Transport in Cement-based Materials. *Cem. Concr. Res.* **2008**, *38*, 1394–1400. [CrossRef]
16. Friedmann, H.; Amiri, O.; Aït-Mokhtar, A. Shortcomings of geometrical approach in multi-species modelling of chloride migration in cement-based materials. *Mag. Concr. Res.* **2008**, *60*, 119–124. [CrossRef]
17. Bourbatache, K.; Millet, O.; Aït-Mokhtar, A. Ionic transfer in charged porous media. Periodic homogenization and parametric study on 2D microstructures. *Int. J. Heat Mass Transf.* **2012**, *55*, 5979–5991. [CrossRef]
18. Bourbatache, K.; Millet, O.; Aït-Mokhtar, A.; Amiri, O. Modeling the Chlorides Transport in Cementitious Materials By Periodic Homogenization. *Transp. Por. Med.* **2012**, *94*, 437–459. [CrossRef]
19. Andrade, C.; Prieto, M.; Tanner, P.; Tavares, F.; Andrea, R. Testing and modelling chloride penetration into concrete. *Constr. Build. Mater.* **2013**, *39*, 9–18. [CrossRef]
20. Nguyen, P.T.; Amiri, O. Study of the chloride transport in unsaturated concrete: Highlighting of electrical double layer, temperature and hysteresis effects. *Constr. Build. Mater.* **2016**, *122*, 284–293. [CrossRef]
21. Xia, J.; Li, L. Numerical simulation of ionic transport in cement paste under the action of externally applied electric field. *Constr. Build. Mater.* **2013**, *39*, 51–59. [CrossRef]
22. Fenaux, M.; Reyes, E.; Gálvez, J.C.; Moragues, A. Modelling the transport of chloride and other ions in cement-based materials. *Cem. Concr. Compos.* **2019**, *97*, 33–42. [CrossRef]
23. Mao, L.; Hu, Z.; Xia, J.; Feng, G.L.; Azim, I.; Yang, J.; Liu, Q.F. Multi-phase modelling of electrochemical rehabilitation for ASR and chloride affected concrete composites. *Compos. Struct.* **2019**, *207*, 176–189. [CrossRef]
24. Steefel, C.I.; Appelo, C.A.J.; Arora, B.; Jacques, D.; Kalbacher, T.; Kolditz, O.; Lagneau, V.; Lichtner, P.C.; Mayer, K.U.; Meeussen, J.C.L.; et al. Reactive transport codes for subsurface environmental simulation. *Comput. Geosci.* **2015**, *19*, 445–478. [CrossRef]

25. Guo, B.; Hong, Y.; Qiao, G.; Ou, J.; Li, Z. Thermodynamic modeling of the essential physicochemical interactions between the pore solution and the cement hydrates in chloride-contaminated cement-based materials. *J. Colloid Interface Sci.* **2018**, *531*, 56–63. [CrossRef] [PubMed]
26. Sanchez, T.; Henocq, P.; Millet, O.; Aït-Mokhtar, A. Coupling PhreeqC with electro-diffusion tests for an accurate de-termination of the diffusion properties on cementitious materials. *J. Electroanal. Chem.* **2020**, *858*, 113791. [CrossRef]
27. Sanchez, T.; Conciatori, D.; Laferriere, F.; Sorelli, L. Modelling capillary effects on the reactive transport of chloride ions in cementitious materials. *Cem. Concr. Res.* **2020**, *131*, 106033. [CrossRef]
28. Yu, Y.; Zhang, Y.X. Coupling of chemical kinetics and thermodynamics for simulations of leaching of cement paste in ammonium nitrate solution. *Cem. Concr. Res.* **2017**, *95*, 95–107. [CrossRef]
29. Tran, V.Q.; Soive, A.; Baroghel-Bouny, V. Modelisation of chloride reactive transport in concrete including thermodynamic equilibrium, kinetic control and surface complexation. *Cem. Concr. Res.* **2018**, *110*, 70–85. [CrossRef]
30. Tran, V.; Soive, A.; Bonnet, S.; Khelidj, A. A numerical model including thermodynamic equilibrium, kinetic control and surface complexation in order to explain cation type effect on chloride binding capability of concrete. *Constr. Build. Mater.* **2018**, *191*, 608–618. [CrossRef]
31. Jensen, M.M.; De Weerdt, K.; Johannesson, B.; Geiker, M.R. Framework for reactive mass transport: Phase change modeling of concrete by a coupled mass transport and chemical equilibrium model. *Comput. Mater. Sci.* **2014**, *92*, 213–223. [CrossRef]
32. Jensen, M.M.; De Weerdt, K.; Johannesson, B.; Geiker, M.R. Use of a multi-species reactive transport model to simulate chloride ingress in mortar exposed to NaCl solution or sea-water. *Comput. Mater. Sci.* **2015**, *105*, 75–82. [CrossRef]
33. Cherif, R.; Hamami, A.E.A.; Aït-Mokhtar, A.; Bosschaerts, W. Thermodynamic equilibria-based modelling of reactive chloride transport in blended cementitious materials. *Cem. Concr. Res.* **2022**, *156*, 106770. [CrossRef]
34. Andersson, K.; Allard, B.; Bengtsson, M.; Magnusson, B. Chemical composition of cement pore solutions. *Cem. Concr. Res.* **1989**, *19*, 327–332. [CrossRef]
35. Bogas, J.A.; Gomes, A. Non-steady-state accelerated chloride penetration resistance of structural lightweight aggregate concrete. *Cem. Concr. Compos.* **2015**, *60*, 111–122. [CrossRef]
36. Castellote, M.; Andrade, C.; Alonso, M.C. Changes in Concrete pore size distribution due to electrochemical chloride migration trials. *Mater. J.* **1999**, *96*, 314–319.
37. Huang, Q.; Wang, C.; Zeng, Q.; Yang, C.; Luo, C.; Yang, K. Deterioration of mortars exposed to sulfate attack under electrical field. *Constr. Build. Mater.* **2016**, *117*, 121–128. [CrossRef]

materials proceedings

MDPI

Proceeding Paper

The Effect of Preconditioning Temperature on Gas Permeability of Alkali-Activated Concretes †

Patrycja Duży [1,2,*], Marta Choinska Colombel [2], Izabela Hager [1] and Ouali Amiri [2]

1 Chair of Building Materials Engineering, Faculty of Civil Engineering, Cracow University of Technology, 31-155 Cracow, Poland
2 Research Institute in Civil and Mechanical Engineering GeM—UMR CNRS 6183, IUT Saint-Nazaire, Nantes University, 44600 Saint-Nazaire, France
* Correspondence: patrycja.duzy@doktorant.pk.edu.pl
† Presented at the 10th MATBUD'2023 Scientific-Technical Conference "Building Materials Engineering and Innovative Sustainable Materials", Cracow, Poland, 19–21 April 2023.

Abstract: Alkali-activated materials (AAMs) are a group of environmentally friendly binders considered alternatives to conventional cementitious binders. They utilise industrial wastes such as slag and fly ash to reduce cement production and related CO_2 emissions. Despite the strong interest of researchers, the application of alkali-activated concrete (AAC) in constructions is still very limited. Given the difference in the process of producing the AAC and ordinary Portland cement concrete (OPCC), some of the testing methods need to be adjusted to a new type of binder. The increased sensitivity of AAM to high temperatures leads to discussions on the results achieved in the gas permeability tests that require the samples to be dried first. In this paper, the influence of drying temperature applied to the samples on the gas permeability will be presented. The binders' precursors are blends of fly ash (FA) and ground granulated blast furnace slag (GGBFS) in slag proportions of 5%, 20%, and 35%, expressed by the mass of FA. Materials are denoted AAC5, AAC20, and AAC35, respectively. Measurements of the gas permeability of concretes were conducted by the RILEM–CEMBUREAU method, with lab adaptation for gas flow measurements. The comparison of results obtained shows the increase in gas permeability values with the temperature. However, the corresponding effect of temperature on permeability is driven by, on the one hand, the binder composition, and on the other hand, the aggregate's nature.

Keywords: gas permeability; geopolymer concrete; temperature impact; drying

Citation: Duży, P.; Choinska Colombel, M.; Hager, I.; Amiri, O. The Effect of Preconditioning Temperature on Gas Permeability of Alkali-Activated Concretes. *Mater. Proc.* **2023**, *13*, 38. https://doi.org/10.3390/materproc2023013038

Academic Editors: Katarzyna Mróz, Tomasz Tracz and Tomasz Zdeb

Published: 20 February 2023

1. Introduction

Concrete, as one of the most commonly used construction materials, is exposed to plenty of aggressive environmental factors [1]. Transport of external media is conducted by a connected pore structure [2]. This is the reason why the description of the pore network deserves special attention in the analysis of concrete's durability [3,4].

Gas permeability is one of the parameters that describes the pore structure of materials, therefore, it is considered one of the most popular parameters to determine their quality and durability [3,5]. Among the variety of methods used to determine the gas permeability, the most common are the CEMBUREAU method [6] and the oxygen permeability index. The principle of these tests is to measure the amount of gas passing through the sample with a known cross-sectional area per unit of time. The most commonly used gases are oxygen, nitrogen, or dry air. These methods of measurement refer to laboratory tests. In situ testing by Figg's method [7] or others methods described in the literature [8,9] is acceptable but the main disadvantage of them is the lack of possibility to control the impact of temperature and of moisture on the material. These parameters strongly affect the result of permeability measurements [10,11].

Gas permeability measurements are basic to the determination of the durability of constructions and modelling the service life [12]. The CEMBUREAU method is considered a reference method for air permeability for other methods. However, laboratory methods require temperature preconditioning of the specimens [6,13]. The temperature of drying ranges from 50 °C to 105 °C. It is clear that the results of measurements will not be the same for variable conditions. The definition of the temperature impact on the results obtained may allow researchers to adjust the condition to the tested materials and compare it with others.

The temperature impact on geopolymers' microstructure is the subject of many investigations [14,15]. Many of them focus on behaviour in high temperatures up to 1000 °C. [16,17]. However, it is a well-known fact that the temperature of curing strongly affects geopolymer properties, especially alkali-activated fly ashes [18,19]. In addition, for alkali-activated slag, temperature of curing influences its microstructure and porosity [20]. Taking into consideration the influence of preconditioning of specimens on gas permeability, the question about its applicability to geopolymer concretes is an unsolved problem.

The temperature that activates fly ashes has been investigated by many researchers in recent years [21,22]. Palomo et al. [23] presented the evolution of properties of alkali-activated fly ash cured at temperatures between 65 °C and 85 °C. The ingress of compressive strength with high temperature reached 60 MPa. With these changes, it is obvious to assume that all the properties connected with porosity were also affected by curing temperature. These studies only confirmed the theories made much earlier [24]. Similar research was conducted to define the influence of the activator used [25] for mixes and their behaviours in temperatures between 75 °C and 95 °C. Unfortunately, there is no possibility to specify one general rule to describe the influence of curing temperature on porosity and microstructure of alkali-activated materials. It depends on the type of precursor and activator and on the temperature applied. In the case of gas permeability measurement, it is more necessary to compare the results obtained for materials with different compositions than to obtain accurate values of permeability. In this paper, the influence of preconditioning temperature on gas permeability measurements and comparison of this influence on different compositions of alkali-activated concretes are presented.

2. Materials

The studies were conducted on six compositions of alkali-activated concretes based on fly ash (Połaniec powerplant, Poland) with ground granulated blast furnace slag (Ekocem Dąbrowa Górnicza, Poland) replacement by mass. Three levels of GGBFS dosage were examined: 5%, 20%, and 35%. Two types of coarse aggregate were used to detect its influence on gas permeability. The tested blends were denoted AAC5B, AAC20B, and AAC35B for basalt aggregate and AAC5D, AAC20D, and AAC35D for dolomite aggregate with 5%, 20%, and 35% of slag, respectively. According to ASTM C618 Standard Specification for Coal Fly Ash and Raw or Calcined Natural Pozzolan for Use in Concrete [26], the FA was categorised as class F, and is mainly composed of silicon dioxide and aluminium oxide. In the case of the slag used, it mainly consists of calcium oxide and silicon oxide. The chemical compositions of precursors are presented in Table 1. For activation, the diluted sodium silicate solution Geosil® 34417, supplied by Woellner, was used.

Table 1. Oxide composition of fly ash and ground granulated blast furnace slag.

wt.%	SiO_2	Al_2O_3	Fe_xO_y	CaO	MgO	SO_3	K_2O	Na_2O	P_2O_5	TiO_2	Mn_3O_4	Cl^-
FA	52.30	28.05	6.32	3.05	1.71	0.28	2.51	0.76	0.69	1.35	0.07	-
GGBFS	39.31	7.61	1.49	43.90	4.15	0.51	0.36	0.47	-	-	-	0.04

The mixing procedure was performed based on prior experience of the research group [27]. The first step was to prepare pastes. Fly ash was mixed with diluted alkaline solution until a homogenous consistency was obtained. Slag was added with constant stirring. At that time, a mixture of coarse aggregate (basalt or dolomite) and quartz sand was prepared in a larger mixer. When the paste was ready, it was added to the mixer with the aggregate and distributed evenly throughout the mass. The samples were cast and compacted on a shaking table in cylindrical moulds with diameter 11 cm and height 22 cm. They were stored in ambient conditions at 20 ± 2 °C and protected from water evaporation by plastic film. For the prepared samples, no thermal curing was applied.

3. Methods

The CEMBUREAU method is still one of the most popular methods of gas permeability measurements [28]. The values of inlet and outlet pressures are steady during the measurement. Identification of gas permeability is based on stabilised permanent gas flow. The atmospheric pressure and temperature, which may affect the results, are also taken into account for permeability estimation. Lateral pressure of 8 bars is applied to the rubber gum surrounding the specimen to exclude gas leaking around the sample.

Measurements in accordance with this standard were initially carried out within samples which had never been dried before. The value of gas permeability started to stabilise after 240 days and this continued for up to 360 days, as given in Figure 1.

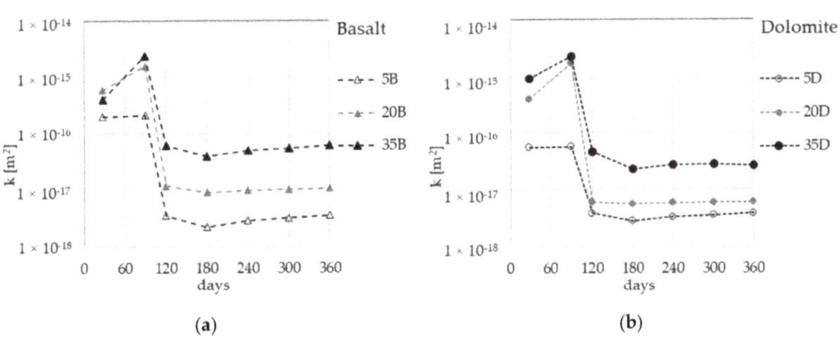

Figure 1. Gas permeability development in time for samples stored in room conditions: (**a**) basalt aggregate; (**b**) dolomite aggregate.

In order to exclude test disturbance caused by the age of the materials, the impact of preconditioning temperature started to be investigated once the values of permeability stabilised (after 1 year). AAMs are highly sensitive to external temperature application. Referring to this fact, three temperatures of preconditioning were taken into consideration: 40 °C, 80 °C, and 105 °C.

After the first step, corresponding to the measurement performed on the natural state of the material (denoted as 20 °C, corresponding to data at 360 days, see Figure 2), further stages of research were run. For each step of temperature, the drying process was carried out in order to reach a stable mass of samples (relative mass loss Δm less than 0.1% during 24 h). In order to cool down the specimens, they were placed in a desiccator fulfilled with hydrophobic pellets to reduce moisture. After 24 h of cooling down, the mass of each material was verified. Then, the permeability test was performed. For each material, two specimens were tested. Just after measurements, the specimens were placed in a dryer at the next level of temperature and the procedure was repeated. All of the tests were conducted on the same specimens to avoid impact of heterogeneity of concretes. The total drying process and changes in specimens' mass are shown in Figure 2.

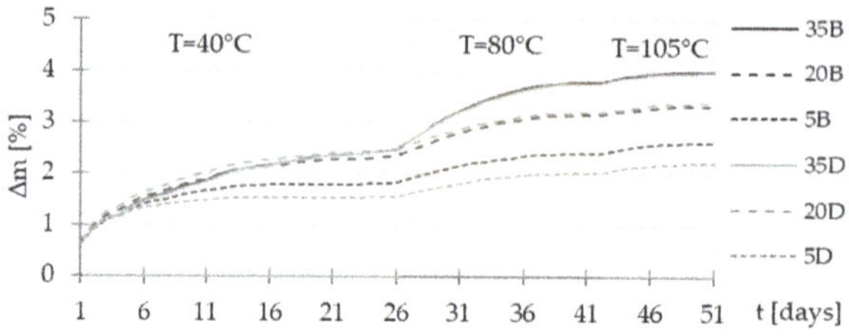

Figure 2. Scheme of drying process.

The values of gas permeability coefficients were calculated according to modified Darcy's law. The analysis of the obtained results consists of influence of temperature on the values of apparent and intrinsic gas permeability and the occurrence of boundary slippage effect related to the Klinkenberg effect [29].

4. Results and Discussion

Apparent permeabilities as a function of inversed mean pressure show permeability increased with temperature and followed Klinkenberg's law. The pore fineness increased, probably because of the creation of newly accessible fine pores during drying and cooling. In parallel, a competitive action is normally due to expansion of already present and accessible pores, caused by water evaporation from internal surfaces [30,31].

Table 2 presents intrinsic permeability values for initial, not oven-dried, specimens (20°C) of all the tested concretes. These intrinsic permeabilities were calculated following Klinkenberg's law, using apparent permeabilities. The values for low slag content materials (AAC5) are the lowest, independently of aggregate type, in comparison to other materials. Globally, there is no tendency in permeability evolution as a function of aggregate type.

Table 2. The intrinsic permeability values for specimens preconditioned at 20 °C.

	AAC5B	AAC20B	AAC35B	AAC5D	AAC20D	AAC35D
k_{20} [m^2]	3.53×10^{-18}	1.08×10^{-17}	6.07×10^{-17}	3.81×10^{-18}	6.00×10^{-18}	4.65×10^{-17}

The analysis presented below is based on intrinsic gas permeability values and the relation between them for appropriate drying temperatures. The ratio proposed for the further analysis is k_i/k_y, where i, y mean drying temperatures. Values of k_i/k_{20} are shown in Figure 3.

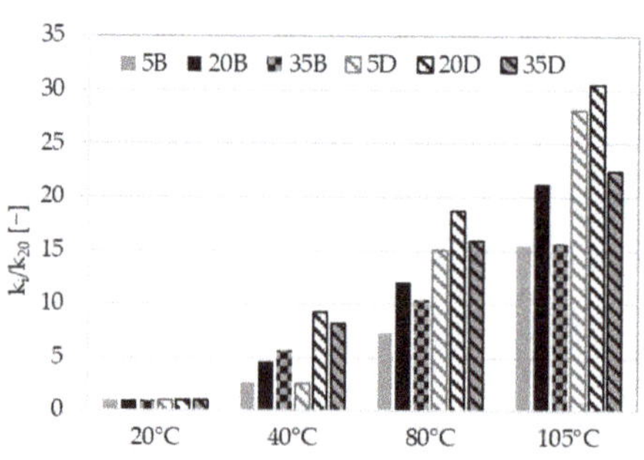

Figure 3. Values of k_i/k_{20} ratios for temperatures of preconditioning.

For almost all of the temperatures (40, 80, 105 °C), the values of permeability for dolomite-based concrete increased more than for basalt-based concrete. This behaviour is probably due to physical changes in the microstructure of the zone between paste and aggregate, as the latter is not intrinsically impacted at these temperatures. The impact of the drying process at 40 °C is much higher for AAC35B and AAC35D than for AAC5B and AAC5D, which is clearly visible in Figure 4. For AAC35B, the value of permeability increased from 6.07×10^{-17} m^2 to 34.8×10^{-17} m^2 for 20 °C and 40 °C, respectively (the k_{40}/k_{20} ratio was equal to 5.74). For AAC35D, permeability at these temperatures reached 4.65×10^{-17} m^2 and 38.2×10^{-17} m^2, making the k_{40}/k_{20} ratio equal to 8.22. However, low slag content materials show lower sensitivity when heated to a temperature of 40 °C. The values of the k_{40}/k_{20} ratio for AAC5B and AAC5D were 2.71 and 2.64, respectively. The impact of the drying process at over 40 °C is related to complex physical and chemical changes and will be discussed below.

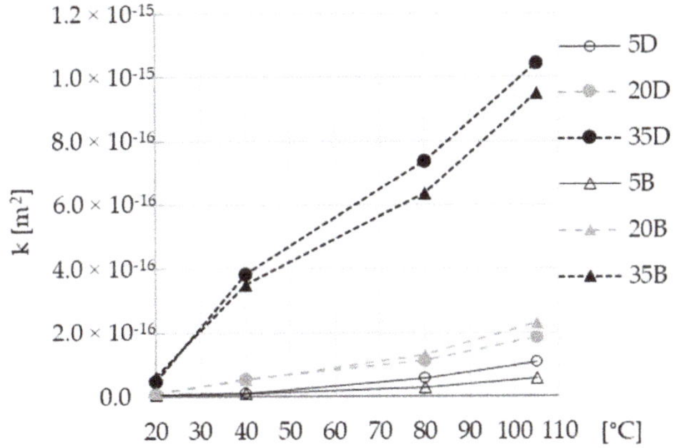

Figure 4. Impact of drying temperature on gas permeability of AAC.

Materials with the highest amount of fly ashes reveal a significant response to temperatures between 40 °C and 105 °C. As was explained in the Introduction, these materials are sensitive to temperature treatment. On the one hand, this may be due to the activation of the FA and therefore the creation of new phases, but on the other hand, it may be due to a rapid release of water from the pores of the material and generation of microcracks in the material, amplifying gas flow. All the absolute values of gas permeability are summed up in Figure 4.

As the results presented before are mainly related to aggregates' nature, the results presented in Figure 4 are strongly focused on slag content impact. Three families of results, depending on slag amount, may be highlighted. In the case of 5% and 20% slag content, the influence of the drying temperature is slight in comparison to 35% slag. Thermal expansion strain differences between high-content slag paste and aggregate should be more significant than for other pastes. It provides geometrical incompatibilities and therefore probably microcracking, causing an increase in permeability of up to two orders of magnitude.

The analysis carried out has brought many valuable observations. The difference in drying temperature impact shows distinct sensitivity of materials to drying. Furthermore, a meaningful impact of slag addition is highlighted.

5. Conclusions

The tests carried out were aimed at comparing the influence of the sample preconditioning temperature on the result of the gas permeability test. The variety of standards used for measurements may significantly affect the obtained result, which has been proven in this paper. Based on the obtained experimental results, the following conclusions can be drawn:

- The drying temperature of the samples significantly affects the obtained permeability values.

- The effect of temperature on AAC permeability is strongly related to the precursor used and especially to the slag content.
- All permeability test reports have to include sample preconditioning conditions to enable correct interpretation of results.
- The gas permeability value of materials with the same binder can be compared under the same conditions, taking into account the impact of the aggregate and the quality of the zone between aggregate and paste.
- The results of permeability tests carried out on AAC samples with different binders under different sample moisture conditions should not be compared.

Experimental research has clarified the possibility of comparing the results obtained for materials with different binders and aggregates. Conclusions drawn from these studies may provide guidance for further analysis and interpretation of permeability test results performed on alkali-activated concrete.

Author Contributions: Conceptualisation, I.H., M.C.C. and O.A.; methodology, M.C.C. and P.D.; validation, I.H. and O.A.; formal analysis, M.C.C. and I.H.; investigation, P.D.; data curation, P.D. and M.C.C.; writing—original draft preparation, P.D. and M.C.C.; writing—review and editing, M.C.C., I.H., O.A. and P.D.; visualisation, P.D. and M.C.C.; supervision, M.C.C.; project administration, I.H.; funding acquisition, I.H. and M.C.C. All authors have read and agreed to the published version of the manuscript.

Funding: The authors of this research were financially supported by SMART-G Smart Geopolymers project financed by National Centre for Research and Development (NCBR)–ERA-MIN2-3/SMART-G/1/2022.

Data Availability Statement: The data presented in this study are available on request from the corresponding author.

Acknowledgments: The publication cost of this paper was covered with funds of the Polish National Agency for Academic Exchange (NAWA): "MATBUD'2023–Developing international scientific cooperation in the field of building materials engineering" BPI/WTP/2021/1/00002, MATBUD'2023.

Conflicts of Interest: The authors declare no conflict of interest.

References

1. Bernal, S.A.; Bílek, V.; Criado, M.; Fernández-Jiménez, A.; Kavalerova, E.; Krivenko, P.V.; Winnefeld, F. Durability and testing—Degradation via mass transport. In *RILEM State-of-the-Art Reports: Alkali Activated Materials*; Provis, J.L., van Deventer, J.S.J., Eds.; Springer: Dordrecht, The Netherlands; Heidelberg, Germany; New York, NY, USA; London, UK, 2014; Volume 13, pp. 223–276.
2. Li, K.F. *Durability Design of Concrete Structure: Phenomena, Modeling and Application*; Wily & Sons: London, UK, 2016.
3. Beushausen, H.; Luco, L.F. *Performance-Based Specifications and Control of Concrete Durability, RILEM State-of-the-Art Reports RILEM TC 230-PSC*; Springer: Dordrecht, The Netherlands, 2016.
4. Stanish, K.D.; Hooton, R.D.; Thomas, M.D. *Testing the Chloride Penetration Resistance of Concrete: A Literature Review*; FHWA Contract DTFH61-97-R-00022; Department of Civil Engineering University of Toronto: Toronto, ON, Canada, 2001.
5. AKatz, J.; Thompson, A. Quantitative prediction of permeability. *Phys. Rev.* **1986**, *34*, 8179–8181.
6. Kropp, J. Permeability of Concrete as a Criterion of its Durability. *Mater. Struct.* **1999**, *32*, 5.
7. Cather, R.; Figg, J.W.; Marsden, A.F.; O'Brien, T.P. Improvements to the figg method for determining the air permeability of concrete. *Mag. Concr. Res.* **2015**, *36*, 241–245. [CrossRef]
8. Kropp, J.; Hilsdorf, H.K. *Performance Criteria for Concrete Durability, State-of-the-Art Report of RILEM TC116-PCD*; E&FN Spon: London, UK, 1995.
9. Yang, K.; Basheer PA, M.; Bai, Y.; Magee, B.J.; Long, A.E. Development of a new in situ test method to measure the air permeability of high performance concretes. *NDT Int.* **2014**, *64*, 30–40. [CrossRef]
10. Romer, M. Effect of moisture and concrete composition on the Torrent permeability measurement. *Mater. Struct.* **2005**, *38*, 541–547. [CrossRef]
11. Zhang, D.; Li, K. Concrete gas permeability from different methods: Correlation analysis. *Cem. Concr. Compos.* **2019**, *104*, 103379. [CrossRef]
12. Alexander, M.G. Service life design and modelling of concrete structures–background, developments, and implementation. *Revista Alconpat* **2018**, *8*, 224–245. [CrossRef]
13. *XP P18-463*; French Standards, Concretes–Gas Permeability Test on Hardened Concrete. AFNOR: St. Denis, France, 2011.

14. Sitarz, M.; Hager, I.; Kochanek, J. Effect of High Temperature on Mechanical Properties of Geopolymer Mortar. *MATEC Web Conf.* **2018**, *163*, 06004. [CrossRef]
15. Criado, M.; Palomo, A.; Fernández-Jiménez, A. Alkali activation of fly ashes. Part 1: Effect of curing conditions on the carbonation of the reaction products. *Fuel* **2005**, *84*, 2048–2054. [CrossRef]
16. Dudek, M.; Sitarz, M. Analysis of changes in the microstructure of geopolymer mortar after exposure to high temperatures. *Materials* **2020**, *13*, 4263. [CrossRef]
17. Poornima, N.; Revathi, T.; Sivasakthi, M.; Jeyalakshmi, R. Effect of curing on mechanical strength and microstructure of fly ash blend GGBS geopolymer, Portland cement mortar and its behavior at elevated temperature. *Mater. Today Proc.* **2021**, *47*, 863–870. [CrossRef]
18. Runci, A.; Serdar, M. Effect of curing time on the chloride diffusion of alkali-activated slag. *Case Stud. Constr. Mater.* **2022**, *16*, e00927. [CrossRef]
19. Noushini, A.; Castel, A. The effect of heat-curing on transport properties of low-calcium fly ash-based geopolymer concrete. *Constr. Build. Mater.* **2016**, *112*, 464–477. [CrossRef]
20. Chi, M. Effects of dosage of alkali-activated solution and curing conditions on the properties and durability of alkali-activated slag concrete. *Constr. Build. Mater.* **2012**, *35*, 240–245. [CrossRef]
21. Swanepoel, J.C.; Strydom, C.A. Utilisation of fly ash in a geopolymeric material. *Appl. Geochem.* **2022**, *17*, 1143–1148. [CrossRef]
22. Fernandez-Timenez, A.; Palomo, A. Alkali-activated fly ashes: Properties and characteristics. In Proceedings of the 11th ICCC 3, Durban, South Africa, 11–16 May 2003; pp. 1332–1340.
23. Palomo, A.; Grutzeck, M.W.; Blanco, M.T. Alkali-activated fly ashes, a cement for the future. *Cem. Concr. Res.* **1999**, *29*, 1323–1329. [CrossRef]
24. Jiang, W.; Roy, D. Hydrothermal processing of new fly ash cement. *Ceram. Bull.* **1990**, *71*, 642–647.
25. Bakharev, T. Geopolymeric materials prepared using Class F fly ash and elevated temperature curing. *Cem. Concr. Res.* **2005**, *35*, 1224–1232. [CrossRef]
26. *ASTM C618-19*; Standard Specification for Coal Fly Ash and Raw or Calcined Natural Pozzolan for Use in Concrete. ASTM International: West Conshohocken, PA, USA, 2019.
27. Hager, I.; Sitarz, M.; Mróz, K. Fly-ash based geopolymer mortar for high-temperature application–Effect of slag addition. *J. Clean. Prod.* **2021**, *316*, 128168. [CrossRef]
28. Kollek, J.J. The determination of the permeability of concrete to oxygen by the Cembureau method—A recommendation. *Mater. Struct.* **1989**, *22*, 225–230. [CrossRef]
29. Lide, D.R.; Frederikse, H.P.R. *CRC Handbook of Chemistry and Physics*, 84th ed.; CRC Press LLC: New York, NY, USA, 2003; p. 1004.
30. Zhang, J.; Jin, T.; He, Y.; Wang, Y.; Gao, Y.; Zhang, Y. The slippage effect of concrete gas permeability and the influence of its microstructure. *Constr. Build. Mater.* **2022**, *333*, 127384. [CrossRef]
31. Choinska, M. *Effets de la Température, du Chargement Mécanique et de Leurs Interactions sur la Perméabilité du Béton de Structure*; Matériaux. Ecole Centrale de Nantes (ECN); Université de Nantes: Nantes, France, 2006; p. 185.

materials proceedings

MDPI

Proceeding Paper

Thermal Performance Evaluation of a PCM-Integrated Gypsum Plaster Board †

Katarzyna Nowak [1,*], Tomasz Kisilewicz [1], Umberto Berardi [2] and Anna Zastawna-Rumin [1]

1 Chair of Building Design and Building Physics, Cracow University of Technology, Warszawska 24 Street, 31-155 Cracow, Poland
2 Faculty of Engineering and Architectural Science, Toronto Metropolitan University, 325 Church Street, Toronto, ON M5B 2K3, Canada
* Correspondence: knowak@pk.edu.pl
† Presented at the 10th MATBUD'2023 Scientific-Technical Conference "Building Materials Engineering and Innovative Sustainable Materials", Cracow, Poland, 19–21 April 2023.

Abstract: In order to design and optimize building materials containing phase changing material (PCM), it is important to accurately characterize the thermal properties of these composites: the enthalpy curve and its hysteresis. This paper presents the initial tests of these properties for a composite panel: gypsum plasterboard with an evenly distributed layer of PCM. Performance testing of the selected material was carried out by means of the dynamic method. This method (dynamic heat flow meter apparatus (DHFMA)) involves the measurement of non-stationary heat flow to determine the stored energy (enthalpy change) as a function of temperature using a stationary heat flow measurement apparatus (HFMA). This method allows for the measurement of the sensible and latent heat capacity of the products containing phase change material. In addition to presenting the obtained results, the study will discuss the practical aspects of this test method, recently introduced in the standard ASTM C 1784-20. The preliminary experiments described in the article were aimed at learning a new research technique, determining the required conditions for conducting research and the capabilities of the possessed apparatus in this regard.

Keywords: phase change material; PCM; dynamic thermal properties; differential scanning calorimetry; DSC method; DHFMA method

Citation: Nowak, K.; Kisilewicz, T.; Berardi, U.; Zastawna-Rumin, A. Thermal Performance Evaluation of a PCM-Integrated Gypsum Plaster Board. *Mater. Proc.* **2023**, *13*, 39. https://doi.org/10.3390/materproc2023013039

Academic Editors: Katarzyna Mróz, Tomasz Tracz, Tomasz Zdeb and Izabela Hager

Published: 20 February 2023

1. Introduction

Modern buildings are characterized by significantly improved thermal insulation and the other solutions aimed at minimizing energy demand. A frequently used techno- logy for erecting buildings is a technology based on durable structural elements forming the skeleton filled with thermal insulation. Heat transfer through such partitions is minimized, but the disadvantage of this technology is the low thermal mass. In addition, oversized glazing often causes a disproportion between the solar gains and the limited thermal capacity of the building. The inability to counteract significant temperature fluctuations in the rooms has a very negative effect on the microclimate and the feelings of people staying in them, as well as on the efficiency of solar energy use.

The ability to store excess heat energy and release it when the room cools down results in the better use of solar radiation in the winter and reduces the risk of space overheating in the winter and summer. The occurrence of a too-high temperature is usually connected with the use of expensive and energy-intensive air-conditioning equipment. In low energy buildings, the ability to store energy becomes particularly important.

The lack of a time coincidence of the demand and possible supply of energy gains contribute to the development of research on heat and cold storage. One of the effective ways of passive accumulation of thermal energy in buildings is the use of materials that change the state of aggregation, usually marked with the abbreviation PCM (phase changing material).

During the change of aggregation phase, a large amount of heat is accumulated or released. Particularly in buildings erected using light construction technologies, the application of additional energy storage in the form of PCM proves to be very effective [1]. PCM can contribute to reducing the peak demand on cooling and allow for more even operation of the air conditioning system, also reducing the need for heating at night. Thanks to the use of PCM, it is possible to avoid or at least reduce the currently frequent effects of the overheating of buildings with excessive glazing [1,2].

Phase change materials offered for building use can take the form of:

- Microcapsules added to plasters and mortars, concrete screeds or plasterboards;
- Small containers in the form of flexible mats with cavities filled with Bio-PCM or aluminum sheet plates filled with phase change material (Dupont);
- Fibrous insulation with the addition of PCM in the form of capsules [3–5].

This article presents the results of preliminary tests of the thermal capacity of a plasterboard integrated with PCM. The purpose of these studies was to test a new research technique, the necessary conditions for conducting research and a completely different than designed application of the FOX 314 apparatus for measuring the thermal conductivity of building materials. The apparatus and its original software were designed to conduct tests in stationary conditions, while thermal capacity tests require dynamic conditions.

2. Measurement Methods of PCM Thermal Characteristics

The proper selection of the PCM type and its quantity requires the precise experimental determination of the thermal characteristics of heterogeneous products.

Differential scanning calorimetry (DSC) is one of the most widely used PCM measurement methods because of the ease with which different thermodynamic data can be obtained. DSC measures the amount of heat needed to raise the temperature of the test sample in a given temperature range [4]. On this basis, the phase change temperature, enthalpy, heat capacity and specific heat can be determined. However, the DSC method is applicable to millimeter-scale samples with weight in the order of a few milligrams. The DSC method also requires relatively homogeneous samples. Due to the large heterogeneity of composites on a small-scale sample, determining the average PCM content in a mass-produced product requires many tests, which is troublesome and expensive, and may be subject to a large error of sample randomness [3]. In the case of composite materials (PCM—enhanced building products), with unevenly distributed PCM in the structure, testing according to the DSC method does not make sense.

In search of a possibly fast and at the same time precise method that can be used in determining the thermodynamic parameters of large samples of PCM-reinforced building components, a dynamic method using a plate apparatus was developed.

In the Oak Ridge National Laboratory, a dynamic method using a plate apparatus was developed to measure the content of phase-change material in composite samples [6,7]. The plate apparatus is basically used to measure the thermal conductivity of materials in a steady state of heat conduction. However, it can also be used to measure the dynamic thermal properties of samples of the tested materials, in which the heat flux stabilization is relatively slow due to the ongoing phase change. The developed method is referred to as dynamic HFMA or DHFMA. Since the beginning of its development, the DHFMA method has been subject to constant modifications. The theory of the method and test procedure has been described by the creators of the method [8].

In an isothermal process such as a phase transition, the enthalpy change (ΔH) is equal to the amount of heat absorbed (heat input) or released (heat output) during the process (ΔQ). The heat capacity (c_p) is the temperature derivative of the enthalpy (H), i.e., $c_p = dH/dT$.

Assuming there is a linear relationship between enthalpy and temperature for small temperature increments (ΔT):

$$c_p = dH/dT \approx \Delta H/\Delta T = \Delta Q/\Delta T, \tag{1}$$

where:

ΔH—enthalpy change [J/m^2];

ΔQ—energy amount absorbed or released during a given process [J/m^2];

ΔT—temperature increase [K].

In the DHFMA method, the temperature of the sample is changed in small steps and the resulting heat flow into or out of the sample is measured during this process. The heat capacity at an average temperature is then determined using Equation (1). The DHFMA method uses a conventional HFMA instrument containing at least one heat flow sensor on each of the isothermal plates. The sample is placed parallel to the plates while the sides are thermally insulated to achieve near-adiabatic conditions. Both isothermal plates are kept at the same temperature. A temperature step is then applied to both plates and the heat flows through the plates are measured until a thermal equilibrium is reached where the heat flow values become negligible. In the case of a step change in temperature, the change in enthalpy per unit area of the sample is calculated by integrating the heat flow rates over time (2):

$$\Delta H = \Delta Q = \sum[\{(qU_i - qU_{final}) + (qL_i - qL_{final})\}\tau] \quad (2)$$

where:

qU_i—heat flux flowing through the upper plate, recorded at time intervals τ [W/m^2];

qL_i—heat flux flowing through the bottom plate, recorded at time intervals τ [W/m^2];

qU_{final}—the value of the remaining heat flux of the upper plate, caused by side heat losses in the equilibrium state [W/m^2];

qL_{final}—the value of the remaining heat flux of the lower plate, caused by side heat losses in the equilibrium state [W/m^2];

τ—time step between readings [s].

The volumetric heat capacity of the PCM-enhanced component is determined as follows (3):

$$c_V = (1/l)\cdot(\Delta Q/\Delta T), \quad (3)$$

where l is the thickness of the sample. It should be noted that the DHFMA method can be used to measure the heat capacity of any solid or liquid material. Recently, ASTM introduced the test standard C1784 for measuring the thermal storage properties of phase change materials and products, based on the DHFMA method [9].

3. Thermal Capacity Testing of Micronal® PCM SmartBoard™ 23

The composite material Micronal® PCM SmartBoard™ 23, in the form of a plasterboard with the addition of a phase change material (approximately 3 kg of dry Micronal per 1 m^2 of the board), was tested. According to the authors of [10], this microencapsulated paraffin avoids leakage during the liquid phase and makes heat transfer easier by increasing the contact area. The authors of [10] analyzed gypsum boards with a mass PCM content up to 45% [11].

3.1. DSC Test

Before starting the dynamic DHFMA testing of the building material, thanks to the kindness of Netzsch, it was possible to perform a calorimetric study of the Micronal itself. It is an organic material which, according to the manufacturer's data, has a melting point of 23 °C. PCM makes up about 30% of the weight of the SmartBoard panel. The phase change material added to the plasterboard matrix is in the form of microcapsules (with a diameter of 2–20 μm) filled with paraffin.

The thermal characteristics of the PCM materials used, obtained as a result of DSC tests, are shown in Figure 1. The tests performed by Netzsch were carried out in the DSC 214 Polyma calorimeter with an aluminum melting pot, for the range from −50 °C to +50 °C. Calorimetric studies allowed the value of the heat of the phase change to be determined: PCM 23: 127.7 J/g. This value is similar to the properties of PCM described in article [9], where Micronal with a heat of fusion of 110 J/g was used and was significantly lower than the paraffin presented in [11] of 151 J/g. The phase transition range is a wide temperature range from 17.8 °C to 31.5 °C, practically identical in the case of heating and cooling of the material. However, the peak heat flux was achieved at 25.1 °C, not 23 °C. During the DSC test, there was no sub-cooling effect, but there was a noticeable difference between the heat capacity in the entire heating and solidification process.

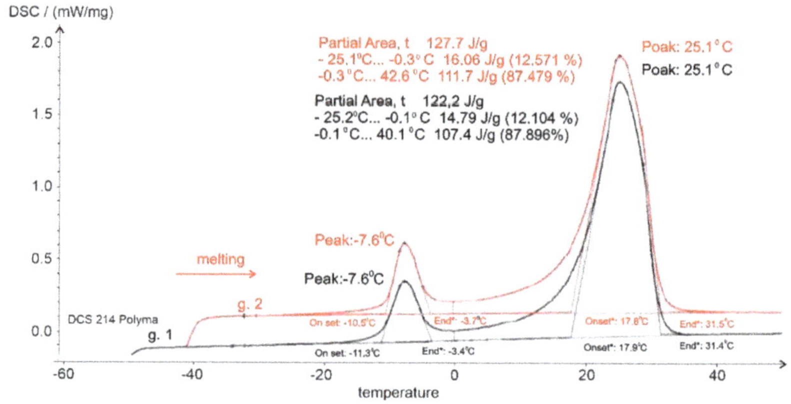

Figure 1. Results of DSC test.

3.2. *FOX 314 Tests*

The equipment used for dynamic tests of thermal conductivity and heat capacity is a device called FOX 314, manufactured by LaserComp USA, serial number: 1043. The design of the FOX 314 device is based on the method of measuring thermal resistance using heat flux sensors in stationary conditions and is compatible with the following standards:

- EN ISO 8301:1998: Thermal insulation, determination of steady-state thermal resistance and related properties and heat flow meter apparatus;
- EN 12667:2002: Thermal performance of building materials and products, determination of thermal resistance by means of guarded hot plate and heat flow meter methods and products of high and medium thermal resistance;
- ASTM C518-91: Standard test method for steady-state heat flux, and measurements and thermal transmission properties by means of the heat flow meter apparatus.

The characteristics of the sample tested in the FOX 314 apparatus are:

geometrical dimensions: 300 × 301 × 14.8 mm; volume: 0.001336 m^3; weight: 1089.36 g; density: 815.12 kg/m^3.

3.2.1. Thermal Conductivity Tests

The values of the thermal conductivity coefficient shown in Table 1 were tested for two sets of temperature differences:

- 0 °C to 20 °C, the average temperature of the sample during this test equal to 10 °C, corresponds to the condition of full PCM solidification;
- In the range of 10 °C to 30 °C, partial liquefaction of the PCM in the cross-section of the plate should be expected. Such average conditions occur in reality during the summer period in Polish climatic conditions.

Table 1. Thermal conductivity coefficients of Micronal® PCM SmartBoard™ 23 plasterboard in two temperature ranges: from 0 to 20 °C and from 10 to 30 °C.

T_1/T_2, [°C]	λ, [W/mK]
0/20	0.1658
10/30	0.1651

3.2.2. Thermal Capacity Testing

Due to the results of the Micronal tests obtained by the DSC method, a very wide range of temperatures in the DHFMA tests of the plate containing this material was adopted. Since this type of research is extremely time-consuming, a temperature change step of 2K was assumed for the preliminary testing.

The heating stage started at 10 °C and finished at 32 °C. The cooling of the material was carried out immediately after the heating stage, so the first measurement value obtained after reversing the process was the temperature of 30 °C, and the last one was 8 °C. Table 2 shows the temperature values that were actually obtained in the apparatus during the measurements and mass heat capacity. Temperature values given in Table 2 differ insignificantly from the measurement conditions set in the FOX 314 apparatus. Figure 2 and Table 2 summarize the results of the measurement obtained during the heating stage, while Figure 3 and Table 3 summarize the results during the cooling stage.

Table 2. Mass thermal capacity of Micronal® PCM SmartBoard™ 23, melting stage: 10–32 °C.

Melting	
Temperature [°C]	**c_p [J/kgK]**
10.01	1518.801
12.01	1544.036
14.01	1553.112
16.03	1581.835
18.01	1707.86
20.02	1970.278
22.02	3789.865
24.02	8795.741
26.01	1380.011
28.01	1384.403
30.02	1396.335
32.02	1401.394

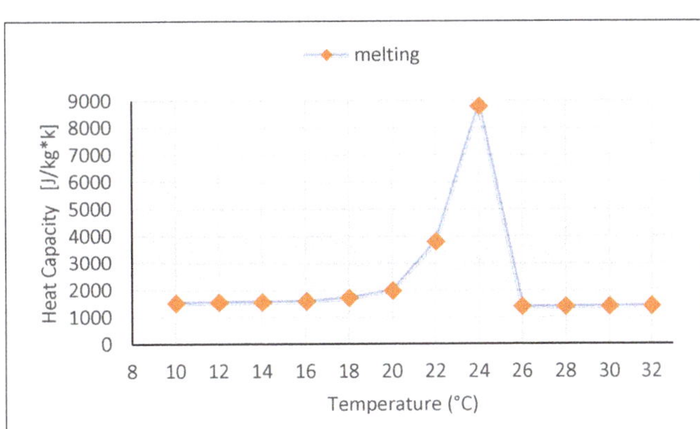

Figure 2. Mass thermal capacity of Micronal® PCM SmartBoard™ 23, melting stage: 10–32 °C.

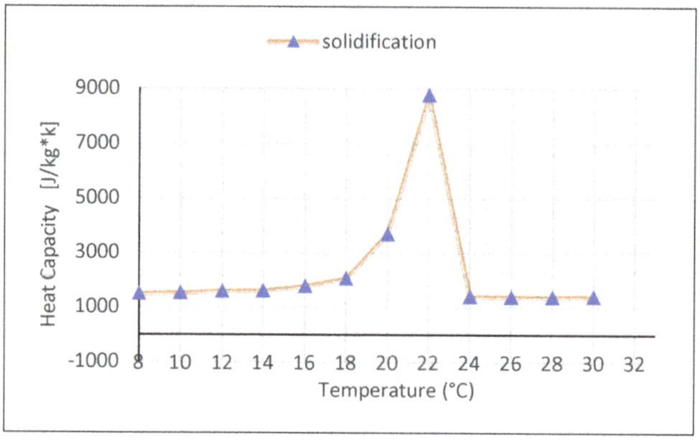

Figure 3. Mass thermal capacity of Micronal® PCM SmartBoard™ 23, cooling stage: 30–8 °C.

Table 3. Mass thermal capacity of Micronal® PCM SmartBoard™ 23, cooling stage: 30–8 °C.

Solidification	
Temperature [°C]	**c_p [J/kgK]**
8	1500.192
10.01	1537.904
12.01	1596.831
14.01	1607.684
16.03	1778.953
18.01	2060.286
20.02	3689.62
22.02	8771.898
24.02	1412.638
26.01	1396.558
28.01	1399.615
30.02	1405.134

Based on the measurement in the FOX apparatus, the volumetric heat capacity [J/m^3·K] was obtained. On this basis, after conversion, the value of the mass heat capacity [J/kg·K] was calculated

The apparatus used for the tests described here does not have specialized or very expensive software for dynamic tests. Therefore, obtaining the results was an extremely time- and labor-intensive process. It was necessary to manually control the course of the research and observe the process of stabilization after changing the boundary conditions. The measurement step of the apparatus is approximately 0.7 s. The tested sample is a relatively thin gypsum board, for which the average stabilization time was close to 4 h. For this period, almost 21,000 measurement records were obtained, which had to be further copied and processed. With 12 measurement steps used in the research, the calculation sheet counted over 220,000 records for only one series of tests.

Figure 4 shows the combined results obtained during the heating and cooling of the Micronal® PCM SmartBoard™ 23 in one graph.

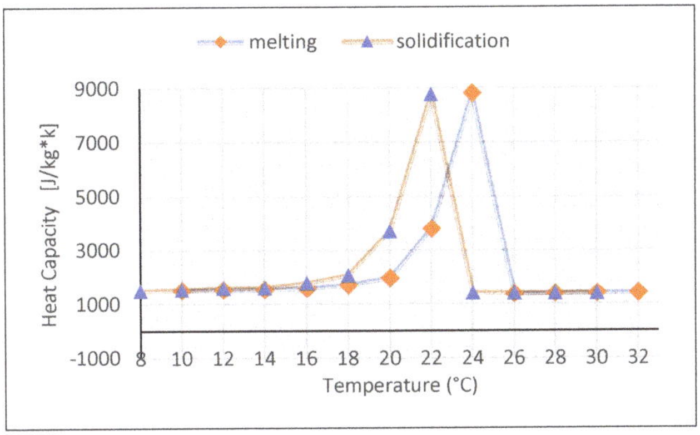

Figure 4. Thermal capacity of Micronal® PCM SmartBoard™ 23 for melting and cooling stages.

The temperature range in which the phase activity of the material can be observed starts at 16 °C and extends up to 26 °C. Due to the long duration and preliminary nature of the tests, a large (2K) step of temperature changes over time was adopted, and unfortunately, the stages of heating and cooling were carried out for the same temperature values, without any shift. Thus, the results and graphs obtained in this way are approximates only; they do not allow a reproduction of the real shape of the enthalpy curve or to clearly specify the phase transition temperature. Even with a 2K step, offsetting the heating and cooling cycles by 1K would yield additional information. According to the ASTM guidelines [9], the step of temperature changes can be as small as 0.5K. However, in order to maintain the accuracy of the measurement, several separate measurement series should be carried out for this purpose with a larger step, but mutually shifted by 1K or 0.5K.

The preliminary results of the PCM tests based on a new procedure and the non-standard use of the FOX 314 apparatus allowed estimations that the phase change in the tested material takes place close to the temperature declared by the manufacturer. The maximum heat capacity of a gypsum board with PCM was 8772 J/kgK, and in terms of volumetric capacity, it was 7.15 MJ/m^{3}K. The authors of the article [8] obtained, in the case of a gypsum board containing 25% paraffin-based PCM, ca. 6 MJ/m^{3}K. The heat capacity values of the tested material during melting and solidification were not identical, showing the hysteresis effect. Solidification took place at a lower temperature than melting, while the maximum heat capacity values obtained were almost identical. Both observations differ from the characteristics of Micronal itself, obtained from the DSC study. The material containing PCM already in the liquid phase has a lower heat capacity than in the solid phase. A similar result was obtained by the authors of [8].

4. Conclusions

The conducted pilot tests of the thermal characteristics of the phase change material allow the following conclusions:

- Using the ordinary FOX 314 apparatus, designed to measure thermal conductivity in stationary conditions, it is possible to conduct dynamic enthalpy studies of PCM;
- Research conducted for a wide range of temperatures is very time-consuming, and without a special software, extending the apparatus application, which is also very laborious, requires the processing of hundreds of thousands of measurement data;
- The measured maximum heat capacity of the plate during the phase transformation, equal to 8772 J/kgK, is close to the results reported in the literature. However, the adopted step change of the test temperature every 2K was too large to faithfully reproduce the thermal properties of the sample and enable detailed comparisons;

- The described dynamic method gives the opportunity to test large samples of composite building materials with non-homogeneous PCM distribution. Such data cannot be obtained by means of DSC method.

Author Contributions: Conceptualization, K.N., T.K., A.Z.-R. and U.B.; methodology, K.N. and T.K.; software, K.N.; validation, K.N. and T.K.; formal analysis, K.N., U.B., T.K. and A.Z.-R.; investigation, K.N.; data curation, K.N.; writing—original draft preparation, K.N., T.K and A.Z.-R.; writing—review and editing, U.B. and A.Z.-R.; visualization, K.N. and T.K. All authors have read and agreed to the published version of the manuscript.

Funding: This research received no external funding.

Institutional Review Board Statement: Not applicable.

Informed Consent Statement: Not applicable.

Data Availability Statement: The data presented in this study are available on request from the corresponding author. The data are not publicly available due to privacy issues.

Acknowledgments: Special thanks to the member companies of Netzsch that provided data for our research. Publication cost of this paper was covered with founds of the Polish National Agency for Academic Exchange (NAWA): "MATBUD'2023—Developing international scientific cooperation in the field of building materials engineering" BPI/WTP/2021/1/00002, MATBUD'2023.

Conflicts of Interest: The authors declare no conflict of interest.

References

1. Berardi, U.; Manca, M. The energy saving and indoor comfort improvements with latent thermal energy storage in building retrofits in Canada, 8th Int. Conference on Sustainability in Energy and Buildings, Turin. *Energy Procedia* **2017**, *111*, 462–471. [CrossRef]
2. Soudian, S.; Berardi, U. Assessing the effect of night ventilation on PCM performance in high-rise residential buildings. *J. Build. Phys.* **2019**, *43*, 229–249. [CrossRef]
3. Kossecka, E.; Kośny, J. Dynamiczna metoda pomiaru zawartości materiału fazowo-zmiennego w izolacji włóknistej. *Fiz. Budowli Teor. Prakt.* **2009**, *IV*, 109–112.
4. Kośny, J. *PCM- Enhanced Building Components: An application of Phase Change Materials in Building Envelopes and Internal Structures*; Springer International Publishing: Cham, Switzerland, 2015. [CrossRef]
5. Zastawna-Rumin, A. The Analysis of the Appliaction Efficiency of Phase Change Materials in Partitions in Polish Low-Energy Buildings. Doctoral Thesis, Cracow University of Technology, Cracow, Poland, 2018.
6. Kośny, J.; Yarbrough, D.W.; Miller, W.A.; Wilkes, K.E.; Lee, E.S. Analysis of the dynamic thermal performance of fibrous insulations containing phase change materials. In Proceedings of the 11th International Conference on Thermal Energy Storage, Effstock 2009—Thermal Energy Storage for Energy Efficiency and Sustainability, Stockholm, Sweden, 14–17 June 2009.
7. Kośny, J.; Yarbrough, D.W.; Kossecka, E. Use of a Heat-Flow Meter to Determine Active PCM Content in an Insulation. In Proceedings of the 30th International Thermal Conductivity Conference, Pittsburgh, PA, USA, 29 August–2 September 2009.
8. Shukla, N.; Kosny, J. DHFMA Method for Dynamic Thermal Property Measurement of PCM-integrated Building Materials. *Curr Sustain. Renew. Energy Rep.* **2015**, *2*, 41–46. [CrossRef]
9. *ASTM C174-20*; Standard Test Method for Using a Heat Flow Meter Apparatus for Measuring Thermal Storage Properties of Phase Change Materials and Products. ASTM: West Conshohocken, PA, USA, 2014.
10. Oliver, A. Thermal characterization of gypsum boards with PCM included: Thermal energy storage in buildings through latent heat. *Energy Build.* **2012**, *48*, 1–7. [CrossRef]
11. Sharifi, N.P.; Shaikh, A.A.N.; Sakulich, A.R. Application of phase change materials in gypsum boards to meet building energy conservation goals. *Energy Build.* **2017**, *138*, 455–467. [CrossRef]

materials proceedings

MDPI

Proceeding Paper

Effect of Glass Powder on the Cement Hydration, Microstructure and Mechanical Properties of Mortar †

Magdalena Dobiszewska [1,*], Waldemar Pichór [2], Tomasz Tracz [3], Andrea Petrella [4] and Michele Notarnicola [4]

1 Faculty of Civil and Environmental Engineering and Architecture, Bydgoszcz University of Science and Technology, 85-796 Bydgoszcz, Poland
2 Faculty of Materials Science and Ceramics, AGH University of Science and Technology in Krakow, 30-059 Cracow, Poland
3 Faculty of Civil Engineering, Cracow University of Technology, 31-155 Cracow, Poland
4 Department of Civil, Environmental, Land, Construction and Chemistry, Polytechnic University of Bari, 70125 Bari, Italy
* Correspondence: magdalena.dobiszewska@pbs.edu.pl

† Presented at the 10th MATBUD'2023 Scientific-Technical Conference "Building Materials Engineering and Innovative Sustainable Materials", Cracow, Poland, 19–21 April 2023.

Abstract: Cement production has the largest carbon-footprint since it leads to the release of carbon dioxide and enormous energy consumption. Therefore, scientific research is being conducted worldwide on the possibility of using other materials that can be used as a cement substitute. A potential alternative material is glass powder. This paper present research results on the use of glass powder as a partial cement replacement with a substitution level of 0–20%. The pozzolanic activity and the influence of glass powder on cement hydration were analyzed. The porosity, the microstructure of the interfacial transition zone and the compressive strength of mortar were also investigated.

Keywords: mortar properties; microstructure; compressive strength; porosity; cement hydration; glass powder

Citation: Dobiszewska, M.; Pichór, W.; Tracz, T.; Petrella, A.; Notarnicola, M. Effect of Glass Powder on the Cement Hydration, Microstructure and Mechanical Properties of Mortar. *Mater. Proc.* **2023**, *13*, 40. https://doi.org/10.3390/materproc2023013040

Academic Editors: Katarzyna Mróz, Tomasz Zdeb and Izabela Hager

Published: 22 February 2023

1. Introduction

Concrete is the most-often-used construction material, and in the largest quantities, all over the world. The production of concrete significantly impacts the environment, as it consumes major amounts of fossil fuels and raw materials and is energy-intensive. Cement production has the largest carbon-footprint since it leads to the release of carbon dioxide and enormous energy consumption. The production of one ton of cement leads to emissions of about 0.8 tons of CO_2 into the atmosphere. Therefore, alternative materials that can substitute cement as a binder in concrete production are currently being sought. Researchers are mostly focused on the use supplementary cementitious materials (SCM) as a partial or complete replacement of cement in mortar and concrete production, i.e., fly ash, slag and silica fume. The use of SCMs contributes to improving the material properties and durability and also brings economic and ecological benefits. However, the decarbonization policy, the transformation towards alternative renewable energy resources and the departure from coal combustion significantly reduce the availability of fly ash [1]. The worldwide slag supply is also quite limited compared to the demand for concrete production [2]. Therefore, scientific research is being conducted worldwide on the possibility of using other materials that can be used as an SCM in the production of concrete as a substitute for cement.

One of the potential alternative materials is recycled powder glass. Glass is a very common material used for the production of everyday appliances such as glasses, dishes and windowpanes, as well as for the production of glass packaging: bottles and jars. Due to the limited-service life, these products must be recycled or landfilled. It has been estimated that the annual worldwide volume of waste glass stored in landfills is about 200 million tons [3]. According to the European Container Glass Federation, the average collection for the recycling rate for glass packaging reached the rate of 78% in 2019 [4]. The unrecycled glass is stored in landfills, causing environmental problems. In recent decades, there has been increasing interest in the use of waste glass as a substitute for cement [5–7]. The utilization of ground glass in construction production allows for effective waste management and contributes to the reduction in waste glass volumes in landfills, which is currently a serious problem [8,9]. Most of all, it reduces the consumption of natural resources and greenhouse gas emissions. Previous research related to the assessment of the use of powder glass in cementitious composites revealed that an up to 10–30% replacement of cement with glass powder led to no significant deterioration or even an improvement in the properties and durability of mortar and concrete [10–15]. The effect of glass powder on the properties of cement composites depends mainly on the fineness of the glass and is more pronounced in the later curing time [3,8,11,16,17]. Researchers observed a decrease in the mortar and concrete compressive strength in the early curing time with an increase in the substitution level of cement with glass powder. However, in the later curing time, a comparable compressive strength to that of the control cement composites or even a slight increase in strength were reported [12,18–20]. Compressive strength improvement is associated mostly with the filler effect of glass powder, which leads to the formation of a denser and less permeable cement matrix.

The effect of glass powder on the properties of cementitious composites is related not only to the filler effect but also to the pozzolanic activity [12,13,19,21–26]. The amorphous structure of glass enables its dissolution in a highly alkaline pore solution and acts as a pozzolana in the cement matrix [27–30]. The pozzolanic reactivity of glass powder increases with the decreasing particle diameter [12,16,31]. The pozzolanic reactivity of glass powder contributes to the formation of C-S-H phases, which, in turn, leads to the densification of the cement microstructure. Thus, the compressive strength increases as the glass powder amount increases.

This paper presents research results on the use of glass powder as a partial cement replacement with a substitution level of 0–20%. The pozzolanic activity and influence of glass powder on cement hydration were analyzed. The porosity, the microstructure of the interfacial transition zone and the mechanical properties of mortar were also investigated. Due to the relatively low pozzolanic activity of the glass powder, its effect on cement hydration is small. The replacement of cement with ground glass up to 10% does not lead to the deterioration of mortar properties. In the later curing time, a slight increase in the mortar compressive strength with the glass powder addition was observed. The influence of glass powder on cement hydration and mortar properties is related mainly to the filler effect and the heteronucleation of the C-S-H phase on the glass powder surface.

2. Materials and Methods

2.1. Materials

Mortar samples were prepared with ordinary Portland cement CEM I 42.5 R, which complies with the requirements of European Standard EN 197-1. The Blaine fineness of cement is about 4100 cm^2/g, and the specific gravity is 3.15 g/cm^3. The chemical and mineral compositions of cement are given in Table 1. Chemical analysis of the used raw materials was performed with the XRF method (WDXRF AxiosmAX spectrometer, Malvern Panalytical). The mineral composition was calculated in simple form, taking into consideration the relation between the main oxides in the cement clinker.

Table 1. Chemical and mineral composition of OPC.

Chemical Composition, [%]		Mineral Composition, [%]	
SiO_2	19.33		
Al_2O_3	5.15	C_3S	58.3
Fe_2O_3	2.90		
CaO	64.59	C_2S	12.2
MgO	1.25		
SO_3	3.23	C_3A	8.75
K_2O	0.47		
Na_2O	0.21	C_4AF	8.9
Cl^-	0.05		

CEN Standard quartz sand and de-ionized water were used for all mortar mixtures. The glass powder used in this study is a post-consumer by-product derived from recycled white glass (bottles and jars). The glass, after preliminary crushing in the crusher, was ground in a laboratory mill to a Blaine fineness of about 2700 cm^2/g. The median particle size of the powder glass is about 20 μm, and the specific gravity is 2.31 g/cm^3. The results of the chemical composition of the waste glass analysis are presented in Table 2.

Table 2. Chemical composition of powder glass.

Chemical Composition, [%]			
SiO_2	72.3	K_2O	0.6
Na_2O	12.8	BaO	0.2
CaO	10.2	SO_3	0.1
Al_2O_3	1.5	TiO_2	0.1
Fe_2O_3	1.2	Cl	0.1
MgO	0.9		

The scanning electron microscope images of powder glass are shown in Figure 1.

Figure 1. Scanning electron microscope images of glass powder.

2.2. Mixture Proportion and Preparation

The influence of glass powder on cement hydration was analyzed by preparing cement paste with 10% glass addition. To investigate the mechanical properties and microstructure of mortar, five mixtures were prepared, with different amounts of glass powder used as a partial cement substitution, i.e., 5%, 10%, 15% and 20% by cement mass. The composition of each mixture is presented in Table 3. The control mortar was marked as M0 (mortar without glass powder), and the experimental mortars were named, accordingly, M5, M10, M15 and M20, where the number indicates the cement replacement with the glass powder. All mixtures were prepared in accordance with European Standard EN 196-1, placed into 40 × 40 × 160 steel prim molds and consolidated by external vibration. The mortar samples after demolding were cured in water and at 20 ± 1 °C until the testing time.

Table 3. Mortar mixture proportions for one batch of laboratory mixer, [g].

Mortar	Cement	Water	Sand	Glass Powder
M0	450			0
M5	427.5			22.5
M10	405	225	1350	45
M15	382.5			67.5
M20	360			90

2.3. *Test Methods*

The pozzolanic activity of glass powder was assessed by chemical and mechanical methods. The chemical method is based on the determination of the soluble silicon and aluminum oxides content in the glass powder in the NaOH solution. The physical method is based on the determination of the Strength Activity Index (SAI), which is the ratio of the strength of the experimental mortar (cement mortars with 25% cement replacement with glass powder) to that of the reference mortar, and it was conducted in accordance with EN 450-1.

The total heat of hydration after 72 h of hardening cements with the addition of glass waste was determined by the semi-adiabatic method, in accordance with the PN-EN 196-9:2010 standard. The samples prepared with CEM I 42.5 and the addition of 10% waste and the reference sample of pure cement were used. To determine the compressive strength of the specimens after 28, 90 and 180 days of curing, EN 196-1 was employed. The compressive strength is the average value from six specimens tested at each age for each mixture. The porosity of each specimen was determined by the MIP method after 28 days of curing.

The microstructure analyses were carried out using an electron microscope, FESEM-EDX Carl Zeiss Sigma 300 VP (Carl Zeiss Microscopy GmbH, Jena, Germany), after sputtering the samples with graphite (Sputter Quorum Q150 from Quorum Technologies Ltd., East Sussex, UK).

3. Results and Discussion

3.1. *Glass Powder Pozzolanic Activity*

The study of the pozzolanic activity index of glass powder based on the strength of cement mortars (physical method) indicates the moderate pozzolanic properties of glass. SAI gained 79% and 88% after 28 days and 90 days of curing, respectively. According to EN-450-1, the SAI should not be lower than 75% after 28 days and should not be lower than 85% after 90 days of hardening the mortar. The low pozzolanic activity of the glass powder was also confirmed by the chemical method. Based on the research, it was determined that the content of the reactive ingredients in the glass is about 9%. The content of reactive components in additives with pozzolanic properties, i.e., silica fume and fly ash, is 70–75% and 10–15%, respectively. According to literature reports, glass dust has good pozzolanic properties, and the reactivity increases as the fineness increases [12,16,31]. The lower-than-expected pozzolanic activity of the analyzed glass powder is related to its insufficient fineness.

3.2. *Cement Hydration*

The influence of glass powder on cement hydration was evaluated on the basis of the analysis of the hydration reaction rate and the microstructure of the hardening cement paste. Two samples were investigated, i.e., pure cement paste and paste with 10% glass powder addition. Cement replacement with glass powder leads to a decrease in the heat release rate and the amount of heat released during hydration. This is of course due to the dilution of cement. At the same time, it was found that a 10% reduction in the cement content resulted in a reduction in the amount of heat released by only 6%. Thus, the relative reduction in heat released during hydration is smaller than the reduction in the cement content in the paste. This indicates a slight acceleration of hydration due to the addition of glass powder.

The influence of the glass powder additive on cement hydration was also examined on the basis of the microstructure analysis of the cement matrix in the early period of hydration. It was noticed that the C-S-H phase crystallizes on glass powder particles (Figure 2). Very fine particles of glass powder act as crystallization centers and provide an additional area where C-S-H nuclei can settle. Thus, a slight acceleration of hydration was observed.

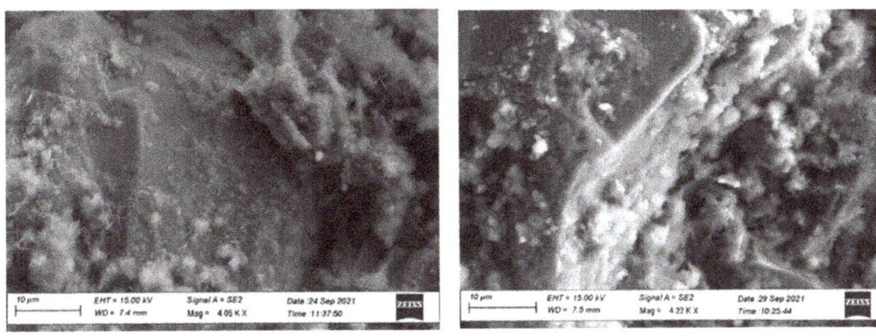

Figure 2. SEM images show the crystallization of C-S-H phases on the glass powder surface after 2 h (**left**) and after 12 h (**right**) of hydration.

The interfacial region visible on the glass grain surface is dense, and it is similar to the matrix far from the surface. Calcium hydroxide crystals or increased local porosity were not observed in the area of the contact zone. In particular, the increased concentration of shrinkage cracks was not visible in this area. Heteronucleated small crystals of hydrated calcium silicates are visible on the surface of glass grains.

3.3. Compressive Strength

The compressive strength results of mortars at various curing ages with respect to the replacement of glass powder are shown in Figure 3. It is seen that, after 28 days of curing, the mortar compressive strength decreases with the increase in the glass powder content. The most visible mortar compressive strength deterioration, i.e., a 16% decrease, was observed in the case of mortar with a 20% replacement of cement with glass powder. This situation reverses at 90 and 180 days of age, when the partial cement replacement with glass powder does not lead to the compressive strength reduction. Even a slight increase is observed, i.e., 3% for mortar with a 5% glass additive and 4% for mortar with a 10% glass addition after 90 and 180 days of curing, respectively.

At an early age, glass powder acts only as an inert material causing the compressive strength reduction. The substitution of cement with glass powder leads to an increase in the water/cement ratio from a value of 0.5 to 0.63 for mortar with a 20% glass powder addition. Thus, due to the dilution of cement, a decrease in compressive strength is observed. The increase in the compressive strength at the later age of the curing of mortar with the glass powder additive may be attributed to the pozzolanic activity of glass powder. The pozzolanic reaction of reactive silica oxide in glass power and calcium hydroxide leads to C-S-H phases formation, which results in cement matrix microstructure densification and, consequently, a mortar compressive strength increase.

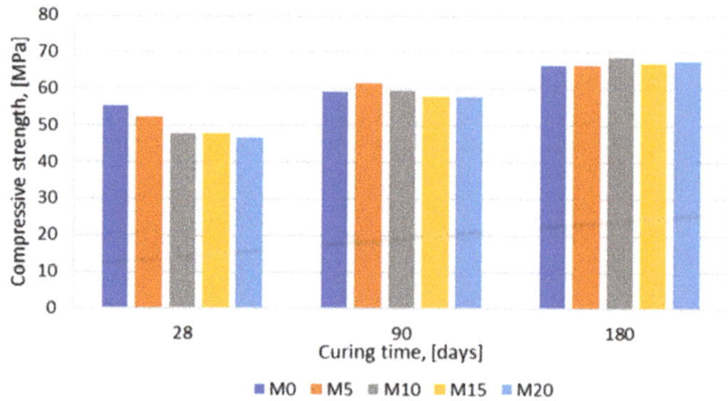

Figure 3. Compressive strength of mortars with glass powder addition.

3.4. Porosity

The influence of glass powder on the microstructure was analyzed by mercury porosimetry. Cement was substituted with 10% and 20% glass powder by mass. The porosity and pore size distribution of mortars after 28 days of curing are shown in Figures 4 and 5. By increasing the amount of glass powder, the porosity increases from 0.58 cm^3/g for the control mortar to 0.60 cm^3/g and 0.66 cm^3/g in the case of mortar with 10% and 20% glass powder addition, respectively (about 7% and 12%).

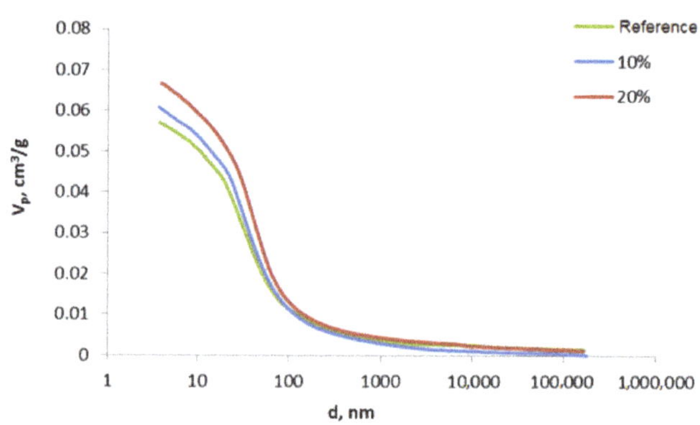

Figure 4. Total porosity of samples after 28 days of hydration.

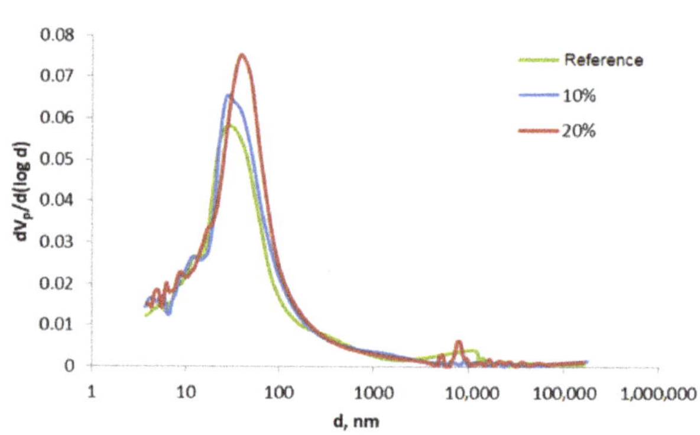

Figure 5. Porosity of samples after 28 days of hydration.

An increase in the amount of the pore in the diameter range of 50–100 μm was observed. There is also a visible shift of the maximum of the pore size distribution towards the pores with larger diameters as the proportion of glass addition increases. This results in a compressive strength decrease. The unfavorable influence of glass powder on the porosity and pore size distribution is mainly due to the substitution of cement with coarser inert glass powder.

4. Conclusions

The following conclusions can be drawn from the research presented in this paper:

- The analyzed glass powder possesses slight pozzolanic reactivity. It is expected that the finer grinding of glass will result in an enhancement of pozzolanic activity.
- The partial replacement of cement with glass powder results in a delay of the hydration rate. However, the reduction in cumulative heat is smaller than the reduction in the cement content, which indicates a slight acceleration of hydration due to the addition of glass powder. The effect of the heteronucleation of the C-S-H phase on the glass powder surface was also observed.
- The addition of glass powder negatively affects the compressive strength of mortar after 28 days of curing. In the latter age, the improvement of the compressive strength of mortar with glass powder addition is observed. This is mainly due to the pozzolanic activity of glass powder, which is revealed in the later time.
- No negative effect on the microstructure in the interfacial region was observed. The glass grain acts as heteronucleation centers of C-S-H crystals.
- Ground waste glass can be used as a replacement for about 10% of cement. In this case, a slight increase in the mortar compressive strength and no significant deterioration of the porosity are observed. In order to use this waste more effectively, it is necessary to grind it finer to a specific surface that is not smaller than that of cement. Then, not only will the pozzolanic activity of the glass be enhanced, but the additional effect of the filler will also be important.

Author Contributions: Conceptualization, M.D. and W.P.; methodology, M.D. and W.P.; validation, M.D., W.P., T.T., A.P. and M.N.; formal analysis, M.D., W.P., T.T., A.P. and M.N.; investigation, M.D., W.P., T.T. and A.P.; resources, M.D., W.P., T.T. and A.P.; data curation, M.D., W.P., T.T. and A.P.; writing—original draft preparation, M.D. and W.P.; writing—review and editing, M.D., W.P., T.T., A.P. and M.N.; visualization, M.D., W.P., T.T. and A.P.; supervision, M.D. and W.P.; project administration, M.D. All authors have read and agreed to the published version of the manuscript.

Funding: This research received no external funding.

Institutional Review Board Statement: Not applicable.

Informed Consent Statement: Not applicable.

Data Availability Statement: Not applicable.

Acknowledgments: The publication cost of this paper was covered by funds from the Polish National Agency for Academic Exchange (NAWA): "MATBUD'2023—Developing international scientific cooperation in the field of building materials engineering" BPI/WTP/2021/1/00002, MATBUD'2023.

Conflicts of Interest: The authors declare no conflict of interest.

References

1. Rehana, S.J. Where's the Fly Ash? Available online: https://www.concreteconstruction.net/producers/wheres-the-fly-ash_o (accessed on 30 September 2021).
2. Bentz, D.P.; Ferraris, C.F.; Jones, S.Z.; Lootens, D.; Zunino, F. Limestone and Silica Powder Replacements for Cement: Early-Age Performance. *Cem. Concr. Compos.* **2017**, *78*, 43–56. [CrossRef] [PubMed]
3. Mirzahosseini, M.; Riding, K.A. Influence of Different Particle Sizes on Reactivity of Finely Ground Glass as Supplementary Cementitious Material (SCM). *Cem. Concr. Compos.* **2015**, *56*, 95–105. [CrossRef]
4. Michael Record Collection of Glass Containers for Recycling Hits 78% in the EU. Available online: https://feve.org/glass_recycling_stats_2019/ (accessed on 2 November 2022).

5. Dong, W.; Li, W.; Tao, Z. A Comprehensive Review on Performance of Cementitious and Geopolymeric Concretes with Recycled Waste Glass as Powder, Sand or Cullet. *Resour. Conserv. Recycl.* **2021**, *172*, 105664. [CrossRef]
6. Nassar, R.-U.-D.; Soroushian, P.; Sufyan-Ud-Din, M. Long-Term Field Performance of Concrete Produced with Powder Waste Glass as Partial Replacement of Cement. *Case Stud. Constr. Mater.* **2021**, *15*, e00745. [CrossRef]
7. Gołek, Ł. Glass Powder and High-Calcium Fly Ash Based Binders—Long Term Examinations. *J. Clean. Prod.* **2019**, *220*, 493–506. [CrossRef]
8. Khan, M.N.N.; Saha, A.K.; Sarker, P.K. Reuse of Waste Glass as a Supplementary Binder and Aggregate for Sustainable Cement-Based Construction Materials: A Review. *J. Build. Eng.* **2020**, *28*, 101052. [CrossRef]
9. Tan, K.H.; Du, H. Use of Waste Glass as Sand in Mortar: Part I—Fresh, Mechanical and Durability Properties. *Cem. Concr. Compos.* **2013**, *35*, 109–117. [CrossRef]
10. Ortega, J.M.; Letelier, V.; Solas, C.; Miró, M.; Moriconi, G.; Climent, M.Á.; Sánchez, I. Influence of Waste Glass Powder Addition on the Pore Structure and Service Properties of Cement Mortars. *Sustainability* **2018**, *10*, 842. [CrossRef]
11. Jiang, Y.; Ling, T.-C.; Mo, K.H.; Shi, C. A Critical Review of Waste Glass Powder—Multiple Roles of Utilization in Cement-Based Materials and Construction Products. *J. Environ. Manag.* **2019**, *242*, 440–449. [CrossRef]
12. Shao, Y.; Lefort, T.; Moras, S.; Rodriguez, D. Studies on Concrete Containing Ground Waste Glass. *Cem. Concr. Res.* **2000**, *30*, 91–100. [CrossRef]
13. Kamali, M.; Ghahremaninezhad, A. Effect of Glass Powders on the Mechanical and Durability Properties of Cementitious Materials. *Constr. Build. Mater.* **2015**, *98*, 407–416. [CrossRef]
14. Du, H.; Tan, K. Waste Glass Powder as Cement Replacement in Concrete. *J. Adv. Concr. Technol.* **2014**, *12*, 468–477. [CrossRef]
15. Kalakada, Z.; Doh, J.H.; Chowdhury, S. Glass Powder as Replacement of Cement for Concrete—An Investigative Study. *Eur. J. Environ. Civ. Eng.* **2019**, *26*, 1–18. [CrossRef]
16. Idir, R.; Cyr, M.; Tagnit-Hamou, A. Pozzolanic Properties of Fine and Coarse Color-Mixed Glass Cullet. *Cem. Concr. Compos.* **2011**, *33*, 19–29. [CrossRef]
17. Lu, J.; Duan, Z.; Poon, C.S. Combined Use of Waste Glass Powder and Cullet in Architectural Mortar. *Cem. Concr. Compos.* **2017**, *82*, 34–44. [CrossRef]
18. Patel, D.; Tiwari, R.P.; Shrivastava, R.; Yadav, R.K. Effective Utilization of Waste Glass Powder as the Substitution of Cement in Making Paste and Mortar. *Constr. Build. Mater.* **2019**, *199*, 406–415. [CrossRef]
19. Carsana, M.; Frassoni, M.; Bertolini, L. Comparison of Ground Waste Glass with Other Supplementary Cementitious Materials. *Cem. Concr. Compos.* **2014**, *45*, 39–45. [CrossRef]
20. Elaqra, H.; Haloub, M.; Rustom, R. Effect of New Mixing Method of Glass Powder as Cement Replacement on Mechanical Behavior of Concrete. *Constr. Build. Mater.* **2019**, *203*, 75–82. [CrossRef]
21. Dyer, T.D.; Dhir, R.K. Chemical Reactions of Glass Cullet Used as Cement Component. *J. Mater. Civ. Eng.* **2001**, *13*, 412–417. [CrossRef]
22. Nassar, R.-U.-D.; Soroushian, P. Strength and Durability of Recycled Aggregate Concrete Containing Milled Glass as Partial Replacement for Cement. *Constr. Build. Mater.* **2012**, *29*, 368–377. [CrossRef]
23. Shayan, A.; Xu, A. Value-Added Utilisation of Waste Glass in Concrete. *Cem. Concr. Res.* **2004**, *34*, 81–89. [CrossRef]
24. Khmiri, A.; Chaabouni, M.; Samet, B. Chemical Behaviour of Ground Waste Glass When Used as Partial Cement Replacement in Mortars. *Constr. Build. Mater.* **2013**, *44*, 74–80. [CrossRef]
25. Shi, C.; Zheng, K. A Review on the Use of Waste Glasses in the Production of Cement and Concrete. *Resour. Conserv. Recycl.* **2007**, *52*, 234–247. [CrossRef]
26. Najduchowska, M.; Pabiś, E.; Rolka, G.; Baran, T. *Właściwości Betonu z Zastosowaniem Stłuczki Szklanej*; Stowarzyszenie Producentów Cementu: Wisła, Poland, 2016; Volume 2, pp. 627–648.
27. Liu, S.; Xie, G.; Wang, S. Effect of Curing Temperature on Hydration Properties of Waste Glass Powder in Cement-Based Materials. *J. Therm. Anal. Calorim.* **2015**, *119*, 47–55. [CrossRef]
28. Matos, A.M.; Sousa-Coutinho, J. Durability of Mortar Using Waste Glass Powder as Cement Replacement. *Constr. Build. Mater.* **2012**, *36*, 205–215. [CrossRef]
29. Omran, A.; Tagnit-Hamou, A. Performance of Glass-Powder Concrete in Field Applications. *Constr. Build. Mater.* **2016**, *109*, 84–95. [CrossRef]
30. Maraghechi, H.; Maraghechi, M.; Rajabipour, F.; Pantano, C.G. Pozzolanic Reactivity of Recycled Glass Powder at Elevated Temperatures: Reaction Stoichiometry, Reaction Products and Effect of Alkali Activation. *Cem. Concr. Compos.* **2014**, *53*, 105–114. [CrossRef]
31. Shi, C.; Wu, Y.; Riefler, C.; Wang, H. Characteristics and Pozzolanic Reactivity of Glass Powders. *Cem. Concr. Res.* **2005**, *35*, 987–993. [CrossRef]

materials proceedings

MDPI

Proceeding Paper

Alkali-Activated Metashale Mortar with Waste Cementitious Aggregate: Material Characterization †

Petr Hotěk [1,*], Lukáš Fiala [1], Wei-Ting Lin [2], Yi-Hua Chang [2] and Robert Černý [1]

1 Department of Materials Engineering and Chemistry, Faculty of Civil Engineering, Czech Technical University in Prague, Thákurova 2077/7, 166 29 Prague, Czech Republic

2 Department of Civil Engineering, National Ilan University, Ilan 26047, Taiwan

* Correspondence: petr.hotek@fsv.cvut.cz; Tel.: +420-2-2435-7102

† Presented at the 10th MATBUD'2023 Scientific-Technical Conference "Building Materials Engineering and Innovative Sustainable Materials", Cracow, Poland, 19–21 April 2023.

Abstract: The design of sustainable construction materials is continuously gaining increasing importance in civil engineering. Geopolymers are attractive alternatives to cementitious materials in terms of environmental impact and specific material properties, such as durability, an initial increase in mechanical properties, or chemical and thermal resistance. Such favorable properties can be advantageously utilized within various applications involving the design of materials for heavily stressed industrial floors. The research presented in the paper was focused on the design of a geopolymer composite based on metashale MEFISTO L05 and waste metashale RON D460HR binders. The 1:4 raw/waste mix of binders activated by potassium hydroxide/silicate was supplemented by 0.11 wt.% of graphite fibers to optimize electrical properties and bestow on it some new material functions, such as self-heating. The further utilization of fine waste aggregate (crushed defective concrete products, waste concrete from auto-mixers) resulted in an ~85% utilization of input waste materials. An acceptable mechanical performance of the mortar for particular civil engineering applications was observed (28d: R_f ~ 2.5 MPa, R_c ~ 15 MPa), as well as favorable thermal and DC/AC electrical properties, predicting the self-heating potential.

Keywords: alkali activation; waste metashale mortar; waste aggregate; low environmental impact

Citation: Hotěk, P.; Fiala, L.; Lin, W.-T.; Chang, Y.-H.; Černý, R. Alkali-Activated Metashale Mortar with Waste Cementitious Aggregate: Material Characterization. *Mater. Proc.* **2023**, *13*, 41. https://doi.org/10.3390/materproc2023013041

Academic Editors: Katarzyna Mróz, Tomasz Tracz, Tomasz Zdeb and Izabela Hager

Published: 1 March 2023

1. Introduction

The construction industry, with its almost 30% share of the total industrial activity within the European Union (EU), is one of its largest industrial sectors. Besides the construction of new buildings, it also involves renovation, demolition, and the disposing of waste building materials. In 2020, more than 800 million tons of the produced construction and demolition waste in the EU had to be recycled or landfilled [1]. Due to the negative environmental impact of landfilled construction materials polluted by hazardous agents, the design of sustainable materials leading to a reduction in landfilling gained higher importance in recent years [1–3].

With over 25 gigatons produced per year, concrete is the most widely used building material worldwide and undoubtedly also a major component of construction waste [4]. Therefore, its reuse in the construction industry, e.g., as a recycled waste aggregate for new composites, is beneficial for the reduction in landfilled material as well as the protection of natural resources [5,6]. It should be noted that the use of recycled concrete aggregate has its limits and rules. Nováková and Mikulica [6] emphasized that in the case of landfilled concrete, special attention needs to be paid to the removal of impurities coming from other materials, such as asphalts, bricks or plastics. They also noted that the European standard EN 12 620+A1 requirements for the use of aggregate in concretes slightly differs from country to country [7].

The degree of the environmental impact of the construction composite production significantly depends on the used binder and filler. Although alternative binders can positively influence the total environmental impact of the composite, the choice of aggregate plays the decisive role since it is present in a significantly higher amount than the binder. Alkali-activated composites (geopolymers) based on aluminosilicate waste/industrial product precursors (slag, brick dust, fly ash, metakaolin, metashale, etc.) offer a good environmental performance as far as the binder is concerned [8,9]. Despite the use of alkali activators (potassium/sodium hydroxides and water glasses), which are characterized by a higher negative environmental impact, the amount necessary for activation is low in comparison with the amount of binder and filler [10,11]. Geopolymers have a huge potential in civil engineering due to their various beneficial properties, e.g., high initial mechanical properties and chemical and fire resistance, ensuring durability and a low environmental impact, which can be further improved by replacing natural products with recycled aggregate [12–14].

This paper is focused on the design of a geopolymer mortar based on the mix of the newly produced and waste metashale binder, potassium hydroxide/silicate activator and recycled concrete aggregate filler. The composite was supplemented with a small amount of carbon fibers to examine their influence on electrical properties, which would be crucial for the new functional properties, such as self-sensing, self-heating or energy harvesting. The composite was characterized in terms of the basic physical, thermal and electrical properties.

2. Experimental Stage

2.1. Materials

The designed composite was based on the mix of two aluminosilicate precursors, Mefisto L05 and RON D460 HR (Figure 1). Mefisto L05 is a very fine metashale powder (d_{50} = 3 μm, d_{90} = 10 μm) originating from the thermal and granulometric treatment of clays and float kaolins. RON D460 HR is an industrial waste from the production of Mefisto L05, consisting of some unburnt particles that are of slightly higher size than in Mefisto L05 (d_{50} = 4.5 μm and d_{90} = 23 μm). The chemical composition and particle size distributions of the precursors are summarized in Table 1 and Figure 2.

(**a**) (**b**)

Figure 1. (**a**) Mefisto L05, (**b**) RON D460 HR.

Table 1. XRF analysis of Mefisto L05 and RON D460 HR precursors (%).

Precursor	SiO_2	Al_2O_3	Fe_2O_3	TiO_2	MgO	CaO	K_2O
Mefisto L05	49.58	41.04	2.76	2.39	1.08	0.9	1.03
RON D460 HR	49.1	47.3	0.9	1.6	0.1	0.2	0.5

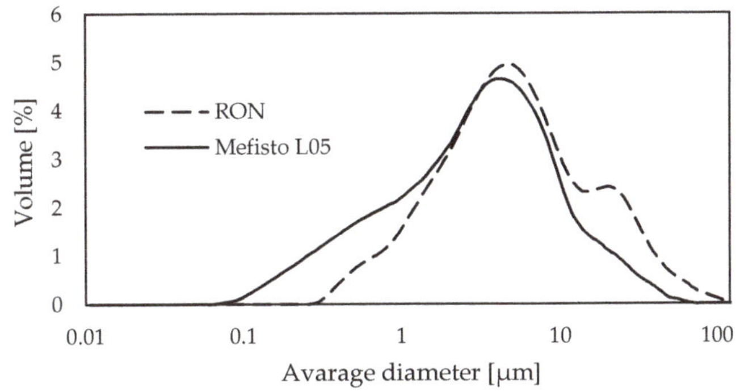

Figure 2. Particle size distribution of Mefisto L05 and RON D460 HR.

Metashale precursors were activated by a mix of potassium alkalis: potassium water glass with a K_2O/SiO_2 ratio = 1.7 and potassium hydroxide to avoid efflorescence, which occurs upon activation by sodium water glass/hydroxide [15]. The filler—a recycled cementitious aggregate—was of a grain size within the range of 0.063 to 2 mm, prepared by crushing damaged structural formworks. The H-shaped structural formwork (Figure 3a) was first crushed into smaller pieces using an EDB 400 mechanical press. Subsequently, smaller pieces were crushed using a crusher with 2.2 mm maximum jaw distance. The resulting aggregate was sieved and sorted into individual fractions of 0.063–0.5 mm, 0.5–1 mm and 1–2 mm, following the pattern of standardized quartz sand fractions (Figure 3b). The fresh mixture was supplemented with a small amount of carbon fibers to examine their influence on the electrical properties of the designed composite.

(**a**) (**b**)

Figure 3. (**a**) H-shape structural formwork, (**b**) fractions of waste aggregates.

2.2. Sample Preparation

First, Mefisto L05 and Ron D460 HR precursors were mixed with three waste aggregate fractions. Pellets of carbon fibers were dispersed in water with non-ionic surfactant Triton X-100 and siloxane-based air-detraining agent Lukosan S to reduce the surface tension of the carbonaceous admixture and defoam the suspension. The vessel with the suspension was then treated in an ultrasonic bath for 10 min to effectively crumble the pellets and disperse individual fibers. The well-prepared suspension was poured into a dry mixture of precursors, aggregate, and alkali activator and mixed for 10 min. Fresh mortar was finally placed into molds (Figure 4): 160 × 40 × 40 mm^3—mechanical properties, 100 × 100 × 100 mm^3—electrical properties, 70 × 70 × 70 mm^3—thermal properties. The 100 × 100 × 100 mm^3 samples were additionally embedded with copper-grid electrodes using a 3D-printed plastic board for precise positioning. After 24 h of curing in laboratory conditions (22 °C, 50% RH), samples were unmolded and left in equal conditions for a further 28 days. The composition of the studied geopolymer is summarized in Table 2.

Figure 4. RMCF1 mortar samples.

Table 2. Composition of RMCF1 mortar.

Component	RMCF1
Mefisto L05 [g]	272
RON D460 HR [g]	1088
Carbon fibers [g]	1.5
Potassium water glass [g]	474
Potassium hydroxide [g]	51
Aggregate 0.063–0.5 mm [g]	1700
Aggregate 0.5–1 mm [g]	850
Aggregate 1–2 mm [g]	1020
Water [g]	500

2.3. Methods

The bulk density ρ_v [kg·m^{-3}] was determined on the 40 × 40 × 160 mm^3 samples using the gravimetric method (Equation (1)). The matrix density ρ_{mat} [kg·m^{-3}] was evaluated by helium pycnometry (Pycnomatic ATC EVO). The total open porosity ψ [%] was then calculated using Equation (2).

$$\rho_v = \frac{m}{V} \tag{1}$$

$$\psi = \left(1 - \frac{\rho_v}{\rho_{mat}}\right) \cdot 100 \tag{2}$$

The dynamic modulus of elasticity E_{dyn} [MPa] was determined on the 40 × 40 × 160 mm^3 samples via a non-destructive method using the Pundit ultrasonic device according to Equation (3).

$$E_{dyn} = \rho_v \cdot v^2 \tag{3}$$

where ρ_v [kg·m^{-3}] is the bulk density of the material, and v [m·s^{-1}] is the speed of ultrasonic wave propagation through the sample.

Mechanical properties represented by the flexural and compressive strength were determined according to the ČSN EN 196-1 [16] using FP 100 and ED60 presses on the 40 × 40 × 160 mm^3 samples after 7 and 28 days.

The thermal properties, the thermal conductivity λ [W·m^{-1}·K^{-1}] and the specific heat capacity c_p [J·kg^{-1}·K^{-1}] were determined on the 70 × 70 × 70 mm^3 samples via non-stationary measurements using the ISOMET 2114 device equipped with a flat surface probe.

The DC electrical conductivity was determined on the 100 × 100 × 100 mm^3 electrode-embedded samples using a GW Instek GPR-11H30D power source and two Fluke 8846A multimeters for the voltage and current measurements. The electrical conductivity σ [$S \cdot m^{-1}$] was determined for three different input voltage levels according to Equation (4).

$$\sigma = \frac{I}{U} \cdot \frac{l}{S} = \frac{1}{R} \cdot \frac{l}{S} \quad (4)$$

where I [A] is the electric current, U [V] is the voltage, S = 0.0072 [m^2] is the area of electrodes, l = 0.07 [m] is the distance between electrodes and R [Ω] is the resistance of the material.

AC electrical properties represented by the magnitude of the impedance Z [Ω] and the phase shift θ [°] were determined in the range of 10 Hz–10 MHz on the 100 × 100 × 100 mm^3 samples using a GW Instek 8210 LCR bridge.

3. Results and Discussion

The summary of the material properties of the geopolymer mortar is given in Table 3. The total open porosity of ~17%, corresponding to the bulk and matrix density of 1954 $kg \cdot m^{-3}$ and 2362 $kg \cdot m^{-3}$, was mainly affected by the amount of mixing water (water/binder = ~0.37) due to the use of waste aggregates and carbon fibers [17]. The thermal conductivity and specific heat capacity were of 1.18 $W \cdot m^{-1} \cdot K^{-1}$ and 884 $J \cdot kg^{-1} \cdot K^{-1}$ which are typical values for such composites. Thermal conductivity was reasonably high for an effective spreading of the heat, which is an important presumption for self-heating ability.

Table 3. Basic physical, thermal and electrical properties of RMCF1 mortar.

Material Property [Unit]	RMCF1
Bulk Density [$kg \cdot m^{-3}$]	1954
Matrix Density [$kg \cdot m^{-3}$]	2362
Total Open Porosity [%]	17.3
Dynamic Modulus of Elasticity (28 days) [MPa]	7.5
Thermal Conductivity [$W \cdot m^{-1} \cdot K^{-1}$]	1.18
Specific Heat Capacity [$J \cdot kg^{-1} \cdot K^{-1}$]	884
Electrical Conductivity [$S \cdot m^{-1}$]	$3.95 \cdot 10^{-2}$

The flexural and compressive strengths at 7 d and 28 d are summarized in Figure 5. An increase in flexural strength from 2 MPa to 2.5 MPa and 12.5 MPa to 15 MPa in compressive strength was observed. The relatively low strength values were mainly due to the use of a significant amount of waste metashale binder, as well as fine waste aggregate (input waste materials ~85 wt.%). The decrease in the strength of cementitious and geopolymer composites due to the substitution of natural aggregate with waste counterparts is well known and mentioned in the literature, e.g., Nuaklong et al. [18], who investigated fly ash geopolymers with limestone and recycled aggregate, observing that the recycled aggregate led to a reduction in strength from 40 MPa to 30.6 MPa. Zaid et al. [19] studied natural aggregate replacement with recycled aggregate in steel-fiber-reinforced concrete and concluded that with an increasing amount of the recycled aggregate, the strength of the composites decreased. It was justified by a higher porosity of the recycled aggregate, resulting in an increase in material drying shrinkage and the formation of microcracks.

AC electrical properties represented by the frequency-dependent magnitude of the impedance and phase shift are presented in Figure 6. The magnitude of the impedance, involving both the resistive and capacitive component (resistance, capacitance), decreased from the initial ~175 Ω (10 Hz) to ~100 Ω (10 MHz), which due to the low values, revealed the potential for new functional properties, such as self-heating. The phase shift was from −3° to −10° in the tested frequency range and a noticeable decrease was observed at higher frequencies of 1–10 MHz. The phase shift close to 0° confirmed the resistive nature of the mortar and its potential for the self-heating function in an AC electric field. Since the DC electrical conductivity of the mortar (σ = 3.9·10^{-2} S·m^{-1}) was reasonably high, the self-heating potential was confirmed also in a DC electric field.

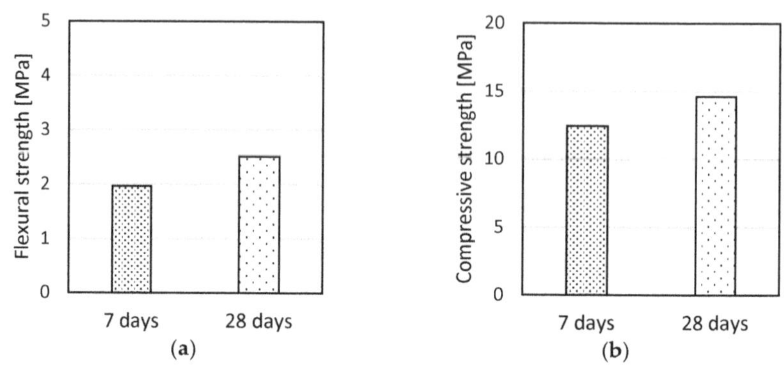

Figure 5. (**a**) Flexural strength, (**b**) Compressive strength of RMCF1 mortar.

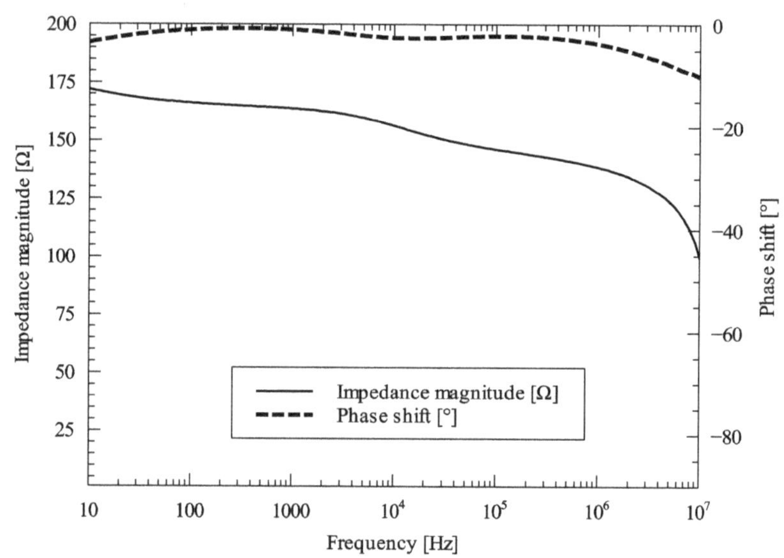

Figure 6. AC characteristics of RMCF1 mortar.

4. Conclusions

The study was focused on the design and basic material characterization of a geopolymer mortar with a special emphasis on the utilization of a significant amount of waste input materials. The waste metashale binder and waste cementitious aggregate originating from crushed defective cementitious products and cement mix used within the design of the geopolymer composite ensured an ~85 wt.% waste origin of input materials.

The mechanical properties of the mortar (28 d: R_f ~ 2.5 MPa, R_c ~ 15 MPa) are acceptable for some civil engineering applications. Nevertheless, it is important to target further efforts on the optimization of the geopolymer mortar composition, ensuring better mechanical performance. The thermal and electrical properties were favorable for the self-heating function in a DC and AC electric field, even with a low amount of carbon fibers. Nevertheless, it should be noted that the samples were characterized in a partially water-saturated state (curing in laboratory conditions, successive measurements without the preceding drying). Since the porous system in a heterogeneous geopolymer matrix is partially filled with water/salt solutions, which is beneficial in view of the thermal and electrical conductivity increase, the evaluated self-heating potential is slightly higher than in the case of dry material.

Author Contributions: Conceptualization, L.F. and R.Č.; methodology, P.H. and Y.-H.C.; validation, L.F. and W.-T.L.; investigation, P.H. and Y.-H.C.; resources, L.F. and W.-T.L.; data curation, P.H. and Y.-H.C.; writing—original draft preparation, L.F. and P.H.; supervision, L.F., W.-T.L. and R.Č.; project administration, L.F. All authors have read and agreed to the published version of the manuscript.

Funding: The outcome has been achieved with the support of M.era-Net Call 2021, Project No. 9262 and financial support from the Technology Agency of the Czech Republic under the project No. TH80020002 and the Grant Agency of the Czech Technical University in Prague under the project No. SGS22/137/OHK1/3T/11. Publication cost of this paper was covered with funds from the Polish National Agency for Aca-demic Exchange (NAWA): "MATBUD'2023—Developing international scientific cooperation in the field of building materials engineering" BPI/WTP/2021/1/00002, MATBUD'2023.

Institutional Review Board Statement: Not applicable.

Informed Consent Statement: Not applicable.

Data Availability Statement: Not applicable.

Conflicts of Interest: The authors declare no conflict of interest.

References

1. Eurostat, Environment and Energy, Generation and Treatment of Waste. Available online: http://ec.europa.eu/eurostat/ (accessed on 11 December 2022).
2. Miguel, M.; Brito, J.; Pinheiro, M.; Bravo, M. Construction and demolition waste indicators. *ISWA* **2013**, *31*, 241–255. [CrossRef]
3. Saca, N.; Dimache, A.; Radu, L.R.; Iancu, I. Leaching behavior of some demolition wastes. *J. Mater. Cycles Waste Manag.* **2017**, *19*, 623–630. [CrossRef]
4. IEA; WBCSD. *Cement Technology Roadmap 2009—Carbon Emissions Reductions up to 2050*; International Energy Agency [IEA], World Business Council for Sustainable Development [WBCSD]: Paris, France, 2009.
5. Wang, B.; Yan, L.; Fu, Q.; Kasal, B. A Comprehensive Review on Recycled Aggregate and Recycled Aggregate Concrete. *Resour. Conserv. Recycl.* **2021**, *171*, 105565. [CrossRef]
6. Nováková, I.; Mikulica, K. Properties of concrete with partial replacement of natural aggregate by recycled concrete aggregates from precast production. *Procedia Eng.* **2016**, *151*, 360–367. [CrossRef]
7. *ČSN EN 12 620+A1*; Aggregates for Concrete. Czech Office for Standards, Metrology and Testing: Prague, Czech Republic, 2008.
8. Castillo, H.; Collado, H.; Droguett, T.; Vesely, M.; Garrido, P.; Palma, S. State of the art of geopolymers: A review. *E-Polymers* **2022**, *22*, 108–124. [CrossRef]
9. Krivenko, P.; Shi, C.; Roy, D. *Alkali-Activated Cements and Concretes*, 1st ed.; Taylor and Francis Group: London, UK, 2006; ISBN 978-0-415-70004-7.
10. Duxson, P.; Fernández-Jiménez, A.; Provis, J.L.; Lukey, G.C.; Palomo, A.; Deventer, J.S.J. Geopolymer technology: The current state of the art. *J. Mater. Sci.* **2007**, *42*, 2917–2933. [CrossRef]
11. Davidovits, J. *Geopolymer: Chemistry and Applications*, 5th ed.; Davidovits, J., Ed.; Institut Géopolymére: SaintQuentin, France, 2020; ISBN 9782954453118.
12. Yan, D.; Chen, S.; Zeng, Q.; Li, H.; Xu, S. Correlating the elastic properties of metakaolin-based geopolymer with its composition. *Mater. Des.* **2016**, *95*, 306–318. [CrossRef]
13. Temuujin, J.; Rickard, W.; Lee, M.; Riessen, A. Preparation and thermal properties of fire resistant metakaolin-based geopolymer-type coatings. *J. Non Cryst. Solids* **2011**, *357*, 1399–1404. [CrossRef]
14. Azimi, E.A.; Abdullah, M.A.A.B.; Ming, L.Y. Review of Geopolymer Materials for Thermal Insulating Applications. *Key Eng. Mater.* **2015**, *660*, 17–22. [CrossRef]
15. Kani, E.N.; Allahverdi, A.; Provis, J.L. Efflorescence control in geopolymer binders based on natural pozzolan. *Cem. Concr. Compos.* **2012**, *34*, 25–33. [CrossRef]
16. *ČSN EN 196-1(722100)*; Methods of Testing Cement—Part 1: Determination of Strength. Czech Office for Standards, Metrology and Testing: Prague, Czech Republic, 2016.
17. Frayyeh, Q.; Swaif, A. Mechanical properties of fly ash geopolymer mortar reinforced with carbon fibers. In Proceedings of the the 3rd International Conference on Buildings, Construction and Environmental Engineering, University of Technology, Bahgdad, Iraq, 7 May 2018. [CrossRef]

18. Nuaklong, P.; Sata, V.; Chindaprasirt, P. Influence of recycled aggregate on fly ash geopolymer concrete properties. *J. Clean. Prod.* **2016**, *112*, 2300–2307. [CrossRef]
19. Zaid, O.; García, R.; Fraile, F.; Alshaikh, I.; Palencia, C. To determine the performance of metakaolin-based fiber-reinforced geopolymer concrete with recycled aggregates. *Arch. Civ. Mech. Eng.* **2022**, *22*, 114. [CrossRef]

materials proceedings

MDPI

Proceeding Paper

About Gas Permeability and Diffusion through Concrete †

Takwa Lamouchi [1,2], Severine Levasseur [3], Ludovic Potier [2], Thierry Dubois [2] and Frédéric Skoczylas [1,2,*]

1 School of Civil Engineering and Architecture, Changzhou Institute of Technology, Changzhou 213032, China
2 CNRS, Centrale Lille, UMR9013—LaMcube—Laboratoire de mécanique multiphysique et multiéchelle, Université de Lille, F-59000 Lille, France
3 ONDRAF/NIRAS, RD&D-Department, Avenue des Arts 14, 1210 Bruxelles, Belgium
* Correspondence: frederic.skoczylas@centralelille.com
† Presented at the 10th MATBUD'2023 Scientific-Technical Conference "Building Materials Engineering and Innovative Sustainable Materials", Cracow, Poland, 19–21 April 2023.

Abstract: Gas production is expected in radioactive-waste storage structures. This will induce a slow increase in gas pressure, which necessitates the study of gas transfer at a low pressure. In this special case, calculations of the flow through storing materials while solely using permeability and Darcy's law are likely to be inadequate, as diffusion may play a crucial role in the process. The gas permeability and gas diffusion coefficient of industrial concrete have then been measured on the dry material. Diffusion tests were performed with a new device, specially designed for this study. The diffusion coefficient was directly measured with the use of the first Fick's law, as the test was analyzed under a steady state. Using some simplified hypotheses, it was then possible to compare the proportion of flow occurring due to diffusion with the one occurring due to permeation. The tendency is very clear and unambiguously shows that diffusion is predominant at a very low injection pressure but becomes negligible as soon as the gas pressure exceeds a moderate value.

Keywords: gas permeability; gas diffusion; dry concrete; injection pressure

Citation: Lamouchi, T.; Levasseur, S.; Potier, L.; Dubois, T.; Skoczylas, F. About Gas Permeability and Diffusion through Concrete. *Mater. Proc.* **2023**, *13*, 42. https://doi.org/10.3390/materproc2023013042

Academic Editors: Katarzyna Mróz, Tomasz Tracz, Tomasz Zdeb and Izabela Hager

Published: 7 March 2023

1. Introduction

Radioactive waste storage at great depths is likely to produce gas, mainly di-hydrogen, due to water radiolysis and corrosion. This will lead to gas transport through storage structures, such as concrete tunnels, or surrounding rocks—often clay rocks. This is why numerous studies were conducted by ONDRAF/NIRAS (in Belgium) or Andra (in France) in order to characterize the gas transfer properties of dry or partially saturated concrete or of its host rock. Gas-permeability tests are often carried out with a significant pressure gradient, i.e., with quite a high injection pressure—a few MPa, for example. It must be nevertheless mentioned that the gas pressure would slowly increase in the storage structure at the beginning of its production. As a consequence, and for the purpose of further simulations, it seems important to take this phase into account and to measure the respective role of permeation and diffusion in the gas-transfer process. Diffusion, which occurs when there is a gas concentration gradient, has in fact to be taken into account during the low-gas-pressure phase. Hence, the scope of the present experimental study was to design a gas-diffusion device and to evaluate the effective diffusion coefficient of the concrete intended to be used by ONDRAF/NIRAS (Organisme national des déchets radioactifs et des matières fissiles enrichies—in Belgium) for its storage tunnels. This measurement, conducted alongside 'traditional' permeability tests, demonstrates that a large proportion of gas transport may be occurring due to diffusion under a very low pressure injection.

2. Material and Sample Preparation

Concrete cylinders (around 140 mm diameter) cored from tunnel voussoirs of the HADES URL (the Belgian Underground research laboratory) were provided by ON-

DRAF/NIRAS. They were then re-cored into small rectified cylinders with a diameter of 65 mm and height of 50 mm. Such a height was considered to be enough for permeability and diffusion measurements, as it is more than three times the maximum aggregate size (14 mm). The concrete composition as provided by ONDRAF/NIRAS is presented in Table 1 below.

Table 1. Concrete composition.

Water kg/m^3	Cement CEM II/B-V kg/m^3	Fly Ash kg/m^3	Coarse Agg. (5–14 mm) kg/m^3	Fine Agg. (0–4 mm) kg/m^3
135	335	115	1252	540

Four samples were prepared, and their porosity was measured with the classical vacuum and weighing techniques [1]. The samples were dried at a temperature of 105 °C until the mass was constant. The results gave a mean porosity of 12.5%.

3. Experimental Setup and Experimental Conditions

Two different setups were used: one for permeability measurements and one for gas diffusion. Gas-permeability measurements are quite usual in our lab, and they did not require any new design. This was not the case with diffusion tests, which required the performance of a new design.

3.1. Gas Permeability

Most of the tests performed in the laboratory are 1D-flow-type tests on cylindrical samples (Figure 1). The gas pressure is P_1 at the upstream sample side and P_0 at the downstream side. Using Darcy's law and steady flow [2], we obtain:

$$P(x) = \sqrt{P_1^2\left(1 - \frac{x}{L}\right) + P_0^2\frac{x}{L}} \quad (1)$$

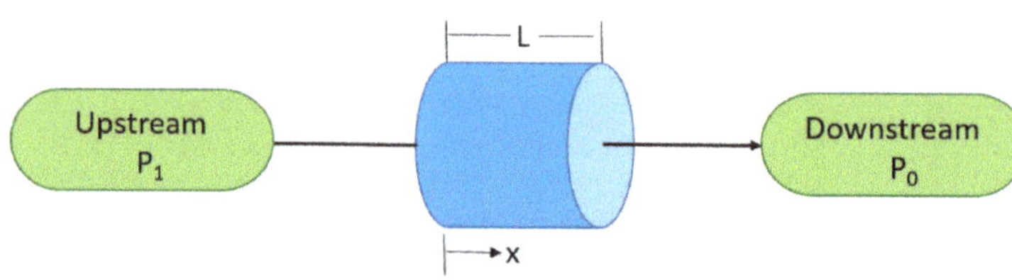

Figure 1. Schematic diagram of the steady-flow test method.

Q_1 is the volumetric gas flowrate at the upstream sample side [2]:

$$Q_1 = \frac{K_{app}A}{2\mu L}\frac{P_1^2 - P_0^2}{P_1} \quad (2)$$

K_{app} is the apparent gas permeability (apparent due to a potential Klinkenberg effect—see further), A is the sample cross-section, L is the sample length, and μ is the gas viscosity.

The flowrate Q_1 has to be measured to find the apparent gas permeability. Different methods can be used for this purpose: direct measurement with flowmeters (for example Brooks or Bronkhorst) or a measurement based on small pressure variation techniques (often used to calibrate the usual mass flowmeters). This second method was specially developed in our laboratory for materials with a very low permeability. Figure 2 presents a scheme of the system designed and used for this purpose.

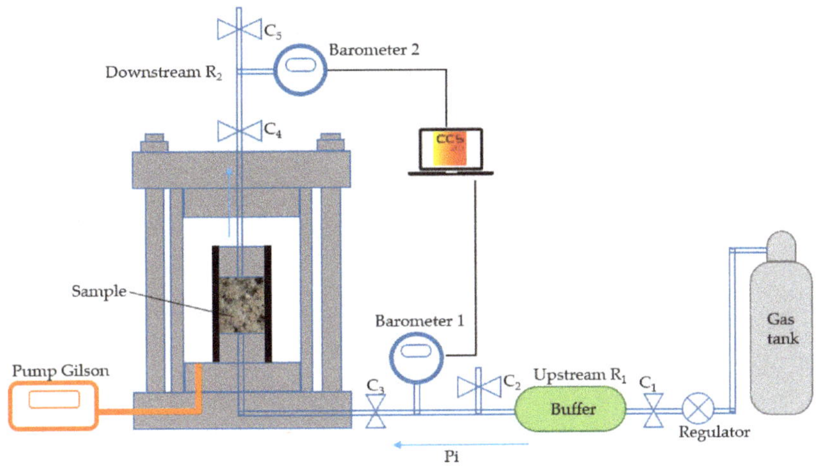

Figure 2. The device used for gas-permeability experiments.

The device is composed of a buffer reservoir R_1 and a tube reservoir R_2, respectively, connected at the upstream and downstream sample sides. The gas is first injected from a big gas tank at constant pressure P_1 (or P_i). The valve C_1 is closed as soon as a steady flow is assumed and R_1 is now feeding the sample with gas. The first possibility is then to measure the incoming flow rate Q_1. It is in fact the mean flowrate Q_1^{mean} during a time Δt for which there is a decrease ΔP_1 of pressure P_1. Assuming that there is a steady flow during Δt at a mean injection pressure $P_1^{mean} = P_1 - \Delta P_i/2$, it can be easily shown [3] that:

$$Q_1^{mean} = \frac{V_1 \Delta P_1}{P_1^{mean} \Delta t} \quad (3)$$

The apparent permeability K_{app} can then be deduced from relation 2 in which $P_1 = P_1^{mean}$ and $Q_1 = Q_1^{mean}$. This method is called the quasi-steady flow method at high pressure because it is applied at the upstream sample side. Experiments were also conducted in the laboratory with electronic mass flowmeters when it was possible. They provided results that were compared to those given upstream by the quasi-steady method. The same results were virtually obtained with a difference in permeability of often less than 1%, as long as the ΔP_1 decrease did not exceed 5% of P_1.

V_1 is the volume of the R_1 reservoir, which includes the tubing volume between R_1 and the sample. This volume is obtained with an accurate calibration.

3.2. *Diffusion Test—Principle of the Method*

This test has been newly designed in our laboratory in order to induce gas transport through a concentration difference at atmospheric pressure on both sample sides. As the use of a gas spectrometer has caused some calibration difficulties, we chose to use a simpler gas analyzer with a 500 ppm resolution. This apparatus was calibrated to detect helium into nitrogen, and it can work in a closed circuit. This means that helium will diffuse into a reservoir containing nitrogen (initially 100% nitrogen) and that the apparatus will analyze a small proportion of the mix (nitrogen + helium) and re-inject this proportion into the reservoir after analysis. The principle of this test is indicated in Figure 3. There is a continuous pure helium flow at the upstream side. This implies that the helium concentration is constant at this side despite nitrogen diffusion.

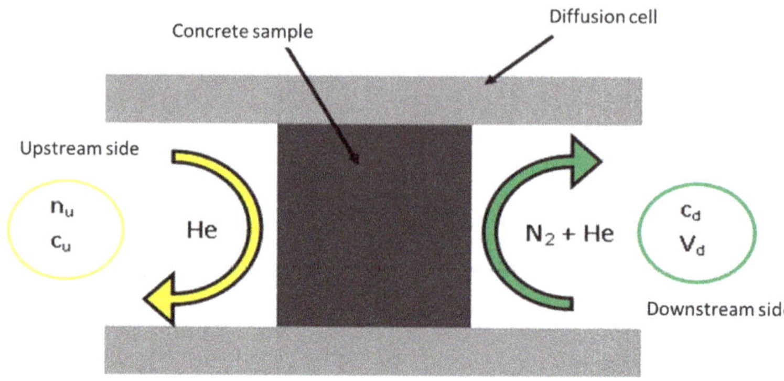

Figure 3. Schematic principle of the diffusion test; n_u is the number of helium moles, c_u and c_d are helium concentrations at the upstream (u) or downstream (d) sides, respectively. V_d is the downstream reservoir volume.

The device designed for diffusion tests is presented in the picture in Figure 4. The balancing device at Patm is realized by dip tubes whose height (into oil to avoid evaporation) can be adjusted. This allows for a pressure regulation with an accuracy that is better than one millibar. The percentage of helium at the downstream side is periodically measured by the gas analyzer.

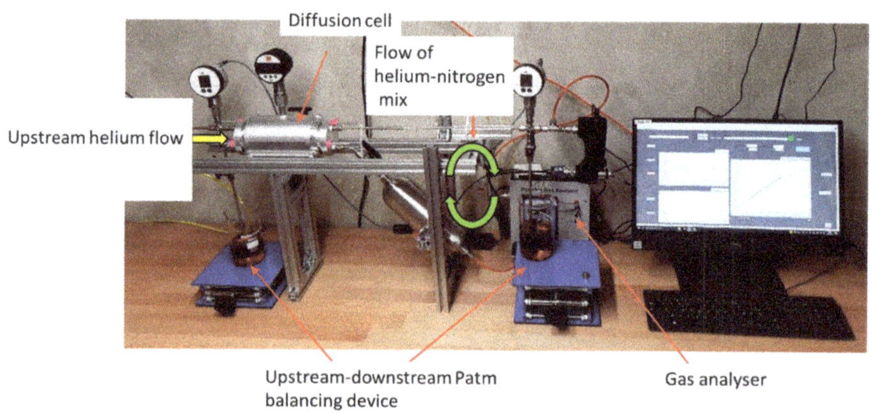

Figure 4. Picture of the device designed for diffusion.

3.3. *Hypotheses and Test Analysis*

By using some hypotheses, it is supposed that the diffusion process is controlled by the first Fick's law (1856), which, with a 1D geometry, gives:

$$J_x(x,t) = -D\frac{\partial c(x,t)}{\partial x} \tag{4}$$

J_x is the molar surface flow in mol·s^{-1}·m^{-2};
D is the effective diffusion coefficient in m^2·s^{-1};
c is the gas concentration in mol·m^{-3}.

The goal is then to measure J through the sample to obtain D under known or measured concentrations at the upstream and downstream sample sides. To be applied, Equation (3) needs a knowledge of the concentration 'c(x,t)' through the sample, which evolves with time. Under a constant molar flow J_x, the downstream concentration increases linearly with time. At this stage, the mass balance equation coupled with Equation (3) leads to a linear concentration profile in the sample. This will be the main hypothesis used in the results.

The room temperature is constant and controlled at 22 °C or 295 K. Assuming that helium is a perfect gas (P_{atm} V = n_uRT), it is found that n_u = 40.8 moles per unit volume (1 m^3); thus, the concentration at the upstream side is c_u = 40.8 mol/m^3.

The molar flux through the sample cross-section will be:

$$\varphi(x,t) = J_x(x,t)\ A \tag{5}$$

φ molar flux through surface A in mol·s^{-1};
A sample cross-section in m^2.

For the calculations, it was supposed that the downstream helium concentration can be neglected in front of the upstream one, i.e., $c_d << c_u$. This leads to:

$$\frac{\partial c(x,t)}{\partial x} \approx \frac{-c_u}{L} \tag{6}$$

L is the sample length.

Under a stationary flow, 'p', the rate of helium particles' increase into the downstream reservoir, is directly linked to φ:

$$\varphi = p\,\frac{V_d}{V_m} \tag{7}$$

V_d downstream reservoir volume in m^3;
V_m molar volume (at P_{atm}) in m^3·mol^{-1};
'p' is in s^{-1}.

The following is thus obtained:

$$D = \frac{pLV_d}{Ac_uV_m} \tag{8}$$

with
V_d = 1.09 × 10^{-3} m^3;
V_m = 24.05 × 10^{-3} m^3·mol^{-1};
c_u = 40.8 mol·m^{-3}.

4. Results

4.1. Gas-Permeability Results

4.1.1. Results with Argon

Gas-permeability tests were performed with argon gas on dry material after the porosity measurements. Three injection pressures were used: 0.5, 1 and 1.5 MPa, in order to quantify the potential Klinkenberg effect [4]. This effect, also known as the 'slipping effect', may occur when the mean gas free path 'λ' is close to or lower than the mean pore size. As 'λ' is increasing when the gas pressure is decreasing, this effect is often visible on material with small pores (like concrete) and/or during tests with a weak injection pressure. As a result, if this effect is present, the measured permeability is apparent and higher than the intrinsic one. A very well-known correction was brought by Klinkenberg [4] in order to take this effect into account:

$$K_{app} = K_{int}\left(1 + \frac{\beta}{P_m}\right) \tag{9}$$

K_{app} is the apparent permeability (m^2);
K_{int} is the intrinsic permeability;
β is the Klinkenberg coefficient and P_m is the mean test pressure:

$$P_m = \frac{1}{L}\int_0^L P(x)dx \tag{10}$$

From relation (8), it can be seen that three different injection pressures are sufficient to assess and correct the Klinkenberg effect.

Two confining pressures (hydrostatic pressures) had been required by our partner ONDRAF/NIRAS: 2.25 and 4.5 MPa. Such a change in confining pressure can induce a significant permeability variation depending on whether the material is (micro-)cracked [3,5]. A typical result, obtained from sample OB-111, can be seen in Figure 5, and the whole set of intrinsic permeability results is presented in Table 2.

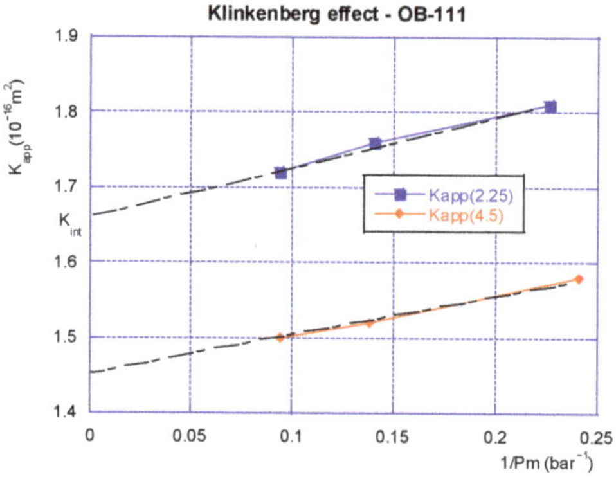

Figure 5. Example of Klinkenberg effect: tests performed at 2.25 or 4.5 confining pressure.

Table 2. Gas permeability results obtained with argon.

Sample	Confining Pressure (MPa)	K_{int} (10^{-16} m^2)	Coef. β (bar)
OB-111	2.25 4.5	1.66 1.45	0.4 0.38
OB-121	2.25 4.5	1.73 1.5	0.31 0.34
OB-321	2.25 4.5	1.57 1.35	0.28 0.35
OB-422	2.25 4.5	1.37 1.16	0.31 0.52

First of all, these results show a good material homogeneity in terms of its gas permeability. The confinement effect is very weak; this means that the material is not significantly cracked, as it is well known [6] that cracks close with a confining pressure, which in return induces a strong (and non-reversible) reduction in permeability. This is not the case here. The Klinkenberg effect is actually present but can be considered as being quite low.

4.1.2. Results with Helium

As mentioned before, the diffusion test will be performed with helium and not with argon. The results obtained will be used to evaluate the respective proportion of gas transfer due, respectively, to diffusion and permeation. It is thus important to compare the 'argon permeability' with the 'helium permeability'. One comparative set of tests was therefore performed on sample OB-422. The Klinkenberg effect is more sensitive with helium, as can be seen in Figure 6. This is consistent with the fact that the helium molecule size is lower than the argon molecule size [7]. 'β' is supposed to vary as 1/r, 'r' being the radius of the molecule. 'r' is three times higher for argon than for helium. This ratio (1/3) is more or less respected by the 'β' coefficients presented in Table 3. On the other hand, it is clear that argon's and helium's intrinsic permeabilities are virtually the same, i.e., both can be used, without significant differences, to compare the respective flow resulting from permeation or from diffusion.

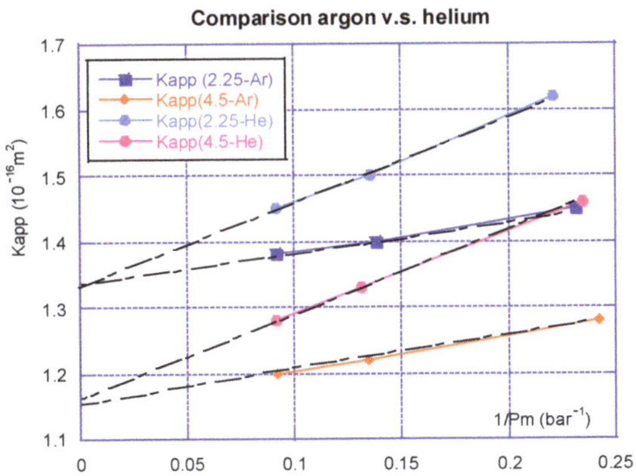

Figure 6. Permeability and Klinkenberg effect: argon vs. helium.

Table 3. Comparison of argon vs. helium.

Sample	Confining Pressure (MPa)	K_{int} (10^{-16} m^2)	β (Bars)
OB-422 Argon	2.25	1.33	0.38
	4.5	1.15	0.48
OB-422 Helium	2.25	1.32	1.02
	4.5	1.16	1.09

4.2. Gas-Diffusion Results

Helium-diffusion tests were carried out right after the permeability experiments. The downstream helium concentration is given in Figure 7. One can observe a very good homogeneity in these results for the four samples, as was also the case for the gas permeability. The four tests were performed at a 2.25 MPa confining pressure, as it was supposed that this pressure did not play a crucial role for permeability measurements. It is clear that, after around 3500 s, a permanent flow rate can be assumed, as the He concentration increases linearly. This provides evidence that for the dry concrete, the diffusion phenomenon is quite rapid. The slope 'p' (presented in §3.3) can be obtained from these results, allowing for the calculation of the diffusion coefficient 'D' with relation (7). The results are presented in Table 4; they lie within the range of the gas-diffusion coefficient often reported for concrete [8].

Table 4. Diffusion coefficient for the four samples.

Référence	p (s^{-1})	D (m^2·s^{-1})
OB-111	2.5×10^{-6}	4.2×10^{-8}
OB-212	2.4×10^{-6}	3.9×10^{-8}
OB-321	2.6×10^{-6}	4.4×10^{-8}
OB-422	2.4×10^{-6}	3.9×10^{-8}

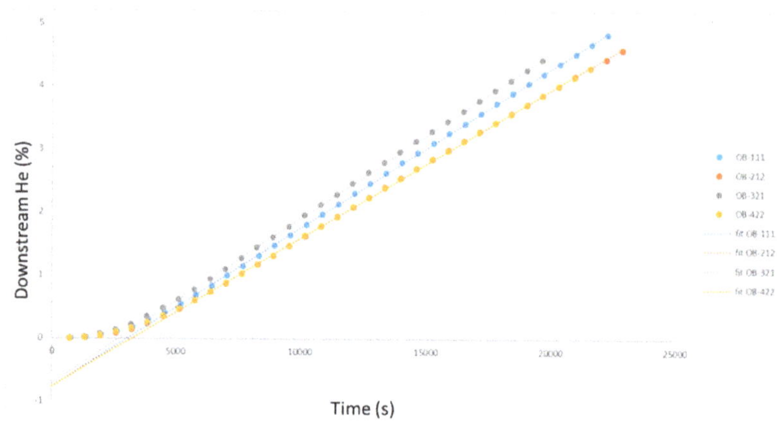

Figure 7. Helium proportion at the downstream side of samples.

4.3. Equivalent Permeability and Discussion

As mentioned before, the pressure due to gas production in a radioactive-waste storage should slowly increase. It is thus interesting to evaluate, in the case of a very low pression gradient, the proportion of gas transfer resulting from permeation and diffusion. This calculation was made within some simplified hypotheses. In particular, it was assumed that the diffusion coefficient does not depend on the gas pressure, which is not the case. The first step is to calculate the equivalent permeability due to diffusion. If an experiment is conducted with the gas pressure (pure helium, for example) at injection pressure P_1 (upstream sample side) and the drainage pressure at P_0 = Patm (air), it is supposed here that the diffusion coefficient helium-air is almost the same as for helium-nitrogen. If the test is interpreted as a permeability test, the downstream volumetric flowrate Q_0 is given by:

$$Q_0 = A\frac{K}{2\mu L}\frac{\left(P_1^2 - P_0^2\right)}{P_0} \tag{11}$$

This is linked to the number of moles per second n_k:

$$n_k = \frac{Q_0}{V_m} \tag{12}$$

In a diffusion test, this quantity is the same as φ mentioned before:

$$\frac{Q_0}{V_m} = n_k = \varphi = J \cdot A = D\frac{c_u}{L}A \tag{13}$$

The helium downstream concentration is still neglected, while the upstream concentration is:

$$c_u = n_u\frac{P_1}{P_0} \tag{14}$$

Then:

$$\frac{Q_0}{V_m} = D\frac{n_u}{L}A\frac{P_1}{P_0} = A\frac{K}{2\mu L}\frac{\left(P_1^2 - P_0^2\right)}{P_0}\frac{1}{V_m} \tag{15}$$

The equivalent permeability K_D can then be extracted from relation (15):

$$K_D = 2D\mu n_u V_m\frac{P_1}{\left(P_1^2 - P_o^2\right)} \tag{16}$$

As can be seen in relation (15), the equivalent permeability depends on the injection pressure P_1. K_D is roughly in the form Cste/P_1 when P_1 is increased. This means that the proportion of flow due to diffusion will be lesser and lesser as P_1 is increased. This is illustrated in Figure 8, which presents the ratio K_D/K, in which K has been chosen as a mean value of 1.5×10^{-16} m^2. This ratio is equivalent to the proportion of gas flow due to diffusion compared to the one due to permeation.

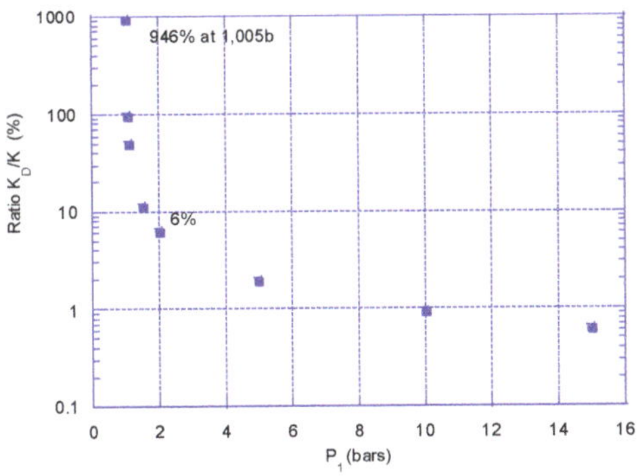

Figure 8. Flow ratio due to diffusion compared to the one due to permeation. P_1 is the absolute injection gas pressure.

5. Conclusions

A special device has been designed in order to measure gas diffusion through concrete (or other porous materials). The main goal of this study was to measure both the gas permeability and gas diffusion of an industrial concrete, which could be employed for tunnels intended for radioactive-waste storage. Gas production due to corrosion and water radiolysis should take place in these structures, and the low rate of production would first induce gas transfer at very low pressures, which are generally not used in gas-permeability experiments. Gas-permeability and diffusion tests were then performed on the same set of samples. They revealed that argon and helium permeability is virtually the same when corrected from a (slight) Klinkenberg effect. On the whole, gas permeability was found to be very homogeneous (order of magnitude of 1.5×10^{-16} m^2). This homogeneity was also verified for the effective diffusion coefficients (around 4×10^{-8} m^2/s). These coefficients were used to calculate an equivalent permeability K_D, which is dependent on the gas injection pressure. This clearly showed that under a low pressure gradient (or injection pressure), diffusion is largely predominant, whereas its induced flow can be neglected as soon as the injection pressure is larger than a few bars. This implies that gas diffusion must be taken into account at the beginning of gas production. Such a study should find a logical extension in the case of partially saturated concrete, which is likely to be encountered in 'in situ' structures.

Author Contributions: T.D., L.P. and F.S. conceived and planned the experiments. T.L. and L.P. carried out the experiments. T.L. and F.S. planned and carried out the analyses. S.L., T.L. and F.S. contributed to the interpretation of the results. F.S. and S.L. took the lead in writing the manuscript. All authors have read and agreed to the published version of the manuscript.

Funding: The publication cost of this paper was covered with funds from the Polish National Agency for Academic Exchange (NAWA): "MATBUD'2023-Developing international scientific cooperation in the field of building materials engineering" BPI/WTP/2021/1/00002, MATBUD'2023.

Acknowledgments: The authors thank ONDRAF/NIRAS for its technical and scientific support.

Conflicts of Interest: The authors declare no conflict of interest.

References

1. Andreola, F.; Leonelli, C.; Romagnoli, M.; Miselli, P. Techniques Used to Determine Porosity. *Am. Ceram. Soc. Bull.* **2000**, *79*, 49–52.
2. Zhang, D.; Agostini, F.; Jeannin, L.; Skoczylas, F. New Insights Brought by Micro-Tomography to Better Understand Gas Transfer Property Variation and Coupling Effects in Salt Rocks. *Rock. Mech. Rock Eng.* **2021**, *54*, 6457–6480. [CrossRef]
3. Chen, X.; Caratini, G.; Davy, C.A.; Troadec, D. Skoczylas, Coupled transport and poro-mechanical properties of a heat-treated mortar under confinement. *Cem. Concr. Res.* **2013**, *49*, 10–20. [CrossRef]
4. Klinkenberg, L.J. The permeability of porous media to liquids and gases. In *Drilling and Production Practices*; American Petroleum Institute: Washington, DC, USA, 1941; pp. 200–213.
5. Chen, W.; Han, Y.; Agostini, F.; Skoczylas, F.; Corbeel, D. Permeability of a Macro-Cracked Concrete Effect of Confining Pressure and Modelling. *Materials* **2021**, *14*, 862. [CrossRef] [PubMed]
6. Ding, Q.; Wang, P.; Cheng, Z. Influence of temperature and confining pressure on the mechanical properties of granite. *Powder Technol.* **2021**, *394*, 10–19. [CrossRef]
7. Degao, H.; Feng, Y.; Zhiguo, S.; Aiwei, Z.; He, Z.; Bin, L. Experimental study about the gas slip flow in Longmaxi shales from the southern Sichuan Basin. *Bull. Geol. Sci. Technol.* **2021**, *40*, 36–41. [CrossRef]
8. Sercombe, R.; Vidal, C.; Gallé, F. Adenot, Experimental study of gas diffusion in cement paste. *Cem. Concr. Res.* **2007**, *37*, 579–588. [CrossRef]

materials proceedings

MDPI

Proceeding Paper

Geopolymers—Base Materials and Properties of Green Structural Materials †

Tarreck Mahaman Manssour Issa [1], Mateusz Sitarz [2,*], Katarzyna Mróz [2] and Marcin Różycki [2]

1 Buildings and Public Works, School of Engineering, École d'ingénieurs, Bachelor et Mastère Spécialisé® (CESI), 76800 Saint-Étienne-du-Rouvray, France

2 Chair of Building Materials, Faculty of Civil Engineering, Cracow University of Technology, 31-155 Cracow, Poland

* Correspondence: mateusz.sitarz@pk.edu.pl; Tel.: +48-12-628-23-71

† Presented at the 10th MATBUD'2023 Scientific-Technical Conference "Building Materials Engineering and Innovative Sustainable Materials", Cracow, Poland, 19–21 April 2023.

Abstract: Nowadays, most industrial solid waste, agriculture waste, and natural minerals are piled up, which not only occupies land resources, but also has a dangerous impact on the environment. The responsible management and recycling of these materials can have significant environmental benefits, while also providing a means of accessing raw materials for the creation of geopolymers. Geopolymers are new, environment-friendly, cementless materials and alternative construction materials to ordinary Portland cement. They not only have excellent mechanical properties, but also have valuable properties, e.g., high-temperature resistance. This paper summarizes the feasibility studies in terms of base materials and properties of today's developing geopolymers. This paper also highlights the significance of developing sustainable materials for civil engineering.

Keywords: geopolymers; precursors; activator solutions; environment

Citation: Issa, T.M.M.; Sitarz, M.; Mróz, K.; Różycki, M. Geopolymers—Base Materials and Properties of Green Structural Materials. *Mater. Proc.* **2023**, *13*, 43. https://doi.org/10.3390/materproc2023013043

Academic Editors: Tomasz Tracz, Tomasz Zdeb and Izabela Hager

Published: 9 March 2023

1. Introduction

Geopolymer composites are a "green" alternative to traditional cementitious materials. It is estimated that the production of geopolymers creates four to eight times less carbon dioxide than cement production [1]. The process requires twice less energy compared to the manufacture of Portland cement [1]. It seems that geopolymers have low emissions of CO_2, SO_2, and NO_X. In spite of the mechanical resemblances, the chemical and morphological differences in this material create challenges when comparing it to ordinary Portland cement (OPC), particularly in light of the current standards.

The term "geopolymers", concerning aluminosilicate binders, was introduced in the late 1970s, and developed by the French scientist and engineering prof. Joseph Davidovits in 1978 [2]. Geopolymers are inorganic [3] and belong to the family of alkali-activated materials, which, unlike cementitious materials, require alkalis to harden. They are also amorphous aluminosilicate materials with three-dimensional frameworks of SiO_4 and AlO_4 tetrahedra that can be produced via alkali activation even at low temperatures (20 and 120 °C, according to Davidovits [1]) and at low pressure. Geopolymers are synthesized via geopolymerization. The mineral raw materials dissolve in the alkaline environment at room temperature or higher. As a result, an amorphous phase and three-dimensional aluminosilicate structure are formed. The mechanisms of geopolymerization are not fully understood at present, but most hold that the geopolymer binding process can be presented in three main stages [4–8]:

- The dissolution of the aluminosilicate precursors to form reactive particles $Si(OH)_4$ and $Al(OH)^{4-}$.
- Restructuring and modification of aluminosilicate structures to a more stable state. The hydrolysis process leads to the release of water from the structure at this stage.

- Gelation/Polycondensation: Polymerization and precipitation of the system, i.e., the formation of a three-dimensional network of silico-aluminate that forms a geopolymer.

The production of a hard geopolymer involves mixing aluminosilicate powder that meets specific requirements for particle size distribution and specific surface area with an alkali solution (such as NaOH, KOH, or waterglass), resulting in the formation of a gel-like substance that solidifies quickly [9–11]. This alkali-activated material boasts exceptional physical, chemical, and mechanical properties, including low density, micro- and nano-porosity, high mechanical strength, thermal stability, fire resistance, and chemical resistance. Geopolymer materials allow non-polluting production and do not emit toxic gases or fumes. They can be reinforced, for example with carbon fibers, which provide excellent adhesive properties to the reinforcement.

2. Base Materials of Geopolymers

2.1. Mineral Precursors

As a geopolymer precursor, any aluminosilicate that is able to dissolve in an alkaline solution is suitable. There is a wide range of reactive products that can be used, including calcined or non-calcined clays [12], industrial or agricultural by-products [13] and artificial pozzolans [14]. However, the choice of raw material may depend on several factors, including availability, cost, and type of application. The raw materials may be distinguished as primary and secondary raw materials.

2.1.1. Primary Raw Materials

According to the literature, Geopolymer production can be achieved using raw materials with high silica and alumina content. The main sources of primary raw materials are natural minerals. [15].

➢ *Clay and clay minerals*

Through thermal treatments, clay can be transformed into various clay minerals. As an aluminosilicate salt consisting of tiny particles (less than 2 µm), clay is a pliant and cohesive earthen material [16]. It is a layered silicate that is composed of alternating silicon–oxygen tetrahedrons and alumina octahedrons, and it is an appropriate precursor for preparing geopolymers [16].

➢ *Metakaolin*

Metakaolin (MK) is a type of clay mineral known as kaolinite that has been dehydroxylated. Kaolinite is a soft, fine, white clay that exhibits excellent plasticity and fire resistance. When kaolinite is subjected to thermal treatment, typically at a temperature range of 600–800 °C, the original structure of the clay is destroyed, leading to the formation of metakaolin. This new substance is an anhydrous aluminum silicate [16]. Metakaolin is a commonly used aluminosilicate source for geopolymerization due to its purity and predictable properties. This material provides a consistent chemical composition, which makes it a preferred starting material for geopolymer research [15]. Despite its benefits, metakaolin-based geopolymers have some drawbacks, such as the need for high amounts of water and sometimes low mechanical strength, which limit their application areas [15]. As a result, the geopolymerization of various Al–Si minerals and clays, particularly kaolinite and metakaolin, has been extensively studied over the past few decades.

2.1.2. Secondary Raw Materials

Within the framework of reducing environmental impact via alternative waste management, the production of geopolymers can be envisaged using industrial and agricultural wastes as raw materials, as follows.

➢ *Fly Ash*

Coal-fired power plants generate fly ash (FA) as a byproduct, which is considered an industrial waste (Figure 1). Its main components are SiO_2 and Al_2O_3, but it can contain small components such as CaO, Fe_2O_3, MgO, etc. However, the composition of these components in this waste material is significantly variable, especially depending on the coal source and burning conditions. FA is generally divided into classes F (FFA) and C (CFF). FFA is a low calcium fly ash and is a by-product of burning bituminous coal; FFC is a high calcium fly ash and is produced by burning lignite and sub-bituminous coal [17]. Geopolymer synthesis can benefit from the unique characteristics of FA, such as its alumina-silicate composition, ability to function with minimal water, high malleability, and widespread availability. Notably, in the United States alone, an estimated 63 million tons of FA are produced annually, and Hungary generates approximately 200 million m^3 of fly ash and slag each year [15].

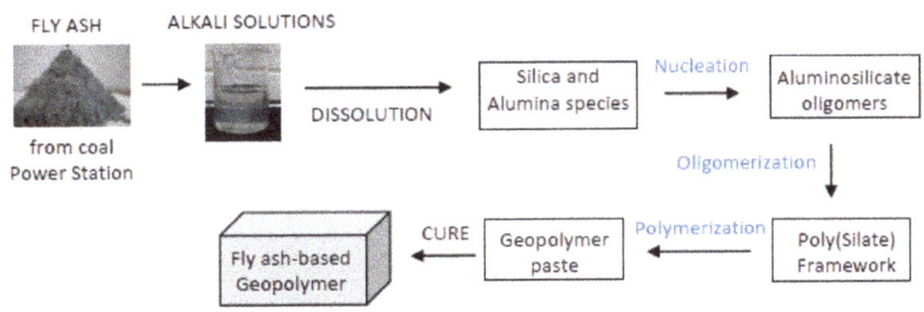

Figure 1. The transformation process of fly ash into a binder for geopolymer materials [10,11].

➢ *Granulated Blast Furnace Slag (GBFS)*

Blast furnace slag (BFS) is a by-product of iron making and is often referred to as slag. It can be obtained at approximately 1500 °C [18]. Following rapid cooling and grinding processes, a granular substance known as granulated blast furnace slag (GBFS) can be formed. This raw material (GBFS) contains S_iO_2, Al_2O_3, CaO, and MgO. GBFS is one of the common geopolymer raw materials, given its strong reactivity towards geopolymer synthesis. Additionally, it is possible to achieve an optimal reaction rate using GBFS at temperatures as low as 0 °C [19]. GBFS is known as a raw material which can reduce porosity. GBFS in geopolymer can improve reactivity of the mix, enhance mineral structure, rise long-term strength and increase resistance to sulfate. It also reduces the water demand, permeability, and hydration heat of a geopolymer.

➢ *Red Mud*

Red Mud (RD) is a by-product of processing bauxite into alumina using the Bayer process. The Bayer process dissolves the soluble part of bauxite using sodium hydroxide under high-temperature and high-pressure conditions. However, a small amount of sodium hydroxide remains in the RD after the Bayer process and allows it to have a high pH value (over 12) [20]. RD contains generally solid and metallic oxides, especially Fe_2O_3, Al_2O_3, S_iO_2, Cao, Na_2O, and TiO_2. The red color of RD comes from iron oxides, which represent more than 60% of red mud mass [21].

➢ *Rice Husk*

Fillers known as rice husk ash (RHA) are obtained by calcining rice husks, which are often considered a form of agricultural waste and a potential environmental hazard. During calcination, most rice husks components disappear, and generally an amorphous silicate remains. Research has shown that every 100 kg of husks burnt in a boiler yields approximately 25 kg of RHA [22]. The temperature of calcination must not exceed 600 °C to obtain reactive pozzolanic ashes and to avoid the risk of forming less reactive crystalline silicas. Well-burnt RHA contains 90% amorphous silica, 5% carbon and 2% K_2O [22]. However, the chemical composition of RHA depends on combustion conditions. RHA is a small, fine material; its particle size ranges from 3 to 75 µm [22]. RHA is greyish-black in color because of unburned carbon.

Some other raw materials, such as steel slag (STS), silica fume (SF), volcanic ash (VA), waste glass (WG), coal gangue (CG), high magnesium nickel slag (HMNS), etc., can be used in synthetic geopolymers because of their abundant silicate and aluminum elements and amorphous structures [16].

A variety of structures based on geopolymers can be manufactured using different precursors. For example, previous research has shown that the mechanical strength and durability of geopolymer materials based on FA can be improved by the addition of metakaolin or other wastes [15]. Geopolymer-based elements can exhibit superior properties and greater durability when compared to structures formed using interconnected Portland cement. Additionally, the microstructure of geopolymer materials is influenced by the composition of the amorphous phase, the content of oxides, and especially, the particle fineness [23].

2.2. Activator Solutions

In the geopolymerization process, the activating solution plays a fundamental role. Depending on its quantity and concentration, it will provide the necessary mixture that not only initiates the reaction, but also determines the ultimate structure of the solidified material. For geopolymerization to occur, the presence of strongly alkaline activators is essential in the solution. Indeed, they accelerate the dissolution of the aluminosilicate source, favoring the formation of stable hydrates with low solubility and the formation of a compact structure with these hydrates. The physical and chemical properties of the activating solutions play an important role in the behavior of the activated material.

2.2.1. Alkali Activators

Numerous studies have demonstrated that the activation of geopolymer materials is achieved through the use of alkali activators. Alkalinity, resistance to chemical attacks, strength development, and durability are all dependent on the type of Alkali activator used. Additionally, the selection of suitable activator solutions is based on the chemical composition and the source of raw materials. Typically, the activation of alkali materials involves the use of sodium- or potassium-based activators (e.g., hydroxides, such as NaOH and KOH), sodium- and potassium-based silicate solutions, and carbonates. Each of these activators has distinct advantages, with sodium- and potassium-based activators being the most commonly used due to their cost-effectiveness and widespread availability. Prior investigations have revealed that, in the case of FA, sodium-based activators exhibit greater activation efficiency than their potassium-based counterparts [24], despite the higher alkalinity in potassium-based solutions [25]. To produce a metakaolin-based geopolymer, sodium waterglass has been used as an activator; the research has shown a compressive strength reaching 63.8 MPa [26].

In addition to conventional activators, ash generated from the incineration of solid waste and organic material can be employed as activators. Moreover, ash sourced from olive oil biomass can serve as an alkali activator for GBFS-based geopolymer production. [16].

2.2.2. Acidic Activators

Although Alkali activators are the most commonly used means of activation in geopolymers, some compositions rely on the use of acidic activators. For instance, MK-based geopolymer has exhibited remarkable compressive strength, up to 93.8 MPa, via the implementation of phosphoric acid as an activator [26,27]. Moreover, research has indicated that acid-based geopolymers possess higher temperature resistance (up to 1450 °C) and superior mechanical properties than alkali-based geopolymers [28]. In addition, recent studies suggest that an innovative phosphoric acid-based geopolymer could function as a dependable and robust fire or heat insulator, as it has a conductivity of no more than 10^{-7} S/cm at elevated temperatures [29].

3. Properties of Geopolymers

Geopolymers are known for their exceptional fire resistance up to 1400 °C, as well as their resistance to heat and acid [30]. These materials have high early compressive strength, excellent fracture toughness, long-term durability, low apparent porosity or nano-porosity, and freeze-thaw resistance. Geopolymer composites and concrete present similar mechanical properties to OPC-based materials, or better [31]. Its Mohs hardness ranges from four to seven [30].

One of the important properties of geopolymer binders is that they can be use to immobilization toxic waste [32]. Geopolymer materials have been observed to exhibit zeolitic and feldspathoid-like behavior, which allows them to immobilize hazardous elemental wastes within their matrix. This property makes them an effective binder for converting semi-solid waste into solid adhesive materials. [32]. Moreover, geopolymers have excellent workability because of the particle shape of raw materials, which allows their fluent production [16]. Previous studies have indicated that the addition of certain materials can enhance the workability of geopolymers. For example, the inclusion of silicon powder, calcium carbonate, or specific amounts of superplasticizer have been shown to improve their workability [16].

Furthermore, a high slag content in geopolymer mixtures can accelerate the initial and final setting time. A decrease in the molar concentration of NaOH can extend the setting time of FA-based geopolymer. [16].

Geopolymers can be strengthened using various types of material, such as carbon, glass, minerals, or steel, which can be used to produce advanced composite materials with improved mechanical properties, such as flexure, tension, shear, and flexural fatigue, particularly at high temperatures. This is due to the excellent adhesive properties of geopolymer binders, making them an attractive option for reinforcement [16].

4. Conclusions

In this study, the ways geopolymer composites are designed and developed were summarized. According to the state-of-the-art, the following conclusions can be drawn:

- Many mineral raw materials can be used in geopolymer technology.
- Especially interesting is the possibility of using waste materials from the industrial, energy, and mining sectors in the geopolymer synthesis process. This solution allows the reduction of the carbon footprint of the geopolymer composite, which makes the geopolymer material more environmentally friendly.
- The most optimal solution is to use locally available waste materials. In this way, it is possible to reduce both the financial and environmental costs of producing geopolymer materials.
- The properties of geopolymer composites indicate that such materials can be an alternative to cement-based materials in some applications (Figure 2).

Geopolymers are a promising, eco-friendly alternative to ordinary Portland cement (OPC). The production of geopolymer binders is associated with lower carbon dioxide emissions compared to the production of Portland cement. [30]. Industrial solid waste and waste incineration bottom ash are commonly piled up, taking up valuable land resources and posing environmental risks. However, these wastes can be recycled and used as raw materials for geopolymer production, which presents significant potential. The possibility of using geopolymer materials as an alternative to OPC binder allows to reduce the consumption of cement, which production is associated with high CO_2 emissions.

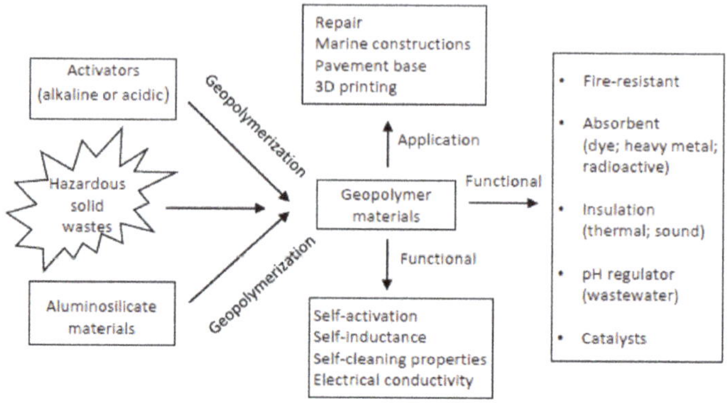

Figure 2. Geopolymerization process and geopolymers applications [16].

Author Contributions: Conceptualization, T.M.M.I., M.S. and K.M.; methodology, K.M. and T.M.M.I.; software, M.S. and K.M.; validation, K..M. and M.R.; formal analysis, T.M.M.I.; investigation T.M.M.I., K.M., M.R. and M.S.; data curation, T.M.M.I., K.M., and M.S.; writing—original draft preparation, K.M. and M.S.; writing—review and editing, T.M.M.I., M.S. and K.M.; visualization, K.M. and M.S.; supervision, K.M.; project administration, K.M.; funding acquisition, K.M. and M.S. All authors have read and agreed to the published version of the manuscript.

Funding: This research was funded by the Cracow University of Technology and the Polish National Agency for foreign students in the framework of the project NAWA Spinaker, POWR.03.03.00-00-PN16/18. This research was supported by the National Centre for Research and Development in Poland in the framework of the project SMART-G Smart geopolymers (ERA MIN2-3/SMARTG/1/2022).

Institutional Review Board Statement: Not applicable.

Informed Consent Statement: Not applicable.

Data Availability Statement: Not applicable.

Acknowledgments: This research was part of the work provided by two interdisciplinary research groups at Building Materials Engineering, Faculty of Civil Engineering, Cracow University of Technology (31-155 Cracow, Poland) and Buildings and Public Works, School of Engineering, CESI (76800 Saint Etienne du Rouvray, France). The publication cost of this paper was paid by the Polish National Agency for Academic Exchange (NAWA), "MATBUD'2023—Developing international scientific cooperation in the field of building materials engineering", BPI/WTP/2021/1/00002, MATBUD'2023.

Conflicts of Interest: There are no conflict of interest stated by the authors.

References

1. Davidovits, J. *Geopolymer Chemistry and Applications*, 5th ed.; Institut Géopolymère: Saint-Quentin, France, 2020.
2. Davidovits, J. Geopolymers and geopolymeric materials. *J. Therm. Anal.* **1989**, *35*, 429–441. [CrossRef]
3. What Is a Geopolymer? Introduction. Institut Géopolymère, Saint-Quentin, France. Available online: http://www.geopolymer.org/science/introduction (accessed on 10 July 2022).
4. Duxson, P.; Fernández-Jiménez, A.; Provis, J.L.; Lukey, G.C.; Palomo, A.; van Deventer, J.S. Geopolymer technology: The current state of the art. *J. Mater. Sci.* **2007**, *42*, 2917–2933. [CrossRef]
5. Provis, J.L.; van Deventer, J.S.J. Geopolymerisation kinetics. 2. Reaction kinetic modelling. *Chem. Eng. Sci.* **2007**, *62*, 2318–2329. [CrossRef]
6. Prud'Homme, E.; Michaud, P.; Joussein, E.; Peyratout, C.; Smith, A.; Rossignol, S. In situ inorganic foams prepared from various clays at low temperature. *Appl. Clay Sci.* **2011**, *51*, 15–22. [CrossRef]
7. Ren, D.; Yan, C.; Duan, P.; Zhang, Z.; Li, L.; Yan, Z. Durability performances of wollastonite, tremolite and basalt fiber-reinforced metakaolin geopolymer composites under sulfate and chloride attack. *Constr. Build. Mater.* **2017**, *134*, 56–66. [CrossRef]
8. De Silva, P.; Sagoe-Crenstil, K.; Sirivivatnanon, V. Kinetics of geopolymerization: Role of Al_2O_3 and SiO_2. *Cem. Concr. Res.* **2007**, *37*, 512–518. [CrossRef]

9. Barbosa, T.R.; Foletto, E.L.; Dotto, G.L.; Jahn, S.L. Preparation of mesoporous geopolymer using metakaolin and rice husk ash as synthesis precursors and its use as potential adsorbent to remove organic dye from aqueous solutions. *Ceram. Int.* **2018**, *44*, 416–423. [CrossRef]
10. Perná, I.; Hanzlíček, T. The solidification of aluminum production waste in geopolymer matrix. *J. Clean. Prod.* **2014**, *84*, 657–662. [CrossRef]
11. Łach, M. Geopolymer Foams—Will They Ever Become a Viable Alternative to Popular Insulation Materials?—A Critical Opinion. *Materials* **2021**, *14*, 3568. [CrossRef]
12. Tchakouté, H.K.; Rüscher, C.H.; Kong, S.; Kamseu, E.; Leonelli, C. Geopolymer binders from metakaolin using sodium waterglass from waste glass and rice husk ash as alternative activators: A comparative study. *Constr. Build. Mater.* **2016**, *114*, 276–289. [CrossRef]
13. Gokhale, C. The Immobilisation of Organic Waste by Geopolymerization. Ph.D. Thesis, Stellenbosch University, Stellenbosch, South Africa, 2001.
14. Diffo, B.K.; Elimbi, A.; Cyr, M.; Manga, J.D.; Kouamo, H.T. Effect of the rate of calcination of kaolin on the properties of metakaolin-based geopolymers. *J. Asian Ceram. Soc.* **2015**, *3*, 130–138. [CrossRef]
15. Mucsi, G.; Ambrus, M. Raw materials for geopolymerization. In Proceedings of the MultiScience—XXXI. microCAD International Multidisciplinary Scientific Conference, University of Miskolc, Miskolc, Hungary, 20–21 April 2017.
16. Cong, P.; Cheng, Y. Advances in geopolymer materials: A comprehensive review. *J. Traffic Transp. Eng.* **2021**, *8*, 283–314. [CrossRef]
17. Guo, Y.; Zhao, Z.; Zhao, Q.; Cheng, F. Novel process of alumina extraction from coal fly ash by pre-desilicating—Na_2CO_3 activation—Acid leaching technique. *Hydrometallurgy* **2017**, *169*, 418–425. [CrossRef]
18. Amran, Y.M.; Alyousef, R.; Alabduljabbar, H.; El-Zeadani, M. Clean production and properties of geopolymer concrete; A review. *J. Clean. Prod.* **2020**, *251*, 119679. [CrossRef]
19. Katarzyna, B.; Le, C.H.; Louda, P.; Michał, S.; Bakalova, T.; Tadeusz, P.; Prałat, K. The Fabrication of Geopolymer Foam Composites Incorporating Coke Dust Waste. *Processes* **2020**, *8*, 1052. [CrossRef]
20. Nie, Q.; Hu, W.; Huang, B.; Shu, X.; He, Q. Synergistic utilization of red mud for flue-gas desulfurization and fly ash-based geopolymer preparation. *J. Hazard. Mater.* **2019**, *369*, 503–511. [CrossRef]
21. Red Mud. Available online: https://en.wikipedia.org/wiki/Red_mud (accessed on 15 July 2022).
22. Singh, B. Rice husk ash. In *Waste and Supplementary Cementitious Materials in Concrete*; Woodhead Publishing: Cambridge, UK, 2018; pp. 417–460.
23. Sitarz, M.; Castro-Gomes, J.; Hager, I. Strength and Microstructure Characteristics of Blended Fly Ash and Ground Granulated Blast Furnace Slag Geopolymer Mortars with Na and K Silicate Solution. *Materials* **2022**, *15*, 211. [CrossRef]
24. Helmy, A.I. Intermittent curing of fly ash geopolymer mortar. *Constr. Build. Mater.* **2016**, *110*, 54–64. [CrossRef]
25. Singh, B.P.; Bhargava, O.N.; Chaubey, R.S.; Kishore, N.; Prasad, S.K. Early Cambrian Trail Archaeonassa from the Sankholi Formation (Tal Group), Nigali Dhar Syncline (Sirmur District), Himachal Pradesh. *J. Geol. Soc. India* **2015**, *85*, 717–721. [CrossRef]
26. Tchakouté, H.K.; Rüscher, C.H. Mechanical and microstructural properties of metakaolin-based geopolymer cements from sodium waterglass and phosphoric acid solution as hardeners: A comparative study. *Appl. Clay. Sci.* **2017**, *140*, 81–87. [CrossRef]
27. Shuai, Q.; Xu, Z.; Yao, Z.; Chen, X.; Jiang, Z.; Peng, X.; An, R.; Li, Y.; Jiang, X.; Li, H. Fire resistance of phosphoric acid-based geopolymer foams fabricated from metakaolin and hydrogen peroxide. *Mater. Lett.* **2020**, *263*, 127228. [CrossRef]
28. Celerier, H.; Jouin, J.; Gharzouni, A.; Mathivet, V.; Sobrados, I.; Tessier-Doyen, N.; Rossignol, S. Relation between working properties and structural properties from 27Al, 29Si and 31P NMR and XRD of acid-based geopolymers from 25 to 1000 °C. *Mater. Chem. Phys.* **2019**, *228*, 293–302. [CrossRef]
29. Sellami, M.; Barre, M.; Toumi, M. Synthesis, thermal properties and electrical conductivity of phosphoric acid-based geopolymer with metakaolin. *Appl. Clay Sci.* **2019**, *180*, 105192. [CrossRef]
30. Janošević, N.; Đorić-Veljković, S.; Topličić-Ćurčić, G.; Karamarković, J. Properties of geopolymers. *FU Arch. Civ. Eng.* **2018**, *16*, 45–56.
31. Kaya, M.; Koksal, F.; Gencel, O.; Munir, M.J.; Kazmi, S.M. Influence of micro Fe_2O_3 and MgO on the physical and mechanical properties of the zeolite and kaolin based geopolymer mortar. *J. Build. Eng.* **2022**, *52*, 104443. [CrossRef]
32. Davidovits, J. Properties of geopolymer cements. In Proceedings of the First International Conference on Alkaline Cements and Concretes, Kiev, Ukraine, 11–14 October 1994.

materials proceedings

MDPI

Proceeding Paper

Influence of Environmental Conditions on Steel Corrosion in Concrete Exposed to Gamma Radiation †

Mariusz Dąbrowski [1,*], Justyna Kuziak [2], Kinga Dziedzic [1] and Michał A. Glinicki [1]

1 Institute of Fundamental Technological Research, Polish Academy of Sciences, Pawińskiego 5b, 02-106 Warsaw, Poland

2 Department of Building Materials Engineering, Warsaw University of Technology, al. Armii Ludowej 16, 00-637 Warsaw, Poland

* Correspondence: mdabrow@ippt.pan.pl

† Presented at the 10th MATBUD'2023 Scientific-Technical Conference "Building Materials Engineering and Innovative Sustainable Materials", Cracow, Poland, 19–21 April 2023.

Abstract: This article examines the problem of the service life of reinforced concrete structures intended for nuclear power plants and radiation waste storage bunkers when exposed to radiation. This research focused on assessing the corrosion resistance of steel bars under conditions of simultaneous exposure to gamma radiation and various environmental conditions affecting the rate of carbonation. Electrochemical measurements of steel bars were carried out on samples in three environmental conditions: in a laboratory–dry; enclosed in a can at RH = 50%; and enclosed in a can at RH = 100%. The durability of the passivation layer of steel on non-irradiated and irradiated specimens after 8 months of exposure to gamma radiation was compared. A lower degradation effect of gamma radiation was visible in fully water-saturated specimens.

Keywords: corrosion of steel; relative humidity variability; carbonation condition; EIS; polarization curve

Citation: Dąbrowski, M.; Kuziak, J.; Dziedzic, K.; Glinicki, M.A. Influence of Environmental Conditions on Steel Corrosion in Concrete Exposed to Gamma Radiation. *Mater. Proc.* **2023**, *13*, 44. https://doi.org/10.3390/materproc2023013044

Academic Editors: Katarzyna Mróz, Tomasz Tracz, Tomasz Zdeb and Izabela Hager

Published: 13 March 2023

1. Introduction

The durability of nuclear power plant constructions, with a service life of 30 years or more, is a desirable feature affecting electricity prices [1]. The corrosion of steel in concrete structures exposed to irradiation is one of the important parameters determining the long term suitability of the structure for further operation [2]. The structural properties of existing and new nuclear power plant buildings should be well established in terms of the expected period of operation.

The irradiation-induced aging of concrete in nuclear power plants, occurring over a long period of time in nuclear environments, should be considered. Common knowledge concerning the influence of gamma radiation shows that a significant deterioration of mechanical properties of concrete occurs when the gamma dose exceeds the threshold of 1×10^8 Gy [3]. However, there are studies that report the deterioration of concrete properties resulting just from a few MGy of gamma radiation [4–6], especially in the early stage of hardening [7,8]. For the reinforced concrete structures of a nuclear power plant, the impact of CO_2 from the atmosphere is very important. Researchers confirm that the main product of carbonation is the formation of calcite and vaterite [9–11] and have suggested an increase in the rate of the carbonation reaction. Normally, reinforcing bars surrounded by concrete exist in high pH (around 13 for Portland cement) conditions, and such an alkaline environment protects against steel corrosion and the further deterioration of the structure. The acceleration of carbonation lowers the pH of the concrete and consequently leads to steel depassivation [12,13]. In addition, gamma radiation causes the decomposition of the aqueous pore solution near the steel reinforcing bars and the formation of oxidizing and reducing compounds [14], which can affect the properties of the passivation layer on the steel surface. Investigations into gamma radiation on corrosion of metals and alloys are generally measured in water solution systems, simulating cement matrix conditions. Studies [15,16] conducted on different types of steel and alloys (AISI 304, carbon steel) and gamma radiation doses (range from 75 kGy to 3 MGy) did not provide a consistent answer regarding the effect of irradiation on steel corrosion. In the case of real cement-based composite systems, the low dose (300 kGy) of gamma radiation [17] does not change the properties of the steel protective layer. A dose of up to 2 MGy, as seen in previous research [18], demonstrated the effect of gamma radiation on the reduction of the protective properties of the passive layer of steel in mortar, located in a sealed can with CO_2 and medium relative humidity (RH = 50%). The effect has not been unequivocally confirmed, due to a lack of studies that have conducted measurements in strong radiation fields. A marginal number of papers (as reported in review paper [12]) consider the influence of humidity conditions on the passivation of steel in gamma radiation field. Humidity conditions are of key importance for the intensity of concrete carbonation [9] and, consequently, for the conditions of the reinforcement steel.

This research is a continuation of considerations concerning the effects of gamma radiation on steel corrosion in cement-based materials [18]. The main purpose of this research is to assess the effect of gamma radiation on reinforcing steel in humid conditions that favour concrete carbonation (approximately RH = 50%) and inhibit carbonation (RH = 100%). The potentiodynamic and electrochemical impedance spectroscopy measurements were used to assess the susceptibility of the steel to corrosion.

2. Experimental Program

2.1. Materials and Specimens

A mortar mix with Portland cement CEM I 52.5R [19] and quartz sand [20] was prepared. A total of 400 g of cement (density 3.1 kg/dm^3 and 4500 cm^2/g Blaine specific surface) and 1350 g of sand was used. The w/c was 0.6. The cement mortar component proportions ensured an increase in open porosity and prevented the segregation of composition.

The specimens with the steel rebar were cast. The steel rebars, S235JR, with a diameter of 6 mm (according to EN 10025-2 [21]) were used. The chemical composition of steel, as declared by the producer, was given in [18]. Wet polishing with SiC paper (4000 grit) followed by sonication was performed before the steel bars were coated with epoxy resin. For each bar, 10 cm^2 of smooth steel surface was left for electrochemical measurements, according to PN-86-B-01810 [22]. Standard curing conditions before irradiation were used (temperature of 20 ± 2 °C, duration 28 days).

2.2. Specimen Conditioning and Gamma Radiation Exposure

Cylindrical mortar specimens with central placed steel rod were divided into three groups after curing time: the first was to be irradiated and second and third to be stored in the laboratory without irradiation. The two groups of specimens were placed in steel cans, as shown in Figure 1a. The third group was conditioned without canning. Subsequently, each group of specimens was placed in a CTS climate-controlled chamber (65% RH and CO_2 concentration of 1% in the air). Specimens were conditioned in open cans in the climate chamber for 24 h and then the cans were closed. The environmental conditions in the cans were designed for two RH—fully water-saturated condition and RH = 50 ± 5% (controlled by superabsorbent polymer). The third group of specimens, after 24 h of conditioning without a can, was cured under laboratory conditions (23 ± 1 °C and RH = 43 ± 5%).

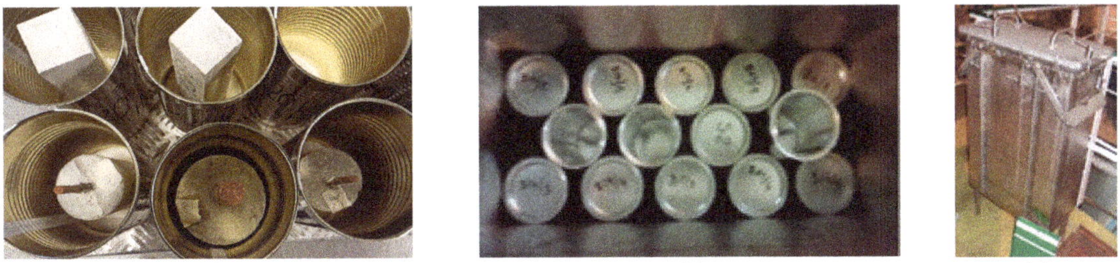

(a) (b) (c)

Figure 1. Location of specimens in the cans, (**a**) the arrangement of sealed cans (**b**), and the view of the prismatic steel box (**c**) immersed in the reactor pool close to spent nuclear fuel rods.

The research reactor of National Centre for Nuclear Research was used for the gamma irradiation of specimens sealed in cans and set in a prismatic steel box (Figure 1) in a manner described in detail in [18]. The absorbed dose was determined as the average value of two dosimeters attached to the walls of the cans. The dose of gamma radiation absorbed by the samples was close to 1.8 MGy. The specimens in the nuclear reactor-spent fuel pool were irradiated for 8 months. The second group of canned specimens was stored in climatic chamber at 38 °C without irradiation for the same period. The temperature in the climatic chamber corresponds with the temperature of irradiation specimen conditions. The measurements of environmental conditions (temperature and humidity) in closed cans confirmed the correctness of the adopted conditions. The results of mortar carbonation are presented and discussed in the research of Jóźwiak-Niedźwiedzka et al. [23].

2.3. Test Methods

The corrosion behaviour of steel in mortar was evaluated by electrochemical measurement (EIS and anodic polarization curves) using a potentiostat/galvanostat Autolab PGSTAT 302N with Nova 2.1 software. A three-electrode system consisted of reinforcement steel (the working electrode), saturated calomel electrode (SCE; the reference electrode), and a stainless-steel plate (counter electrode) was used. Measurements were taken at a temperature of 20 ± 1 °C after 24 h of conditioning in a saturated solution of $Ca(OH)_2$ in water.

The EIS measurements settings:

- frequency range of 5×10^5–5×10^{-1} Hz;
- amplitude of the sine wave perturbation—10 mV.

The equivalent circuit (Figure 2) for the analysis of the obtained EIS spectra was chosen under the assumptions described in a previous study [18]. The following parameters were determined on the basis of the measurements:

- R_{pas}—the ohmic resistance in the defects of passive layer;
- R_{ct}—the charge transfer resistance;
- Y,n—parameters of a constant phase element (CPE_m—mortar constant phase element; CPE_{pas}—passive surface constant phase element; CPE_{dl}—double layer on the steel constant phase element).

The EIS measurement was followed by an anodic polarization curve measurement. The settings for polarization measurements were as follows:

- potential range from −100 mV, with respect to the corrosion potential, to 800 mV, with respect to the reference electrode,
- rate of potential change—1 mV/s.
- Parameters determined based on the recorded polarization curve:
- E_{cor}—corrosion potential,
- E_P—passivation potential,

- E_{tr}—transpassivation potential,
- j_{cor}—corrosion current density,
- j_p—passivation current density.

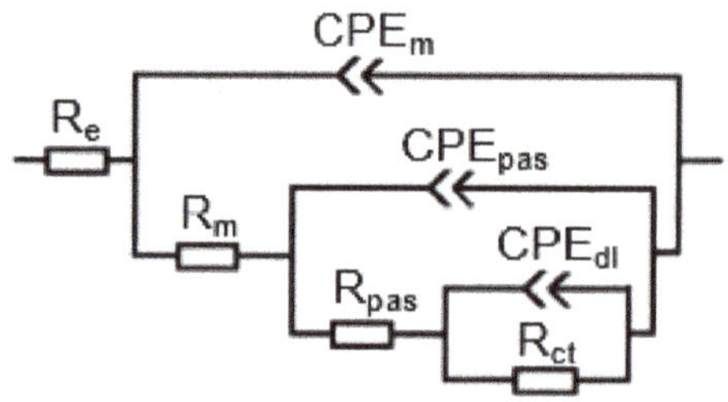

Figure 2. Equivalent circuit for steel in cement mortar (R_e—the electrolyte resistance; R_m—mortar resistance; CPE_m—constant phase element describing mortar; R_{pas}—the ohmic resistance in pits or defects of passive layer; CPE_{pas}—the constant phase element for passive surface; R_{ct}—the charge transfer resistance; CPE_{dl}—constant phase element describing the double layer on the steel) [18].

The criteria for polarization data evaluation are given in [22] and [24].

3. Results and Discussion

All the obtained polarization curves have a course characteristic of steel in the passive state (Figure 3). In all cases, the gamma irradiation of the samples results in a shift of the corrosion potential values towards more negative potentials, indicating an increase in the probability of corrosion. The steel in all mortars is characterized by low j_{cor} values. However, only the non-irradiated mortars in cans in an environment with an RH = 50% j_{cor} value met the requirements for steel in the passive state ($j_{cor} < 0.1$ μA/cm^2; negligible corrosion level). In other cases, the j_{cor} value indicates a low corrosion level. The corrosion current density increased after irradiation (Table 1), which means that the steel in irradiated mortars has a higher corrosion rate. In the case of samples stored in environment with 100% RH, gamma radiation caused a small increase in current density (1.1–1.3 times), while in the case of RH 50%, significantly higher j_{cor} values (4–12 times) for steel in irradiated samples were observed. The effect of RH on the course of the polarization curves of steel in irradiated mortars was small—the courses of polarization curves of steel in samples in environments with 50 and 100% RH were comparable. In contrast, in the case of non-irradiated samples, significantly lower values of j_{cor} and anodic current density were observed for 50% RH than for 100% RH. The highest values of j_{cor} and j_p were obtained for steel in mortar stored under laboratory conditions, which demonstrates the worst protective properties of the passive layer and the highest corrosion rate of steel. It may be related to the unrestricted access of CO_2 to the samples stored in air, while other samples were sealed in cans.

Table 1. Characteristic parameters of polarization curves in plain mortars (50/100—RH value; ir—irradiated; can—in a can in a climatic chamber; air—laboratory conditions).

Parameters	air	50—can	50—ir	100—can	100—ir
E_{cor}, V	−0.52	−0.25	−0.43	−0.31	−0.36
j_{cor}, μA/cm^2	0.57	0.02	0.24	0.20	0.26
E_p, V	−0.47	−0.19	−0.36	−0.25	−0.30
j_p, μA/cm^2	13.9	0.6	4.4	3.8	5.0
E_{tr}, V	0.55	0.57	0.55	0.54	0.54

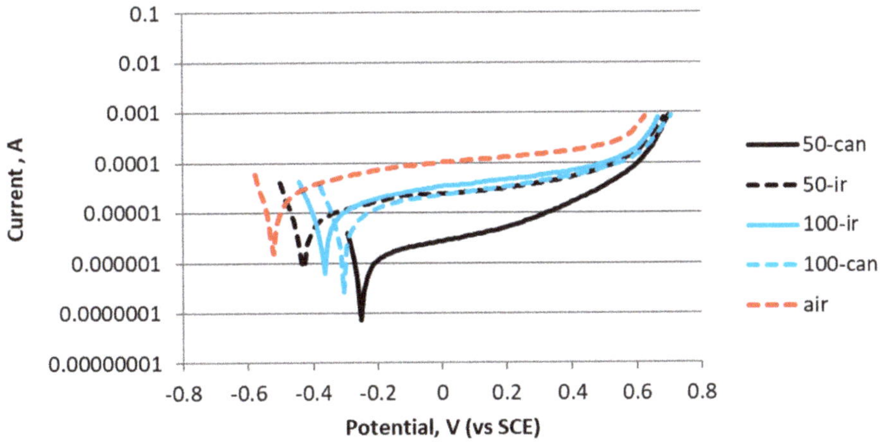

Figure 3. Polarization curves for steel rebar embedded in Portland cement mortar (50/100—RH value; ir—irradiated; can—in a can in a climatic chamber; air—laboratory conditions).

The characteristics of the steel in mortars on the Bode plots of the EIS spectra for samples conditioned under five different conditions are shown in Figure 4. The shape of the EIS spectra (wide phase angle peak) and high impedance at a low frequency indicates the presence of a passivation layer on the steel surface. The gamma radiation is responsible for shifting the impedance and phase angle to smaller values in the low frequency range. The above-mentioned effects are observed for all investigated specimens and indicate a deterioration in the quality of the passivation layer.

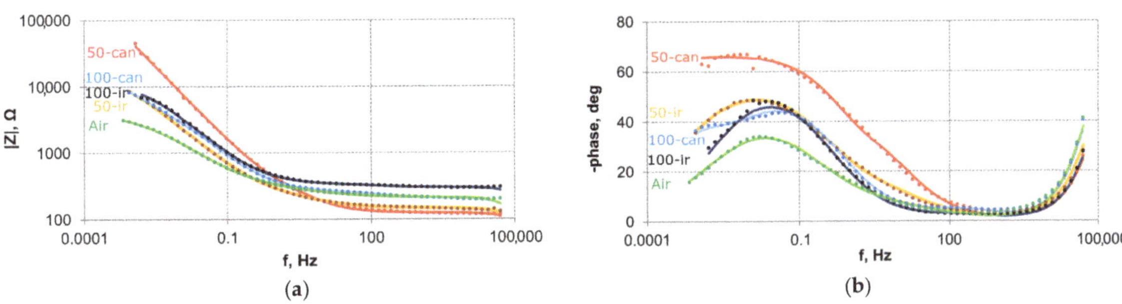

Figure 4. EIS spectra: (**a**) impedance, (**b**) phase shift for steel in mortars (points—measurement data; lines—fitting for equivalent circuit from Figure 2; 50/100—RH value; ir—irradiated; can—in a can in a climatic chamber; air—laboratory conditions).

The parameters determined on the basis of the equivalent circuit and the EIS measurement showed similar conclusions compared to polarization curve parameters (especially j_{cor}). The polarization resistance R_p (R_{ct} + R_{pas}) is the highest for the non-irradiated specimen at 50% RH (Table 2), which should be interpreted as the best protection of the passivation layer among the analysed specimens. In contrast, the smallest R_p values were present in mortar stored in air (air). The gamma irradiation of steel in mortar caused a decrease in R_p values—it was larger for RH = 50% than for RH = 100%—which indicates an acceleration of steel corrosion. It should be emphasized that the interpretation of changes takes place at a low level of steel corrosion. Increasing RH from 50% to 100% decreased R_p and therefore accelerated corrosion. It can be seen that the change in R_p due to an increase in RH was small for irradiated samples and significant for non-irradiated samples.

Table 2. Electrical circuit parameters for steel in mortar (50/100—RH value; ir—irradiated; can—in a can in a climatic chamber; air—laboratory conditions).

Parameters	air	50—can	50—ir	100—can	100—ir
R_{pas}, kΩ·cm^2	1.56	5.15	2.12	0.38	0.48
Y_{pas}, μFs^{n-1}·cm^{-2}	132.0	50.5	111.0	2.4	20.1
n_{pas}	0.55	0.73	0.61	0.90	0.71
R_{ct}, kΩ·cm^2	39	5010	204	207	122
Y_{dl}, μFs^{n-1}·cm^{-2}	201	43	122	173	146
n_{dl}	0.62	0.78	0.67	0.70	0.68

The gamma irradiation of mortars with steel bars caused a three-fold increase in capacity of the double layer (Y_{dl}) compared to non-irradiated specimens stored under 50% RH conditions. The physical interpretation of such a determination is an increasing thickness of the electrical double layer and consequently loses the stability of the passivation layer and iron dissolution [25]. In the case of fully saturated mortar (RH-100), the opposite relationship was observed—Y_{dl} after gamma irradiation decreased by about 15–30%. A 15% decrease in the *n* parameter is also observed for steel in mortar at 50% and exposure to gamma radiation. This effect is due to the reduced uniformity of the passive layer on steel, and thus a degradation of its protective properties. In the case of RH-100 mortar, no effect of gamma radiation on the *n* parameter was observed.

4. Conclusions

The electrochemical parameters of steel rods in mortar conditioning in specific environmental conditions (the dose of gamma radiation reached 1.8 MGy and environmental conditions favoured the progress of carbonation phenomena) result in the following conclusions:

1. The increasing RH from 50% to 100% resulted in the intensification of the deterioration of the protective properties of the passive layer on the steel and the accelerated corrosion of the steel in non-irradiated specimens. The effect of gamma radiation on the corrosion of steel in the mortar is significantly greater in an atmosphere of 1% CO_2 and moderate RH (50%) than 1% CO_2 and fully saturated conditions (RH = 100%);
2. Steel in mortar stored in closed, low-volume cans had a better passive layer and lower corrosion rate than steel in mortar stored in a laboratory atmosphere.

Author Contributions: Conceptualization, M.D.; methodology, M.D. and M.A.G.; validation, M.D.; formal analysis, M.A.G.; investigation, M.D., J.K., K.D.; data curation, M.D. and J.K.; writing—original draft preparation, M.D.; writing—review and editing, M.D., J.K.; visualization, M.D. and J.K.; supervision, M.A.G.; project administration, M.A.G.; funding acquisition, M.A.G. All authors have read and agreed to the published version of the manuscript.

Funding: This research was funded by the National Centre for Research and Development, Poland, grant number Project V4-Korea/2/2018.

Institutional Review Board Statement: Not applicable.

Informed Consent Statement: Not applicable.

Data Availability Statement: Not applicable.

Acknowledgments: Publication cost of this paper was covered with founds of the Polish National Agency for Academic Exchange (NAWA): "MATBUD'2023—Developing international scientific cooperation in the field of building materials engineering" BPI/WTP/2021/1/00002, MATBUD'2023

Conflicts of Interest: The authors declare no conflict of interest.

References

1. Tcherner, J.; Vaithilingam, L.; Han, M. Effective aging management of NPP concrete structures. *J. Adv. Concr. Technol.* **2017**, *15*, 1–9. [CrossRef]
2. Lindqvist, J.A.; Flansbjer, M.; Hansson, E. *Analysis of Irradiated Concrete*; Report 2016:239; Energiforsk: Stockholm, Sweden, 2016.
3. Field, K.G.; Remec, I.; Le Pape, Y. Radiation effects in concrete for nuclear power plants—Part I: Quantification of radiation exposure and radiation effects. *Nucl. Eng. Des.* **2015**, *282*, 126–143. [CrossRef]
4. Soo, P.; Milian, L.M. The effect of gamma radiation on the strength of Portland cement mortars. *J. Mater. Sci. Lett.* **2001**, *20*, 1345–1348. [CrossRef]
5. Robira, M.; Hilloulin, B.; Loukili, A.; Potin, G.; Bourbon, X.; Abdelouas, A. Multi-scale investigation of the effect of γ irradiations on the mechanical properties of cementitious materials. *Constr. Build. Mater.* **2018**, *186*, 484–494. [CrossRef]
6. Mobasher, N.; Bernal, S.; Kinoshita, H.; Sharrad, C.; Provis, J. Gamma irradiation resistance of an early age slag-blended cement matrix for nuclear waste encapsulation. *J. Mater. Res.* **2015**, *30*, 1563–1571. [CrossRef]
7. Glinicki, M.A.; Dąbrowski, M.; Antolik, A.; Dziedzic, K.; Sikorin, S.; Fateev, V.; Povolansky, E. Gamma irradiation sensitivity of early hardening cement mortar. *Cem. Concr. Compos.* **2022**, *126*, 104327. [CrossRef]
8. Dąbrowski, M.; Glinicki, M.A.; Dziedzic, K.; Jóźwiak-Niedźwiedzka, D.; Sikorin, S.; Fateev, V.S.; Povalansky, E.I. Early age hardening of concrete with heavy aggregate in gamma radiation source—Impact on the modulus of elasticity and microstructural features. *J. Adv. Concr. Technol.* **2021**, *19*, 555–570. [CrossRef]
9. Maruyama, I.; Ishikawa, S.; Yasukouchi, J.; Sawada, S.; Kurihara, R.; Takizawa, M.; Kontani, O. Impact of gamma-ray irradiation on hardened white Portland cement pastes exposed to atmosphere. *Cem. Concr. Res.* **2018**, *108*, 59–71. [CrossRef]
10. Potts, A.; Butcher, E.; Cann, G.; Leay, L. Long term effects of gamma irradiation on in-service concrete structures. *J. Nucl. Mater.* **2021**, *548*, 152868. [CrossRef]
11. Rasheed, P.A.; Nayar, S.K.; Barsoum, I.; Alfantazi, A. Degradation of Concrete Structures in Nuclear Power Plants: A Review of the Major Causes and Possible Preventive Measures. *Energies* **2022**, *15*, 8011. [CrossRef]
12. Reches, Y. A multi-scale review of the effects of gamma radiation on concrete. *Results Mater.* **2019**, *2*, 100039. [CrossRef]
13. Jaśniok, M.; Jaśniok, T. Measurements on corrosion rate of reinforcing steel under various environmental conditions, using an insulator to delimit the polarized area. *Procedia Eng.* **2017**, *193*, 431–438. [CrossRef]
14. Macdonald, D.D.; Engelhardt, G.R.; Petrov, A. A Critical Review of Radiolysis Issues in Water-Cooled Fission and Fusion Reactors: Part I, Assessment of Radiolysis Model. *Corros. Mate. Degrad.* **2022**, *3*, 470–535. [CrossRef]
15. Aljohania, T.A.; Geesib, M.H.; Kaibac, A.; Al-Mayoufd, A.M.; Khoshnaw, F. Characterization of gamma-ray irradiation influence on the corrosion behaviour of austenitic stainless steel. *Mater. Today Commun.* **2020**, *24*, 101242. [CrossRef]
16. Liu, C.; Wang, J.; Zhang, Z.; Han, E.-H.; Liu, W.; Liang, D.; Yang, Z.; Cao, X. Effect of cumulative gamma irradiation on microstructure and corrosion behaviour of X65 low carbon steel. *J. Mater. Sci. Technol.* **2018**, *34*, 2131–2139. [CrossRef]
17. Dewynter-Marty, V.; Chomat, L.; Guillot, W.; Amblard, E.; Durand, D.; Cornaton, M.; Bourbon, X. Concrete radiolysis effect on steels corrosion and comparison with non-irradiated material. In Proceedings of the EUROCORR 2017—20th International Corrosion Congress/Process Safety Congress 2017, Prague, Czech Republic, 3–7 September 2017.
18. Dąbrowski, M.; Glinicki, M.A.; Kuziak, J.; Jóźwiak-Niedźwiedzka, D.; Dziedzic, K. Effects of 2 MGy gamma irradiation on the steel corrosion in cement-based composites. *Constr. Build. Mater.* **2022**, *342*, 127967. [CrossRef]
19. *EN 197-1:2012*; Cement: Composition, Specifications and Conformity Criteria for Common Cements. European Committee for Standardization: Brussels, Belgium, 2012.
20. *EN 196-1:2016*; Methods of Testing Cement: Determination of Strength. European Committee for Standardization: Brussels, Belgium, 2016.
21. *EN 10025-2:2019*; Hot Rolled Products of Structural Steels Part 2: Technical Delivery Conditions for Non-Alloy Structural Steels. European Committee for Standardization: Brussels, Belgium, 2019.
22. *PN-B-01810:1986*; Protection Against Corrosion in Building—Protective Properties of Concrete Referring to Reinforcing Steel—Electrochemical Tests. Polish Committee for Standardization: Brussels, Belgium, 1986. (In Polish)
23. Jóźwiak-Niedźwiedzka, D.; Dąbrowski, M.; Dziedzic, K.; Jarząbek, D.; Antolik, A.; Denis, P.; Glinicki, M.A. Effect of gamma irradiation on the mechanical properties of carbonation reaction products in mortar. *Mater. Struct.* **2022**, *55*, 164. [CrossRef]
24. Andrade, C.; Alonso, C.; Gulikers, J.; Polder, R.; Cigna, R.; Vennesland, Ø.; Salta, M.; Raharinaivo, A.; Elsener, B. RILEM TC 154-EMC: Electrochemical Techniques for Measuring Metallic Corrosion. Recommendations. Test methods for on-site corrosion rate measurement of steel reinforcement in concrete by means of the polarization resistance method. *Mater. Struct.* **2004**, *37*, 623–643. [CrossRef]
25. Montemor, M.F.; Cunha, M.P.; Ferreira, M.G.; Simoes, A.M. Corrosion behaviour of rebars in fly ash mortar exposed to carbon dioxide and chlorides. *Cem. Concr. Compos.* **2002**, *24*, 45–53. [CrossRef]

MDPI
St. Alban-Anlage 66
4052 Basel
Switzerland
Tel. +41 61 683 77 34
Fax +41 61 302 89 18
www.mdpi.com

Materials Proceedings Editorial Office
E-mail: materproc@mdpi.com
www.mdpi.com/journal/materproc

www.ingramcontent.com/pod-product-compliance
Lightning Source LLC
Chambersburg PA
CBHW061554220926
51850CB00005B/337